DISCOVERING GIS and ArcGIS

Bradley A. Shellito

Youngstown State University

W. H. FREEMAN
& COMPANY

A Macmillan Education Imprint

DEDICATION

This book is dedicated to my parents, David and Elizabeth, whose encouragement showed me how to roll the dice and make them come up sevens.

ASSOCIATE PUBLISHER: Steven Rigolosi

ACQUISITIONS EDITOR: Bill Minick

ASSOCIATE EDITOR: Mary Walsh

ASSISTANT EDITOR AND MARKET DEVELOPMENT MANAGER: Stephanie Ellis

EDITORIAL ASSISTANT: Carlos Marin

MARKETING MANAGER: Taryn Burns

MARKETING ASSISTANT: Bailey James

EXECUTIVE MEDIA MANAGER: Rachel Comerford

MEDIA EDITOR: Lukia Kliossis

PROJECT EDITORS: Enrico Bruno and Edward Dionne

PRODUCTION SUPERVISOR: Paul Rohloff

PHOTO EDITOR: Sheena Goldstein

ART MANAGER: Matt McAdams

COVER AND TEXT DESIGN: Vicki Tomaselli

COMPOSITION: MPS Limited

PRINTING AND BINDING: RR Donnelley

COVER PHOTOS: Top: Source: jl661227/Shutterstock, Center: Source: Planet Observer/Universal Images Group/Getty Images, Bottom: Source: © Michael DeYoung/Corbis, Map: Source: Kraska/Shutterstock

ISBN-13: 978-1-4641-4520-9

ISBN-10: 1-4641-4520-2

Library of Congress Preassigned Control Number: 2014950584

Printed in the United States of America

First printing

W. H. Freeman and Company

41 Madison Avenue

New York, NY 10010

Houndmills, Basingstoke RG21, 5XS, England

This book was not prepared, approved, or endorsed by the owners or creators of any of the software products discussed herein. The graphical user interfaces, emblems, trademarks and associated materials discussed in this book remain the intellectual property of their respective owners.

ArcGIS for Desktop 10.2, ArcGIS Online

SOURCE: NEAL P. MCNALL

B radley A. Shellito is a geographer whose work focuses on the application of geospatial technologies. Dr. Shellito has been a professor at Youngstown State University (YSU) since 2004, and was previously a faculty member at Old Dominion University. He teaches classes in GIS, remote sensing, GPS, and 3D visualization, and his research interests involve using these concepts within a variety of real-world issues. His first book, *Introduction to Geospatial Technologies,* was also published by Macmillan Education. He also serves as YSU's Principal Investigator in OhioView, a statewide geospatial consortium. A native of the Youngstown area, Dr. Shellito received a bachelor's degree from YSU, a master's degree from the Ohio State University, and a doctorate from Michigan State University.

BRIEF CONTENTS

CONTENTS

Why I Wrote *Discovering GIS and ArcGIS*

My approach to teaching GIS courses involves two goals for the students. First, they should be able to work hands-on with the GIS software. Second, and more importantly, they should know the "how" and "why" behind what they're doing. Software changes quickly; the theory has a longer shelf life. The goal of *Discovering GIS and ArcGIS* is to teach students how to combine GIS concepts with ArcGIS software skills. Students learn to use the software, apply it to real-world tasks, and discover why they are doing things. They learn background and theory subjects, when appropriate, as they work through the hands-on application of ArcGIS in each chapter.

For instance, when students are constructing a map and choosing a data-classification method, an accompanying theory-based "Smartbox" provides background on the different classification methods, how each method classifies data, and which method should be used based on the distribution of the data. Similarly, when students are preparing to publish their data to ArcGIS Online as a service, an accompanying Smartbox provides background and information on publishing, the difference between feature services and tiled map services, and how these tools are set up and used. In this way, students encounter the theory as they work with the software. In short: Rather than reading pages of background before getting to the hands-on application, students encounter the underlying concepts and explanations while working with the software. Students "learn while doing," gaining an understanding of GIS as they are implementing GIS concepts with ArcGIS.

When using a worktext, students can all too often click all the right buttons, input the correct values, and end up with the correct answers without having a clear idea of why they were pushing those buttons or using those particular settings. For instance, students can turn a set of two-dimensional building footprints into a 3D representation of the structures by clicking the mouse a few times and entering values in an ArcGIS dialog. Similarly, they can open the "Create TIN" tool, select some inputs, click "OK," and get a TIN (triangulated irregular network) to work within ArcGIS, but not have to understand what a TIN is or how to create one properly. The goal of *Discovering GIS and ArcGIS* is to provide step-by-step instructions that allow students to use ArcGIS for real-world tasks and applications, while also helping them understand the "hows" and "whys" behind their actions.

The target audience for *Discovering GIS and ArcGIS* is introductory and advanced GIS courses or applied GIS courses. Colleges and universities will often offer more than one GIS course—such as one at an introductory level and another at an advanced level. *Discovering GIS and ArcGIS* would be appropriate for one or both courses. Within the text, instructors will be able to find the topics they want to present whether they are teaching a single intro-level course, a single advanced or applied course, or a two-course sequence.

Through its generous software promotion program, Esri makes it easy for students to access ArcGIS for educational purposes. Instructors at schools with an Esri site license can request a free one-year version of ArcGIS for Desktop to distribute to their students (the online request form is here: http://www.esri.com/industries/apps/education/offers/promo/index.cfm). This is a full version of ArcGIS for Desktop 10.2 with all of the extensions, toolbars, and functions used in this book; students can install it on their own computer or laptop. Instructors can request either a DVD to distribute, or a code for students to use in downloading the software.

Organization

The focus of the 22 chapters and two appendixes of *Discovering GIS and ArcGIS* is as follows:

Chapter 1: How to Use Geospatial Data and ArcGIS. This chapter explains how items in the real world are represented with vector objects, shows that each object has a spatial reference (in terms of real-world geographic or projected coordinates), and explains how those real-world items can be used and viewed in ArcGIS.

Chapter 2: How to Use Tables and Attributes in ArcGIS 10.2. This chapter explains how each vector object has a series of attributes that relate to it and how those attributes can be accessed and queried for analysis.

Chapter 3: How to Create a Map Layout with ArcGIS 10.2. This chapter explains how to take GIS data layers to make a professional-looking map in ArcGIS.

Chapter 4: How to Create a Web Map and Share Data Online with ArcGIS 10.2. This chapter explains how to take GIS data layers and create an interactive Web map using ArcGIS Online.

Chapter 5: How to Obtain and Use Online Data with ArcGIS 10.2. This chapter explains how to obtain GIS data (for free) via the Internet and then use it in ArcGIS, with a focus on obtaining data from The National Map and ArcGIS Online.

Chapter 6: How to Create Geospatial Data with ArcGIS 10.2. This chapter explains how users can create their own GIS vector data to use in ArcGIS.

Chapter 7: How to Edit Data with ArcGIS 10.2. This chapter explains how users can edit their data (whether created or obtained) and assess the quality of the data.

Chapter 8: How to Perform Spatial Analysis in ArcGIS 10.2. This chapter explains the basics of spatial analysis using ArcGIS, including working with different types of spatial queries and spatial joins.

Chapter 9: How to Perform Geoprocessing in ArcGIS 10.2. This chapter introduces how geoprocessing and map overlay concepts are used in ArcGIS and how multiple layers can be combined for analysis.

Chapter 10: How to Perform Geocoding in ArcGIS 10.2. This chapter explains how to take a spreadsheet of addresses, turn it into a point layer, and use it for spatial analysis.

Chapter 11: How to Perform Network Analysis in ArcGIS 10.2. This chapter explains how network analysis (as it is used in transportation studies) is performed using ArcGIS.

Chapter 12: How to Use Raster Data in ArcGIS 10.2. This chapter explains what raster data is and how it is used in ArcGIS (as opposed to the vector data used in the previous 11 chapters).

Chapter 13: How to Use Remotely Sensed Imagery in ArcGIS 10.2. This chapter explains what remotely sensed data (both satellite and aerial imagery) is and how it can be used in ArcGIS as a raster data layer.

Chapter 14: How to Perform Spatial Interpolation with ArcGIS 10.2. This chapter explains how spatial interpolation methods can create a continuous surface from a set of points.

Chapter 15: How to Work with Digital Elevation Models in ArcGIS 10.2. This chapter introduces working with GIS data that has a third dimension, specifically digital elevation models. It examines how to use this type of data including slopes, hillshades, and visibility analysis.

Chapter 16: How to Work with Contours, TINs, and 3D Imagery in ArcGIS 10.2. This chapter explains how TINs are built and used in ArcGIS with other data sources such as contours, and also how high-resolution imagery can be utilized with TINs.

Chapter 17: How to Work with Lidar Data in ArcGIS 10.2. This chapter introduces lidar data and how lidar points can be used to measure heights and elevations of the landscape and objects on the surface.

Chapter 18: How to Represent Geospatial Data in 3D with ArcGIS 10.2. This chapter explains how structures (and other non-landscape features) can be designed and visualized in 3D in ArcGIS.

Chapter 19: How to Utilize Distance Calculations in ArcGIS 10.2. This chapter introduces how distance measurements are made in ArcGIS and how distance layers are treated, as well as how to use cost distances and shortest paths.

Chapter 20: How to Perform Map Algebra in ArcGIS 10.2. This chapter explains how to overlay two or more raster layers using Map Algebra techniques, as well as introducing setting up workflows with ModelBuilder.

Chapter 21: How to Build a Model in ArcGIS 10.2. This chapter explains how to create a separate model (using ModelBuilder) that can be shared or used as a tool in the context of developing a site suitability model.

Chapter 22: How to Use Hydrologic Modeling Tools in ArcGIS 10.2. This chapter explains how hydrologic modeling tools are used in ArcGIS by examining water flow, stream extraction, and watershed delineations (using ModelBuilder).

Appendix A: How to Customize Toolbars in ArcGIS 10.2. This is a short appendix on using the customization tools within ArcGIS to create your own toolbars and add your own buttons, menu items, and shortcuts.

Appendix B: Using Coordinate Systems in ArcGIS 10.2. This appendix provides more technical and computational details on the geographic coordinate systems (GCS) and projected coordinate systems, specifically Universal Transverse Mercator (UTM) measurements and State Plane Coordinate System (SPCS) measurements. It supplements Chapter 1 and provides a resource for using these spatial references in other chapters.

Features of Each Chapter

Each chapter presents the material using the following structure and special features:

An application-based approach. Each chapter focuses on using a variety of ArcGIS tools in a real-world context. At the start of each chapter, a scenario puts the student in a particular role with a number of tasks to accomplish. These scenarios include the role of a park ranger trying to find the best overland route to rescue a group of stranded hikers (Chapter 19), a delivery person finding the shortest driving route between a set of libraries (Chapter 11), and an urban planner trying to determine the minimum height necessary for constructing an observation platform (Chapter 15). Because the early chapters teach basic GIS and ArcGIS skills using data that can be easily replicated for a local area, the scenarios used are simple. The scenarios used in later chapters become more specific (as the book assumes that students have mastered the basic skills and can work with more difficult or involved tasks). In addition, each chapter describes professions and applications in the real world that make use of the chapter's theory and skills. For example, each chapter may discuss how archeologists, geologists, biologists, civil engineers, city planners, realtors, water resource managers, or law enforcement officers can make use of GIS.

The ability to localize each chapter with freely available data. Some students may respond better to working with GIS applications if they get to work with local data. For instance, being able to geocode the locations of nearby libraries or determine the sites for an ecological preserve within their own county may better connect students to these tasks. Though each chapter's data is taken from various locations within Ohio, the data being used is freely available; more importantly, the same kind of data can be obtained for most local areas within the United States. Each chapter contains brief instructions on how to obtain similar local data for performing the chapter's tasks. For instance, Chapter 1 uses five data layers for Mahoning County, Ohio. These layers came from a free download via the National Map and the U.S. Census, which happen to have those same five layers available for all U.S. counties. If you want to perform the chapter's actions in Ingham County, Michigan, you can download the same layers for Ingham via the National Map and the Census Bureau. All of the chapters are designed so that activities can be easily replicated for other locations using free data.

Smartboxes, Troubleboxes, and Questions. The bulk of each chapter consists of hands-on use of ArcGIS 10.2 to complete the tasks involved with the chapter's scenario. Important key terms are highlighted in blue. Chapters are divided into subheadings such as Step x.1, Step x.2, and so forth, to break up specific tasks and skills. Each chapter also contains "Smartboxes." The Smartboxes present theory (and background information) at an appropriate time to help readers understand what their actions are really doing. For instance, one of the steps in Chapter 2 walks the reader through the process of joining two tables in ArcGIS 10.2. A nearby Smartbox succinctly describes the theory behind joining tables, what is necessary to join tables, and the rules needed for an effective join. Similarly, a step in Chapter 1 shows the reader how to project a dataset. A nearby Smartbox provides the theory and information on what projection is, how it works, and what the end result will be.

Some chapters also contain "Troubleboxes," which contain troubleshooting tips for common problems that may arise. For instance, when first working with ArcGIS, students frequently close necessary windows or remove necessary ArcMap items, and then don't know how to get them back. A Troublebox is placed in an appropriate location in Chapter 1 to describe how to easily restore the missing windows. Similarly, when students connect to ArcGIS Online (in Chapter 4), the Hosted Services section may not be available the first time it's used. A Troublebox is placed at that spot to give students the information they need to access the Hosted Services. The Troubleboxes are the result of common student questions that crop up in my classes—often I find myself answering the same questions over and over again to get students back on track.

Each chapter contains a set of questions for students to answer. These are interspersed throughout the chapter and have two goals. First, they are intended to keep students on track as they move through the chapter, connecting the buttons and dialogs they're using to the goals of the project they are working on. (Sometimes these are simple questions related to the number of selected records or the size of a grid cell.) Second, many questions ask students to think about what they're doing, such as evaluating the output of a shortest path calculation or figuring out why an address didn't geocode properly. By pausing to answer these questions, students will be more connected to the activities they're working on. In addition, the end of each chapter asks students to present their results either in the form of a map layout, a shared Web map on ArcGIS Online, or (in some cases) an interactive Web Scene or a video file.

Related Concepts for each chapter. There are many topics in GIS—more than can be covered in a single chapter without making that chapter seem like a very contrived cookbook. The Related Concepts section at the end of each chapter presents important information about related topics that weren't covered in the chapter. For example, in Chapter 4, students create a Web map using ArcGIS Online. The Related Concepts section for Chapter 4 describes creating Web applications and the use of ArcGIS for Server. Similarly, in Chapter 18, students create a 3D version of a campus using extruded polygons. The Related Concepts section for Chapter 18 describes how to create 3D versions of structures using multipatches and refining them using the free SketchUp software. The Related Concepts section can be presented to the class as part of a lecture; alternatively, students may choose to read these sections on their own for further expanded information about the chapter's topics.

For More Information: This is the reference section for the chapter, highlighting the materials used in compiling it. These sources were largely drawn from either ArcGIS Help or the Esri Resource Center (these two items together constitute a tremendous resource for GIS theory and ArcGIS application). The lists of items in the For More Information section provide the exact phrase or word string you can type in to the Help or the Esri Resource Center utility to get the necessary information back. This section also lists some specific articles that may be of particular use (or were used to write the chapter), although they are limited to a handful, and only as necessary.

Online and Supplementary Materials

Instructors can download supplements to accompany *Discovering GIS and ArcGIS* from the Macmillan catalog at http://www.macmillanhighered.com/Catalog/product/discoveringgisandarcgis-firstedition-shellito (or visit http://www.macmillanhighered.com/Catalog and search by author name or textbook title). From the catalog page the following resources can be downloaded by instructors:

• Instructor's Manual/Test Bank – An online instructor's manual has been set up as a supplement to *Discovering GIS and ArcGIS*. It includes a solutions manual for all of the questions from the various chapters, as well as separate PDFs of answer sheets for questions that can be distributed to students. It also includes copies of the book's graphics for instructors to use in lectures, along with an extensive test bank of questions for each chapter.

• ArcGIS Installation Guide – There is also a PDF of step-by-step directions for installing the student version of ArcGIS 10.2 so that students in a GIS class can utilize a copy of ArcGIS for Desktop on their home computer or laptop.

Students can download the folders containing the data required for each chapter from the Macmillan catalog at http://www.macmillanhighered.com/ Catalog/product/discoveringgisandarcgis-firstedition-shellito (or visit http://www. macmillanhighered.com/Catalog and search by author name or textbook title).

ACKNOWLEDGMENTS AND THANKS

Books like this don't just spring out of thin air—I owe a great deal to several people who have provided inspiration, help, and support for what would eventually become this book:

• Steven Rigolosi, Bill Minick, and Chuck Linsmeier of Macmillan Education, for invaluable help, advice, patience, and guidance throughout this entire project. I would also very much like to thank Ed Dionne, Sheena Goldstein, Anthony Calcara, Enrico Bruno, Stephanie Ellis, Vicki Tomaselli, Matt McAdams, and Lukia Kliossis of W. H. Freeman for all of their help on the production side, as well as all of their extensive "behind the curtain" work that shaped this book into a finished product.

• Neil Salkind and Stacey Czarnowski, for great representation and advice.

• Sean Young, for an extensive expert review of the text and ArcGIS methods used in this book.

• The students who contributed to the development of the YSU 3D Campus Model: Rob Carter, Ginger Cartright, Paul Crabtree, Jason Delisio, Sherif El Seuofi, Nicole Eve, Paul Gromen, Wook Rak Jung, Colin LaForme, Sam Mancino, Jeremy Mickler, Eric Ondrasik, Craig Strahler, Jaime Webber, Sean Welton, and Nate Wood. The 3D data utilized and presented in Chapter 18 comes directly from their hard work.

• Also, a very special thanks to all of my professors, instructors, colleagues, and mentors past and present (who are too numerous to list) from Youngstown State University, The Ohio State University, Michigan State University, Old Dominion University, OhioView, and everywhere else for the help, knowledge, notes, information, skills, and tools they've given me over the years. I am also deeply indebted to the work of Tom Allen and Nicole Eve for some of the concepts and methods used in various chapters.

I would also like to thank the following colleagues, who reviewed an early draft of the manuscript:

• Shivanand Balram, Simon Fraser University

• W.B. Clapham, Jr., Cleveland State University

• Alison E. Feeney, Shippensburg University

• Arlene Guest, Naval Postgraduate School

• John McGee, Virginia Tech

• Rafael Moreno, University of Colorado Denver

• Thomas Mueller, California University of Pennsylvania

• Reed Perkins, Queens University of Charlotte

I'd very much like to hear from you regarding any thoughts or suggestions you might have for the book. You can follow me on Twitter @GeoBradShellito or you can reach me via email at bashellito@ysu.edu.

Bradley A. Shellito
Youngstown, Ohio

Accessing Folders and Datasets for Each Chapter

Each chapter in this book requires you to download a folder for use within ArcGIS. To download these free folders, visit http://www.macmillanhighered. com/Catalog/product/discoveringgisandarcgis-firstedition-shellito

How to Use Geospatial Data and ArcGIS 10.2

Introduction

Here's something to think about—when a wastewater treatment plant takes in raw sewage, its goal is to separate out water from the rest of the material. The treated water flows back to local streams or rivers, while the remainder is pressed into sludge. This sewage sludge is treated and referred to as biosolids, which has the look and feel of dirt or potting soil. Often, the final destination of biosolids is a landfill or incinerator, but under certain conditions, landowners use this treated sludge as a fertilizer (Figure 1.1). Because biosolids sometimes contain pathogens, the Environmental Protection Agency regulates their application as fertilizer.

FIGURE 1.1 Land application of biosolids on an agricultural field.

Still, the practice of applying biosolids to the land should raise many questions, among them:

• What are the present and past locations of the agricultural fields that have been fertilized with biosolids?

• How close is your home to an agricultural field where application is taking place?

• How close to these agricultural fields are the wells from which you get your water?

• If land application of biosolids took place in the past at a particular site, how is that land being used today?

All of these questions involve examining locations (such as the past and present land-applied fields), qualities or attributes of those areas (such as the type of land use at the sites), and how one location relates to another (such as the distance from your home or water well to a land-applied field). Answering these questions involves working with large-scale location-based processes, thousands of location data points, and multiple types of data. For this reason, you're going to need a powerful tool to help you conduct your analysis.

Geographic information systems (GIS) are composed of the hardware and software that allow for computer-based analysis, manipulation, visualization, and retrieval of location-related data. In GIS terms, we refer to the data associated with real-world locations as **geospatial** data (or often simply as **spatial** data). GIS provides a means of solving problems and answering questions related to geospatial data and information; it is used for countless different applications (like the issues surrounding the biosolids scenario). GIS is part of a larger field of technologies, referred to as **geospatial technologies**, which encompasses many types of methods and techniques for the collection, analysis, modeling, and visualization of geospatial data.

Several types of GIS software are available today, ranging from expensive commercial products to freely available open-source alternatives. The industry leader in

geographic information systems (GIS) The hardware and software that allow for computer-based analysis, manipulation, visualization, and retrieval of location-related data.

geospatial or **spatial** Referring to items that are tied to a specific real-world location.

geospatial technologies A term that encompasses many types of methods and techniques for the collection, analysis, modeling, and visualization of geospatial data.

Esri Environmental Systems Research Institute—the market leader in GIS software and the developer of ArcGIS.

ArcGIS for Desktop 10.2 The version of ArcGIS that runs on a personal computer.

ArcGIS 10.2 The version of the ArcGIS software package released in 2013 and the basis for the activities in this book.

ArcGIS Basic The version of ArcGIS for Desktop with the fewest functions.

ArcGIS Standard The version of ArcGIS for Desktop with a moderate number of functions.

ArcGIS Advanced The version of ArcGIS for Desktop with the greatest number of functions.

GIS is **Esri** (Environmental Systems Research Institute), a company based in Redlands, CA. Esri was founded in 1969 and released its first version of its Arc/Info GIS software in 1982. Arc/Info has evolved into Esri's current ArcGIS software. A recent salary survey conducted by Geospatial Training Services showed that Esri's GIS products were among the most popularly used software packages in the industry.

ArcGIS for Desktop 10.2 is the version of Esri's GIS software for use on a personal computer released in 2013. **ArcGIS 10.2** comes in three varieties: **ArcGIS Basic**, **ArcGIS Standard**, and **ArcGIS Advanced**. Think of these as different levels of the same software. ArcGIS Basic contains fewer features than ArcGIS Standard, which in turn contains fewer features than ArcGIS Advanced. The version we will be using in this book is ArcGIS for Desktop Advanced 10.2, but for simplicity we will use the terms "ArcGIS" and "ArcGIS 10.2" interchangeably to refer to the version of ArcGIS for Desktop that contains the full functionality.

The goal of this first chapter is to introduce you to some key GIS and geospatial concepts; these will be used throughout the book as important building blocks for using GIS. You'll be using several components of ArcGIS to see how GIS data represent real-world items such as roads, airport locations, and water bodies. You'll also examine the geospatial characteristics of these data (such as the coordinate system the data is set up in) and how ArcGIS handles these data. Last, you'll learn how to present these data effectively by using various symbols and colors. Throughout the text (and in many other GIS applications you may work with), you will use many of the ArcGIS skills and tools you learn in this chapter. "Smartboxes" within this chapter explain many of these primary concepts.

Chapter Scenario and Applications

This chapter puts you in the role of a GIS analyst for a county engineer's office. Your first task will be to assemble a basic set of data for the county's many features (including airports, roads, bodies of water, and locations of county structures) within ArcGIS for a presentation to the general public. You will compile this dataset with an eye to creating future presentations of this material (such as making a map of the data or creating a basic Web mapping application).

The following are examples of other real-world applications of this chapter's theory and skills:

- An EMT operator has been assigned to give a presentation regarding the impact of current road closures on a county's hospitals. She will use ArcGIS to show the locations of home injuries, the locations of hospitals, the county's road network, and the locations of road closures. She will then display these data for the presentation.

- To prepare for a presentation to the board of trustees, a manager of a local outdoor recreation area wants to examine the property boundaries between the lands he manages and the neighboring properties. He'll use ArcGIS to present the rec area's property boundaries, the local road network, walking paths in and out of the area, and the locations of "No Trespassing" signs posted on the property.

- A local government planner needs to present information to a group of volunteers about the new routes the Labor Day parade will take. Using ArcGIS, he'll be able to show the route boundaries, roads, and locations of emergency facilities to the volunteers.

ArcGIS Skills

In this chapter, you will learn:

• How to use the ArcGIS 10.2 environment, including basic navigation and tool use in ArcMap.

• How to use the Catalog to connect to folders and move GIS data from one place to another.

• How to preview data in Catalog and retrieve basic information about GIS data.

• How to add point, line, and polygon GIS data to ArcMap and manipulate their data layer properties, including their appearances and symbologies.

• How to examine the geospatial characteristics of data layers, such as their coordinate system, datum, and projection information.

• How to change a data layer from one projected coordinate system to another.

• How to define a projection for data layers that are missing this information.

• How to export your work in ArcMap to a PDF format.

• How to save your work in ArcMap as both a map document and a map package.

Study Area

• For this chapter, you will be working with data from Mahoning County, Ohio.

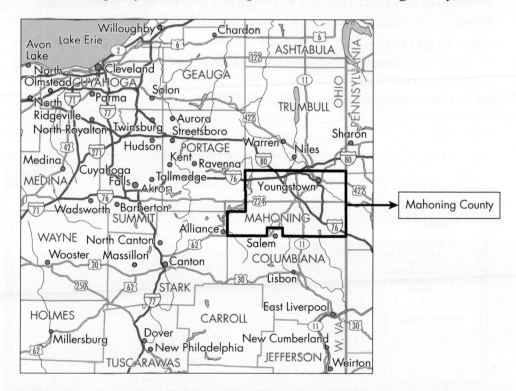

Data Sources and Localizing This Chapter

This chapter's data focus on features and locations within Mahoning County, Ohio. However, this chapter can be easily modified to use data from your own county or local area instead.

The water bodies' data were downloaded from the TIGER/Line files of the Census Bureau Website here: ftp://ftp2.census.gov/geo/tiger/TIGER2013. (We'll do more with TIGER files from the Census in Chapters 2, 3, and 4.) From the AREAWATER folder, download the zip file (containing the water polygons as a shapefile) for your county. The files are sorted by county FIPS code (for instance, Mahoning County, Ohio, is FIPS code 39099). For example, if you were using data for this chapter from Ingham County, Michigan, you would download the AREAWATER file for FIPS code 26065). A full list of state and county FIPS codes is available from the EPA online here: http://www.epa.gov/envirofw/html/codes/state.html. For this chapter, after downloading the AREAWATER shapefile, delete the .prj file before importing the shapefile into a geodatabase—doing so will remove the projection information that you will create in Step 1.7.

The remaining data layers were downloaded (for free) from the National Map Website here: http://nationalmap.gov. Step-by-step details on downloading and extracting GIS data from the National Map are in Chapter 5. When downloading data for another county (for instance, Ingham County, Michigan), the feature classes to use for this chapter are struct_point (the structures layer), trans_roadsegment (the roads layer), trans_airportpoint (the airports layer), and GU_countyorequivalent (the boundary layer).

STEP 1.1   Getting Started

- This "Getting Started" step of each chapter will instruct you on which data you'll be using and which toolbars and extensions you'll use during the hands-on portion of the chapter. Because this is the first time through, this step of this chapter will be more thorough than usual to introduce these concepts.

- ArcGIS for Desktop 10.2 consists of a main component (called **ArcMap**) and several sub-components and extensions. ArcMap is used to work with and create maps of geospatial data for analysis and visualization. To begin, start ArcMap (its icon is a magnifying glass looking at the world):

ArcMap The main component of ArcGIS for Desktop.

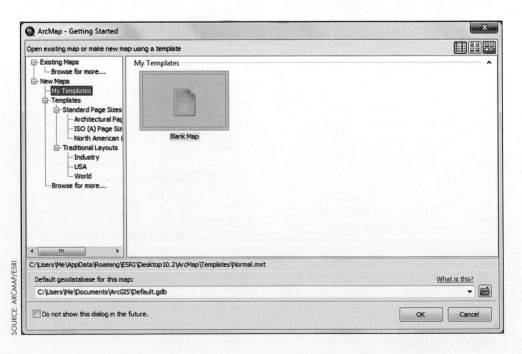

• After the software opens, ArcMap will start asking you questions about "Maps." When prompted, select the option for a new "Blank Map" to begin (see **Smartbox 1** for more information about the "maps" you're working with) and click **OK**.

Smartbox 1

What are these "maps" in ArcGIS 10.2?

ArcMap treats the overall project you are working with as a "map" or "map document." A "map document" is a file that ends with the .mxd extension. When you save your work to a **map document (.mxd)** file, note that you are *not* saving all of your work into a single file. The .mxd file contains reference information as to where your data are, what format they are in, what colors your map is, and so on. It does *not* contain the actual data you are using, just references to where your data can be found.

> **map document (.mxd)** File that contains information about where all of the data layers used in a session are located, as well as their appearance and settings.

What this all means is that if you save your work on a GIS project, copy the .mxd file to a flash drive, go to a different computer, and then try and open the map again, it will not open properly and you will not have your data. The layers and files you used will appear in the Table of Contents but with a red "!" next to them, which indicates that the pathway to that particular layer was not found. The map document holds information about the path to the files that are located on your computer but it does *not* hold the files themselves.

For instance, if you save your map document and it is using all of the files on your flash drive (which is mapped on the computer as a G: drive) and you move to a different computer and attempt to reopen the files (and this second computer maps your flash drive as an H: drive), ArcGIS will be unable to load all of the data. It will be searching for files with a pathway to a particular folder on a G: drive, when those files are now inside a folder on an H: drive. You'll see your files in the map document with the red "!" symbol next to them. In essence, ArcMap knows those files should be in that map document, but it can't find the correct pathway to get to them. When the map document was saved, it understood that those files were on a G: drive. However, when the map document was opened again, ArcMap does not see them on a G: drive (because the flash drive is now mapped as the H: drive on the new computer) and thus cannot find the path to those files. When you move your maps from machine to machine, you will have to take all of your files with you and place them on the new computer in the same path that the map document is looking for (or rebuild the links and references stored in the .mxd file on the new machine).

Thus, when you start ArcMap, you can either open a previously saved map document (.mxd) or start with a new, blank map to work with. Think of a blank map as a blank workspace to which you can add data in order to start creating your map or to start doing data analysis.

• ArcMap works like a standard Windows interface with buttons and pull-down menus. A few things to note at this beginning stage (Figure 1.2):

   • The area marked with a "1"—labeled "Untitled" in Figure 1.2: This is the current name of the map document/.mxd file. Until you save your data, the .mxd file has no name (and is called "Untitled") and your work is not saved.

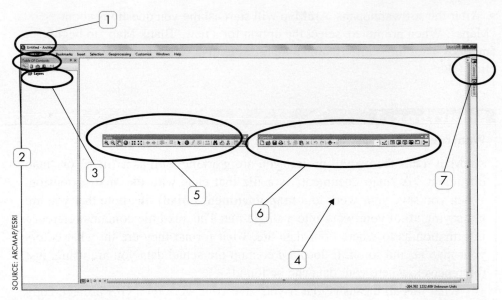

SOURCE: ARCMAP/ESRI

FIGURE 1.2 The basic layout of ArcMap.

- The area marked with a "2"—labeled "Table of Contents": This entire box on the left side of the map is the **Table of Contents**. It will list all layers and Data Frames available for use in the current map document. The Table of Contents is a moveable window in ArcMap. It can be "pinned" in place or hidden to the side of the ArcMap window in a shrunken tab form by clicking on the pushpin symbol in the TOC's window.

- The area marked with a "3"—labeled "Layers": This is the name of the **Data Frame** that will contain any data that you add to the map. Any data you use will be referenced in this area. ArcGIS refers to different kinds of GIS data as **layers**. Thus, this Data Frame will contain any layers you add to the map.

- The large empty area marked with a "4": This is the Data Frame itself, where the visualization of the GIS data layers you use will be shown. We'll also refer to this empty space as the View.

- The area marked with a "5"—a toolbar full of icons: This represents the **Tools toolbar** for ArcMap. You can move the toolbar to "snap" it onto the gray area where the other icons are (and probably where it will be located when you first start ArcMap). Several other toolbars and add-ons are available for ArcMap, but this set represents the default options.

- The area marked with a "6"—another toolbar of icons: This represents the **Standard toolbar** for basic ArcMap usage. Like the Tools toolbar, these are basic utilities for use in ArcMap. The Standard toolbar is also a dockable window.

- The area marked with a "7"—the pinned windows: These are additional windows that can be opened and pinned in place in ArcMap but stay shrunk as tabs on the right side until they need to be used. These include the Catalog and the Search features.

- For further information on what's available via the Tools and Standard toolbars and how they operate, see **Smartboxes 2 and 3**. You can add any toolbar to ArcMap by selecting the **Customize** pull-down menu, then selecting **Toolbars**, then selecting the specific toolbar you want to add. Future chapters will instruct you on which toolbars to add for that particular chapter's steps.

Table of Contents Component of ArcGIS that shows all layers being used in a map document.

Data Frame Component of ArcMap that contains and displays all data layers.

layer A single dataset used in ArcGIS.

Tools toolbar ArcMap toolbar that contains functions such as zooming in or out, and selecting or identifying features.

Standard toolbar ArcMap toolbar that contains functions such as opening, saving, and printing map documents, as well as options to open other windows such as ArcToolbox or Catalog.

Smartbox 2

What are the functions of the Tools toolbar in ArcGIS 10.2?

ArcMap's **Tools toolbar** provides several tools to navigate ArcMap and work with the data you've added to the map (Figure 1.3). Selecting a tool from the toolbar will often change the cursor to a new shape. The cursor will remain in this new shape (with a new set of properties) until the cursor icon is selected again from the toolbar.

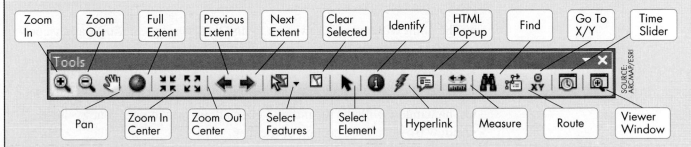

FIGURE 1.3 The ArcGIS 10.2 Tools toolbar.

• The magnifying glass with the plus sign will **zoom in** to a portion of the map. You can select a specific point to zoom in on or click and drag a box to zoom in to.

• The magnifying glass with the minus sign will **zoom out** of a portion of the map. You can select a specific point to zoom out from or click and drag a box to zoom out from.

• The hand (the **pan** icon) allows you to "grab" the map and move it around. Clicking with the hand makes it "clutch" the map, and then dragging the map allows you to move the map to a different point on the screen.

• The globe restores the zoom and the view to a "global view" (that is, the **full extent** of all data layers).

• The arrows pointing inward will **zoom in to the center** of the map.

• The arrows pointing outward will **zoom out from the center** of the map.

• The arrows pointing left and right allow you to restore to a previous extent (or move to the next extent) after moving it with a zoom or pan.

• The arrow next to the box allows you to interactively **select** map features. Selected features are displayed in cyan (light blue) on the map display (see Chapters 2, 8, and 9 for more about selection).

• The all-white box (with no cyan) allows you to **clear** features that have been selected.

• The **Select Element** tool (which looks like a black cursor) restores the icon to the default cursor instead of the other icons. It also allows you to select or resize any graphics (not data layers) that have been added to the maps. (Graphic items are things like shapes, text, and labels that are added to a map.)

• The "i" with a circle is the **Identify** tool. Clicking this tool on a specific map element will return information about that element. Whatever information is stored for that element in the layer's attribute table will be displayed.

• The lightning bolt is used to trigger **hyperlinks** from a map feature if they have been created (see Chapter 2).

• The word balloon can be used to trigger an **HTML pop-up** from the data. Clicking this icon will open a box containing the attributes of a feature.

• The ruler is the **Measure** tool, which will allow you to measure the distance between map features (see Chapter 8).

• The binoculars activate the **Find** tool. By selecting this icon, you can query a layer's attributes and locate a particular item on the map (see Chapter 5).

• The set of connected points allows you to find a new **route** (a network tool; see Chapter 11).

• The point with the **xy** allows you to find locations by entering *x* and *y* coordinates (see Chapter 10).

• The icon with the clock represents the **Time Slider** tool, which is used for visualizing data from multiple time periods (when these data are available).

• The window with the magnifying class allows you to open a new **viewer window** by selecting a section of the map.

Smartbox 3

What are the functions of the Standard toolbar in ArcGIS 10.2?

ArcMap's **Standard toolbar** (Figure 1.4) allows you to create new maps, save your work, print, add data, and copy and paste—all standard functions. It also allows you to access ArcMap's other components and windows.

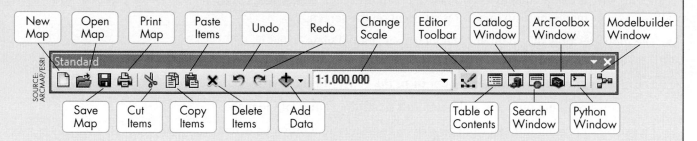

FIGURE 1.4 The ArcGIS 10.2 Standard toolbar.

The icons on the left side of the toolbar should be familiar from other Windows-based applications:

• The blank page icon allows you to create a **new** map document.

• The folder icon allows you to **open** an existing map document.

• The disk icon allows you to **save** a map document.

• The printer icon will activate the **print** dialog.

• The scissors, dual pages, clipboard, and x allow you to **cut**, **copy**, **paste**, and **delete** items, respectively.

• The curved arrows allow you to **undo** and **redo** your most recent actions.

• The black and yellow "plus" icon allows you to **add data** to ArcGIS (more on this shortly).

• The pull-down menu allows you to **change the scale** at which your data are being displayed (see Chapter 3 for more about scale).

- The icon with the pencil and the lines opens the **Editor** toolbar (you'll use the Editor toolbar in Chapters 2, 6, 7, and 15).

- The icon with the lines opens the **Table of Contents** window.

- The icon with the yellow "catalog" symbol opens the **Catalog** window.

- The icon with the globe symbol opens the **Search** window.

- The icon with the red box symbol opens the **ArcToolbox** dockable window.

- The icon with the > prompt opens the **Python** scripting window. This is used in programming using the Python language for customizing ArcGIS (see Chapters 9 and 21 for more about Python).

- The icon with the four color boxes opens the **Modelbuilder** utility, which is used for creating models and chaining several tasks together at once (we'll use Modelbuilder in Chapters 20, 21, and 22).

• In ArcMap, be sure to have the **Tools** and **Standard** toolbars available to use. See Troublebox 1 for information on accessing the toolbars if they aren't available (and also how not to display them if you aren't working with them).

Troublebox 1

What happened to the toolbars I was using in ArcGIS 10.2?

If you close a toolbar in ArcGIS, getting it back again is easy. From the **Customize** pull-down menu, select **Toolbars**. ArcGIS will display all of the available toolbars (including Standard and Tools). Select the one you want to display (a checkmark next to its name means it's displayed). All of the toolbars are dockable windows, so if you close one, just select it from this menu. Likewise, you can hide or display any ArcGIS toolbars using this menu.

• The first step is to access the data you'll be using in this chapter. The **Catalog** is the ArcGIS component for organizing, moving, and copying GIS data. For more about the Catalog and its functions, see Smartbox 4.

Catalog The component of ArcGIS used for copying, moving, and organizing GIS data.

Smartbox 4

How can I access folders and data through the Catalog?

In ArcGIS 10.2, the Catalog is a window that can be pinned to the right-hand side of the screen. Clicking on the tabbed version will expand it, while clicking on the expanded window's pushpin symbol will "pin" the window in place. Clicking the pin once more will hide the window in the tab on the side of the screen (Figure 1.5).

The Catalog can be used to examine all of the available folders (whether local on the computer's hard drive, on a USB drive, or on a network drive or place somewhere) and the data they have available.

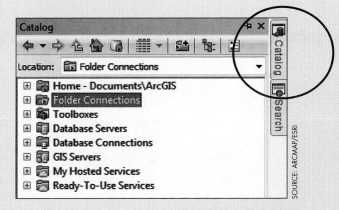

FIGURE 1.5 The Catalog and its tabbed version.

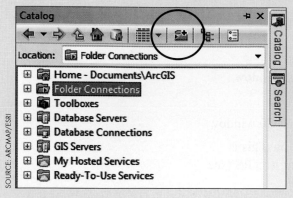

FIGURE 1.6 The Catalog and the Connect to Folder icon.

The **Connect to Folder** option (Figure 1.6) can be used to select new folders (such as your computer's C: or D: drives, a mapped network drive, a data server, or your USB drive). Pressing this button will allow you to select a new folder, which will be added to the available Catalog options.

By expanding the folder options you'll be able to see the data contained in the folders. You can then copy, paste, and remove items like in Windows. In this way you can move geographic data from one location to another. Moving files in this fashion instead of using utilities like Windows Explorer will ensure that any extra or ancillary files associated with your data are properly moved or deleted.

- Open the Catalog tool either by pushing its button on the Standard toolbar or expanding the pinned version on the right-hand side of the screen.

- Press the Connect to Folder button in Catalog.

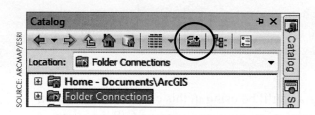

- In the new dialog that appears, set up a Folder Connection to the location of the chapter data. ***Important Note:*** For all chapters in this book, it is assumed that the book's data are located on a C: drive in a folder called GISBookdata. If the data are stored somewhere else on your machine or network, substitute that location for C:\GISBookdata\.

- Also set up a Folder Connection to the local area where you will be working with a copy of the book's data. ***Important Note:*** For all chapters in this book, it is assumed that the copy of the book's data you're working with is located on a computer's D: drive in a folder called GIS. If the copy of the data is stored somewhere else on your machine, a flash drive, or a network space, substitute that location for D:\GIS\.

- Once you have the two folders connected, expand the folder structure and copy the **Chapter1** folder from the C:\GISBookdata\ folder and place it in your D:\GIS\ folder. Be sure to copy the entire folder, not just the files within it.

  - Your D:\GIS\ drive should now have a folder called Chapter1.

- In Catalog, navigate to the D:\GIS\ folder on your machine.

- From there, expand the Chapter1 folder.

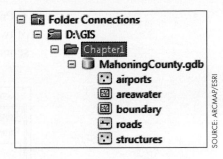

- Chapter1 contains a grey "barrel" icon (called a geodatabase) named MahoningCounty.gdb that you can expand. It contains five different GIS layers (each shown with its own grey icon). Check to be sure you have all of the following:

  - Airports: The locations of airports in the county

  - Areawater: The locations of lakes and rivers in the county

  - Boundary: The outline of Mahoning County

  - Roads: The roads within the county

  - Structures: The locations of many different kinds of buildings (schools, colleges, fire stations, bridges, and so on) in the county

STEP 1.2 Previewing and Examining GIS Data

- In Catalog, right click on the areawater layer and select **Item Description**.

- A new window will open with information about the areawater dataset. In this new window, select the **Preview** tab. You should see a set of shapes representing the county's water bodies.

- In the Item Description box, select the **Identify** tool from its toolbar (the white "i" inside a blue circle). Click on each of the two large lakes in the southeastern portion of the county and you'll get information back about that lake. Answer Question 1.1.

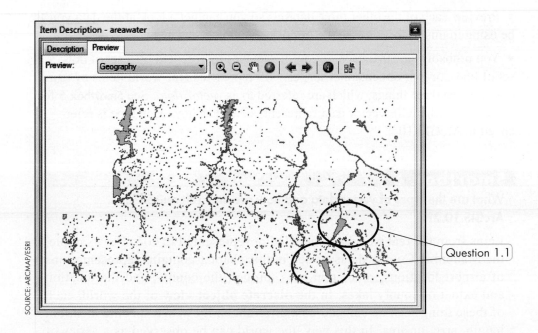

QUESTION 1.1 What are the full names of these two lakes, according to this dataset?

• Preview the airports dataset. You'll see that the airports are represented by points of two different colors. One color represents airport complexes and the other represents airport runways. Keeping this information in mind, answer Question 1.2.

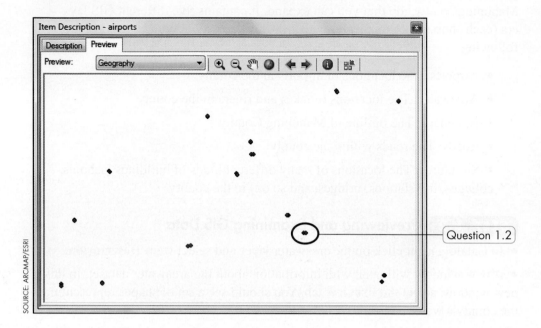

SOURCE: ARCMAP/ESRI

QUESTION 1.2 What is the name of the point representing an airport complex (not a runway), according to this dataset?

• Preview each of the other three datasets so you can get a look at the data you'll be using in this chapter. Close the Item Description box when you're done.

• You probably saw that each of your layers was represented by a set of dots, a set of lines, or two-dimensional shapes. GIS represents real-world items by using one of these three things, which are referred to as *vector data*. See **Smartbox 5** for more about what GIS vector data represent, as well as how each type is referenced in ArcGIS 10.2.

Smartbox 5

What are the types of vector data and how are they referenced in ArcGIS 10.2?

Items from the real world are modeled or represented in GIS. For example, the Mahoning County dataset used in this chapter contains representations of airport locations, roads that pass through the county, and the location and extent of county lakes. In the **discrete object view** of the world, each of these things can be considered a different entity with a specific location, length, size, or area. In this way, the world can be observed as a series of objects. For instance, each lake in the county has a fixed boundary, just like each road has a definite starting and stopping point. If there are multiple lakes in the county, then each one is considered its own object. The same is true with each leg of a county road.

discrete object view The conceptualization of the world that all items can be represented by objects.

Each of these objects can be represented in Arc-GIS in one of three ways (Figure 1.7): as a **point** (such as the airport locations), a **line** (such as the roads), or a **polygon** (such as the lakes). A point is a zero-dimensional object and represents just a simple set of coordinate locations. A line is a one-dimensional object created by connecting the starting and stopping points (and any points in between that give the line its shape). A polygon is a two-dimensional object that forms an area from a set of lines (or that define an area from a line that creates a boundary).

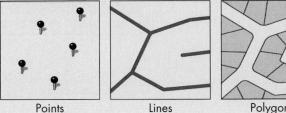

FIGURE 1.7 Representations of the three vector objects: points, lines, and polygons.

These points, lines, and polygons are referred to as **vector objects** and make up the basis of the GIS **vector data model**. By using these three vector objects, any real-world phenomena can be represented. For instance, the locations of stop signs, water wells, or groundwater testing sites can be modeled as points; streams, railroads, or power lines can be modeled as lines; and residential neighborhoods, congressional districts, or watershed boundaries can be modeled as polygons.

In ArcGIS, there are different data storage formats for the vector data model (and we'll go into more depth about them in Chapter 6). For now, however, two common ones (that we'll use in this book) are as follows.

• A **shapefile** is a simple data structure composed of several computer files that can store points, lines (referred to as *polylines*), or polygons. A shapefile can hold only one type of data. Thus, a line shapefile would hold roads data, while a polygon shapefile would hold county boundary data.

• A **geodatabase** is a single item (a file or a folder) that can hold one or more feature classes, and each feature class can hold a different type of data. For instance, a line feature class could hold walking path data, while a point feature class could be used to mark a trailhead.

In this chapter, the Mahoningcounty.gdb file that you are using is a file geodatabase, and each of the layers you're using is a feature class. The airports and structures layers are point feature classes, the roads layer is a line feature class, and the boundary and areawater layers are polygon feature classes.

You'll be working exclusively with vector data in Chapters 1 through 11 of this book. There is a second type of data storage, called *raster data,* that is completely different from vector data. You'll begin incorporating raster data into ArcGIS starting with Chapter 12.

point Zero-dimensional vector object.

line One-dimensional vector object.

polygon Two-dimensional vector object.

vector objects Points, lines, and polygons that are used to model real-world phenomena using the vector data model.

vector data model Model that represents geospatial data with a series of vector objects.

shapefile A series of files that make up one vector data layer.

geodatabase A single item that can contain multiple datasets, each as its own feature class.

STEP 1.3 **Adding Data to the Data Frame**

• It's time to add actual data to ArcMap instead of just previewing it.

• You can add data to the map by doing any of the following:

  • Pressing the "Add Data" button (the yellow and black "plus" button on the Standard toolbar)

  • Selecting **Add Data** from the **File** pull-down menu.

  • Dragging and dropping the files from the Catalog into either the View or the Table of Contents.

- Choosing one of the "add data" options will bring up a new dialog that allows you to select which layers you want to add to the map. The dialog will show you the drives and locations available from which to add data.

  - Note: The circled area in the graphic below shows a "**Connect to Folder**" option, which works here just like it works in the Catalog if you need to add data from another source (such as a flash drive or a data server) that's not set up as a Folder Connection in the Catalog.

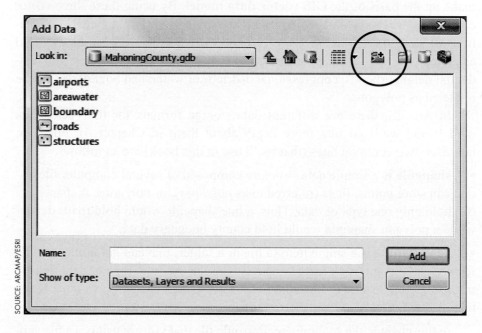

- To start, navigate to the D:\GIS\Chapter1\ folder and double click on the Mahoningcounty.gdb geodatabase to access the feature classes it contains.

- Select the **boundary** feature class to click **Add** to place it on the map.

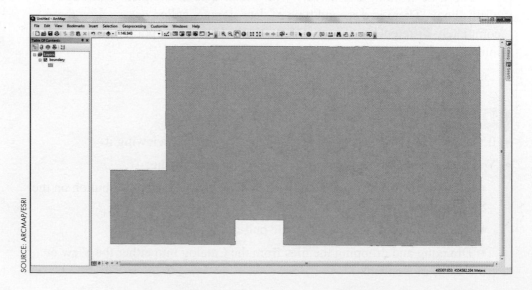

- Note that there is a checkmark in the box next to the layer's name in the Table of Contents. By clicking in this box, you can turn the checkmark off and on. When the box has a check in it, the data layer will be displayed. Without the checkmark, the data are not displayed.

- Note that your layer has been placed into the default **Data Frame** (named "Layers") in the Table of Contents. For more information about the Data Frame and what it does, see **Smartbox 6**.

Smartbox 6

What is the Data Frame and how does it work in ArcGIS 10.2?

The Data Frame contains all of the data being used in a single map document. Whatever layers you've added to the Table of Contents for this particular Data Frame can be displayed, changed, and worked with. You can have more than one Data Frame in the Table of Contents, but you can show the contents of only one Data Frame at a time in the View. By selecting the **Insert** pull-down menu and choosing **Data Frame**, you can place an additional Data Frame into the Table of Contents. You can drag and drop layers back and forth from one Data Frame to another. To select which Data Frame has its contents displayed, right click on a Data Frame and choose **Activate**. The Data Frame with its name in bold letters is activated and its contents are available to display and work with (Figure 1.8).

The Data Frame's properties will default to those of the first layer that is added to it. For instance, the coordinate system and units of measurement used by the Data Frame will match those of the first layer added to the Data Frame. If the first layer added to the Data Frame is a Universal Transverse Mercator projection with measurements made in meters, then the Data Frame will try to display all layers using this projection and will allow you to measure things in meters. You can change these Data Frame properties by right clicking on the Data Frame in the Table of Contents, selecting the Data Frame's **Properties**, then selecting the **Coordinate System** tab. You can then choose a new coordinate system for all layers in the Data Frame to be displayed in.

The Data Frame also controls the **extent** (geographic boundaries) of the display. By pressing the full extent button on the Tools toolbar, you will change the display to the full geographic boundaries of the layers in the Data Frame. The default extent will be the largest of the layers that is in the Data Frame itself. However, you can set your own custom extent by right clicking on the Data Frame and selecting **Properties**. Then, within the **Data Frame** tab, you can set the extent boundaries you wish to display.

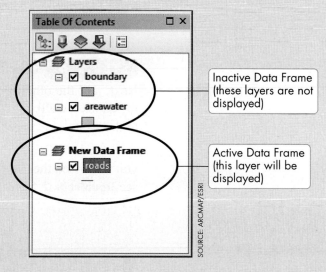

SOURCE: ARCMAP/ESRI

FIGURE 1.8 Inactive and active Data Frames in the Table of Contents.

extent The viewable dimensions of a layer.

- You'll start by examining the properties of the Data Frame itself. Right click on the Data Frame, then select **Properties**. This will bring up the Data Frame Properties dialog.

- Click the **General** tab and type a new name for the Data Frame instead of just "Layers." Call it "Mahoning County."

- Also under the General tab will be information about the units of the map and the units of display. Map Units represent the units of measurement that the data were created in, such as decimal degrees, feet, meters, or miles. Display Units represent how you are measuring things on the map. The Map Units should be shaded out so you can't change them. Answer Question 1.3.

QUESTION 1.3 What measurement units were the Mahoning County data made in?

- Select **Kilometers** for Display Units and click **Apply**. Now when you make measurements on the map, you will receive distances back as kilometers. Note that the only thing that has changed is the way ArcMap reports results back to you. It does not change what units the data were created in.

  - Return to the map and move the cursor around the screen. In the lower-right portion of the screen, you will see distances and coordinates being reported in kilometers.

- Go back to the Data Frame Properties and change the Display Units to **Miles**. Go back to the map and you will see measurements in miles.

STEP 1.4 Displaying Multiple Layers in the Table of Contents

Next, add the other four feature classes to the map (just click through any error messages—we'll deal with them in Steps 1.6 and 1.7). You'll see the layers added to the Table of Contents.

- Pin or hide Catalog back to the side of the screen. If during any of the actions you've closed the Catalog, Table of Contents, or other windows on the screen, see Troublebox 2 for an easy way to get them back.

Troublebox 2

What happened to the Catalog, Table of Contents, or other windows in ArcGIS 10.2?

It's easy to close a window accidentally when you meant to minimize it or pin it. If the Catalog, Table of Contents, or other windows disappear from ArcMap, don't panic. You can easily get them back. From the Windows pull-down menu, you can show (or not show) the Table of Contents, Catalog, and Search windows. Also, the Standard toolbar (see **Smartbox 3**) includes icons for opening (but not closing) these three windows, along with ArcToolbox (which we'll use in Steps 1.6 and 1.7) and the Python window (see Chapter 9).

• Data layers are displayed in ArcMap by their placement in the Table of Contents, or TOC (see **Smartbox 7**).

Smartbox 7

How can data layers be displayed in the Table of Contents (TOC) in ArcGIS 10.2?

There are four ways of presenting your data in the Table of Contents (TOC) (Figure 1.9):

• List by drawing order: This is the default setting. The layer at the top of the stack will be drawn on top of all other layers, then the layer underneath it drawn next, and so on. By using this option, you can drag layers from all sources to different positions in the TOC.

• List by source: Layers are listed according to their source. For instance, all the layers from one folder will be listed, then all of the layers from a second folder, and so on.

• List by visibility: Layers will be listed in two categories: those layers that are turned on and can be seen, and those layers that are turned off and are invisible.

• List by selection: Layers will be listed in two categories: those layers that can have objects selected from them (and how many objects are selected) and those layers that are unavailable for selection (see Chapters 2, 8, and 9 for more about selection).

FIGURE 1.9 The four different ways of displaying your data layers in the TOC.

• Select the **List by Drawing Order** option. The layer at the top of the TOC is displayed first. The layer displayed second from the top is displayed next, and so on. The top layer will be displayed "on top of" the other layers. Thus, if you have two layers that cover the same area, the one at the top of the TOC list will be displayed on top of the others.

> • You can move a layer by clicking on the name of the layer and dragging it to another place in the TOC.

> • Rearrange the layers on your map so that water bodies are displayed at the top, then the structures next, then the airports, then the roads, and finally the county boundary at the bottom.

STEP 1.5 Examining GIS Data and Working with Coordinate Systems, Datums, and Projections

• When using geospatial data in ArcGIS, it's a good idea to keep all of the data that you are using in the same projection, coordinate system, and datum to avoid any computational or analysis problems. For more information about working with these topics, see **Smartbox 8**, and for more in-depth technical details, see Appendix B.

Smartbox 8

What are datums, coordinate systems, and projections in ArcGIS 10.2?

Each layer of data used in ArcGIS is a piece of geospatial data—that is, it's tied to a specific location on Earth's surface. To create geospatial data, you need to have a **datum**, which is a reference surface or a model of Earth that can be used to determine locations around the globe. Earth is not perfectly round—it's more of an ellipsoid (or spheroid) shape, meaning it's slightly larger at its center than at the poles. Thus, a datum approximates this spheroid, determines where the center of the model is relative to the center of Earth, and allows us to reference coordinates across Earth. The problem is that there's no one universal datum used for all measurements. Literally hundreds of different datums exist, some of which reference the entire globe, while others are used for more local references. Because the datum is the starting point for determining coordinates, measurements made in one datum won't necessarily line up with measurements made from another datum. If you're determining the coordinates of a manhole cover using a system derived from one datum and a friend is measuring the same location but using a different datum, the two sets of coordinates (for the same object) will be different.

Some common datums that you'll encounter with geospatial data in GIS are:

• **NAD27**: The North American Datum of 1927. This datum was used for large-scale mapping of the United States and North America. It placed its center at Meades Ranch in Kansas.

• **NAD83**: The North American Datum of 1983. This datum is used for much of the current data measured in the United States and across North America today. Measurements from NAD83 are very different than measurements from NAD27 (differing by up to a couple hundred meters).

• **WGS84**: The World Geodetic System of 1984. The Global Positioning System (GPS) uses this datum for its measurements. Its coordinates are very similar, but not exactly the same, as those you'd get using NAD83.

Having a reference model allows you to determine the coordinates of locations. ArcGIS references coordinate systems in two different ways (and see Appendix B for further information):

• **Geographic coordinate system (GCS)**: This is the system of measurements made in latitude and longitude and measured in degrees. In ArcGIS, this reference model will commonly be listed as GCS with the datum that was used when the data were created.

• **Projected coordinate system (PCS)**: Often, GCS data will be transformed (or projected) into a flat two-dimensional grid system that allows you to make measurements in units such as feet or meters (rather than degrees) with consistent angles or lengths. Because a PCS is a flat grid system, coordinates are measured as an x (such as an east and west measurement) and y (such as a north and south measurement). As is the case with datums, there are a multitude of different projections, and measurements of distance, area, or perimeter made in one projection will not necessarily match up with the measurements made of the same locations in another projection. Chapter 4

datum A reference surface of Earth.

NAD27 North American Datum of 1927.

NAD83 North American Datum of 1983.

WGS84 World Geodetic System datum of 1984.

geographic coordinate system (GCS) A set of global latitude and longitude measurements used as a reference system for finding locations.

projected coordinate system (PCS) A set of measurements made on a flat grid system, initially derived from a GCS.

offers more information about the kind of projections used with online maps, but some common projected coordinate systems you'll likely encounter with geospatial data and GIS are:

- **UTM**: The Universal Transverse Mercator. This system divides the world into 60 zones, with each zone being 6 degrees of longitude wide. It covers all of Earth except for the poles. Measurements made in each zone are independent of measurements in other zones, so coordinates for zone 17 cannot be compared with coordinates from zone 16. UTM uses meters for its units of measurement.

- **SPCS**: The **State Plane** Coordinate System. This system is used only for the United States. It divides a state into different zones that correspond to county boundaries or similar lines. For instance, Ohio is divided into a north zone and a south zone, while Arizona is divided into an east zone, central zone, and west zone. Measurements made in each zone are independent of those made in another (so Ohio north zone measurements won't match up with Ohio south zone measurements). There are two types of State Plane coordinates—an old one based on NAD27 that uses feet and a new one based on NAD83 that uses either feet or meters.

- **Transverse Mercator**: This projection is commonly used for data with a north-south orientation, and UTM uses this projection as its basis. Units are usually measured in meters.

- **Lambert Conformal Conic**: This projection is commonly used for data that have an east-west orientation at middle latitudes (such as the United States). It keeps the shapes of locations the same as what they are in the real world, and its units are usually measured in meters.

- **Albers Equal Area Conic**: Like Lambert, this projection is used for east-west data at mid-latitudes, so it's commonly used with U.S. data. It keeps the areas (sizes) of locations the same as what they are in the real world, and its units are usually measured in meters.

UTM Universal Transverse Mercator projected coordinate system that divides the world into 60 reference zones with measurements in meters.

State Plane The SPCS (State Plane Coordinate System) projected coordinate system that divides states into zones with measurements made in feet or meters.

Transverse Mercator A projected coordinate system used for north-south data.

Lambert Conformal Conic A projected coordinate system used for east-west data.

Albers Equal Area Conic A projected coordinate system used for east-west data at mid-latitudes.

- Each data layer you're using in this chapter comes with information about the geographic coordinate system and the datum it was created in, as well as the projected coordinate system that each layer was projected to (and the units of measurement that this projected coordinate system uses). To examine this information about your data, do the following:

  - Right click on the layer and select **Properties**.

  - Click on the **Source** tab.

- Look at the coordinate system information for each of your five feature classes, then answer Question 1.4. (*Hint:* Think carefully about what information you're getting about the datum and the coordinate system.)

QUESTION 1.4 Fill out the following chart with the information on the datum, projected coordinate system, and units of measurement of each of the five layers.

Layer	GCS Datum	Projected Coordinate System	Units of Measurement
Airports			
Areawater			
Boundary			
Roads			
Structures			

• As you go through the layers, you'll find that one of them (the airports) is in a different projected coordinate system than the rest of your data and another one has no defined coordinate system at all.

• To make sure that all of your data properly align, you'll want to make these two data layers match the rest of your data in terms of their datum and projected coordinate system. This involves two different tasks: changing the projection of the airports layer (in Step 1.6) to match the others and establishing a coordinate system for the areawater layer (in Step 1.7).

STEP 1.6 **Changing a Data Layer to a Different Projection**

• To get the airports data to match up with the other layers, you'll have to change its projection to match your other data sources. See **Smartbox 9** for more information about changing projections.

Smartbox 9

How do you change projections in ArcGIS 10.2?

When you **project** a layer, you're creating a new layer that has a new datum, coordinate system, or projection. For instance, if you want to change a roads layer set up in GCS NAD83 measured in degrees and turn it into a UTM Zone 20 layer measured in meters, you'll have to project the data. Projection is a mathematical process that takes your data and sets them up in a new projected system (Figure 1.10). You would project your data to change them from their existing coordinate system (whether it's a geographic coordinate system or a projected coordinate system) to a different one. When changing from one datum to another, ArcGIS may ask you to specify or select a particular datum transformation method so that the newly projected layer is set up correctly.

When you add data layers in two different projections to the Data Frame, ArcMap has the capability to "Project On The Fly" and will properly line up the two datasets instead of showing how they don't fit with each other. Note that ArcMap does this for display purposes only. ArcMap doesn't actually change the properties of the layer; it simply alters its appearance to match the other data in the Data Frame. The projection that is displayed in ArcMap is controlled by the coordinate system that the Data Frame is set to. You can set

project To change a layer from one coordinate system into another.

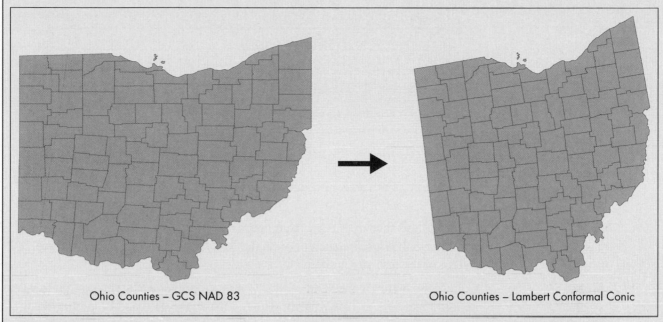

FIGURE 1.10 Ohio county boundary files in a GCS NAD 83 projection and then projected into the Lambert Conformal Conic projection.

Ohio Counties – GCS NAD 83

Ohio Counties – Lambert Conformal Conic

the coordinate system by right clicking on the Data Frame in the Table of Contents, selecting the Data Frame's **Properties**, then selecting the **Coordinate System** tab.

> **ArcToolbox** A separate window in ArcGIS containing multiple tools and functions.

• Projection tools are located in **ArcToolbox**, an ArcGIS utility that contains multiple tools for many different purposes. Open ArcToolbox by clicking on the red "Toolbox" icon on the Standard toolbar.

SOURCE: ARCMAP/ESRI

• ArcToolbox will open in a new window that you can move around, pin, or dock.

• Expand the **Data Management Tools** toolbox (the red icon that looks like a box) and select the **Projections and Transformations** toolset (the red and gray icon that looks like a hammer lying on top of the box). From there, select the **Feature** toolset, then the **Project** tool (the hammer icon itself).

• This will bring up the **Project** tool's dialog.

• For your input dataset, select the airports feature class. Use the browse button to navigate to your D:\GIS\Chapter1\ folder and select airports from inside the MahoningCounty file geodatabase.

• For your output dataset, you will create a new feature class called airports_project, which you should put in the MahoningCounty file geodatabase inside your D:\GIS\Chapter1\ folder.

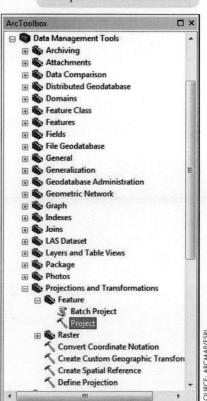

SOURCE: ARCMAP/ESRI

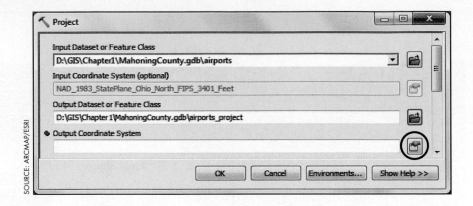

- You'll now have to select what coordinate system you want to reproject your data into so that it will match up with the NAD83 UTM Zone 17 coordinate system of the rest of your data.

- Press the button to the right of the "Output Coordinate System" option to select a coordinate system. This will bring up the "Spatial Reference Properties" dialog.

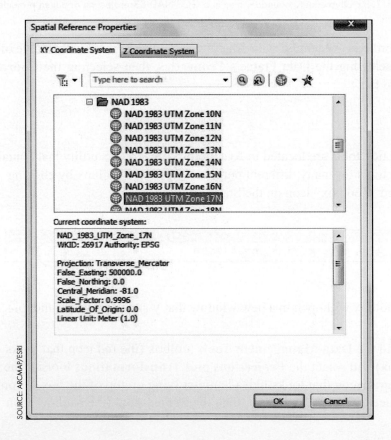

- Select the **XY Coordinate System** tab.
- From the new options choose:
  - **Projected Coordinate Systems**
  - **UTM**

- **NAD83**

- Lastly, choose: **NAD 1983 UTM Zone 17N**

- You should see all the information for your newly selected coordinate system.

- Click **OK** to return to the Project dialog.

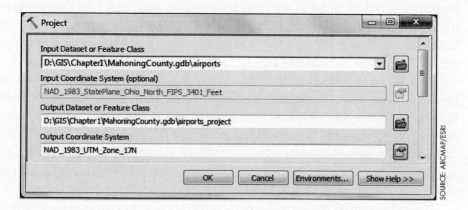

- When all options are properly set, click **OK**. ArcToolbox will reproject your data, creating a new feature class called airports_project in your Mahoning County file geodatabase.

- After a few seconds, you'll see scrolling text across the bottom of ArcMap indicating that the Project tool is in progress. A pop-up box will appear at the bottom of the screen to let you know when the tool has completed the process.

- You'll see that an airports_project layer has been added to ArcMap's TOC. You can delete the old airports layer at this point. Right click on it and choose **Remove**.

STEP 1.7 Defining a Projection for a Layer that is Missing this Information

- The next step is to take care of the areawater feature class that has no coordinate system information attached to it. Unfortunately, you cannot simply project this layer like you did with the interstates layer. Answer Question 1.5.

QUESTION 1.5 Why can't you just use the "project" option to fix the coordinate information for the areawater layer like you did for the airports layer? (*Hint:* Think carefully about what "project" is really doing.)

- Because the areawater layer is missing its coordinate system and projection information, we will need to define a projection for it. See **Smartbox 10** for more background information about defining projections.

Smartbox 10

How do you define a projection in ArcGIS 10.2?

Sometimes, you'll receive geospatial data to work with (or create it yourself) and its coordinate system will be listed as "undefined." The word "undefined" often indicates that the creator of a dataset did not explicitly tell ArcGIS what coordinate system was used in making the layer. In the shapefile data format, a separate .prj is required to hold this information, and other GIS data formats often have something similar—related information that goes with the layer and holds the coordinate information. If this file or information is missing (or not created), then you will have to supply ArcGIS with this information yourself.

The Define Projection tool in ArcGIS allows you to do this. Think of it like a person showing up at a party but forgetting to put on a name tag—he knows what his name is, but nobody else does. Using Define Projection is like putting a name tag on a person to properly identify him to everybody else—in this case, making sure that ArcGIS understands which coordinate system the layer has to try to match with other datasets. In short, Define Projection allows you to tell a layer that is missing the coordinate information exactly what its coordinate system is.

Be very careful to provide the correct projection information when using this tool. If you define the wrong projection, you will encounter a lot of problems. Think of name tags again: If Bob arrives at the party but wears Dave's name tag, everybody at the party will assume Bob is Dave, has Dave's job, drives Dave's car, and has Dave's characteristics. The same concept holds true for geospatial data. If a data layer should be defined as UTM but you incorrectly define it as State Plane, ArcGIS will try to make measurements of the UTM layer using the State Plane system and will try to project the layer "on the fly" using its information about State Plane. The layer will not match the others because ArcGIS is working with incorrect information. Often, you'll see all of your regular data shown in one place on the map, while the incorrectly identified layer is shown in a completely different location.

Note: If you receive a layer that has no geospatial reference whatsoever (such as a scanned photograph, scanned map, or architectural diagram not drawn with real-world grounding), then Define Projection won't solve the problem (because the dataset isn't just missing its coordinate system, it also had no coordinate system to begin with). In this case, a new solution—georeferencing—will have to be used (see Chapter 13 for more on this topic).

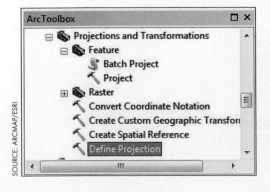

- In ArcToolbox, expand the **Data Management Tools** toolbox and select the **Projections and Transformations** toolset. From there, select the **Define Projection** tool.

- Select the **areawater** layer for the input dataset.

- Right now, the areawater layer has an unknown coordinate system assigned to it. To explicitly select a coordinate system to assign to this layer, press the button next to the Coordinate System pull-down menu.

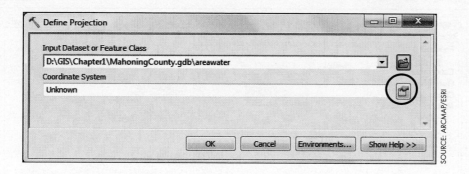

- Select the **XY Coordinate System** tab.
- From the new options choose:
  - **Projected Coordinate Systems**
  - **UTM**
  - **NAD83**
  - Lastly, choose: **NAD 1983 UTM Zone 17N**
- You should see all the information for your newly selected coordinate system.
- Click **OK** to return to the Define Projection dialog.
- Click **OK** again, then wait as ArcMap defines the projection. The pop-up window in the lower right corner of the screen will let you know when it's done.
- From the TOC, bring up the Layer Properties of the areawater layer again and look at the Source. You should see the proper projection information in place.

  - ***Important Note:*** Luckily, the coordinate system that the areawater layer was missing happened to be the same one as the other layers. If it had been different (for instance, if areawater had been missing State Plane coordinate information), you would then have to use the Project tool to change it to match the coordinate and projection information of the other layers.

STEP 1.8 Changing the Appearance of a Layer

- Now that all of your data are properly set with their spatial reference, it's time to make them look good. Since you're compiling these data for future presentations, the first thing you want to do is change the color and appearance of the data. When layers are added to ArcMap, they're assigned a random color, so you could end up with orange lakes or green roads. To alter the appearance of a layer, you'll want to change its **symbology**. All of the information about a layer (including its symbology) is stored in the Layer Properties of each layer. To bring up the Layer Properties for a single layer, you can either double click on the name of the layer in the TOC, or you can right click on a layer and select **Properties** to do the same thing.
- Bring up the Layer Properties of the areawater layer.
- We'll start by altering the color of a layer, so select the **Symbology** tab from the Layer Properties.

symbology The appearance of a data layer.

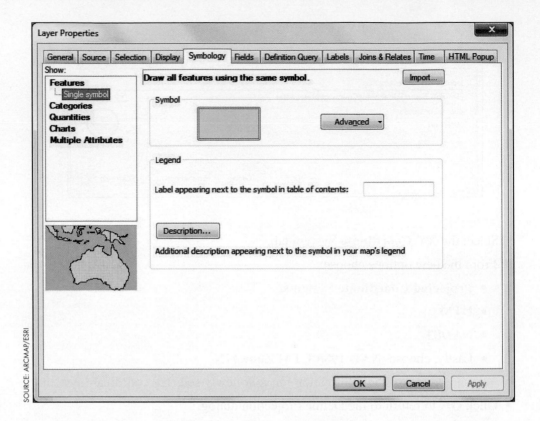

SOURCE: ARCMAP/ESRI

- Under the Show: box (in the left hand corner of the dialog), select **Features** and **Single symbol**. (This means that all of the objects in this layer, no matter what they represent, will be drawn with the same color, outline, or symbol.)

- Click in the Symbol options (the large colored box) to bring up a choice of new colors for display.

  - From here, you can select new colors to fill in the polygons, a new color for the outline of the polygons, and the width of that outline.

  - Select a new appropriate color and look for Mahoning County's water bodies. After selecting the new symbology, you can return to the layer properties by clicking **OK**.

- Next, change the symbology of the roads layer to select a different color for the roads. You may want to adjust the width of the roads as well as use a different type of symbol instead of the default thin line.

- However, you'll note that the areawater and roads layers contain several different types of water bodies or road types. ArcMap allows you to display these layers according to these different types instead of displaying all objects with the same symbol—for instance, showing interstates with a different symbol than residential roads. The structures layer contains several different types of buildings, such as schools, bridges, and hospitals, and you would want to show the layer in ArcMap by giving each building type its own symbol.

  - To do this, bring up the Layer Properties of the structures layer and go to the **Symbology** tab. Under the Show: options, select **Categories**, then select **Unique values**.

  - For the Value Field, choose **Ftype** and then click on **Add All Values**. This will show the different categories of points, each grouped by its own symbol.

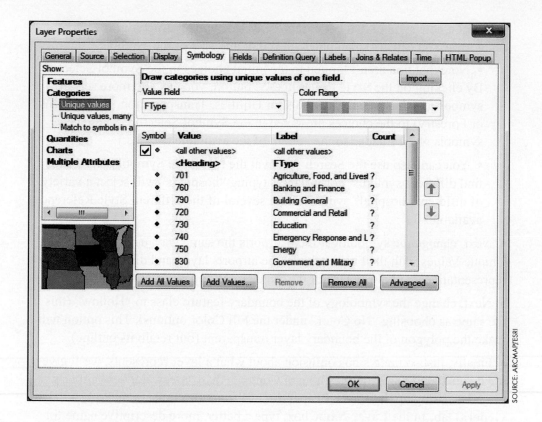

- Click on one of the point symbols; this will bring up the Symbol Selector, which has a variety of different symbols to use for points.

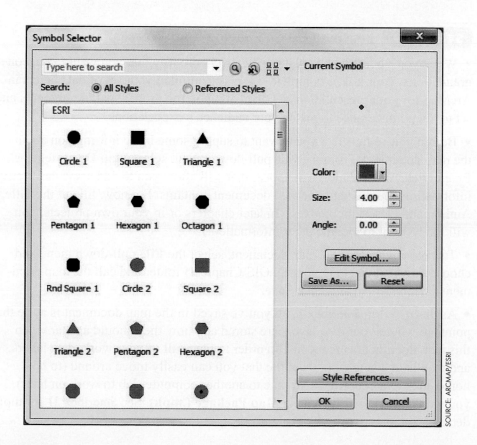

- Select a new symbol that better represents the type of structure for each of the categories. You can also manipulate the color and size of the symbols.

  - Note that you aren't limited to the initial choices in the Symbol Selector. By clicking on the **Style References...** button, you can add more sets of symbols (such as special symbols for Utilities, Transportation, Public Signs, or Forestry) to the choices presented in the Symbol Selector. (These new symbols will be added to the bottom of the available choices.)

  - You can also use the Search option at the top of the Symbol Selector to find different symbols. For instance, typing "hospitals" will select a variety of different "hospital" symbols from several of the different Style References available.

- Next, change the symbology of the airports the same way, using all of the Unique Values with the Ftype value in the airports layer, and display a new, more representative symbol for each of the three airport types.

- Next, change the symbology of the boundary feature class to "Hollow" (this is the same as choosing "No Color" under the Fill Color options). This option will make the polygon of the boundary layer transparent (but retain its outline).

- Finally, just so there's no confusion about what a layer represents, we'll give the layers better names in the Table of Contents than "areawater" or "airports_ project." Bring up the **Layer Properties** of the areawater layer and select the **General** tab. In the **Layer Name** box, type a better, more descriptive name for this layer (such as County Water Bodies), then click **Apply** and **OK**. You'll see the name changed in the TOC. Give each of your other four layers better names than the file name defaults.

STEP 1.9 Finishing Up and Saving Your Work

- When you've finished displaying the layers the way that you want them, congratulations, your task is complete—you've compiled your set of data layers in ArcMap for your presentation, set them all up in the proper spatial reference, and set up their appearance to match your audience's expectations.

- Before you're finished, you'll want to supply some basic information about the map document. From the **File** pull-down menu, select **Map Document Properties**. A new dialog will appear that will allow you to add some descriptive information about what your map document contains. For now, fill out the Title, Author, and Description boxes. (In later chapters or in your own projects, you will likely want to include other information.)

- To save your work as a map document, select the **File** pull-down menu and choose **Save**. Navigate to your D:\GIS\Chapter1\ folder and call the map document Chapter1.mxd and click **Save**.

Map Package (.mpk) A single file that contains the map document, all data layers used in the map document, as well as their appearance.

- As discussed in **Smartbox 1**, all you've saved in the map document is a file that points to where your data layers are stored and how they should appear when this map document is reopened. In order to place all of your work, data layers, and results in a single compact file that you can easily move around (to take home, take to a conference, or take to another computer lab to work on later), you would want to also create a **Map Package (.mpk)**. See **Smartbox 11** for more details.

Smartbox 11

How do I save my work as a Map Package in ArcGIS 10.2?

A Map Package will collect your map document plus all of the data layers that you've used or created in your map document (as well as their assigned symbology and appearance) and place them into a single compressed file. This makes a map package much more portable than a map document because, in a map package, all of your related files are in one file. In addition, a map package can be reopened on another computer that does not have the same path names or folder structure as the original. For instance, if you saved your Chapter 1 map document in your D:\GIS\Chapter1 folder and took the map document to another computer, you would have to place it on the computer's D: drive in a folder of the same name to properly reopen it so that all of the links work correctly. With a Map Package, you could reopen the map document from a different drive (for instance, a flash drive that your computer identifies as a J: drive) and all links will reopen correctly. A Map Package can also be uploaded to the cloud and then downloaded to a different computer and opened there.

To create a Map Package, do the following. From the **File** pull-down menu, select **Share As,** and then select **Map Package**. From here, you can either choose to save your package to your own computer (as a .mpk file) or share your Map Package via ArcGIS Online (see Chapter 4 for more about this option).

Click **Analyze** and ArcGIS will check your work for any outstanding issues (such as layers with undefined coordinate systems). You'll have to fix or address these issues before ArcGIS will allow the Map Package to be created. For instance, you will need to add an Item Description and tags before the Map Package can be created. Also, all layers in the Map Package must have their spatial reference defined. Once the data are validated, click **Share** and the Map Package will be created (Figure 1.11).

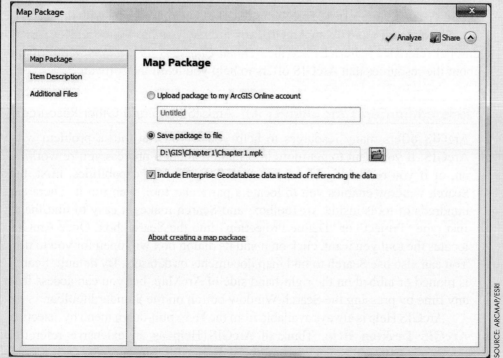

FIGURE 1.11 Creating a Map Package in ArcGIS 10.2.

SOURCE: ARCMAP/ESRI

To reopen a Map Package again, locate the file using Catalog, right-click on the Map Package file, and select **Unpack**. All of the files from the Map Package will open and load into ArcMap.

- In future chapters, you will usually either create a professional looking map layout to print (see Chapter 3) or digitally share your finished product with others via ArcGIS Online (see Chapter 4). For now, however, you can take your completed display and save it as a PDF file that you can print or share with others. From the **File** pull-down menu, select **Export Map**. Give it a File Name of **Chapter1** and for Save as Type select **PDF**. Place this PDF file in your D:\GIS\ Chapter1\ folder. When you're done, answer Question 1.6.

> **QUESTION 1.6** Show your final saved results or your PDF file to your instructor, who will check your work for completeness and accuracy to make sure you get credit for this question.

Closing Time

This chapter introduced the basic principles of working with geospatial data in ArcGIS—that things in the real world can be represented by one of three objects (points, lines, or polygons), that these objects must have a defined spatial reference and coordinate system, and that they can be visualized in different ways in ArcMap. In the next chapter, we'll start looking beyond just the locations and geospatial coordinates of these objects and see how descriptive information can be tied to these locations as well. For instance, each of the points in the structures layer had information about the type of building that point represented (education, industry, health, and so on). It's these types of attributes that we'll begin working with in Chapter 2.

As you learn more about ArcGIS, you are likely to have questions or seek additional reference materials. See the *Related Concepts for Chapter 1* information about the resources that ArcGIS offers to help you learn the software.

Related Concepts for Chapter 1   ArcGIS Help and Other Resources

ArcGIS offers many resources to help you if you run into a problem with ArcGIS, if you want to get more information about a process you're working on, or if you're looking to expand ArcGIS beyond its capabilities. First, the Search window enables you to locate a particular tool, then run it. There are hundreds of tools inside ArcToolbox, and Search makes it easy to find them. Just type "Project" or "Define Projection" into the Search box. Once ArcGIS locates the tool you want, click on it and its dialog box will open for you to use. You can also use Search to find map documents or datasets. By default, Search is pinned or tabbed on the right-hand side of ArcMap, but you can access it at any time by pressing the Search Window button on the Standard toolbar.

ArcGIS Help is always available from the **Help** pull-down menu by selecting **ArcGIS Desktop Help**. Think of ArcGIS Help as an extensive reference library available at the click of a mouse. Help contains information on using tools, syntax for commands in the Python programming language, and information on the theory behind the GIS functions you're working with. Pressing the **Resource Center** button in the Help dialog box will access an entire online repository of additional GIS and ArcGIS 10.2 information for you. Each chapter in this book contains a "For More Information" section that lists the exact phrases to type into Help (or the Resource Center) to get additional info on a particular topic. Topics accessible solely through the Resource Center are indicated by (RC).

The same Help topics and Resource Center information for ArcGIS 10.2 are also available online at http://resources.arcgis.com/en/help/main/10.2/index.html, which includes information from guidebooks and tutorials as well. In fact, Esri maintains an online repository of additional resources beyond the ArcGIS 10.2 Help features at http://resources.arcgis.com/en/home/. From here, you can watch videos that walk you through a certain topic, ask questions of the GIS community in the many ArcGIS forums, follow ArcGIS blogs, and access scripts, add-ons, and other utilities created by ArcGIS users.

You can also customize ArcGIS Desktop to make it look the way you want it to. You can also create your own toolbars, as well as create new buttons or menus for certain tools or tasks. See Appendix A for more information on customizing these options in ArcGIS. Finally, you can use social media to keep up with the latest ArcGIS news and updates from Esri. Find Esri on Facebook at http://www.facebook.com/esrigis or follow them on twitter at @Esri.

For More Information

For in-depth information about the topics presented in this chapter, use the ArcGIS Help feature (in the software or online) to search for the following items:

- Creating a Map Package
- Datums
- Define projection (data management)
- Georeferencing and coordinate systems (RC)
- North American datums
- Project (data management)
- Using Data Frames
- Using the Table of Contents
- What are geographic coordinate systems?
- What are projected coordinate systems?
- What is the Catalog window?

Also, for more information about land application of biosolids, see Czajkowski, K. P., A. Ames, B. Alam, S. Milz, R. Vincent, W. McNulty, T. W. Ault, M. Bisesi, B. Fink, S. Khuder, T. Benko, J. Coss, D. Czajkowski, S. Sritharan, K. Nedunuri, S. Nikolov, J. Witter, and A. Spongberg. 2010. "Application of GIS in Evaluating the Potential Impacts of Land Application of Biosolids on Human Health." *Geotechnologies and the Environment*, 3 (4): 165–186.

And for further information concerning the Geospatial Training Services salary survey, see here: http://www.gisuser.com/content/view/25731/222/

Key Terms

geographic information systems (GIS) (p. 1)

geospatial or spatial (p. 1)

geospatial technologies (p. 1)

Esri (p. 2)

ArcGIS for Desktop 10.2 (p. 2)

ArcGIS 10.2 (p. 2)

ArcGIS Basic (p. 2)

ArcGIS Standard (p. 2)

ArcGIS Advanced (p. 2)

ArcMap (p. 4)

map document (.mxd) (p. 5)

Table of Contents (p. 6)

Data Frame (p. 6)

layer (p. 6)

Tools toolbar (p. 6)

Standard toolbar (p. 6)

Catalog (p. 9)

discrete object view (p. 12)

point (p. 13)

line (p. 13)

polygon (p. 13)

vector objects (p. 13)

vector data model (p. 13)

shapefile (p. 13)

geodatabase (p. 13)

extent (p. 15)

datum (p. 18)

NAD27 (p. 18)

NAD83 (p. 18)

WGS84 (p. 18)

geographic coordinate system (GCS) (p. 18)

projected coordinate system (PCS) (p. 18)

UTM (p. 19)

State Plane (p. 19)

Transverse Mercator (p. 19)

Lambert Conformal Conic (p. 19)

Albers Equal Area Conic (p. 19)

project (p. 20)

ArcToolbox (p. 21)

symbology (p. 25)

Map Package (.mpk) (p. 28)

How to Use Tables and Attributes in ArcGIS 10.2

Introduction

You can represent all types of geospatial data using GIS. For instance, a set of lines can represent the length of roads, while several polygons can represent the area and dimensions of parcel boundaries. However, there's a lot more to capturing these features in GIS than just specifying their spatial properties such as lengths or co-ordinates. A transportation engineer is going to need information about the road's name, speed limit, covering type, and condition, while a county auditor would need information about the owner, zoning, property valuation, and taxes on the land parcels. All of these kinds of data are **non-spatial data**, which are data that are descriptive and not location based. For instance, a spreadsheet could contain the information about the address of a house, the owner's name, and the house's tax-assessed value without having any ties to geospatial data.

Fortunately, GIS easily allows you not only to handle non-spatial data, but also to connect non-spatial data to geospatial locations. (After all, having accurate geospatial representations of land parcels separate from tabular data about those parcels wouldn't be very helpful for analysis.) In ArcGIS, all of the information (spatial and non-spatial) about a layer is stored in that layer's **attribute table**. This table consists of a series of rows (referred to as **records**) and columns (referred to as **fields**). The table's records represent each separate object. For instance, if a layer contains 100 polygons representing the boundaries of lots in a new subdivision, then its attribute table would contain 100 records, one for each polygon/land lot. If you have a point layer representing the locations of 150 different water wells, then its attribute table will have 150 records.

The table's fields contain all of the **attributes** (or related information) about each record, such as any non-spatial data. For instance, each record in the lots layer would have a field indicating the name of the lot's owner, the zoning code for the lot, the monetary value of that lot, a parcel ID, and so on. Figure 2.1 shows an example of an ArcGIS attribute table of polygons representing the boundaries of each county in the United States.

Each record is a separate county polygon, while the fields represent the at-tributes of each county (county name, the state that county is in, the year 2000 population, the population per square mile, and so on). This chapter introduces you to working with attribute data in ArcGIS, building queries of GIS data based on attributes, and joining non-spatial tabular data to geospatial feature class data.

non-spatial data
Descriptive information that does not have location-based qualities.

attribute table A spreadsheet-style form where the rows consist of individual objects and the columns are the attributes associated with these objects.

records The rows of an attribute table.

fields The columns of an attribute table.

attributes The non-spatial data that can be associated with a geospatial location.

Fields

NAME	STATE NAME	POPULATION 2000	POPULATION (00) PER SQMI	POP2010	POP10_SQMI	WHITE	BLACK
Autauga	Alabama	43671	72.3	54571	90.3	42855	9643
Baldwin	Alabama	140415	85.6	182265	111.1	156153	17105
Barbour	Alabama	29038	32.1	27457	30.4	13180	12875
Bibb	Alabama	20826	33.3	22915	36.6	17381	5047
Blount	Alabama	51024	78.4	57322	88.1	53068	761
Bullock	Alabama	11626	18.6	10914	17.5	2507	7666
Butler	Alabama	21399	27.5	20947	26.9	11399	9095
Calhoun	Alabama	112249	183.3	118572	193.6	88840	24382
Chambers	Alabama	36583	60.7	34215	56.7	20112	13257
Cherokee	Alabama	23988	40	25989	43.3	24081	1208
Chilton	Alabama	39593	56.5	43643	62.3	36713	4230
Choctaw	Alabama	15922	17.3	13859	15	7731	6012
Clarke	Alabama	27867	22.3	25833	20.6	14070	11336
Clay	Alabama	14254	23.5	13932	23	11380	2066
Cleburne	Alabama	14123	25.2	14972	26.7	14079	498
Coffee	Alabama	43615	64.1	49948	73.4	37330	8359
Colbert	Alabama	54984	88.4	54428	87.5	43789	8768
Conecuh	Alabama	14089	16.5	13228	15.5	6788	6149

Records

SOURCE: ARCMAP/ESRI

FIGURE 2.1 An ArcGIS attribute table.

Chapter Scenario and Applications

This chapter's scenario puts you in the role of an education coordinator for the state of Ohio. As part of a study related to higher education in the state, your job is to assess the state's population distribution and then examine the locations of the colleges and universities in a set of counties in northeast Ohio. In the end, you'll be presenting your results in an interactive format that shows which colleges are located in which cities and includes graphics of the colleges themselves.

Aside from this chapter's scenario, the following are examples of other real-world applications of this chapter's theory and skills:

• A county auditor is updating the GIS database following county-wide property tax assessment. Rather than retyping the database, she would join the new assessment spreadsheet to the geospatial data of the land parcel locations.

SOURCE: JERRY CLEVELAND/ THE DENVER POST VIA GETTY IMAGES

• A marketing analyst for a department store is working on a focused advertising campaign. To determine where shoppers are traveling from, he can pose a query to the attribute table to find certain ZIP codes and examine their location and boundaries on a map.

• An environmental engineer is performing a soil analysis in GIS. By posing a query to the attribute table, she can quickly select all areas that contain a certain type of soil. She can then export those features to their own layer for further analysis.

ArcGIS Skills

In this chapter, you will learn:

• How to open a GIS data attribute table.

• How to add a table (such as an Excel spreadsheet) to ArcMap.

• How to join a spreadsheet table to a GIS layer's attribute table.

- How to sort an attribute table.

- How to perform both simple and compound Select By Attributes queries.

- How to export the results of a selection to a new feature class.

- How to compute statistics for a field.

- How to select an attribute field to label and how to apply labels to features.

- How to add a new field to an attribute table and how to add data to the new field.

- How to set up and activate a hyperlink in an attribute table as an interactive function.

 ## Study Area

- For this chapter, you will be working with data from Ohio counties, focusing on Mahoning and nearby counties: Trumbull, Columbiana, Stark, and Portage.

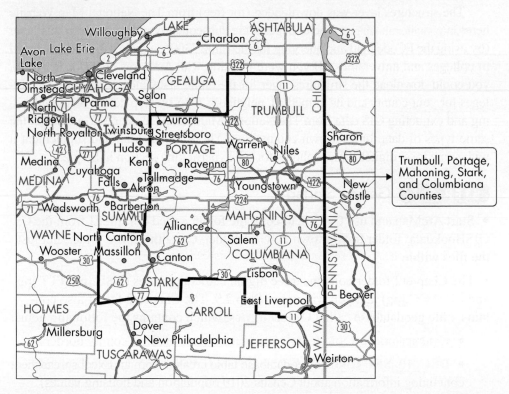

 ## Data Sources and Localizing This Chapter

This chapter's data focuses on features and locations within Ohio's counties. However, this chapter can be easily modified to use county data from your own state by using data from the U.S. Census Bureau (see the *Related Concepts for Chapter 2* for more about using Census data in ArcGIS).

The counties dataset was downloaded from the TIGER/Line files of the Census Bureau Website here: ftp://ftp2.census.gov/geo/tiger/TIGER2013. From the COUNTY folder, download the available shapefile. This contains all of the counties within the United States—you can query the shapefile and extract those counties into a separate feature class for your state based on the STATEFP attribute. This attribute references each state with a two-digit FIPS code. For instance, Ohio is referenced with state FIPS code 39. If you were using this chapter in (for instance) Arkansas, you would use FIPS code 05. A full list of state and county FIPS codes is available from the EPA online here: http://www.epa.gov/envirofw/html/codes/state.html. Note that the U.S. Census Bureau's county boundaries extend to the national boundaries when they border water features (such as on the Great Lakes area between the United States and Canada).

Note however that the TIGER files provide only the geographic boundaries. Attribute data must be downloaded separately and joined to the shapefiles. The attribute data (the population statistics) for each county in the state were downloaded from the Census Bureau's American FactFinder Website available here: http://factfinder2.census.gov. The table used in this chapter was G001: Geographic Identifiers (2010 SF1 100% data). FactFinder attribute data were downloaded as a CSV file, edited and cleaned up, and converted to an Excel file for use in this chapter. The Census Bureau provides an online guide to obtaining data from American FactFinder and preparing them for use in ArcGIS here: http://www.census.gov/geo/education/pdfs/tiger/Downloading_AFFData.pdf.

The structures layer was downloaded (for free) from The National Map Website here: http://nationalmap.gov. The colleges were subsequently extracted from this layer (by using the FCode field and querying for the value 73006, which is the code assigned to colleges and universities). If you were working with Arkansas data (for instance), you could download the structures layer for the entire state and extract only the colleges for your county and its immediate neighbors. Step-by-step details on downloading and extracting GIS data from The National Map are in Chapter 5. You can get the same types of data for your own county from The National Map and supplement it with some local digital camera photos for the hyperlinking section in Step 2.9.

STEP 2.1 Getting Started

• Start ArcMap and use Catalog to copy the folder called **Chapter2** from the C:\ GISBookdata\ folder to your own D:\GIS\ folder. Copy the entire folder, not just the files within it.

• The Chapter2 folder contains three digital camera pictures (**YSU.jpg**, **ITT.jpg**, and **MCTCC.jpg**) that you will use in Step 2.9. The Chapter2 folder also contains a file geodatabase called **Ohiocounties**, which contains the following items:

> • **CensuscountiesOhio** (a polygon feature class of Ohio county borders)

> • **DEC_10_SF1_G001** (a geodatabase table created from an Excel spreadsheet containing information about Census 2010 population and housing values)

> • **Colleges** (a point feature class of colleges and universities within a five-county region of Ohio)

• Add the **CensuscountiesOhio** feature class to ArcMap.

• The Coordinate System being used in this chapter is **GCS NAD 83**, using Map Units of **Decimal Degrees**. Check the Data Frame to verify that this coordinate system is being used. (See Chapter 1 for more about examining the coordinate system and map units of the Data Frame.)

• Add the **Editor toolbar** to ArcMap. From the **Customize** pull-down menu, select **Toolbars**, then select **Editor** (the same way you use or hide any toolbar in ArcGIS, as explained in Chapter 1). The Editor toolbar is moveable and dockable; place it out of the way (or dock it)—you won't be using it until Step 2.9.

STEP 2.2 Examining the Records and Fields of a Layer's
Attribute Table

• You have the geospatial data of the Ohio counties displayed on the screen,
but first we'll look at the counties' attribute table. In the TOC, right-click on the
CensuscountiesOhio layer and select **Open Attribute Table**.

• A new pinnable window called Table will open. For further information about
what you're examining in this attribute table, see Smartbox 12 .

Smartbox 12

What are the components of an attribute table in ArcGIS 10.2?

When an attribute table (or any type of table, for that matter) is opened in
ArcGIS, a single dockable window called Table is added to ArcGIS. You'll
see a tab at the bottom of the Table window indicating the name of the attribute table (CensuscountiesOhio). As you open more attribute tables, you'll see
them added to the tabs at the bottom. No matter how many attribute tables you
have open, ArcGIS will just have one Table window and each attribute table
is accessible as a different tab (similar to how you would use multiple tabs for
different pages in a Web browser).

There are several buttons and controls available for use in the Table window (Figure 2.2):

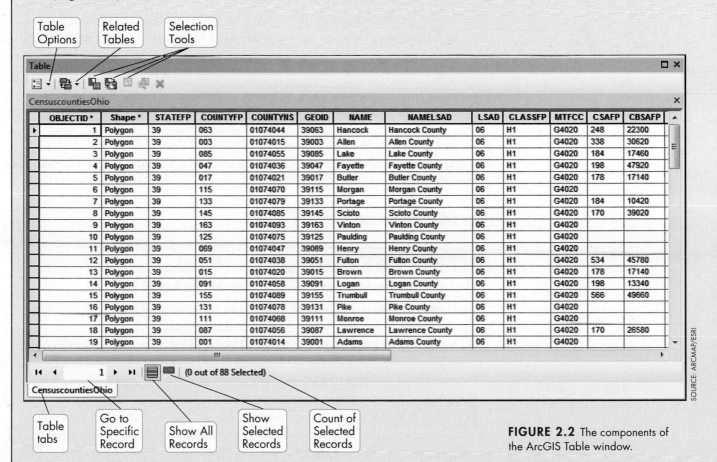

FIGURE 2.2 The components of
the ArcGIS Table window.

- The **Table Options** menu contains several commands related to attribute tables in ArcGIS.

- The **Related Tables** button allows you to see which (if any) tables have been joined or are related to this one (see Step 2.3 and **Smartbox 14** for more information).

- The **Selection Tools** are used when querying or selecting records (and also clearing the results of a query or selection). See Step 2.4 and **Smartbox 15** for more information.

- The **Table tabs** show which of the opened attribute tables are being displayed in the Table window.

- The **"Go to a Specific Record"** buttons allow you to examine a particular record and step through records from that point.

- The **Show All Records** and **Show Selected Records** buttons are used to switch what is being displayed in the Table window. When records are selected (see Step 2.4 and **Smartbox 15**), pressing Show Selected Records will show only those records in the Table. Pressing Show All Records will display all of the records, whether they are selected or not.

- The **Count of Selected Records** will tell you the total number of records in the attribute table, as well as how many have been selected.

- Each polygon in the feature class is shown as a different record. The ObjectID field is used internally by ArcGIS to identify object number 1, object number 2, and so on. From examining the count of selected records at the bottom of the Table window, you'll see that there are a total of 88 records in this attribute table. Answer Question 2.1.

> **QUESTION 2.1** What do each of the records in the attribute table correspond to (in terms of the geospatial data you can see in the View)?

- From examining the fields in the CensuscountiesOhio attribute table, you'll see that there are several different types of data available, including different numerical values and text strings. Each field of attributes can be placed into one of four types of data classifications. Now read **Smartbox 13** for more information about these different data types.

Smartbox 13

What are the different types of GIS attribute data in ArcGIS 10.2?

Fields of non-spatial data values in attribute tables represent one of four types of data: nominal, ordinal, interval, or ratio data (Figure 2.3).

    Nominal data represent some type of unique identifier, such as your phone number, bank account number, or Social Security number. Mathematical operations with nominal data that are numerical just don't make sense—for instance, you can't add two phone numbers together in order to get a mutual friend's phone number. Names or descriptive information associated with a location or phenomenon are also nominal data. Even if the name isn't unique (for example, several points on a map may represent different franchises of a national coffee

nominal data A type of data that is a unique identifier of some kind. If numerical, the differences between numbers are not significant.

FIGURE 2.3 Examples of the four data types: a) nominal data (names and numbers in a phone book), b) ordinal data (gold, silver, and bronze medal winners at the Olympics), c) interval data (temperatures Fahrenheit and Celsius), and d) ratio data (the box office grosses from Hollywood movies).

shop, such as Starbucks, which all have the same name), the fact that it is descriptive or name data classifies it as nominal.

Ordinal data are used to represent a ranking system of some kind with respect to the data. If a field represents a phenomenon where something is in first place, something else is in second place, and another item is in third place, then you're dealing with ordinal data. In the Olympic Games, the first-place winner of the event gets the gold, the second-place winner gets the silver, and the third-place gets the bronze—this ranking represents ordinal data. Here's another example: If you have a map of the addresses of the winners of a horse-racing event and the data indicate who placed first, who placed second, and so on, these would be ordinal data. Ordinal data deal solely with the rankings, and not with anything else associated with these values (such as each horse's finishing time, or the differences in finishing times between competitors). Ordinal data don't tell us anything about how much faster the first-place horse was than the second-place horse, only that one is first and one is second.

Interval data are used when the difference between numbers is significant, but there is no fixed zero point. With interval data, the value of zero is just another number on the scale and does not represent the bottom of the scale. Celsius temperature is an example of interval data—in the Celsius scale, zero degrees represents the freezing point of water, but Celsius measures temperatures below zero (and thus zero is not the bottom of the Celsius scale). Because

ordinal data A type of data that refers solely to a ranking of some kind.

interval data A type of numerical data in which the difference between numbers is significant but there is no fixed non-arbitrary zero point associated with the data.

ratio data A type of numerical data in which the difference between numbers is significant, but there exists a fixed, non-arbitrary zero point associated with the data.

there is no fixed zero point, it's fine to subtract numbers, but dividing numbers makes no sense. For instance, if the temperature is 15 degrees yesterday and 30 degrees today, we can subtract the two values and say that today is 15 degrees warmer than yesterday. However, we cannot divide the numbers and say that today is twice as warm as yesterday (since 30 degrees is not twice as warm as 15 degrees).

Ratio data are used when the difference between numbers is significant, and there exists a fixed and non-arbitrary zero value as the bottom of the measurement scale. For instance, your age and weight are both ratio data because you can't be less than zero years old or weigh less than zero pounds. In terms of temperature, the Kelvin scale would be considered ratio data because zero degrees Kelvin is considered the bottom of the scale (such that zero degrees Kelvin is the coldest that something can be measured). With ratio data, we can subtract values, but we can also multiply and divide them to get meaningful information. For instance, the gross box office receipts of one movie may be $100 million, while another film's gross box office receipts may be $300 million. We can say that the second film took in $200 million more than the first film, but we can also divide the numbers and say that the second film took in three times the amount of money that the first film did (since $100 million is one-third of $300 million).

- Examine the different fields of data and answer Questions 2.2 and 2.3.

QUESTION 2.2 What type of data classification (nominal, ordinal, interval, or ratio) does the Shape_Area field represent and why?

QUESTION 2.3 What type of data classification (nominal, ordinal, interval, or ratio) does the COUNTYFP field represent and why?

STEP 2.3 Joining a Table to a Layer's Attribute Table

- You'll see that there are several attributes for each record in the counties' attribute table, but as part of your project, you're to look into population distributions related to colleges. You'll also see that there are no population attributes for the counties in the attribute table. While you could manually look up the population attributes and type them in, there's a better approach available—you can take another spreadsheet and join its data and attributes to their corresponding records in the counties' attribute table. See **Smartbox 14** for more information about joining tables in ArcGIS.

Smartbox 14

How are tables joined in ArcGIS 10.2?

ArcGIS gives you the capability to link two or more data tables together. Often, non-spatial data (such as a tabular dataset or a spreadsheet) are linked to geospatial data (using a layer's attribute table). For instance, a geospatial layer may represent the polygon boundaries of each building on a college campus.

A separate spreadsheet of non-spatial data indicates information such as the building's name, number of classrooms, number of computers, and so on. By linking these two tables together, you could then examine the geospatial layer, find a record for a particular building, and then access all of the building's non-spatial data as well.

In ArcGIS, tables are connected by a **join**, or the merging of two (or more) tables into one. A join can be performed only if the two tables have a field in common (referred to as a **key**). It's on the basis of this key that data from one set of records can be linked to another set of records. Figure 2.4 shows an example of a join—one of the tables is a layer's attribute table while the second is a tabular spreadsheet. Their common field (the key) is the "COUNTYFP" attribute in the first table and the "County*" attribute in the second table. These two tables can be joined, using these two fields as their key, and the records of the attribute table can thus be linked to their corresponding records in the tabular spreadsheet.

Keep in mind that the common field must be of the same type (that is, both must be numbers, or both must be text strings). If the fields are represented differently (for instance, if the values in the key are represented with numerical values in one table and represented with text strings in the second table), then the field cannot be used as a key and the join will not work. Table values that are left-justified represent text strings (even if they are shown as numbers). Table values that are right-justified represent actual numerical values (not strings of text).

> **join** A method of linking two or more tables together.
>
> **key** The field that two tables have to have in common with each other in order for the tables to be joined.

Key

OBJECTID*	Shape*	STATEFP	COUNTYFP	COUNTYNS	GEOID	NAME	NAMELSAD	LSAD
1	Polygon	39	063	01074044	39063	Hancock	Hancock County	06
2	Polygon	39	003	01074015	39003	Allen	Allen County	06
3	Polygon	39	085	01074055	39085	Lake	Lake County	06
4	Polygon	39	047	01074036	39047	Fayette	Fayette County	06
5	Polygon	39	017	01074021	39017	Butler	Butler County	06
6	Polygon	39	115	01074070	39115	Morgan	Morgan County	06
7	Polygon	39	133	01074079	39133	Portage	Portage County	06
8	Polygon	39	145	01074085	39145	Scioto	Scioto County	06
9	Polygon	39	163	01074093	39163	Vinton	Vinton County	06
10	Polygon	39	125	01074075	39125	Paulding	Paulding County	06
11	Polygon	39	069	01074047	39069	Henry	Henry County	06
12	Polygon	39	051	01074038	39051	Fulton	Fulton County	06
13	Polygon	39	015	01074020	39015	Brown	Brown County	06
14	Polygon	39	091	01074058	39091	Logan	Logan County	06
15	Polygon	39	155	01074089	39155	Trumbull	Trumbull County	06
16	Polygon	39	131	01074078	39131	Pike	Pike County	06
17	Polygon	39	111	01074068	39111	Monroe	Monroe County	06
18	Polygon	39	087	01074056	39087	Lawrence	Lawrence County	06
19	Polygon	39	001	01074014	39001	Adams	Adams County	06

+

OBJECTID*	Id	Id2	State	County*	Population	Housing
1	0500000US39001	3900	39	001	28550	12978
2	0500000US39003	3900	39	003	106331	44999
3	0500000US39005	3900	39	005	53139	22141
4	0500000US39007	3900	39	007	101497	46099
5	0500000US39009	3900	39	009	64757	26385
6	0500000US39011	3901	39	011	45949	19585
7	0500000US39013	3901	39	013	70400	32452
8	0500000US39015	3901	39	015	44846	19301
9	0500000US39017	3901	39	017	368130	148273
10	0500000US39019	3901	39	019	28836	13698
11	0500000US39021	3902	39	021	40097	16755
12	0500000US39023	3902	39	023	138333	61419
13	0500000US39025	3902	39	025	197363	80656
14	0500000US39027	3902	39	027	42040	18133
15	0500000US39029	3902	39	029	107841	47088
16	0500000US39031	3903	39	031	36901	16545
17	0500000US39033	3903	39	033	43784	20167
18	0500000US39035	3903	39	035	1280122	621763
19	0500000US39037	3903	39	037	52959	22730

=

OBJECTID*	Shape*	STATEFP	COUNTYFP	COUNTYNS	GEOID	NAME	NAMELSAD	LSAD	OBJECTID*	Id	Id2	State	County*	Population	Housing
1	Polygon	39	063	01074044	39063	Hancock	Hancock County	06	32	050000	39063	39	063	74782	33174
2	Polygon	39	003	01074015	39003	Allen	Allen County	06	2	050000	39003	39	003	106331	44999
3	Polygon	39	085	01074055	39085	Lake	Lake County	06	43	050000	39085	39	085	230041	101202
4	Polygon	39	047	01074036	39047	Fayette	Fayette County	06	24	050000	39047	39	047	29030	12693
5	Polygon	39	017	01074021	39017	Butler	Butler County	06	9	050000	39017	39	017	368130	148273
6	Polygon	39	115	01074070	39115	Morgan	Morgan County	06	58	050000	39115	39	115	15054	7892
7	Polygon	39	133	01074079	39133	Portage	Portage County	06	67	050000	39133	39	133	161419	67472
8	Polygon	39	145	01074085	39145	Scioto	Scioto County	06	73	050000	39145	39	145	79499	34142
9	Polygon	39	163	01074093	39163	Vinton	Vinton County	06	82	050000	39163	39	163	13435	6291
10	Polygon	39	125	01074075	39125	Paulding	Paulding County	06	63	050000	39125	39	125	19614	8749
11	Polygon	39	069	01074047	39069	Henry	Henry County	06	35	050000	39069	39	069	28215	11963
12	Polygon	39	051	01074038	39051	Fulton	Fulton County	06	26	050000	39051	39	051	42698	17407
13	Polygon	39	015	01074020	39015	Brown	Brown County	06	8	050000	39015	39	015	44846	19301
14	Polygon	39	091	01074058	39091	Logan	Logan County	06	46	050000	39091	39	091	45858	23181
15	Polygon	39	155	01074089	39155	Trumbull	Trumbull County	06	78	050000	39155	39	155	210312	96163
16	Polygon	39	131	01074078	39131	Pike	Pike County	06	66	050000	39131	39	131	28709	12481
17	Polygon	39	111	01074068	39111	Monroe	Monroe County	06	56	050000	39111	39	111	14642	7567
18	Polygon	39	087	01074056	39087	Lawrence	Lawrence County	06	44	050000	39087	39	087	62450	27603
19	Polygon	39	001	01074014	39001	Adams	Adams County	06	1	050000	39001	39	001	28550	12978

SOURCE: ARCMAP/ESRI

FIGURE 2.4 Joining two tables together on the basis of their common field (key).

one-to-one join When a single record is linked to another single record during a join.

many-to-one join When many records in an attribute table can have one other record from another table matched to them during a join.

relate A type of join that establishes a connection between tables but does not append fields from one table to another table.

In addition, the field used as the key should not have any missing or null values since records from the two tables will not link together properly without the value in the key field.

A join is considered a **one-to-one join** if there is only a single record from each table being joined to a single record in another table. For instance, in our campus building example, the geospatial layer has a single record for each building, while the campus spreadsheet also has a single entry for each building.

A different type of join is a **many-to-one join**, in which many of the records in an attribute table are joined to the same one record in the other table. An example of this type of join would be a geospatial layer of land cover for a county. Each polygon in the land-cover layer would be tagged with a particular code indicating if that piece of land represented water, forests, agriculture, urban, barren, or wetland land cover. However, the county is covered with tens of thousands of polygons, each with one of six land-cover codes. A separate spreadsheet with more detailed information about each of the six land covers (such as a description and attribution) will be joined to the geospatial layer. In this case, each of the thousands of records in the geospatial layer will be linked to only one of the six possible records in the spreadsheet, resulting in a many-to-one join. As in a one-to-one join, the result is a single table.

When other types of linkages (such as a one-to-many join, in which several different records can be linked to a single record in another table—this is the reverse of a many-to-one join) need to be performed, a **relate** operation is used. A relate links the information in the two tables rather than merging them together into a single table.

- Add the **DEC_10_SF1_G001** table to ArcMap (you can add a table the same way you add other data).

  - Note: This geodatabase table was derived from an Excel spreadsheet and thus is only one table. If you had the Excel table itself with multiple sheets, you would have to select the sheet you wanted to add.

- In the TOC, right-click on the **DEC_10_SF1_G001** table and select **Open**. In the Table window, you'll see a second tab (corresponding to this spreadsheet) added next to the CensuscountiesOhio attribute table. Select the DEC_10_SF1_G001 tab to examine this table.

- The DEC_10_SF1_G001 table has several more attributes. We can use one of them (the attribute called **COUNTYFP** in the CensuscountiesOhio attribute table and the attribute called **county** in the DEC_10_SF1_G001 table) for the join. Answer Question 2.4.

> **QUESTION 2.4** Each of the two tables has another field with the same information (in the CensuscountiesOhio table, it is called STATEFP, and in the DEC_10_SF1_G001 table, it is called State). Why can these fields not be used as the key when joining the population values to the county feature-class records?

• When joining two tables, you begin with the table you want to join attributes to. Because we want to join the non-spatial attributes in DEC_10_SF1_G001 to the geospatial features in CensuscountiesOhio, we'll begin with the geospatial layer. In the TOC, right-click on the **CensuscountiesOhio** layer, then select **Joins and Relates**, then select **Join**.

• From the **What do you want to join to this layer** pull-down menu, select **Join attributes from a table**. This will allow you to perform the necessary join.

• From the **Choose the field in this layer that the join will be based on** pull-down menu, select **COUNTYFP**. This is the field in the geospatial layer that will be used as the key.

• From the **Choose the table to join to this layer, or load the table from disk** pull-down menu, select **DEC_10_SF1_G001**. This is the table you wish to join to the geospatial layer.

• From the **Choose the field in the table to base the join on** pull-down menu, select **County**. This is the field in the non-spatial layer that will be used as the key.

• Select the radio button for **Keep all records**.

• Click **OK** to perform the join.

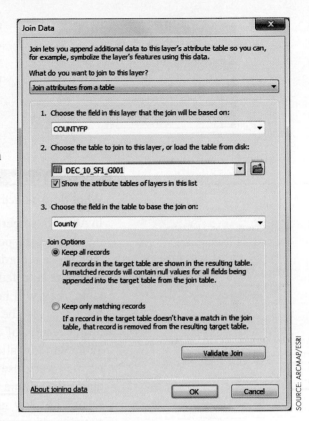

SOURCE: ARCMAP/ESRI

• Back in the Table window, select the **CensuscountiesOhio** tab (the one you joined the table to) and scroll across it. You'll see that the fields from the DEC_10_SF1_G001 table have been added via the key. Each record in CensuscountiesOhio now also has its corresponding data from DEC_10_SF1_G001 properly in place.

STEP 2.4 Selecting Records in an Attribute Table

• Now that the tables are joined, we'll find out which county in Ohio has the largest population. Scroll over to the Population attribute in the CensuscountiesOhio. Right-click in the header of the Field (the grey "Population" name itself) and a new menu of actions you can take to that particular field will appear.

• Select **Sort Descending**. This will sort the entire table using the values in the Population field.

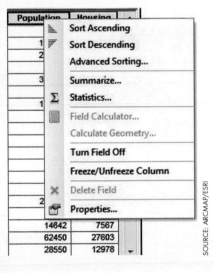

SOURCE: ARCMAP/ESRI

• With the table sorted, the largest population value will now be at the top of the table. Scroll across to find the name field to see which county has the largest population and answer Question 2.5.

QUESTION 2.5 Which Ohio county has the largest population?

• Look at the far left-hand side of the attribute table and click the mouse in the small box to the left of the first field. You'll see the record highlighted in a cyan

Table

CensuscountiesOhio

OBJECTID *	Shape *	STATEFP	COUNTYFP
77	Polygon	39	035
71	Polygon	39	049
58	Polygon	39	061
82	Polygon	39	153

Click Here

SOURCE: ARCMAP/ESRI

color—this indicates that the first record is now selected in ArcGIS. Minimize the table and you'll also see the corresponding county polygon outlined in cyan as well. For more information about selection in ArcGIS, see **Smartbox 15** .

Smartbox 15

What does selection mean in ArcGIS 10.2?

selection When certain records or features are chosen and set aside from the remainder of the records or features.

In ArcGIS, **selection** refers to choosing records from an attribute table or choosing their corresponding features from a layer and holding them aside to work with those records or features separately. By default, ArcGIS uses a cyan color to denote selected features or records. When records are selected in an attribute table, they will be highlighted in cyan and their corresponding features in the View will also be highlighted in cyan. For instance, the records that correspond to the results of a query (see Step 2.5) are considered selected records. The Table window allows you to examine all records or only selected records—this is often very useful for viewing only those records that were the result of a query. The Table window will also give you a count of how many records are selected at any given time. Often, actions that are taken (such as computing statistics or exporting data to another layer) are performed only on the selected features. Thus, it's advisable to check which items (if any) are selected before performing an action. If necessary, you can clear all of the selected features before proceeding with another task.

There are several ways of selecting features and records within ArcGIS. A Select By Attributes query (see Step 2.5) is very commonly used to select records through a query. However, features can also be selected through spatial queries (see Chapter 8) or interactively by the user (see Chapter 9). When performing a selection, you can choose from the following options:

- **Create a new selection**: The query results will wipe any existing selection and specify a new set of selected features or records.

- **Add to current selection**: The query results will be selected and any existing selection will be retained as well.

- **Remove from current selection**: The pool of selected results will be retained except for the results of the query.

- **Select from current selection**: The query results will be drawn only from the pool of selected records, not any un-selected records.

- We'll do much more with selection in the next steps, but for now, clear the one selected record by choosing the **Selection** pull-down menu and then choosing **Clear Selected Features**. You can also press the Clear Selected Features button on the Tools toolbar:

SOURCE: ARCMAP/ESRI

STEP 2.5 Performing a Simple Query (Selection By Attributes)

- Often, you'll find you want to obtain more information from attribute tables than simply sorting fields and digging through the attributes. For instance, in our joined table, there's information about population and housing characteristics. If you wanted to find the counties with a certain level of population *and* a minimum number of housing units, you would need to do a lot of cross-checking of columns, numbers, and names. To obtain this kind of information from a table in GIS, you'll want to perform a **query** of the table. For more information about building simple queries in ArcGIS, see Smartbox 16 .

query The conditions used to retrieve data from a database or table.

Smartbox 16

What is the structure of a simple query in ArcGIS 10.2?

A query in ArcGIS is done in **SQL** (Structured Query Language) format and works like a mathematical function. A **simple query** takes the form of choosing one attribute, then choosing an operator that will affect that attribute, and lastly choosing a value. For instance, as a result of the join, your Ohio counties layer now contains an attribute field called Population. If you wanted to find all counties that have a population of more than 100,000 persons, you would first select the proper attribute (in this case, "Population"), then select an operator (in this case a "greater than" > operator), and lastly choose a value to use in the query (in this case, the number "100000"). Your SQL query expression would then be: **Population > 100000**. After ArcGIS evaluated this simple query, all records that have a value for the Population attribute that is greater than a value of 100000 will be selected.

In ArcGIS, you'll have a choice of **relational operators** to use. The six available are:

SQL The Structured Query Language—a formal setup for building queries.

simple query A query that only contains one operator.

relational operators One of six connectors (=, <>, >, <, >=, <=) used to build a query.

- Equal (=)
- Not Equal (<>)
- Greater Than (>)
- Greater Than or Equal To (>=)
- Less Than (<)
- Less Than or Equal To (<=)

These operators work like their mathematical equivalents, and are useful when selecting population values greater than or less than a certain amount. The Equal and Not Equal operators are useful when selecting text-based attributes as well. For instance, your Ohio counties layer also has an attribute field called NAME that contains the names of each county as a string of text. If you wanted to select the record that corresponded with a particular county (such as Lake County), you could build a query using the NAME field, the Equals operator, and a string of text ('Lake') that you are looking for. In this case, your query expression would read: **NAME = 'Lake'**. When ArcGIS evaluates this query, the record whose NAME field contains the string 'Lake' would be selected. Note that single quotes are necessary when building a query for a text string.

To reverse this operation and instead find all counties that are *not* named Lake, you could use the Not Equal operator and build a query expression that would read: **NAME <> 'Lake'**. When this query is evaluated, all counties except for the record whose NAME field contains the string 'Lake' would be selected.

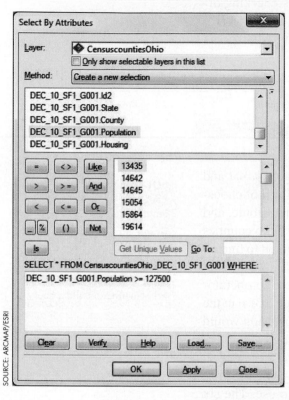

SOURCE: ARCMAP/ESRI

- To start, we'll find out which Ohio counties have high population values, and more importantly, where they're located. The information you want to get back from the table is how many Ohio counties have a population of greater than or equal to 127,500 persons. To build a simple query and select these counties, from the **Selection** pull-down menu, choose **Select By Attributes**.

- For the Layer, choose **CensuscountiesOhio** (this is the layer containing the attributes that will be queried).

- For the Method, choose **Create a new selection**.

- Next, we'll build the SQL query. First, you'll pick the attribute to query. Double-click on **DEC_10_SF1_G001. Population**. In the box at the bottom of the dialog, you'll see that DEC_10_SF1_G001.Population appears as the first part of the query.

- Next, you'll pick the operator. Press the greater than or equal to button (>=). You'll see it appear in the query box after the attribute.

- Last, you want to add the value for the query expression. Press the **Get Unique Values** button—this will show all of the values in the attribute table for the chosen attribute (in this case, DEC_10_SF1_G001.Population). Scroll through the list; you'll see that there is no value that exactly matches the 127500 persons you're querying for (otherwise, you could have chosen it from the box and completed the query). You'll have to type in the value instead. In the query box, type **127500** after the >= sign to complete the query expression.

- Your query expression should read: DEC_10_SF1_G001.Population >= 127500. Click **Apply** to process the query. If you've completed the query incorrectly, press **Clear** to wipe the box clean and build the query fresh.

- When the query is successfully run, click **OK** to close the Select By Attributes dialog.

- You'll see the results of the query highlighted in cyan on both the View (the polygons whose attributes matched the query results) and in the Table (the records that matched the query). Press the Show Selected Records button in the Table to show only the selected records. Answer Questions 2.6, 2.7, and 2.8.

QUESTION 2.6 How many Ohio counties have a population of more than 127,500 persons?

QUESTION 2.7 Where (in general terms) are these higher population counties located in Ohio?

QUESTION 2.8 Do any of the following counties have more than 127,500 persons (and if so, which ones): Stark, Portage, Trumbull, Mahoning, or Columbiana?

• Clear the selection. This will clear the selected polygons from the View and the corresponding selected records from the Table (and restore everything to "un-selected" status again).

• Build a new query to find which counties in Ohio have the lowest overall population—perform a selection by attributes to find all counties with a population of fewer than 40,000 persons. Answer Questions 2.9, 2.10, and 2.11.

QUESTION 2.9 How many Ohio counties have a population of fewer than 40,000 persons?

QUESTION 2.10 Where (in general terms) are these lower-population counties located in Ohio?

QUESTION 2.11 Do any of the following counties have fewer than 40,000 persons (and if so, which ones): Stark, Portage, Trumbull, Mahoning, or Columbiana?

• Clear the selection again.

STEP 2.6 Performing a Compound Query (Selection By Attributes)

• Often, when using queries for analysis or to examine subsets of your data, a simple query is not going to provide enough flexibility. For instance, if you wanted to know something about both population and housing characteristics, a simple query wouldn't be useful since you could build an expression of only one attribute at a time. The same holds true if you wanted to simultaneously determine both the counties with both the largest and smallest populations in a single query—you could not accomplish this with one simple query. To perform these more complicated actions, you'll need to build a compound query (see **Smartbox 17** for more about writing compound queries).

Smartbox 17

What is the structure of a compound query in ArcGIS 10.2?

A simple query will allow you to evaluate only *one* attribute using *one* relational operator (such as NAME = 'Lake'). However, there are going to be plenty of times when you're going to want to perform a query that has multiple criteria (such as selecting both NAME = 'Lake' and NAME = 'Mahoning'

compound query A query that contains more than one operator.

Boolean operator (Query) One of the connectors used in building a compound query.

AND (Query) The Boolean operator wherein the chosen features are those that meet both criteria in the query.

OR (Query) The Boolean operator wherein the chosen features are those that meet one or the other (or both) of the criteria in the query.

NOT (Query) A Boolean operator used to negate the function or query.

simultaneously). A **compound query** can take two (or more) simple queries and combine them. For instance, your Ohio counties layer has attribute fields for both Population (the number of persons in the county) and Housing (the number of homes in the county). If you wanted to find all counties that had both a population of more than 100,000 persons and also had more than 100,000 homes, you would need to create the first simple query (**Population > 100000**) and connect it with the second simple query (**Housing > 100000**) so that you could effectively query both attribute fields at the same time.

A compound query relies on a special kind of connector called a **Boolean operator** that explains how the multiple queries are tied together. In a Select By Attributes query, the Boolean operators AND and OR are used to build compound queries. The **AND** operator is used when finding the commonalities (or the intersection) of both parts of the query. For instance, in the Population and Housing example above, using the AND operator to make the query read **(Population > 100000) AND (Housing > 100000)** would select all records whose value for Population is above 100,000 *and* whose value for Housing is above 100,000. If only one of these two criteria is met (for instance, a record with a Population value of 250,000 but a Housing value of only 99,000), then that record would not be selected. Using AND as the Boolean operator indicates that *both* parts must be valid.

The **OR** operator is used when either one part or the other (or both) of a query is valid. To continue with the same example, suppose the query was constructed as follow: **(Population > 100000) OR (Housing > 100000)**. Any record whose value for Population is above 100,000 would be selected, in addition to any record whose value for Housing is above 100,000. If a record had a Population value of 250,000 and a Housing value of 99,000, it would still be selected, because one part of the query is met (the Population value). Using OR as the Boolean operator indicates that if at least one part of the expression is valid, that record is selected.

Another type of Boolean function available in a query is the **NOT** operator, which is used to effectively reverse the results that would normally be selected. For instance, building a query that reads **NOT (NAME = 'Lake')** would select all records that do not have the text string 'Lake' in their NAME field (likely all records except for the one representing 'Lake' county).

Compound queries can be used to create more complex types of query expressions. For instance, if you wanted to determine all records with either a Population of greater than 100,000 or less than 50,000 and to exclude Lake County from your analysis, your query expression would read: **((Population > 100000) OR (Population < 50000)) AND NOT (NAME = 'Lake')**.

- You'll now want to make a compound query that determines which Ohio counties have a population of more than 127,500 persons but also contain more than 100,000 housing units. Bring up the Select By Attributes dialog again.

- Build the first part of the expression like you did before, to find all values from the DEC_10_SF1_G001.Population field that are greater than or equal to 127,500.

- Next, press the **And** button to add a Boolean operator. You'll see AND added to the query box.

• Next, build the second part of the compound expression to find all values from the DEC_10_SF1_G001.Housing field that are greater than or equal to 100,000.

• Make sure that there are spaces placed between each part of the expressions and the AND for proper syntax.

• When the query is ready, click **Apply**. After the compound query is processed, click **OK** to close the dialog box. Answer Question 2.12.

> **QUESTION 2.12** How many Ohio counties have a population of more than 127,500 persons and also have more than 100,000 housing units? Which counties are they?

• Clear the selection.

• As noted in the Introduction to this chapter, part of your project is to examine the characteristics of a small set of counties in northeast Ohio (Mahoning, Columbiana, Stark, Portage, and Trumbull). Rather than working with all 88 Ohio counties, what you'll do next is select these five counties to work with as a subset of the county data. To do so, you'll build a new compound query to select only these five.

• Bring up the Select By Attributes dialog again. Clear any existing queries that are in it.

• Your compound query will be looking for five different counties, and thus you'll be making the equivalent of five simple queries, each joined by a Boolean operator. For this compound query, you'll be using the CensuscountiesOhio.NAME field, as it contains text strings of county names.

• Build the first part of the compound query by finding the CensuscountiesOhio.NAME equal to 'Columbiana'. First, choose **CensuscountiesOhio.NAME** from the list of attributes, then press the = button, then press the **Unique Values** button and choose **'Columbiana'** from the list of values.

• Next, add the first Boolean operator—press the **Or** button. You'll see OR added to the query box.

• Next, build the second part of the compound query by finding the CensuscountiesOhio.NAME equal to 'Mahoning'. Do this the same way as you selected 'Columbiana'.

• Add the second Boolean connector by pressing the **Or** button.

• Continue until you have the compound query properly built where you are finding CensuscountiesOhio.NAME = 'Columbiana' OR CensuscountiesOhio.NAME = 'Mahoning' OR CensuscountiesOhio.NAME = 'Portage' OR CensuscountiesOhio.NAME = 'Stark' OR CensuscountiesOhio.NAME = 'Trumbull'. This will select all five counties based on their NAME attribute. Answer Question 2.13.

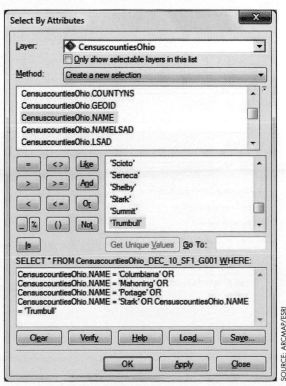

SOURCE: ARCMAP/ESRI

QUESTION 2.13 Why are you using OR as the Boolean connector in all four places in the compound query? What would be the result of using AND in place of OR all four times?

SOURCE: ARCMAP/ESRI

STEP 2.7 Exporting the Results of a Selection to a New Layer

• Now that these five counties are selected, we want to create a new feature class that contains only these counties. Because all further analysis is limited to these five, we don't need the other 83 counties in the layers we're working with. To take selected features and create a new dataset from them, right-click on the layer in the TOC that has the selected features (**CensuscountiesOhio**), then select **Data**, then **Export Data**.

• In the **Export:** pull-down menu, choose **Selected features**.

• For **Use the same coordinate system as:**, choose the radio button for **this layer's source data**.

• For **Output feature class**, you'll need to specify what folder or geodatabase you want to place the exported features in, and also the format. Press the **Browse** button next to the Output feature class box.

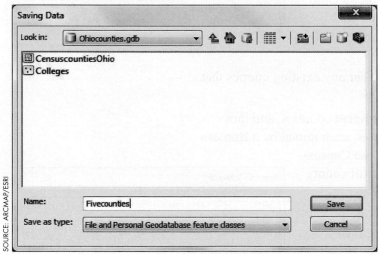

SOURCE: ARCMAP/ESRI

• Navigate to your D:\GIS\Chapter2 folder, then navigate inside the **Ohiocounties.gdb** geodatabase. For Save as type, select **File and Personal Geodatabase feature classes**. For the Name, type **Fivecounties** and click **Save**.

• Back in the Export Data dialog box, you'll see the new path set up. This will take the selected counties and place them into a new feature class called Fivecounties and save it in your Ohiocounties geodatabase. Click **OK** to proceed.

• Choose **Yes** when prompted to add the exported data to the map as a layer.

• A new polygon layer of only the five selected counties will be placed in the TOC. You will not be using the old CensuscountiesOhio layer, so right-click on it in the TOC and select **Remove**. This will delete it from the TOC.

• Click on the **Full Extent** button on the Tools toolbar to adjust the View to only the five counties.

• Open the attribute table for the Fivecounties layer. You'll see that only five records are now in the attribute table but all of the joined attributes have been retained.

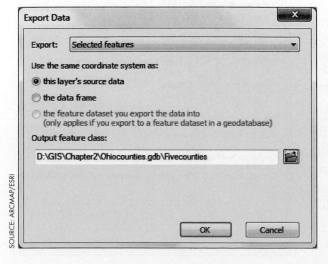

SOURCE: ARCMAP/ESRI

To get information about the population characteristics of only this five-county region, right-click on the **Population** field and select **Statistics**.

• A new window will open with computed statistics about the numerical values of this attribute field, including the count of records and whether any records have null values, the mean and standard deviation of the values, the sum of all values, and the maximum and minimum values. Answer Question 2.14.

> **QUESTION 2.14** What is the total population of this five-county area?

• Close the Statistics window and the Fivecounties attribute table when you're done.

STEP 2.8 Using Queries for Analysis

• The next part of your project involves examining the locations of colleges and universities within this five-county region. Add the **colleges** feature class to the TOC. This layer contains points showing the locations of those places tagged as a college or university by the USGS. Place this layer on top of the counties so that the points are visible. Change the symbology of the layer to a single symbol using something more prominent than a simple point (see Chapter 1 for directions on how to change a layer's symbology).

• Open the colleges attribute table and answer Question 2.15.

> **QUESTION 2.15** How many places in this five-county region are considered a college or university?

• For ease of use, we'll now place labels next to each of the points to help identify them. Right-click on the **colleges** layer in the TOC and select **Properties**. In the Layer Properties dialog box, select the **Labels** tab.

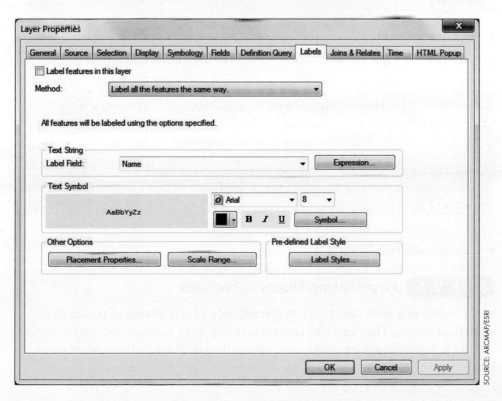

SOURCE: ARCMAP/ESRI

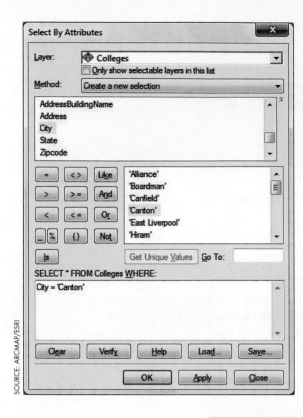

• ArcGIS will assign labels to features based on the contents of an attribute field. You could choose any field you like for a label, but for this chapter, under the **Label Field:** pull-down menu, choose **Name**.

• Place a checkmark in the box that says **Label features in this layer**. Click **OK**. You'll see text placed next to each point—this is the same text string that is in the Name field of the colleges attribute table.

   • Note that you can turn the labels on and off at any time by right-clicking on the **colleges** layer in the TOC and selecting **Label Features**.

• We'll now examine some major cities within this five-county region and see how many colleges and universities are located within each one. We can do this by performing a Select By Attributes query using the colleges layer and its City attribute.

• Bring up the Select By Attributes dialog and build a query to find all records where the City attribute is equal to 'Canton'. Answer Questions 2.16 and 2.17.

QUESTION 2.16 How many colleges or universities are in the city of Canton?

QUESTION 2.17 In which county is Canton located? (*Hint:* Use the Identify tool to examine the attributes of the Fivecounties layer to answer this question.)

• Clear the selected features and build a new query to find all records where the City attribute is equal to 'Youngstown'. Answer Questions 2.18 and 2.19.

QUESTION 2.18 How many colleges or universities are in the city of Youngstown?

QUESTION 2.19 In which county is Youngstown located? (*Hint:* Use the Identify tool to examine the attributes of the Fivecounties layer to answer this question.)

• Clear the selected features.

STEP 2.9 Hyperlinking Photos to Features

• Attributes in a table don't have to consist only of text strings of names or numerical values. They can also contain text that links to other non-spatial media, such as a digital image or a URL. By linking the attributes to these other media, you can set up features such that, by clicking on them, you can access digital photos or Websites connected to those features. This process is called hyperlinking. (For more information, see **Smartbox 18**.)

Smartbox 18

What is a hyperlink and how is it used in ArcGIS 10.2?

A **hyperlink** is used in ArcGIS to connect geospatial features to other items, such as documents, files, photos, URLs, or scripts. Just as clicking on a hyperlink on a Web page will open a new Web page URL or open a photo or download a file, a hyperlink in ArcGIS will trigger a similar type of event. For instance, you could set up the Ohio counties layer to have hyperlinks so that when you click on the polygon of a particular county, the URL for the county government's Web page will open in the computer's Web browser. Similarly, if you have a point layer representing points of interest along a hiking trail, you could set up hyperlinks so that, whenever you click on one of the points, a digital photo will appear. URLs, digital photos, and scripts are all examples of non-spatial data, but hyperlinks allow you to **geotag** them, or link them to spatial locations.

> **hyperlink** A function that allows the user to connect non-spatial media such as documents, files, photos, URLs, or scripts to a geospatial feature and interactively trigger them.
>
> **geotag** A process whereby non-spatial media are linked to geospatial features.

- For presenting your project, you'll be creating a set of interactive hyperlinks using the colleges attribute table. When you click on a point, a photo of that college will appear.

- As you create hyperlinks for features in ArcGIS, you must go through several steps to ensure that your hyperlinks will work properly. Through hyperlinking, you want to be able to click on a point and open a photo of that object. To ensure that this will happen, ArcGIS first has to know where those photos are stored on the computer.

- To set this path, from the **File** pull-down menu, select **Map Document Properties**.

- In the box labeled **Hyperlink base:**, type the full path of the folder containing the photos (in this chapter, it is D:\GIS\Chapter2\). Click **OK** when done. This step ensures that you don't have to create a path to each photo, just one path to where all photos are being stored.

  - *Important Note:* If you want to change this location later, you can do so (for instance, if you move the photos to a different folder, flash drive, or location).

- Next, the point layer contains multiple records, each one being a different point. Thus, each point can have its own photo linked to it. To do this, a new field (column) must be added to the layer. This new field will contain the name of the photo that will be hyperlinked to that point. The following steps will allow you to do this:

- Open the attribute table of the **colleges** layer.

- From the **Table Options** pull-down menu, select **Add Field**.

Map Document Properties — General tab

File:	D:\GIS\Chapter2\Chapter2.mxd
Title:	
Summary:	
Description:	
Author:	
Credits:	
Tags:	
Hyperlink base:	D:\GIS\Chapter2\
Last Saved:	11/20/2013 12:14:09 PM
Last Printed:	
Last Exported:	
Default Geodatabase:	C:\Users\Me\Documents\ArcGIS\Default.gdb
Pathnames:	☐ Store relative pathnames to data sources
Thumbnail:	Make Thumbnail / Delete Thumbnail

OK | Cancel | Apply

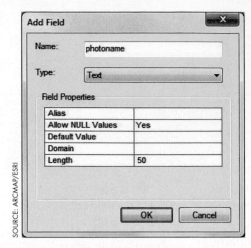

- In the Add Field dialog box that appears, give the new Field a name (call it **photoname**).

- Select **Text** as the Type.

- A Length of **50** should be more than sufficient (this is the maximum number of characters that the field—i.e., the name of the photo—can contain).

- Click **OK** when done. You'll see a new Field (called photoname) has been added to the table.

- Next, you'll need to specify the name of the photo to be linked to each point. You'll do this by editing your new field in the attribute table by adding the photo's file name to that field.

- To perform edits to an attribute table, you'll need to use a separate set of tools—the Editor toolbar. From the Editor toolbar's **Editor** pull-down menu, select **Start Editing**. If prompted, select **Colleges** as the item to edit.

- We'll start by setting up a hyperlink for the point representing Youngstown State University. Locate the record for Youngstown State University and navigate to the photoname field for that record.

- Click in the empty **photoname** field of the Youngstown State University record. Your cursor will switch to a typing icon.

- Type the name of the photo (including its extension, such as .jpg) for the photo you want to hyperlink to this specific point. The photo corresponding to Youngstown State University is called **YSU.jpg**.

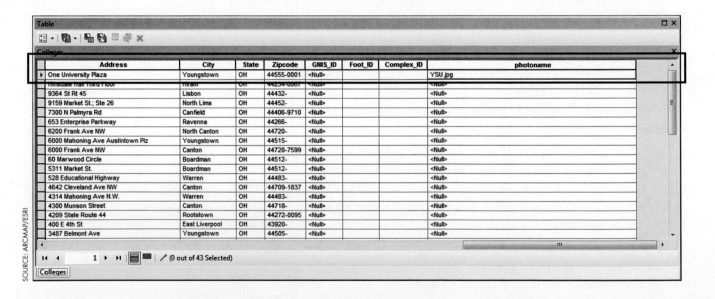

- When you have the name of the photo in place for the point, from the Editor toolbar's **Editor** pull-down menu, select **Save Edits**, then select **Stop Editing**.

- Next, you will tell ArcGIS which field (in this chapter, the field called photoname) contains the information used in hyperlinking.

- Right-click on the **colleges** layer in the TOC and select **Properties**.

- Select the **Display** tab.

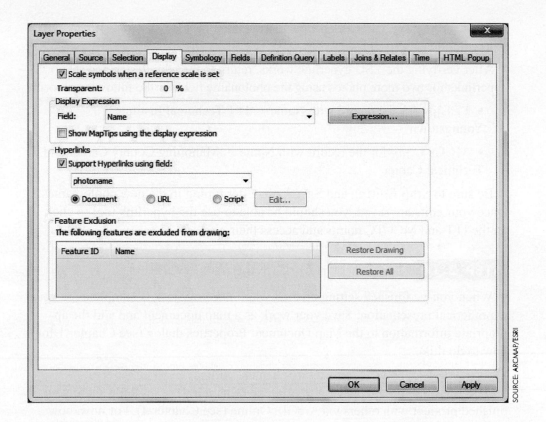

SOURCE: ARCMAP/ESRI

• Under the Hyperlinks heading, place a checkmark in the **Support hyperlinks using field:** box and from the pulldown menu, select the name of the attribute field containing the names of the photos (in this chapter, **photoname** is the name of the field).

• Select the **Document** radio button (to access documents or files such as downloaded photos).

• Click **Apply**, then click **OK** when done.

• Last, to access a hyperlink, select the hyperlink icon from the Tools toolbar. Once you select the hyperlink icon, the points that have hyperlinked photos will change to a different display color.

SOURCE: ARCMAP/ESRI

• Note that you can change this display color with the following steps:

  • From the **Customize** pull-down menu, select **ArcMap Options**.

  • Select the **General** tab.

  • In the Tools heading, put a checkmark in the box for **When the hyperlink tool is selected, highlight features containing clickable content:**. Then, from the pull-down menu, select whatever color you want, then click **Apply** and **OK**. The points that contain hyperlinks will change to this new color.

• When you click on the hyperlink icon, the cursor will change to a lightning bolt shape. Click on the hyperlinked Youngstown State University point and the hyperlinked picture will then open using the default photo viewing software on your computer.

• After verifying the YSU hyperlink works, return to the editing step and set up hyperlinks for two more photos using the photoname field for the following records:

> • **ITT.jpg** for the record with Name = 'ITT Technical Institute – Youngstown'

> • **MCCTC.jpg** for the record with Name = 'Mahoning County Career and Technical Center'

• Be sure to **Stop Editing** and **Save Edits** after typing in the new photo names. Once your edits are saved, you should be able to use the hyperlink tool to click on the ITT and MCCTC points and access their photos.

STEP 2.10 Finishing Up and Saving Your Work

• When you've finished setting up the hyperlinks, your task is complete for your hypothetical presentation. Save your work as a map document and add the appropriate information to the Map Document Properties dialog (see Chapter 1 for how to do this).

• As noted in Chapter 1, most future chapters will have you either create a professional-looking map layout to print (see Chapter 3) or digitally share your finished product with others via ArcGIS Online (see Chapter 4). For now, however, you can take your completed display of the five counties and the labeled points and save it as a PDF file that you can print or share with others. From the **File** pull-down menu, select **Export Map**. Give it a File Name of **Chapter2**, and for Save as Type, select **PDF**. Place this PDF file in your D:\GIS\Chapter2\ folder. When you're done, answer Question 2.20.

> **QUESTION 2.20** Show your final saved results or your PDF file to your instructor, who will check over your work for completeness and accuracy to make sure you get credit for this question. You should also demonstrate your three working hyperlinks.

Closing Time

The ability to link non-spatial data to geospatial locations makes GIS different from other types of information systems. Linking attribute data to geospatial features is a common process in GIS. A good source for large amounts of different types of data is the U.S. Census. The Census Bureau doesn't provide data such as population demographics or housing characteristics in a geospatial format, but rather as a series of tables that can then be used in ArcGIS by joining them to features. See the *Related Concepts for Chapter 2* for more information about how to access and use these kinds of data in ArcGIS.

Handling attribute information, querying attribute tables, and joining tables together are all common tasks in GIS, and throughout this book you'll often find yourself working with attributes tied to geospatial data. As noted in Chapter 1, a different kind of GIS data (called *raster data*) doesn't have the same format as the vector data we've been working with, and thus handles its attributes a bit differently, but we'll get to that in Chapter 12. Starting in the next chapter, you'll be taking your knowledge of geospatial data and attributes and applying it toward creating a professional-looking map of your results.

Related Concepts for Chapter 2 Using U.S. Census Data and Attributes in ArcGIS

The U.S. Census Bureau conducts a census every ten years, collecting a vast amount of data about the U.S. population. This rich dataset includes information on population demographics, as well as the country's social, economic, and housing characteristics. Different characteristics are available depending on the unit of analysis being examined (for example, the state, county, Census Tract, or Block Group level). The Census Bureau makes all of these datasets available to the public (for free) via the **American FactFinder** (AFF) Website (http://factfinder2. census.gov) (Figure 2.5). Using the AFF, you can specify the characteristics you wish to examine and the geographic level, then access all of this information in tabular format. These tables of data can also be downloaded to use as attributes in GIS, as each data table is structured similar to an attribute table. Each unit being examined (such as each state or each county) is a record, while the various Census values (such as total number of houses or total population) are stored as fields. However, even though this attribute data is organized geographically, it is still non-spatial in nature. While you could examine characteristics for a particular county, you could not perform spatial analysis with it. To do so, you would have to join these tables to geospatial data layers within ArcGIS.

The U.S. Census Bureau also provides separate Cartographic Boundary files and TIGER/Line files (in shapefile format) for a variety of different areal units

American FactFinder A Web resource run by the U.S. Census that allows for access to U.S. Census data in tabular form.

FIGURE 2.5 The American FactFinder Website.

SOURCE: AMERICAN FACTFINDER

(such as state, county, congressional district, Census Tract, Block Group, or county subdivision) through its TIGER/Line Website (http://www.census.gov/geo/maps-data/data/tiger.html). You can download the Cartographic Boundary files for the areas you want (such as your state's counties, or your state's Census Tracts) from the U.S. Census TIGER/Line resources, then compile and download the tabular data you wish to examine through AFF. Then, in ArcGIS, you can join the non-spatial AFF data to the geospatial layers of the Cartographic Boundary files and be able to examine U.S. Census data in ArcGIS. This is how we used the data in this chapter (the county boundaries file was downloaded from TIGER / Line resources, while the non-spatial data on population and housing came from AFF).

The data from AFF requires some editing and formatting using a program like Microsoft Excel before it can be properly utilized in ArcGIS and joined to a geospatial layer. The Census Bureau provides some online notes for using attribute data from AFF with ArcGIS here: http://www.census.gov/geo/education/pdfs/tiger/Downloading_AFFData.pdf. Note also that the U.S. Census Bureau also provides a limited set of files for download that combine both Cartographic Boundaries with some Census attribute data already joined together (see here for more information: http://www.census.gov/geo/maps-data/data/tiger-data.html).

For More Information

For in-depth information about the topics presented in this chapter, use the ArcGIS Help feature to search for the following items:

- About joining and relating tables
- Add Join (Data Management)
- Building a query expression
- Deciding between relationship classes, relates, and joins
- Essentials of joining tables
- Setting map document properties
- Using hyperlinks
- Using Select By Attributes
- What are tables and attribute information?

Key Terms

non-spatial data (p. 33)	key (p. 41)	compound query (p. 48)
attribute table (p. 33)	one-to-one join (p. 42)	Boolean operator
records (p. 33)	many-to-one join	(Query): (p. 48)
fields (p. 33)	(p. 42)	AND (Query) (p. 48)
attributes (p. 33)	relate (p. 42)	OR (Query) (p. 48)
nominal data (p. 38)	selection (p. 44)	NOT (Query) (p. 48)
ordinal data (p. 39)	query (p. 45)	hyperlink (p. 53)
interval data (p. 39)	SQL (p. 45)	geotag (p. 53)
ratio data (p. 40)	simple query (p. 45)	American FactFinder
join (p. 41)	relational operators	(p.57)
	(p. 45)	

How to Create a Map Layout with ArcGIS 10.2

Introduction

One of the key uses of GIS is to make a map of your data or the results of your analysis. A **map** is a visual representation of geospatial data that conveys a message about location-based concepts. For instance, GIS can create a map to illustrate property boundaries, hotspots of criminal activity, or the extent of a watershed. This is a unique feature of geospatial data: It can be visualized using a map, rather than a chart or a graph (Figure 3.1, page 60).

However, there's more to creating an effective map than taking your data, adding a title and a legend, and printing it out. **Cartography** is the art and science of mapmaking. Learning good map-design (cartographic) skills will help you create maps that are more useful for your target audience. ArcGIS 10.2 allows you to quickly make professional-looking maps from your geospatial data, while also allowing you the freedom to design the map the way you want to.

Several types of maps can be created using GIS. **Reference maps** convey location information or highlight various features of an area. All of the following are examples of reference maps: a map of the roads and highways between Cleveland and Pittsburgh, a map of tourist locations in San Diego, a map showing the location of abandoned housing in a city, and a map of the location of water wells in a rural area. **Topographic maps** are specific kinds of reference maps that show factors such as landforms, land cover types, and other built or natural features (see Chapter 16 for more about topographic maps).

Thematic maps illustrate a specific theme. The following are examples of thematic maps: a map of the United States showing whether or not a state has a smoking ban in public buildings, a map that shows whether a state's electoral votes went to Barack Obama or Mitt Romney in the 2012 presidential election, and a map showing which countries are members of the European Union. **Choropleth maps** are thematic maps that show multiple values (rather than, say, the two choices of "red state" or "blue state"). One example of a choropleth map is a map showing the percentage of the popular vote in each state for Obama or Romney. Another example is a map showing a state's federal income tax rate.

In this chapter, you will take a value from a layer's attribute table and create a choropleth map from it. Then you'll take the choropleth map and create a professionally designed map layout to print out. Specifically, you'll be making design choices to create an effective map, choosing color and scale, and adding map elements such as a scale bar, legend, north arrow, title, and descriptive text. Unlike most of the other chapters in this book, this chapter does not provide a series of questions for you to answer, but rather a checklist of items to aid you in making sure your map is complete.

Many later chapters will have you compose a map layout at the end of your analysis in order to present your results. While you won't be specifically using

map A visual representation of geospatial data that conveys some sort of message to its reader.

cartography The art and science of making maps.

reference map A map that serves to show the location of features, rather than thematic information.

topographic map A map created by the USGS to show landscape and terrain as well as the location of features on the land.

thematic map A map that displays a particular theme or feature.

choropleth map A type of thematic map in which data are displayed according to one of several different classifications.

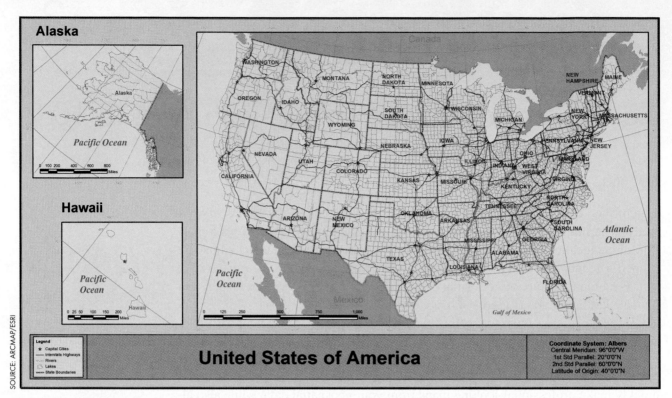

FIGURE 3.1 A map of U.S.
counties created in ArcGIS 10.2.

a choropleth map in other chapters, the skills and knowledge you gain from this chapter will enable you to create professional-quality maps from other types of geospatial data.

Chapter Scenario and Applications

In this chapter, you will take the role of a realtor who is involved with the sales of vacation homes in Ohio. Your goal is to create an effective map that shows seasonal homes in each Ohio county (as the percentage of the total housing stock for that county). You can then distribute this map to colleagues within your agency and to potential home buyers to show them which counties have the highest percentage of seasonal homes.

Here are some other real-world applications of this chapter's theory and skills:

• A researcher at the Centers for Disease Control and Prevention needs to prepare a map showing the rates of bird flu in each township in the state of Georgia. The map needs to be colorful and well designed so that it can be distributed to the general public.

• An analyst at a police department is preparing a map showing the number of 911 calls in a city per household. She is designing the map at the Census block level for distribution to other police analysts.

• An economist is examining population and income demographics for several states and is mapping these at the county level. He is creating several maps of these factors using Census data prior to performing more detailed analysis.

ArcGIS Skills

In this chapter, you will learn:

• How to symbolize data by multiple values using graduated colors.

• How to create a choropleth map and examine different types of data classifications.

• How to create a map layout from your GIS data.

• How to create and edit a map legend.

• How to add various map elements (scale bar, scale text, and north arrow) to a map layout.

• How to add different kinds of text and typology to a map layout.

• How to design a map with artistic and aesthetic merit.

• How to print and export a completed map layout.

Study Area

• In this chapter, you will be working with data for each county in the state of Ohio.

Data Sources and Localizing This Chapter

This chapter's data focus on seasonal homes in the Ohio counties. However, you can easily modify this chapter to use seasonal home data and the counties from your own state by using data from the U.S. Census Bureau (see the *Related Concepts for Chapter 2* for more about using Census data in ArcGIS).

The counties dataset was downloaded from the TIGER/Line files of the Census Bureau Website here: ftp://ftp2.census.gov/geo/tiger/TIGER2013. From the COUNTY folder, download the available shapefile. This contains all of the counties within the United States—you can query the shapefile and extract those counties into a separate feature class for your state based on the STATEFP attribute. This attribute references each state with a two-digit FIPS code. For instance, Ohio is referenced with a state FIPS code of 39. If you were using this chapter to map seasonal home percentages in Florida, you would use FIPS code 12. A full list of state and county FIPS codes is available from the EPA online here: http://www.epa.gov/envirofw/html/codes/state.html.

Note that the U.S. Census Bureau's county boundaries extend to the national boundaries when they border water features, such as on the Great Lakes on the U.S.–Canadian border. Also note that the county boundary file was projected into the *Web Mercator (auxiliary sphere)* projection for purposes of this chapter. There may be another projection for your own state (such as the *Lambert Conformal Conic* projection) that you may want to use instead (see Chapter 1 for more about projecting data).

The TIGER files provide only the geographic boundaries; attribute data must be downloaded separately and joined to the shapefiles (see Chapter 2). The attribute data (the housing statistics) for each county in the state was downloaded from the Census Bureau's American FactFinder Website available here: http://factfinder2.census.gov. The tables used in this chapter were the 2010 SF1 versions of H5: Vacancy Status and H003: Occupancy Status.

FactFinder attribute data was downloaded as a CSV file, underwent minor editing and cleanup in Microsoft Excel, and was converted to an Excel file for use in this chapter (and then imported to the geodatabase as a table). The Census Bureau provides an online guide to obtaining data from American FactFinder and preparing it for use in ArcGIS here: http://www.census.gov/geo/education/pdfs/tiger/Downloading_AFFData.pdf.

STEP 3.1 Getting Started

- Start ArcMap (with a new blank map) and use Catalog to copy the folder called **Chapter3** from the C:\GISBookdata\ folder to your own D:\GIS\ folder. Be sure to copy the entire folder, not just the files within it.

- The Chapter3 folder contains a file geodatabase called **OhioInfo**, which contains the following:

    - Ohiocountyhousing (a polygon feature class of Ohio county borders— housing data from the U.S. Census SF1 H003 and SF1 H5 information has been joined to this feature class)

- Add the Ohiocountyhousing feature class to the Table of Contents.

- Right-click the Ohiocountyhousing layer and select **Properties**. Under the **General** tab, change the name of the layer to **Ohio Seasonal Homes**.

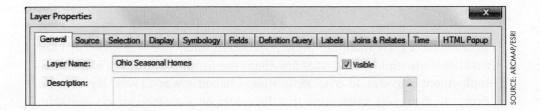

SOURCE: ARCMAP/ESRI

STEP 3.2 Setting Up a Choropleth Map

• Right now, each Ohio county is shown with the same color and no attribute values are displayed in the Table of Contents. Each county contains different attributes, and what you want to do is display the one that shows the percentage of the county's housing stock that is seasonal homes.

• Open the layer's attribute table and scroll across it. You'll see that there are several fields, but two in particular will be useful: (1) a field called *seasonal* that represents the number of seasonal homes in each county and (2) a field called *perseas* that represents the percentage of the total number of houses that are seasonal homes. To create a choropleth map, you'll be using the *perseas* attribute instead of the *seasonal* attribute because the *perseas* attribute has been normalized. For more about using normalized data on a map, see **Smartbox 19** (and close the attribute table).

Smartbox 19

What are normalized data and how are they used in ArcGIS 10.2?

When you're dealing with data values to be presented on a choropleth map, you must first normalize those data values before displaying them on the map. To **normalize** values, you must convert them to a consistent level of data representation, such as a percentage. Normalized values can be compared to each other independent of the size of the polygons (or areal units) that contain those values. A choropleth map that shows count values (the actual numbers constituting whatever is being mapped) will be different from another choropleth map showing the normalized version of those values.

    For instance, if you were creating a choropleth map of unemployment for each state in the United States, chances are that larger states with higher populations like California would also have very high numbers of employment; smaller or less populated states like Rhode Island would have lower numbers. In fact, according to U.S. Bureau of Labor Statistics data for unemployment in 2011 (found here: http://www.bls.gov/lau/table14full11.pdf), California had an estimated 2,138,000 unemployed persons, while Rhode Island had 62,000 unemployed persons. Thus, your choropleth map using these count values would show a massive amount of unemployment in California and very little unemployment in Rhode Island. However, California's size and population are much larger than those of Rhode Island, and a greater number of people will likely equate with a similarly higher number of unemployed persons. Thus, your choropleth map is showing very skewed results. A better mapping strategy would be to normalize the data and instead

normalize Altering count data values so that they are at the same level of representing the data (such as using them as a percentage).

of showing the number of unemployed persons, show the percentage of those employed (by dividing the number of unemployed persons by the total civilian labor force). By mapping these normalized data, you will show that the two states' unemployment rates are about the same—California's 2011 unemployment rate was 11.6%, while Rhode Island's was 11.4%. By normalizing your data, you're ensuring that the values on your map are comparable.

In this chapter, you're working with the percentage of the Ohio county housing stock that is seasonal, which is normalized data. Figure 3.2 shows the seasonal home data in two ways. One map contains the raw count values and the other contains the normalized version of the data (dividing the number of seasonal homes by the total number of houses). The maps show two very different versions of seasonal home distribution in Ohio. In mapping count values, more highly populated counties (with more houses), such as Lucas, Hamilton, or Cuyahoga Counties, have correspondingly more seasonal homes. The choropleth map with the normalized data, however, shows each county mapped as a percentage of the total housing stock and thus keeps the results on the same level (and is the map that should be used).

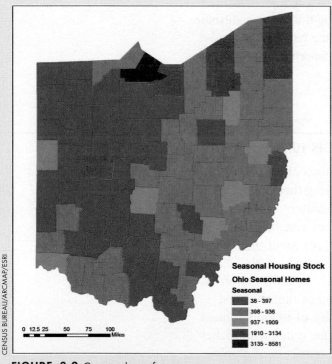

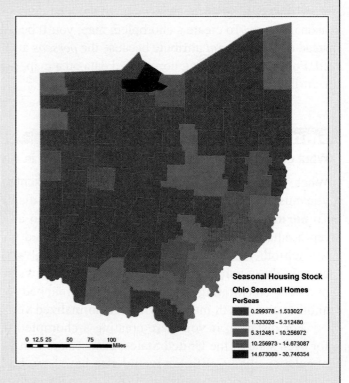

Seasonal Housing Stock
Ohio Seasonal Homes
Seasonal
- 38 - 397
- 398 - 936
- 937 - 1909
- 1910 - 3134
- 3135 - 8581

Seasonal Housing Stock
Ohio Seasonal Homes
PerSeas
- 0.299378 - 1.533027
- 1.533028 - 5.312480
- 5.312481 - 10.256972
- 10.256973 - 14.673087
- 14.673088 - 30.746354

0 12.5 25   50   75   100
Miles

CENSUS BUREAU/ARCMAP/ESRI

FIGURE 3.2 Count values of Ohio counties' seasonal homes vs. normalized percentage values for the same seasonal homes.

- Back in Chapter 1, you changed the colors and symbology for individual items, but that won't work with a choropleth map. You could display the unique value for each county, but you'd then end up with 88 different values (or different colors) in the Table of Contents, which would be really difficult to work with or effectively display on a map. Instead, what you want to do is group several data values together into one category and display a single color for the entire category. To display the perseas attribute for each state, bring up the Ohiocountyhousing layer's properties and click on the Symbology tab.

- In the Show: box, select **Quantities** and **Graduated colors**.

- In the Fields box, for Value select **PerSeas** and for Normalization select **none**. (Since the PerSeas attribute is already normalized, you don't need to normalize it a second time.)

- In the Classification box, use the default of **Natural Breaks (Jenks)** but for Classes: choose **5**.

- A default color ramp will be provided, so use that for now.

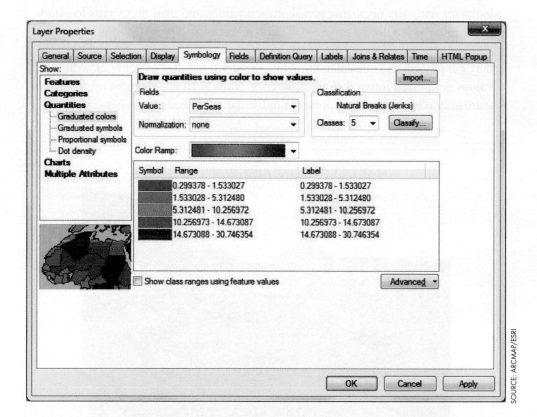

SOURCE: ARCMAP/ESRI

- Click **Apply** and **OK**. You'll see that each county is now displaying the PerSeas attribute, but the values have been grouped into five categories, with each category given its own color. You're using the Natural Breaks method of separating the data into the categories, but there are several different methods of classifying data, depending on the nature of the data being separated. See **Smartbox 20** for more information.

Smartbox 20

What are the differences among the data-classification methods in ArcGIS 10.2?

When setting up a choropleth map, you'll be taking multiple values and dividing them into smaller groupings. For instance, the 88 seasonal home values you're working with in this chapter can get grouped into three categories (highest, lowest, and average) or five categories (highest, high, average, low, and lowest).

data classification Various methods used for grouping together (and displaying) values on a choropleth map.

Natural Breaks A data-classification method that selects class break levels by searching for spaces in the data values.

However, there are several different, more scientific, ways of doing this kind of **data classification** when setting up a choropleth map. ArcGIS has many different types of data-classification methods available, and each will produce a different map, depending on the method used. The way the values are distributed will help determine which of the methods you should apply when making a map. For instance, are all of the seasonal home values evenly separated (are there the same number of high values as low values?), or are they mostly clustered around an average value with only a few outlying high and low values?

There are four main data-classification methods available, with options for setting up your own customized classification. The first of these is the **Natural Breaks** method (also called the Jenks Optimization method). This method looks for naturally occurring gaps between values and uses these gaps to establish the start of each category into which the values will be placed. As Figure 3.3 shows, there are a lot of Ohio counties with low seasonal home percentages, so they get placed into one category, while the small handful of counties with high values get placed into a different category. Natural Breaks is most useful when there are nicely defined break points in the data values.

FIGURE 3.3 A choropleth map of Ohio counties' seasonal home percentages as shown with the Natural Breaks data classification method.

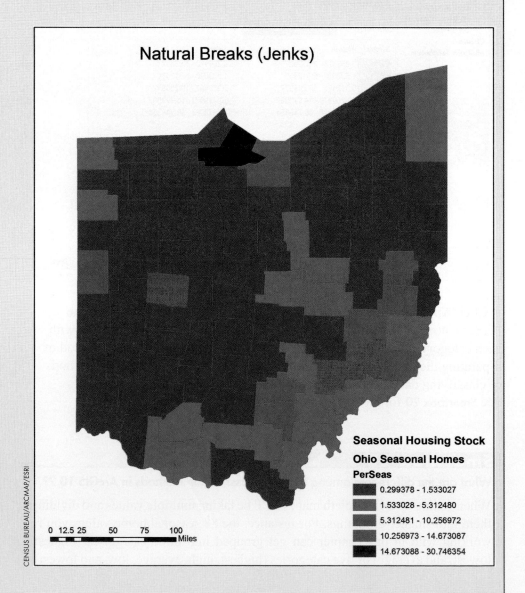

Natural Breaks (Jenks)

CENSUS BUREAU/ARCMAP/ESRI

0 12.5 25   50   75   100
Miles

Seasonal Housing Stock

Ohio Seasonal Homes

PerSeas

■	0.299378 - 1.533027
■	1.533028 - 5.312480
■	5.312481 - 10.256972
■	10.256973 - 14.673087
■	14.673088 - 30.746354

The second classification is the **Equal Interval** method (see Figure 3.4) which creates the category ranges of equal sizes. The category size is dependent on the range of values in the dataset. In this case, seasonal home percentages for Ohio counties are between 0.299% (in Fayette County) and 30.746% (in Ottawa County). The difference between these values is 30.447, and since we're using five categories, each category will contain 6.089%. Thus, all values between 0.299% and 6.388% will be placed in the first category. All values between 6.388% and 12.478% will be placed into the second category, and so on. As Figure 3.4 shows, because there are so many Ohio counties with low percentages of seasonal homes, most of them fall into the first category (less than 6.388%) and only a handful fall into the other defined categories. Equal Interval is most useful when the data values are continuous without a large number of very high or very low values.

The third methods is **Quantiles** (see Figure 3.5), in which the categories each contain an equal (or near-equal) number of entries, regardless of their actual values. Because there are 88 counties in Ohio and we are mapping five different categories, each category will contain 17 or 18 different

Equal Interval A data-classification method that selects class break levels by taking the total span of values (from highest to lowest) and dividing by the number of desired classes.

Quantiles A data-classification method that attempts to place an equal number of values in each class.

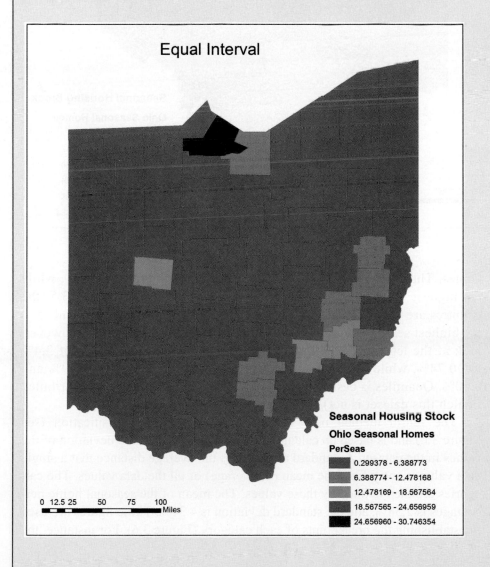

FIGURE 3.4 A choropleth map of Ohio counties' seasonal home percentages as shown with the Equal Interval data classification method.

FIGURE 3.5 A choropleth map of Ohio counties' seasonal home percentages as shown with the Quantiles data classification method.

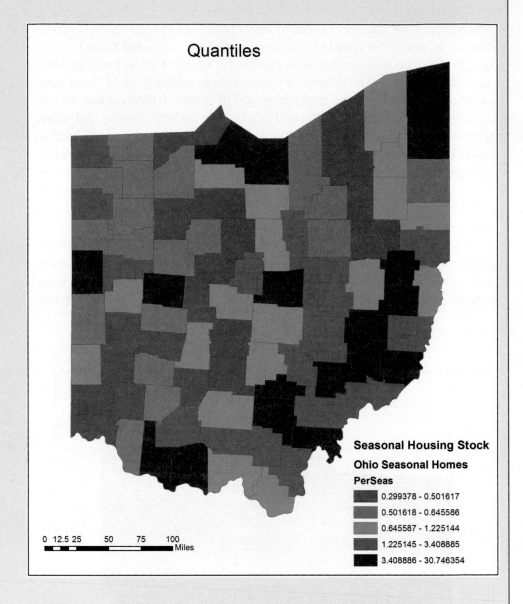

Quantiles

Seasonal Housing Stock

Ohio Seasonal Homes

PerSeas

- 0.299378 - 0.501617
- 0.501618 - 0.645586
- 0.645587 - 1.225144
- 1.225145 - 3.408885
- 3.408886 - 30.746354

0 12.5 25   50   75   100
                                  Miles

Standard Deviation A data-classification method that computes break values by using the mean of the data values and the average distance a value is away from the mean.

values. Thus, the lowest 17 values get placed into the first category, while the highest 17 values get placed into the last category. In Figure 3.5, the counties are very evenly distributed—there are nearly the same number of the highest seasonal home percentages as there are of the lowest. However, look at the legend for the map. The highest category has values of 3.4% to 30.74%, while the lowest category shows values between 0.299% and 0.50%. Quantiles is best used when your data values are evenly distributed (which this dataset is not).

The fourth method uses the **Standard Deviation** classification (see Figure 3.6) and is based on calculating the mean and standard deviation of the values in the dataset. A standard deviation is the average distance that a single data value is away from the mean (or average) of all the data values. The categories are then defined by these values. The mean of the seasonal home percentages is 2.58%, and the standard deviation is 4.31%. These figures are used in establishing the breakpoints of each category (Figure 3.6). For instance, the

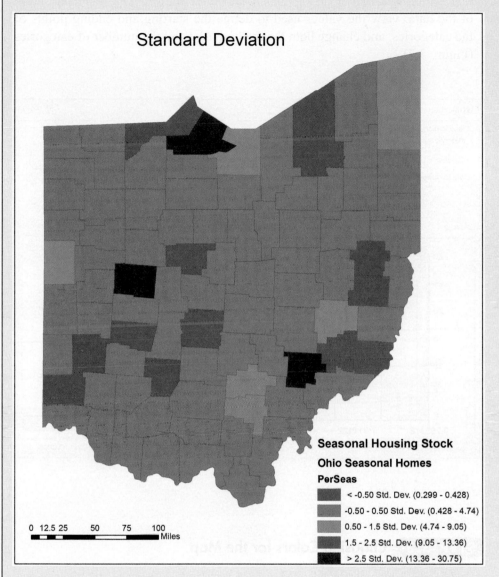

Standard Deviation

Seasonal Housing Stock

Ohio Seasonal Homes

PerSeas

■	< -0.50 Std. Dev. (0.299 - 0.428)
■	-0.50 - 0.50 Std. Dev. (0.428 - 4.74)
■	0.50 - 1.5 Std. Dev. (4.74 - 9.05)
■	1.5 - 2.5 Std. Dev. (9.05 - 13.36)
■	> 2.5 Std. Dev. (13.36 - 30.75)

0 12.5 25   50   75   100
━━━━━━━━━━━━━━━━━━━━━ Miles

FIGURE 3.6 A choropleth map of Ohio counties' seasonal home percentages as shown with the Standard Deviation data classification method.

highest category has values of greater than 2.5 times the standard deviation greater than the mean ((2.5*4.31) + 2.58), or 13.355%. Similarly, the lowest category has values of less than 0.50 times the standard deviation less than the mean ((−0.50*4.31) + 2.58) or 0.425. The Standard Deviation method produces the best results when the data values follow a normal distribution (where the values are well distributed around the mean).

A few other data-classification methods available in ArcGIS allow a greater degree of customization. The Manual method allows you to define your own breakpoints for categories. The Defined Interval method works like Equal Interval, except you can specify how large each of the categories should be. Geometrical Interval attempts to have the same number of values in each category but be useful for continuous data (it's a blend of the Natural Breaks, Equal Interval, and Quantiles methods).

Under the Symbolization tab of the Layer Properties of the layer for which you are setting up the data classification, you'll see a Classify button. Pressing this will bring up a new dialog that allows you to see a chart of the distribution of the data, view the values used to define the starting and ending points of the categories, and change both the method used and the number of categories (Figure 3.7).

FIGURE 3.7. The dialog in ArcGIS containing information for data classification.

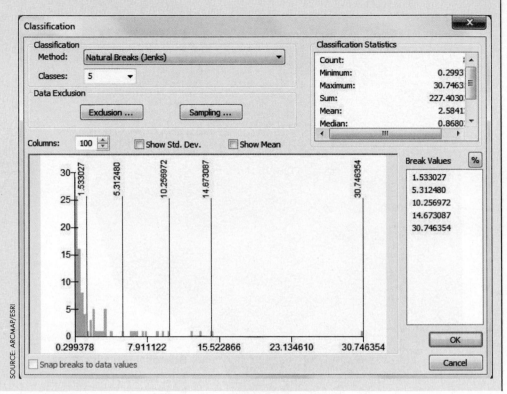

SOURCE: ARCMAP/ESRI

STEP 3.3   Choosing Colors for the Map

• In Chapter 1, you displayed data layers (for example, airports, water bodies, roads, and different types of structures) using different sizes, symbols, and colors, such as an airplane symbol to denote airports. If you were creating a reference map, those symbols you displayed in ArcMap would be the same symbols displayed on a printed map. However, with a choropleth map, you will need to choose different colors for each of your categories. When you set up the choropleth map, you used the default given by ArcGIS, but you can easily change those colors. To adjust the color choices, bring up the Ohiocountyhousing layer's properties and again select the **Symbology** tab.

• From the pull-down menu next to Color Ramp, select a different set of colors for each of the categories on the choropleth map. The **color ramp** will assign a different color to each category on the map, often with a light shade as the first one and a darker shade as the last one, but there are many different options. If none of the color ramps appeal to you, you can select an individual color for a category by changing its display color (as you did in Chapter 1). However, several factors influence the colors you should select. See **Smartbox 21** for more information about color choices in map design.

color ramp A range of colors that are applied to the thematic data of a map.

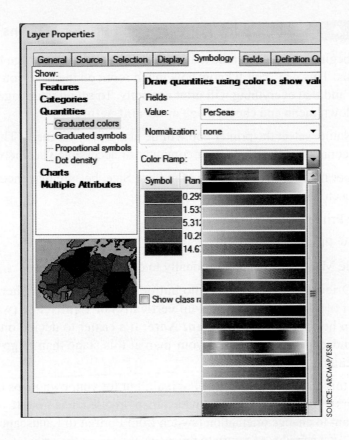

SOURCE: ARCMAP/ESRI

Smartbox 21

Why is color choice important when designing a layout in ArcGIS 10.2?

The colors used on the map are important to the overall map design, especially when it comes to presentation. However, colors may appear one way on a computer monitor, a different way when printed out, and an entirely different way when projected onto a screen from an LCD projector. Shades of green and blue that look great on your monitor will not necessarily look the same when printed on an inkjet printer, or they may appear "washed out" when projected on a screen.

ArcMap allows you to select from a variety of graduated colors when selecting a color ramp. **Graduated colors** are often different shades of a single color (such as several shades of blue, such as light blue, medium blue, dark blue, and an even darker shade of blue); they may also take on multiple hues from shades of green through shades of blue.

Another option for symbology is the use of **graduated symbols**, in which different sizes of the same symbol are used (but the color for all of them remains the same). Figure 3.8 provides examples of graduated colors and graduated symbols.

graduated colors The use of various hues in representing ranges of values on a map.

graduated symbols The use of different sized symbology to convey thematic information on a map.

Symbol	Range
	0.299378–1.533027
	1.533028–5.312480
	5.312481–10.256972
	10.256973–14.673087
	14.673088–30.746354

Symbol	Range
•	0.299378–1.533027
●	1.533028–5.312480
●	5.312481–10.256972
●	10.256973–14.673087
●	14.673088–30.746354

FIGURE 3.8 The use of graduated colors and graduated symbols in ArcMap.

STEP 3.4 Setting Up the Page Size and Printer Margins

• Before beginning to lay out the map, you will want to set the page boundaries and margins. This way, you'll know how much room on the page you have to work with and that everything will print correctly. To set up your page, go to the **File** pull-down menu and choose **Page and Print Setup**.

• Under Name: select the printer you will be using for your map. (The printer should be connected directly to your computer or accessible via a network.)

• For proper map formatting, under Map Page Size, be sure that checkmarks are placed in each of the following boxes:

> • Use Printer Paper Settings
>
> • Show Printer Margins on Layout
>
> • Scale Map Elements proportionally to changes in Page Size

• Under Orientation, select the radio button that indicates whether your map will be **Portrait** (with the page set up vertically) or **Landscape** (with the page set up horizontally). ***Important Note:*** It's easier to decide on a Portrait or Landscape orientation for your map at this stage than to go back and change it later.

> • For this chapter's data, select **Portrait**, but for your own maps (and other chapters), you'll have to decide which map orientation is best. If you later find you want to change orientation (switch from Portrait to Landscape or vice versa), you can always return to this page and adjust the orientation settings.

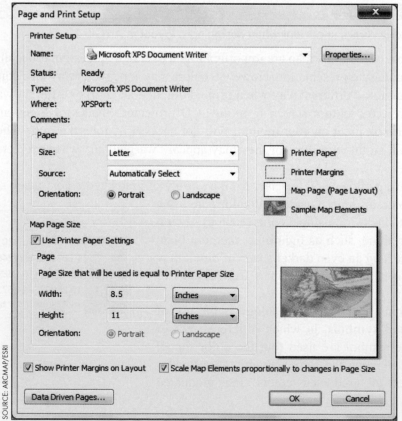

SOURCE: ARCMAP/ESRI

- Also note the sample layout in the lower right of the dialog. It represents the margins that the printer will print and what the page layout will look like.

- When all settings are correct, click **OK**.

layout The assemblage and placement of various map elements used in constructing a map.

STEP 3.5 Working with the Layout View and Data Frames

- When dealing with data in ArcMap, there are two views for examining it: the Data View and the Layout View. These are available under the **View** pull-down menu. The default in ArcMap is the Data View. Switch to the **Layout View**.

- A second method for switching between Data View and Layout View is a set of buttons at the bottom of the map.

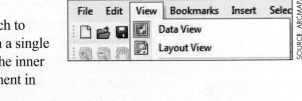

- Think of the ArcMap **layout** as a blank piece of paper on which to design a map. In Layout View you'll see what will be printed on a single page. The outer boundary shows the printed page boundaries. The inner page shows the outline of the Data Frame. The Data Frame element in the layout contains only the layers being displayed from the Table of Contents. The actual boundary of the printed page is NOT the data frame. Rather, the layout shows that page boundary by the grey outline on the layout itself.

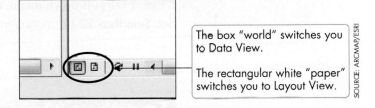

The box "world" switches you to Data View.

The rectangular white "paper" switches you to Layout View.

- To resize data frames, select the cursor tool from the Tools toolbar, click on the data frame to select it, click on the data frame outline (the dashed line), and resize by dragging one of the corners or sides.

- To move data frames, click on the data frame itself, hold down the mouse button, and drag the frame to another position.

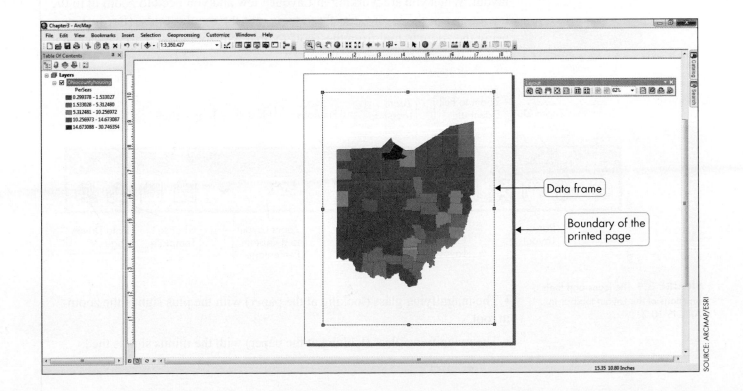

Data frame

Boundary of the printed page

• Note that the Data Frame will show the layers in the Table of Contents. We won't do it in this chapter, but if you wanted to show two separate maps on the same layout, you would need to add a second Data Frame to the layout (by selecting the **Insert** pull-down menu and choosing **Data Frame**). A second data frame would appear on the layout. In the TOC, you could add or copy layers into it, thus showing two different maps on the same layout.

• You'll also see that by default the Data Frame has a border around it. If you want to adjust the Data Frame's border (to change its appearance), right-click on the Data Frame itself (either on the map layout or in the Table of Contents) and choose **Properties**. In the Data Frame Properties dialog, click on the **Frame** tab. A number of options will be available. These include adjusting the color and thickness of the Data Frame's border (including making the border invisible by choosing "no color"), adding a background color to the Data Frame, and adding a drop shadow to the Data Frame.

• When you switch to Layout View, a new Layout toolbar will appear. Like all other toolbars, you can toggle the Layout toolbar on and off by choosing the **Customize** pull-down menu, then selecting **Toolbars**, and then choosing **Layout**. See **Smartbox 22** for more about the functions of the Layout toolbar.

Smartbox 22

What are the functions of the Layout toolbar in ArcGIS 10.2?

The Layout toolbar (Figure 3.9) provides a set of tools for composing map layouts. Some of the tools on the Layout toolbar look the same as their Tools toolbar equivalents, except in the Layout toolbar these tools will affect only the layout. When you are working in Layout View and you need to zoom in to the layout or pan the layout, use these tools, not the regular tools. (For instance, using the zoom in tool from the Tools toolbar will zoom you in to the data, not to how they're shown on the layout.)

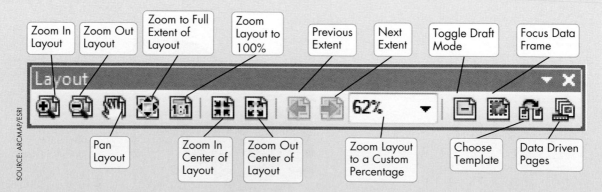

FIGURE 3.9 The icons and their functions of the Layout toolbar in ArcGIS 10.2.

• The magnifying glass (looking at the paper) with the plus sign is the zoom-in tool.

• The magnifying glass (looking at the paper) with the minus sign is the zoom-out tool.

- The hand (over the paper) is the pan tool.

- The paper with arrows in the four cardinal directions will zoom to the whole page.

- The paper labeled 1:1 will zoom the layout to 100%.

- The paper with the arrows pointing in will zoom in to the center.

- The paper with the arrows pointing out will zoom out from the center.

- The papers labeled with the arrows pointing right and left will zoom to previous extents.

- The pull-down menu with the % will select how closely you are zoomed in on the layout (that is, how much of the layout is shown on the screen at one time).

- The page with a line through it allows you to toggle draft mode on and off. When draft mode is on, none of the content of data frames is drawn on the map, allowing you to work with the position of map elements and other features instead.

- The green and blue icon allows you to put a data frame in focus when adding map elements.

- The arrow between two pages allows you to select a default **map template** for your data. Map templates are pre-made layouts in which all elements are already in place and the data get plugged into their previously identified location, size, and format.

Important Note: Because this chapter focuses on creating a map layout from scratch, the use of map templates in this chapter is not allowed, so don't use them this time. However, you may find templates useful for creating maps for future chapters.

- The last button allows for creation of data-driven map items. Data-driven pages are used when making several layouts from one map document (showing different extents on each map) and when creating a map book.

map template A pre-made arrangement of items in a map layout.

STEP 3.6 Adjusting the Display Scale and Adding Scale Information

The next step is to fix the map scale of the data in the layout. While in layout mode, select the Map Scale pull-down menu on the Standard Toolbar and type 1:3000000 and click the enter key on the keyboard. This will adjust the map scale of the layout. Try a few other scale values (such as 1:2500000 or 1:3500000) to see which one best allows you to map the whole state for your layout. For more information about map scale and its effect on layout design, see **Smartbox 23** on page 76.

SOURCE: ARCMAP/ESRI

Smartbox 23

How is scale represented on a layout in ArcGIS 10.2?

In GIS, there are a couple of different ways of thinking about scale.

geographic scale The real-world size or extent of an area.

map scale A metric used to determine the relationship between measurements made on a map and their real-world equivalents.

Representative Fraction (RF) A value indicating how many units of measurement on the map are the equivalent of a number of units of measurement in the real world.

small-scale map A map with a lower value for its representative fraction. Such maps will usually show a large geographic area.

large-scale map A map with a larger value for its representative fraction. Such maps will usually show a smaller geographic area.

Geographic scale refers to the real-world size or area of something. If you're making a map of the entire United States, that map would cover a very large geographic scale, while a map of your property boundaries would reflect a very small geographic scale.

Map scale reflects how many units of measurement on the map are equal to a number of units in the real world. For instance, one inch measured on a map might be equivalent to 5000 inches in the real world. Map scale is usually measured as a **Representative Fraction (RF)**, such as 1:150000. This ratio indicates that one unit of measurement on the map (such as one inch or one centimeter) represents 150,000 of the same units in the real world.

Depending on the Representative Fraction and map scale, maps are considered **small-scale maps** or **large-scale maps**. Small-scale maps often show a larger geographic area but have a smaller RF value (such as 1:250000). Large-scale maps show a smaller geographic area and have a larger RF value (such as 1:4000). The largest-scale map you could make would be 1:1, in which one unit of measurement on the map would reflect one unit of real-world measurement (in other words, a 1:1 map would be the same size as the area that it represented).

The choice of map scale is very important when designing a map, as this choice will affect how much area can be displayed, as well as the types of symbology that can be used. For instance, on a small-scale map (such as a 1:1000000 map), cities would be represented as points. Individual features of cities (such as park boundaries) could not be properly represented at that scale. Major roads could be represented, but individual streets could not. However, on a larger-scale map (such as a 1:24000 map), individual park boundaries could be shown as polygons, and many residential roads could be represented as lines.

In the chapter, you initially set the map scale of the layout to 1:3000000, so you will be producing a very small scale map. At this scale, a distance of one inch measured on the map of Ohio would be equal to 3,000,000 inches (or 250,000 feet) in the real-world Ohio. By zooming in or out of the map (either in the data view or the layout view), the scale will change accordingly. You can also type a new RF into the Standard Toolbar to manually change the display scale for the layout.

- Even though you've set the map scale in ArcMap, you need to communicate this information to the map reader. A graphical device for displaying the map scale is a **scale bar**. To add a scale bar to the map, from the **Insert** pull-down menu, select **Scale Bar...**.

- Select the scale bar you want to add to the map. A preview of it will be shown in the dialog box. Click **OK** to add the scale bar to the layout.

scale bar A graphical device used to represent scale on a map.

- Place the scale bar where you want it on the layout. You can change the appearance of the scale bar by right-clicking on the scale bar itself and selecting **Properties...**. In the properties dialog box for the scale bar, you can make adjustments to the number of divisions or frequency of the numbers on the scale bar.

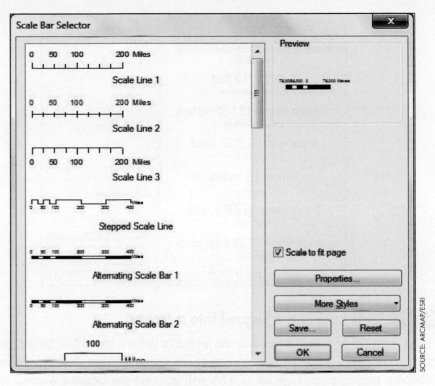

SOURCE: ARCMAP/ESRI

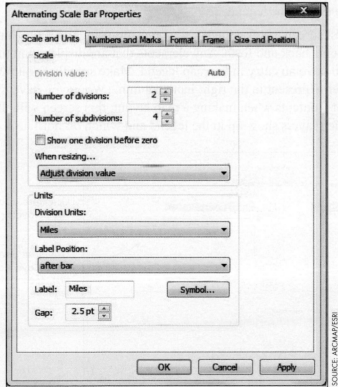

SOURCE: ARCMAP/ESRI

• You can also add **scale text** (written information about the map scale itself) to the map by choosing the **Insert** pull-down menu and selecting **Scale Text…**.

• Select the option for **Absolute Scale** and click **OK**. The RF of the map scale will be added to the map as a text box. Move it to an appropriate place on the map. To adjust the font, color, and other properties of the scale text, right-click on the text and select **Properties**.

scale text Descriptive information used to represent scale on a map.

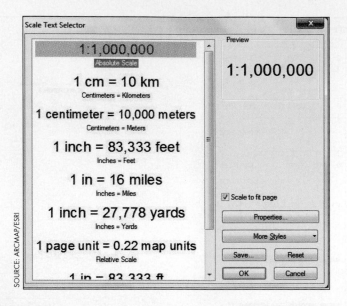

SOURCE: ARCMAP/ESRI

STEP 3.7   **Inserting a Legend into a Layout**

• Next, you'll add a map **legend** to the layout, a device to explain to the map reader what the various symbols and colors on the map represent. From the Insert pull-down menu, select **Legend....** This will activate the Legend Wizard, which will let you design the legend.

• In the first screen, choose the layers you want, using the arrow buttons to move elements back and forth. Any elements that are displayed in the right-hand column will have an entry in the map legend. Make sure that your Ohio Seasonal Homes layer is present in the right-hand column. (When you have more layers in the Table of Contents when making a map layout, this screen will allow you to control which layers show up in the legend and which do not.)

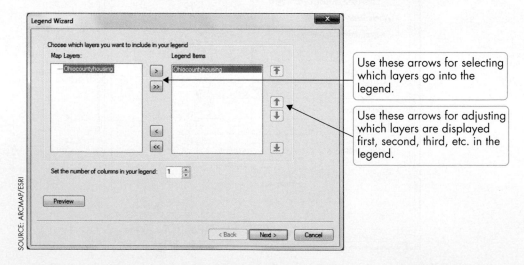

SOURCE: ARCMAP/ESRI

• You can also select the number of columns for your legend. You may want to experiment with legend options—for example, one column, two columns, three columns. With only one item to show in the legend, you'll need only one column, but when you create a map that displays several layers in the legend, multiple columns will probably be useful. Press the **Preview** button to look at the results of using different options. (***Important Note:*** Be sure to de-select the Preview button when you're done.)

legend A graphical device used on a map as an explanation of what the various map symbols and colors represent.

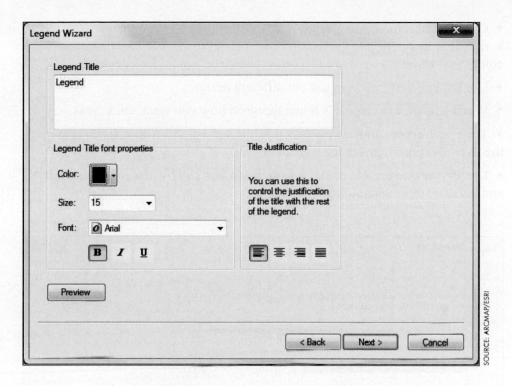

SOURCE: ARCMAP/ESRI

- When you have your layers and columns set for the legend, click **Next**.

- In the second screen, type in the title for the Legend. Give your legend a better name than "Legend."

- You can also adjust the color, size, and font for the title, as well as the placement of the title in relation to other map elements.

- Use the **Preview** option to test out some different design options (but de-select Preview again before moving on).

- Click **Next** when you have the legend's title set up how you want it.

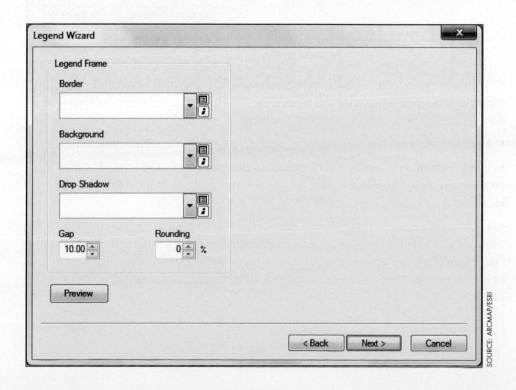

SOURCE: ARCMAP/ESRI

- On the third screen, you can select options for the color and thickness of the border around the legend, the background color of the legend (the default is no color), and whether or not to add a drop shadow design to the legend.

- Use the **Preview** button to test out different designs.

- When you have the legend's frame designed how you want, click **Next**.

- The fourth screen allows you to change the symbol patch, in essence changing the size of various items in the legend.

- Use **Preview** to test out some options, then click **Next** when you have the legend items' patches set how you want them.

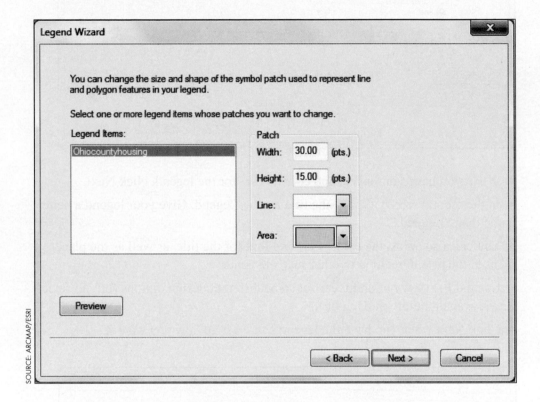

SOURCE: ARCMAP/ESRI

- The last screen allows you to change the spacing between items and entries in the legend, including spacing between items, the distance between columns, or the distance between the legend and the title.

- Like before, **Preview** will show you the results of any changes you make here.

- Click **Finish** to place the final legend on the map. Once the legend is placed on the map, you can select it with the cursor and move it around to another location.

- Each layer you use in the legend has a set of layer properties. Anything changed in the Layer Properties will be reflected in changes to the layout. For instance, if you change the name of a layer or the color used to display it in its Layer Properties, the same change will be made in the legend on the layout.

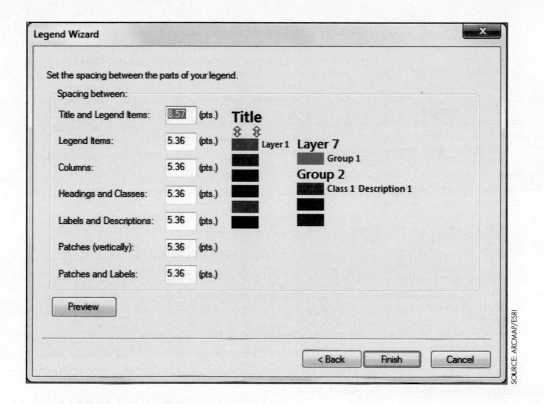

SOURCE: ARCMAP/ESRI

STEP 3.8 **Inserting a North Arrow**

> **north arrow** A graphical device used to show the orientation of the map.

• A **north arrow** is a graphical indicator of what direction north is facing on the layout. Sometimes when making a map (depending on the map projection or the map design), north is not always straight up, so a north arrow helps orient the map reader.

• To add a north arrow to the layout, from the **Insert** pull-down menu, select **North Arrow…**.

• Select the north arrow you want to add to the map. A preview of it will be shown in the dialog box.

• You can access other, more distinctive, north arrow options by selecting the **More Styles** option.

• Once you have the north arrow you want, click **OK** and it will be added to the layout. Move it around to its correct place on the map. You can change the color, size, and appearance of the north arrow by right-clicking on the north arrow itself, selecting **Properties…** and changing the options in the dialog that opens.

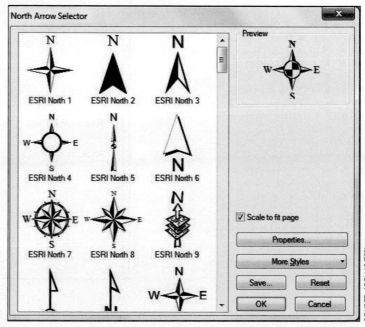

SOURCE: ARCMAP/ESRI

STEP 3.9 **Inserting Text into a Layout**

• Next, you'll want to give your map a title. To add a title, from the **Insert** pull-down menu, select **Title**.

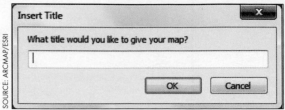

• In the Insert Title dialog box, type in an appropriate map title. When naming a map, don't use the word "map" in the title, as that's pretty self-evident (it would be like naming your dog "Dog" or your cat "Cat"). Choose a more descriptive title that sums up your map of seasonal homes in Ohio counties. Click **OK** when your title is set.

• The title will be added to the top of the map. Click on the title and drag it to an appropriate place in the layout.

• If you want to adjust the font, size, or color of the title, double-click on it. A Properties dialog box will open, allowing you to adjust these options. In the Properties box, select the **Change Symbol** button to select a different font, make the title bold or underlined, and change the size of the text.

• Next, you'll want to add text to the map. Specifically, you want to add information about your name (as the designer of the map), today's date, and the source of the data. To insert this information (or any other text into the layout), from the **Insert** pull-down menu, select **Text**.

• *Important Note:* When text is inserted to a layout, a small box called "text" is placed in the center of the map.

• Double click on this "text" box and a Properties dialog box will open so that you can change the size and font of the text. In the Properties box, select the **Change Symbol** button to select a different font, make the text bold or underlined, and change the size of the text. For more information about the use of type and fonts in a layout, see **Smartbox 24**.

Smartbox 24

Why is the choice of fonts important when designing a layout in ArcGIS 10.2?

Although ArcMap gives you a wide variety of choices when it comes to selecting fonts for text, titles, and labels, usually a map should contain only two different fonts, carefully chosen to complement each other (and the map as a whole). Using too many different fonts on a map, or selecting several of the more esoteric fonts, is likely to make the map more difficult to read (and thus less helpful to its intended audience). By selecting the **Change Symbol** option, you can choose from several pre-made fonts that Esri has set up as standard fonts for map items such as historic regions coastal areas, or oceans (Figure 3.10).

AaBbYyZz	A a B b Y y Z z	A a B b Y y Z z	AaBbYyZz	A a B b Y y Z z
Coastal Region	Ocean	Physical Region	Historic Region	Sea

SOURCE: ARCMAP/ESRI

FIGURE 3.10 Examples of designated fonts available in ArcMap.

- You can then select the text box and move it to another position on the map (like you could with all of the other map elements).

- Last, you'll add some further information about the projected coordinate system being used for this layout. In ArcGIS, this takes the form of **dynamic text**, or text that updates itself as changes are made to the layout. Thus, if you change the coordinate system used in the Data Frame, the corresponding coordinate system text will automatically update itself. There are several other dynamic text options available, including the current date, current time, document name, and document path. To insert this coordinate system information, from the **Insert** pull-down menu, select **Dynamic Text**, then select **Coordinate System**. The text box with the data's coordinate system information will be added to the layout and you can move it and adjust its font and style.

dynamic text Descriptive information about map properties that will change as those properties change or are updated.

STEP 3.10 Evaluating the Map: Moving and Resizing Map Elements

- You can arrange the elements on your map to produce the best designed map possible. For example, you can resize and move map elements by selecting them with the mouse and resizing them like you would the data frame. You can delete map elements by selecting them with the cursor and pressing the delete key on the keyboard. See **Smartbox 25** for further information about map design.

Smartbox 25

What are some strategies for designing an effective map in ArcGIS 10.2?

Designing a good, effective map to communicate its message is crucial when setting up a layout. You don't want to add a data frame, throw on a legend and north arrow, and then print out the map. The map would look sloppy and unprofessional.

A map layout should follow a **visual hierarchy**, with some items more prominently displayed than others. The purpose of the map will help you determine which map elements should be the most prominent in the map's visual hierarchy. For instance, if you're creating a map of a local park, is the purpose to show the park's location in relation to its surroundings (such as the roads needed to reach the park) or is it to highlight the locations and dimensions of the walking paths? Each of these maps should be designed differently and should have different elements prominently displayed. Without a clear hierarchy of its items, the map is likely to look cluttered or be confusing to the reader. A clear visual hierarchy also aids in balancing the "empty spaces" on the map.

ArcMap allows you a great deal of flexibility when putting together a layout. Each element (such as the north arrow, the text boxes, or the legend) you place on the map can have its borders prominently highlighted or made transparent. To access these options, select the element, right-click on it, select **Properties**, and select the **Frame** tab in the dialog that opens (Figure 3.11). In the option for Border, you can change the thickness of the lines, or by selecting **<None>** you can make the element's border transparent. In the options for Background, you can change how the color behind the element appears, while

FIGURE 3.11 The various options available for the Frame around the legend (and other map elements).

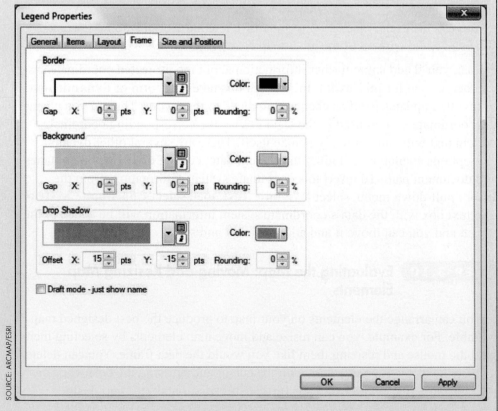

the Drop Shadow options can add extra effects (if desired) to the element's appearance. In some cases, you may find that placing (or removing) borders around the boxes of each element will enhance the map's design.

You can place other borders (referred to as **neatlines**) around a selected group (or all) of the map elements. To add a neatline, choose the Insert pull-down menu and then select **Neatline...** to access the many options available. Neatlines can also be re-sized or changed like the data frame if created as separate elements. Make sure your neatlines stay within the boundary of the printed page.

> **neatline** A border that can be placed around some or all map elements in a layout.

Checklist

Use the following checklist to make sure that you have put together your map layout for this chapter in a complete, visually appealing fashion that conveys information effectively.

_____ The percentage of the seasonal housing stock in all 88 Ohio counties displayed with an appropriate color scheme

_____ Ohio counties displayed at an appropriate scale

_____ Appropriate scale bar

_____ Appropriate scale text

_____ Appropriate map legend, with all items listed with normal names (The legend should not be called "legend," and it should be appropriate with regard to size, font, number of columns, and so on.)

_____ A north arrow of appropriate size and position

_____ Appropriate title (don't use the words "map" or "title" in your title) in appropriate size and font

_____ Type (your name, the date, the source of the data) in an appropriate size and font

_____ Coordinate system information in an appropriate size and font

_____ Overall map design (Check for appropriate borders, color schemes, balance of items and placement, and so on.)

- When you have constructed the map the way you want, you can Preview the map. From the **File** pull-down menu, choose **Print Preview**. Your map will be displayed as it should print on the printer you have selected.

- **Important Note:** If some borders don't appear, if borders are cut off, or if portions of the map are off the screen, you'll need to go back and re-adjust these items before printing the map.

STEP 3.11 Printing and Exporting the Layout

- Save your map document. When everything looks the way you want to print, from the **File** pull-down menu select **Print**. In the Print dialog, click **OK**.

• If you want to export the map layout to a digital format for distribution instead of printing, from the **File** pull-down menu, select **Export Map…**. As in Chapter 1, you can save your final map layout as a PDF or several other graphical formats (such as a JPEG or TIFF).

Closing Time

This chapter explored how to use GIS data to create a professional-looking map. ArcMap provides a wide variety of options for designing and customizing maps for all manner of purposes. While this chapter had you produce a choropleth map, the layout tools in ArcMap can create many other types of maps, such as thematic maps or reference maps. Features on choropleth maps are not commonly labeled (for instance, you wouldn't be identifying individual Ohio cities or interstates on the housing stock map you created in this chapter), but labeling features on a reference map (such as the names of streets or subdivisions) would be critical. See the *Related Concepts for Chapter 3* for more about ArcMap's map labeling properties.

A drawback to creating a layout is that it's a static image—once it's produced, users can't do any of the usual GIS tricks like zooming in or out of the map, turning layers on or off, or querying a particular polygon to find out the percentage of housing stock for a particular county. Also, once the map is produced and distributed (as a printout or as a PDF or a posted graphic), new maps will need to be created and distributed if the map needs to be updated. What might be useful is the ability to design a Web-based interactive map that could be easily updated. In Chapter 4, you'll be doing just that—taking the data you worked with in this chapter, but creating a Web-based map instead of a printed one.

Related Concepts for Chapter 3   Map Labels and Annotation

Although you didn't do it for the map you created in this chapter, chances are good that you'll want to label features when you're creating a reference map. For instance, if you're designing a map of your college campus, you would likely want to provide labels for the names of the buildings, parking lots, roads, and athletic fields shown on the map. In Chapter 2, you labeled the features you were working with by selecting a field in the attribute table; you assigned this field to be displayed in conjunction with a feature in the View. Labels in ArcMap are all displayed with the same font and style, and ArcMap automatically places these labels. While this characteristic of ArcMap does quickly provide names or values, it gives you a limited amount of flexibility as a map designer. For instance, you can't adjust the position or size of each individual label, and you are unable to move an individual label if you don't like its location.

ArcGIS 10.2 gives you access to the **Maplex Label Engine**, a method that allows for more flexibility with placing labels on maps. The Maplex Label Engine provides a variety of labeling tools—for instance, fitting labels within polygons, allowing labels to curve, determining the best placement for labels, and emphasizing label placement for features such as land parcels, rivers, or contours. Figure 3.12 compares (a) a map labeling the Great Lakes with the default label placement with (b) a map created by using some of the options available through the Maplex Label Engine. You can enable Maplex by adding the **Labeling Toolbar**, then selecting **Use Maplex Label Engine** from the **Labeling** pull-down menu.

Maplex Label Engine A toolset in ArcMap that allows for a greater degree of flexibility and options for labeling features.

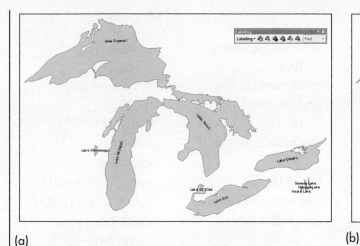
(a)

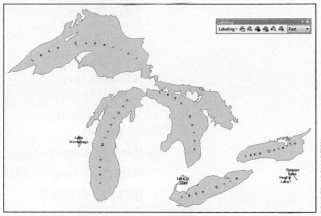
(b)

FIGURE 3.12 The Great Lakes mapped (a) with the default label placement options and (b) using different labeling options with the Maplex Label Engine enabled.

In addition, ArcMap provides you with the ability to create **annotation**, or text that will appear on a map. Annotation is a special type of feature class in a geodatabase (see Chapter 6) that allows information (such as the name of a lake or the designation of a transformer box) to be linked with an object in a separate feature class. For instance, if the owner attribute of a land parcel is changed in a feature class, its linked annotation (the text that accompanies it on the map) will be updated as well. Unlike labels, annotation properties can be adjusted individually.

> **annotation** A method of adding text to a map that allows for individual pieces of text to be edited separately.

For More Information

For further in-depth information about the topics presented in this chapter, use the ArcGIS Help feature to search for the following items:

- A quick tour of page layouts
- About map printing
- Classifying numerical fields for graduated symbology
- Data classification
- Design principles for cartography (RC)
- Essential annotation and graphic text concepts
- Essential labeling concepts
- Spreading the words of a label inside a polygon
- Using map templates
- What is a page layout?
- What is annotation?
- What is the Maplex Label Engine?
- Working with dynamic text
- Working with the Maplex Label Engine

For more about map design with GIS, see Brewer, C. 2005. *Designing Better Maps*. Redlands, CA: Esri Press, 202 pp.

Key Terms

map (p. 59)

cartography (p. 59)

reference map
   (p. 59)

topographic map
   (p. 59)

thematic map (p. 59)

choropleth map (p. 59)

normalize (p. 63)

data classification
   (p. 66)

Natural Breaks (p. 66)

Equal Interval (p. 67)

Quantiles (p. 67)

Standard Deviation
   (p. 68)

color ramp (p. 70)

graduated colors
   (p. 71)

graduated symbols
   (p. 71)

layout (p. 73)

map template (p. 75)

geographic scale
   (p. 76)

map scale (p. 76)

Representative
   Fraction (RF)
   (p. 76)

small-scale map
   (p. 76)

large-scale map
   (p. 76)

scale bar (p. 76)

scale text (p. 77)

legend (p. 78)

north arrow (p. 81)

dynamic text (p. 83)

visual hierarchy
   (p. 84)

neatline (p. 85)

Maplex Label Engine
   (p. 86)

annotation (p. 87)

How to Create a Web Map and Share Data Online with ArcGIS 10.2

Introduction

Being able to make a map of your data is a fundamental trait of GIS. However, a static map (whether a printed map or an exported digital version of the same) isn't always the best method of presenting your data or results. Consider the following situation: A police GIS analyst has put together a set of layers showing the locations of various types of crimes (robbery, assault, and so on) for the county and has mapped them at the census-block level for several different years. To distribute this information, she would have to make maps for each year and perhaps for different combinations of crimes. The result is a stack of well-designed maps, all at the same scale, showing different information. While this set of maps may convey the necessary themes, there is a more flexible option: creating an interactive map that allows users to change scales, select which layers to display, or quickly compare different years by turning layers on and off. An interactive map like this could allow its creator to correct errors on the fly or easily update the map with fresh information as it becomes available, rather than reprinting and redistributing maps.

The idea of sharing and distributing GIS maps and data in an easy-to-use and interactive format is at the heart of **ArcGIS Online**. The resources of ArcGIS Online are part of a **cloud** structure in which your data, maps, and applications are stored on Esri's data servers. Users can then access these materials via the Internet as needed. For instance, you can save your data as a map package and share it via ArcGIS Online. Other users can then access and download your map package directly into ArcGIS Desktop, giving them the ability to quickly utilize your data and map document.

ArcGIS Online also gives you the ability to take your GIS data in a map document and share it online as a **Web map**, which will allow users to turn layers on and off, zoom in and out of the map, make measurements, and identify features, as well as other interactive tasks (Figure 4.1). A big advantage of creating Web maps from GIS data is this: End users don't need ArcGIS 10.2 installed on their computer to use your map; ArcGIS Online Web maps simply run in a Web browser, whether on a desktop or a mobile device or tablet. This ease of distribution can be useful for rapidly distributing geospatial information in emergency situations. For example, a rescue team on the ground after a natural disaster can immediately identify areas of concern on the map, upload those data to a Web map on ArcGIS Online, and immediately distribute that information over the Internet.

ArcGIS Online Esri's cloud-based GIS platform where data and Web mapping software can be accessed via the Internet.

cloud A computing structure where data, content, and resources are all stored at another location and served to the user via the Internet.

Web map An interactive online representation of GIS data, which can be accessed via a Web browser.

SOURCE: ARCMAP/ESRI

FIGURE 4.1 The ArcGIS Online interface for creating Web maps.

This chapter will take you through the steps involved in taking your GIS data in a map document in ArcGIS 10.2 for Desktop and sharing it on the Internet as a Web map via ArcGIS Online. ArcGIS Online works as a **SaaS (Software as a Service)** feature. With a SaaS, all of the software you need is stored on a server and accessed via a Web browser; you do not have to download anything to your own computer. Note that there are two different versions of ArcGIS Online: a free version and a subscription-based version. This chapter assumes you have access to the subscription version (which is included as part of an Esri site license) and can publish maps with it.

Chapter Scenario and Applications

In this chapter, you'll again be taking the role of the realtor from Chapter 3 creating a map of the percentage of seasonal homes in Ohio's counties. However, this time around you'll be taking the same data you used to construct a printed map and instead create an interactive Web map and share it through ArcGIS Online. As in Chapter 3, there are no questions to answer, but instead a checklist of items that will help you make sure your Web map is complete and well designed.

The following are additional examples of other real-world applications of this chapter's theory and skills.

- A park ranger is creating a map of a state park's hiking and driving trails, campsites, and other park features. However, due to the change of seasons and landscape conditions, certain trails and areas are not accessible at all times. She will use ArcGIS Online to create a Web map of current park features and conditions that can be easily updated based on season and other conditions. The Web map will be used by people who want to use the park, as well as park rangers.

- A city's tourist bureau wants to develop a map of tourist attractions, local restaurants, and features of interest for the city's visitors. Rather than printing paper maps, the bureau wants to create an interactive map of the area that shows images of key tourist areas. Visitors can select a location and get information about that

SaaS (Software as a Service) A cloud structure wherein the software being used is stored on a server at another location and accessed on-demand via the Internet.

tourist site, the hours of operation, and photos of the locale. The bureau plans to use ArcGIS Online to develop a map that can be viewed using only a Web browser.

• An archeologist is working at a site in a specific country and is developing a GIS map of the site that can be shared with the local government, as well as other researchers. By sharing this interactive map via the Web, he can easily distribute it to interested parties, as well as make updates as new discoveries are found at the site.

SOURCE: PASQUALE SORRENTINO/ SCIENCE SOURCE

ArcGIS Skills

In this chapter, you will learn:

• How to publish a GIS data layer as a feature service to the cloud structure of ArcGIS Online.

• How to access a feature service within ArcGIS Online.

• How to create a Web map from GIS features uploaded to ArcGIS Online.

• How to change the appearance of layers within ArcGIS Online.

• How to set up the attributes and information that will appear in pop-ups when an item is queried.

• How to view and select a basemap to accompany the features.

• How to create a basic Web application in ArcGIS Online.

• How to share a Web map and a Web application across the Internet.

Study Area

• In this chapter, you will be working with data for each county in the state of Ohio.

Data Sources and Localizing This Chapter

This chapter's data focus on seasonal homes of the counties of Ohio. However, you can easily modify this chapter to use seasonal home data and the counties from your own state by using data from the U.S. Census Bureau (see the *Related Concepts for Chapter 2* for more about using Census data in ArcGIS).

The counties dataset was downloaded from the TIGER/Line files of the Census Bureau Website here: ftp://ftp2.census.gov/geo/tiger/TIGER2013. From the COUNTY folder, download the available shapefile. This contains all of the counties within the United States. You can query the shapefile and extract those counties into a separate feature class for your state based on the STATEFP attribute. This attribute references each state with a two-digit FIPS code. For instance, Ohio is referenced with a state FIPS code of 39. If you were using this chapter to map seasonal home percentages in South Carolina, you would use FIPS code 45. A full list of state and county FIPS codes is available from the EPA online here: http://www.epa.gov/envirofw/html/codes/state.html. Note that the U.S. Census Bureau's county boundaries extend to the national boundaries when they border water features, such as the Great Lakes on the U.S.–Canadian border. Also note that the county boundary file was projected into the *Lambert Conformal Conic* projection for use in this chapter.

The TIGER files provide only the geographic boundaries; attribute data must be downloaded separately and joined to the shapefiles (see Chapter 2). The attribute data (the housing statistics) for each county was downloaded from the Census Bureau's American FactFinder Website available here: http://factfinder2. census.gov. The tables used in this chapter were the 2010 SF1 versions of H5: Vacancy Status and H003: Occupancy Status. The PerSeas attribute was calculated by taking the total number of seasonal homes (the Seasonal attribute) and dividing by the total number of all homes (the TotalHousing attribute). For purposes of this chapter, the percentage of seasonal homes was rounded to two decimal places and saved in a new field called Perseasround. (Note, however, that you may have to slightly adjust the values for the lowest and highest county percentages solely for display purposes in ArcGIS Online.)

FactFinder attribute data were downloaded as a CSV file, underwent some minor editing and cleanup in Microsoft Excel, and were converted to an Excel file for use in this chapter. The Census Bureau provides an online guide to obtaining data from American FactFinder and preparing them for use in ArcGIS here: http://www.census.gov/geo/education/pdfs/tiger/Downloading_AFFData.pdf.

STEP 4.1 Getting Started

- Start ArcMap and use Catalog to copy the folder called **Chapter4** from the C:\ GISBookdata\ folder to your own D:\GIS\ folder. Copy the entire folder, not just the files within it.

- The Chapter4 folder contains a file geodatabase called **OhioWebInfo**. It contains the following:

  - Ohiocountyhousing (a polygon feature class of Ohio county borders; housing data from the U.S. Census SF1 H003 and SF1 H5 information has been joined to this feature class.)

- Add the Ohiocountyhousing feature class to the Table of Contents.

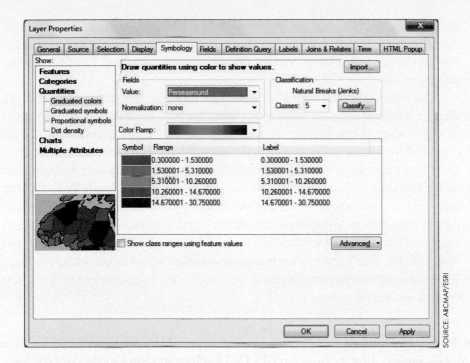

SOURCE: ARCMAP/ESRI

- Like in Chapter 3, change the symbology of the Ohio counties to Quantities and Graduated colors. Use **Perseasround** as the Value to display, using the Natural Breaks (Jenks) method and five classes. Select an appropriate Color Ramp for displaying the data. Click **Apply** and **OK** in the Layer Properties dialog to display the layer. Each Ohio county should now be shown with the percentage of the housing stock that is seasonal homes.

- Before proceeding, save your map document.

STEP 4.2 Signing In to ArcGIS Online

- To use the features of ArcGIS Online, you'll need an Esri Global Account. You can obtain a free account by going to https://www.arcgis.com/home/signin.html. Select the option for **Create a Public Account** and follow the steps to establish your username and password. Write down your username and password.

- Before proceeding, it's critical that you set up your Esri Global Account to access the subscription-level features of ArcGIS Online. If you don't, you won't be able to share your ArcGIS for Desktop work. See **Troublebox 3** for more information.

Troublebox 3

How do I set up my Esri Global Account to access ArcGIS Online subscription-level features?

Your Esri Global Account will allow you to use the free uploading and sharing functions of ArcGIS Online. However, to share your maps directly from ArcGIS for Desktop, you need to access the subscription-level features of ArcGIS Online. When a business purchases a subscription to ArcGIS Online (or when a college or university acquires a subscription as part of an educational site license), one or more individuals at that business or school will be designated the administrator for ArcGIS Online. The administrator will establish the organization for that place. For instance, if subscription-level access to ArcGIS Online is available at your school, an organization should be established. For you to use all of these ArcGIS Online features, you first have to join the organization.

The administrator can allow you to join the organization either as another administrator, as a user, or (what we will be doing in this chapter) as a publisher. When you set up your Esri Global Account, you identified an e-mail address to be associated with the account. The administrator can have an e-mail sent to that account inviting you to join the organization at the publisher level. You must follow the link in the e-mail invitation and carefully follow the steps and instructions. At the end of the process, your Esri Global Account will reflect your role as a publisher within the organization and you'll be able to share your ArcGIS for Desktop data as a service on ArcGIS Online. ***Important Note:*** If you don't know who the administrators are at your school, contact Esri's customer service representatives, who should be able to let you know the contact individual at your school.

- With your Global Account established at the publisher level, you can log in to ArcGIS Online directly from ArcGIS for Desktop, so return to the ArcGIS for Desktop software. From the **File** pull-down menu, select **Sign In…**. A dialog box will appear to allow you to enter your username and password of your Esri Global Account. You will then be signed in to ArcGIS Online.

STEP 4.3   Publishing Your Work to ArcGIS Online

publish Placing data or content onto a cloud server.

- In this next step, you'll **publish** your map to ArcGIS Online. In publishing, you'll be uploading your GIS data as a service on ArcGIS Online. Note that, while you'll be sharing only one layer in this chapter, you do have the ability to share and publish multiple feature layers; you would use the same steps, and all of your layers will be published at the same time as a single service. For more information about publishing and services, see **Smartbox 26**.

Smartbox 26

What is a service and how is it published to ArcGIS Online?

ArcGIS Online is a cloud-based GIS. In the cloud structure, things like GIS data, maps, applications, and geospatial content are stored on servers and can be accessed by users via the Internet (Figure 4.2). In this way, people who generate the data and maps don't have to serve those items out to others or be responsible for distributing them. The benefit of using the cloud is that your maps and content are stored somewhere else and can be accessed on-demand as needed. Thus, by publishing your map with ArcGIS Online, you're making your work accessible to others through a Web browser.

service A format for GIS data and maps to be distributed (or "served") to others via the Internet.

When you're transferring your map and content from ArcGIS Desktop to the cloud, you'll be setting it up as a service in ArcGIS Online. A **service** is a format for your GIS data and maps to be distributed (or "served") to others via the Internet. When publishing to ArcGIS Online, you will be setting up a **hosted service**, so named because someone else (that is, the computers and servers that make up ArcGIS Online) are holding ("hosting") the data and enabling the distribution for you.

You can set up two types of hosted services with ArcGIS Online:

- **Feature service**. In a feature service, the data layers that you publish are directly accessible by others. Depending on how you establish the feature service, other

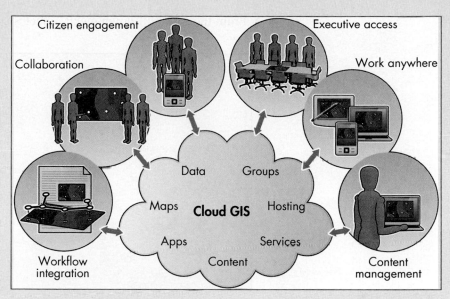

FIGURE 4.2 Using the cloud with GIS.

users will be able to create new features, update your work, make edits, or query a point, line, or polygon to get information about the attribute associated with that object. You will publish your work as a feature service if you want objects to be queried (like you will do in this chapter) or if your work is part of a larger project and others will be working with or updating the layers you upload. For instance, if you are part of an emergency response team, you and the other members of your team could edit and update an online map of a disaster area as new information becomes available.

- **Tiled map service**. In a tiled service (called simply a "**map service**" in ArcGIS Online), your layers are displayed as a set of pre-drawn images (referred to as *tiles*). Thus, your layers will not be set up in a format that will allow you to edit, query, or otherwise access an attribute table. Rather, they will take the form of an image that can only be displayed. Tiled map services are useful for visualizing GIS data, especially large GIS data sets. For instance, a map of population change at the census block level can be displayed as a tiled map service that can be turned on and off to show other layers.

When your layers are published to ArcGIS Online, their initial projection (whether geographic or projected coordinate system) is not maintained. ArcGIS Online services use the **Web Mercator auxiliary sphere** map projection (a common setup used for Web mapping utilities such as Google Earth and Bing Maps) for their content and basemap, so when you publish your services, your data are projected on the fly to match this projection.

Also keep in mind that there is a cost involved in publishing and sharing map services with your ArcGIS Online subscription. Your school or organization receives a certain number of **credits** with its subscription. The sharing of hosted services uses up credits from your organization's account. For instance, publishing and sharing a tiled map service will use up a variable number of credits depending on factors such as the size of the dataset or the scales at which the tiles can be viewed. Your organization's administrator will have information about the number of credits available to you.

hosted services Esri's term for the different types of map services that can be utilized through ArcGIS Online.

feature service A hosted service that allows a user to share GIS data layers that can also be displayed, queried, or edited.

tiled map service A hosted service that sets up GIS data as a series of image tiles that can be displayed, but not queried or edited.

map service Another term for "tiled map service."

Web Mercator auxiliary sphere The projected coordinate system used by ArcGIS Online.

credits A system used by Esri to control the amount of content that can be served by an organization in the ArcGIS Online subscription model.

• From the **File** pull-down menu, select **Share As**, then select **Service…**.

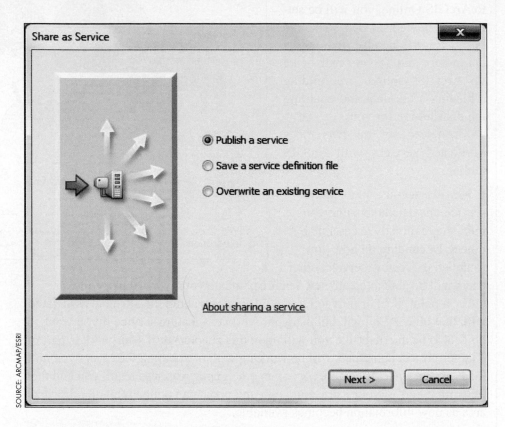

SOURCE: ARCMAP/ESRI

• In the Share as Service dialog that opens, select the radio button for **Publish a service**, then click **Next**.

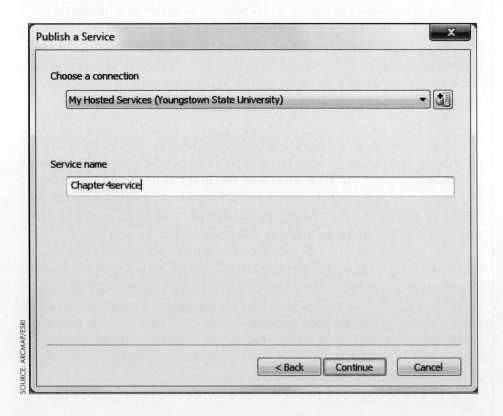

SOURCE: ARCMAP/ESRI

- Under the **Choose a connection** pull-down menu, select the option for **My Hosted Services**. (In the graphic above, it is the connection for Youngstown State University, but yours should have information about your own organization.)

- Under **Service name**, give your map service a descriptive name. (The above example uses "Chapter4service" as the service name, but you can do better than that.)

- If you don't see the option for My Hosted Services, see **Troublebox 4**; you need to have access to your organization's Hosted Services in order to publish.

Troublebox 4

Why is the connection for Hosted Services unavailable?

Your Esri Global Account should be linked to the ArcGIS Online subscription setup for your university or organization. When your Esri Global Account was set up at the publisher level (see **Troublebox 3**), your account was joined to the organization. Be sure that your account is set up at the publisher level and that you previously signed in to ArcGIS Online via ArcGIS for Desktop (see Step 4.2). If you skip either step, your option for Hosted Services will be unavailable.

- When you have all options properly set, click **Continue** to advance.

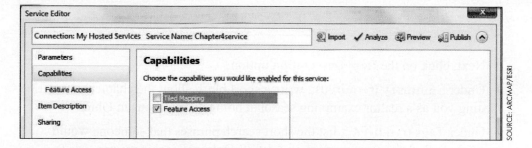

SOURCE: ARCMAP/ESRI

- In the Service Editor dialog, select the **Capabilities** option. For this service, you'll be setting up Feature Access rather than Tiled Mapping (see **Smartbox 26** for the difference between these two settings), so place a checkmark for **Feature Access** and remove the checkmark for **Tiled Mapping**.

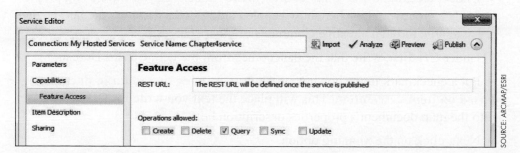

SOURCE: ARCMAP/ESRI

- Next, click on **Feature Access**.

- This is where you can decide what operations are allowed by the users of your Web maps. You can allow them to create new features, delete items, query the features, or provide updates. In many cases, you'd want to enable some of these capabilities, but for now put a checkmark only in the **Query** box.

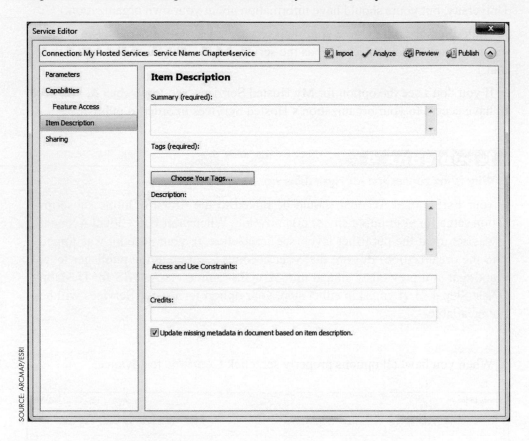

SOURCE: ARCMAP/ESRI

- Next, click on the **Item Description** option.

- Under **Summary (required):**, write a short blurb about the chapter's scenario casting you as a realtor examining seasonal home distributions in Ohio.

- Under **Tags (required):**, list the short search phrases that someone would use to help find your map service on ArcGIS Online. For this service, use the following tags: housing, seasonal homes, Ohio.

- Under **Description:**, write a short blurb about the dataset that you will be sharing on ArcGIS Online.

- Under **Access and Use Constraints:**, write "none." (If you were using proprietary data, you would list here how people could use the data, or whether the data are available through Creative Commons or other sources.)

- Under **Credits:**, write that the data comes from the U.S. Census Bureau.

- Put a checkmark in the box next to **Update missing metadata in document based on item description**. This will place the text you write for Description into the map document's properties description box.

- Next, click on the **Sharing** option.

- In this new dialog, you can choose with whom you want to share your service—for example, the organization hosting your service (in this case, Youngstown State University), anyone using ArcGIS Online (this is the Everyone option), or a specific group (you would probably have this set up for specific

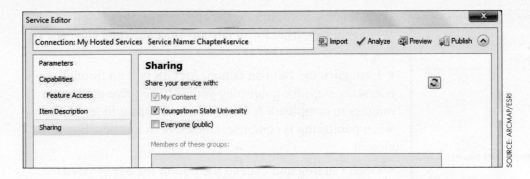

SOURCE: ARCMAP/ESRI

subgroups or development units). For now, put a checkmark in the box next to your organization name as the people with whom to share your results.

• Next, click the **Analyze** button. ArcGIS will examine the data you're using to set up the map service and return warnings about any problems that need to be fixed or changed before the service can be published. A new dialog called **Prepare** will open at the bottom of ArcMap showing the problem areas. In this case, the items in the Prepare dialog will not interfere with publishing this service, but for your own data you may have to make several changes before ArcGIS will allow the service to be published.

• Next, press the **Preview** button. A new window will open, showing what your data will look like in ArcGIS Online.

SOURCE: ARCMAP/ESRI

• Use the pan and zoom tools to examine how your data will look as a map service. If you don't like the color scheme or symbology of any of the features, close the Preview window, return to ArcMap, and change the colors and symbology to match your preferences. Note that this window only shows what your shared data will look like, not what the final Web map will be. (There are a lot of

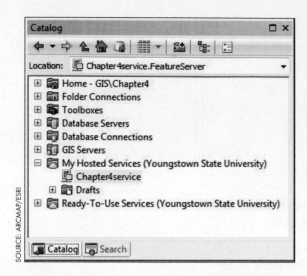

steps to come for designing the appearance of the Web map.) When you're satisfied with the how the data look, close the Preview window.

• Last, click the **Publish** button. ArcGIS will go through the necessary steps for publishing your service (give it a couple minutes to complete). A dialog box will appear to let you know when publishing is complete. Click **OK** in this new box to close it.

• Open Catalog and expand the option for **My Hosted Services**. You should see your newly created service listed there. Right-click on it, select **Properties**, and double-check that all of the options you selected and the text you entered are correct.

STEP 4.4 Creating a Web Map from Your Shared Service

• At this point, you've published your service to ArcGIS Online. If you've shared the data with people in the organization, they could download the feature service into ArcGIS for Desktop. However, to make your housing data more easily accessible to the public (and to use in your realtor's analysis of seasonal homes for individuals looking for houses), you will now take that service and create a Web map from it. A Web map can be viewed and used directly with a Web browser (that is, without using ArcGIS for Desktop).

• Minimize ArcMap and open a Web browser (the activities in this chapter were completed with the latest version of Google Chrome) and go to http://www.arcgis.com. This is the Website for ArcGIS Online.

• On the ArcGIS Online Website, log in with your Esri Global Account username and password. From there, select the **My Content** tab.

• You should see your published Feature Service listed as part of your ArcGIS Online content. If you wanted to, you could make the feature service itself available to others by pressing the Share button. However, in this chapter, we'll be using it to create a Web map instead.

• Put a checkmark in the box next to the **Features** (the Feature Service) version of Chapter4service and click on **Create Map**.

• ArcGIS Online will switch to the Web map interface and display the classified Ohiocountyhousing layer on top of a basemap.

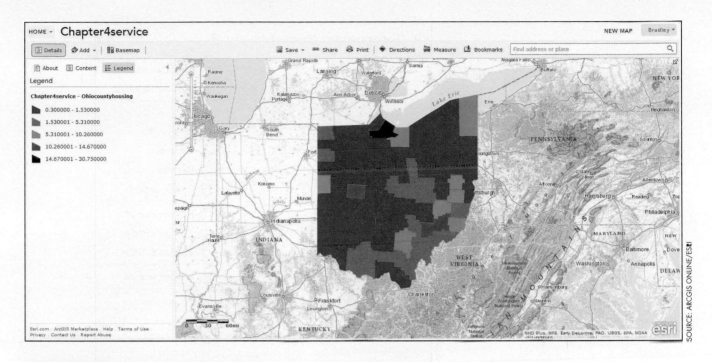

STEP 4.5 **Changing the Appearance of the Web Map**

• Next, we'll look at some different options for changing the appearance of the Web map for a better visualization of the data. To begin, click on the **Show Contents of Map** button (the middle button of the three).

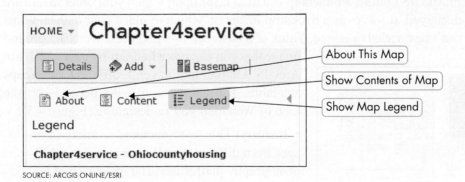

SOURCE: ARCGIS ONLINE/ESRI

• This will change the Table of Contents panel so that you can see what layers are available. To the right of the Chapter4service – Ohiocountyhousing layer is a button to activate a drop-down menu (which becomes visible only when you mouse over it). Click on this button to bring up a list of options for the layer.

• From the pull-down options, select **Transparency**. A slider bar will appear showing how transparent the layer is. Move the slider bar to 50% and the Ohiocountyhousing layer will become partly see-through on the map.

• We'll now select a new basemap to display behind the semi-transparent layer. The default layer is topographic, which doesn't fit well with the theme of seasonal homes. For more about the types of basemaps available in ArcGIS Online, see **Smartbox 27**.

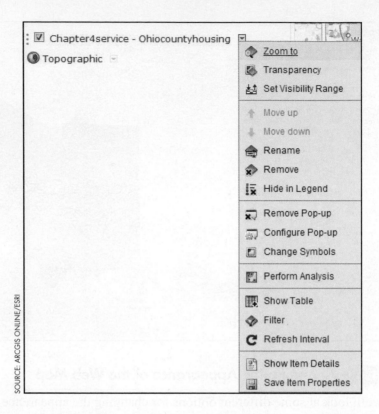

SOURCE: ARCGIS ONLINE/ESRI

Smartbox 27

What kinds of basemaps are available in ArcGIS Online?

> **basemap** An image layer that serves as a backdrop for the other layers used in ArcGIS Online.

In ArcGIS Online, a **basemap** is a tiled layer upon which your other layers are displayed. It serves as a backdrop image on which to place your data and thus can't be queried or edited. Think of a basemap as a georeferenced background image that you can use when setting up a Web map. ArcGIS Online has several different basemaps available. The choice you make depends on the kind of Web map you're designing (Figure 4.3):

• Imagery: This layer contains a variety of images from different satellite sensors (or aerial photography platforms). The imagery that is visible depends on what scale you're zoomed in at or what area of the world you're viewing. See Chapter 13 for more about using aerial and satellite imagery in ArcGIS.

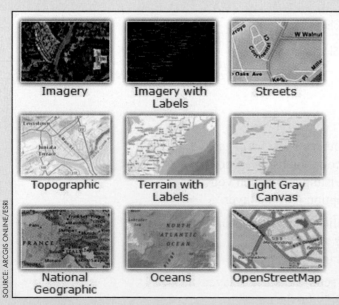

SOURCE: ARCGIS ONLINE/ESRI

FIGURE 4.3 The available basemaps in ArcGIS Online.

• Imagery with Labels: This is the same layer as the Imagery layer, but it also includes labeled country boundaries, as well as labeled state and county boundaries for the United States.

• Streets: This is a world street map that also labels cities, parks, water features, and some building footprints.

- Topographic: This layer covers the world, showing built features as well as land cover and shaded relief.

- Terrain with Labels: This layer provides shaded terrain relief, water features, and bathymetry, as well as country, state, and county borders and labels.

- Light Gray Canvas: This layer is intended to provide geographic reference, but because of its light gray color, it remains in the background and allows for the appearance of your data to "jump" off the screen. As such, it contains minimal labels and features and is intended for use as a neutral backdrop.

- National Geographic: This world map shows a variety of roads, city features, water bodies, and landmarks, along with shaded relief and land cover.

- Oceans: This basemap is intended for use in ocean and marine maps. It features water bodies, coastlines, and bathymetry.

- OpenStreetMap: This layer shows road features from the OpenStreetMap project, an open source online map that can be edited by anyone (see Chapter 5 for more information about OpenStreetMap). OpenStreetMap data can be viewed online at http://www.openstreetmap.org.

- In addition, you can use basemaps of imagery from Microsoft's Bing Maps program by obtaining a Bing Maps key from Microsoft. Bing Maps imagery can also be viewed online at http://www.bing.com/maps. These basemaps include:

    - Bing Maps Aerial: This layer contains high-resolution imagery.

    - Bing Maps Hybrid: This is the same layer as the Bing Maps Aerial, but it also includes labels for features like cities, boundaries, and road names.

    - Bing Maps Road: This layer contains the street map information (but not the imagery) that is also available through Bing Maps.

    As with other tiled ArcGIS Online map services, you can download and access these layers in ArcGIS Desktop (see Chapter 5 for more about how to do this).

- To change basemaps, press the **Basemap** button. For this Web map, select **Imagery with Labels** as the basemap to use.

- Zoom in closer to the various Ohio counties and you'll see the county names and city locations and names visible through the semi-transparent layer. If your initial color ramp doesn't fit well with the satellite imagery basemap, it's easy to change. Select the layer's pull-down menu again and choose **Change Symbols**. The Table of Contents will change, allowing you to select a new classification scheme, a new value to map, a new number of classification breaks, different color ramps, or different labels. For now, select a different color ramp that's more compatible with the appearance of the basemap.

- Close the Change Symbols dialog and click on the **Show Map Legend** button in the Table of Contents (the third button from the left) to see the new color ramp displayed along with the values that each color represents.

STEP 4.6 **Working with Queries and Pop-ups on the Web Map**

• When you initially set up the feature service, you selected Query as an option that users of the Web map would be able to perform. Thus, by clicking on one of the counties, you can display a pop-up dialog box of all the values in the attribute table for that county. However, as a realtor interested in displaying data about seasonal homes, you'll see that there's a lot of unnecessary information that gets displayed in the pop-up. In an ArcGIS Online Web map, it's easy to control what attributes get displayed as the result of a query.

• Click on the **Show Contents of Map** button, then select the pull-down menu next to the Ohiocountyhousing layer. From these options, select **Configure Pop-Up**. A new set of options will appear in the Table of Contents panel. These options will allow you to set the name of the pop-up, what gets displayed in the pop-up (such as custom attributes or a description), what fields of the attribute table are shown in the pop-up, and any other media that you want to display (such as a picture or chart; this function works in a manner similar to the way hyperlinks work, as described in Chapter 2).

• For this Web map, in the **Display:** pull-down menu, select **A list of field attributes**. This option indicates that the pop-up will display values from the attribute table.

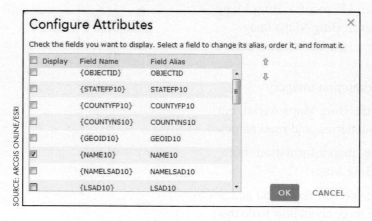

SOURCE: ARCGIS ONLINE/ESRI

• To control which attributes are shown, click on the **Configure Attributes** button.

• A field with a checkmark will be displayed. Make sure only the following fields are checked: NAME10, TotalHouses, Occupied, Vacant, Seasonal, and PerSeas. Click **OK** when you're done.

• Once the dialog closes, click on **Save Pop-Up** button in the Pop-Up Properties box. Your changes to how the pop-ups are displayed will be saved.

• Click on a county now to query it. The pop-up should show only the name and the housing information.

STEP 4.7 **Saving and Sharing Your Web Map**

• To save your work, click the **Save** button and select **Save**. A new dialog will appear. Fill in a descriptive title, useful tags, and a summary of the map, then click **Save Map**. The Web map will now be saved in the My Content section of ArcGIS Online.

• For others to use your map, you need to **share** it. Sharing in ArcGIS Online will assign your Web map a URL so that others can find it and use it as a service via ArcGIS Online. Click on the **Share** button to begin.

• Select the group or organization you want to share the map with—Everyone (meaning anyone using ArcGIS Online) or just the people in your organization.

share Distributing data, content, maps, or applications across the Internet.

- Under Link to this map, a URL (beginning with bit.ly) will be shown. This is the URL of your Web map. You can share the URL via your Facebook or Twitter account if you want. By copying and pasting this URL into a Web browser, you can access your Web map at any time.

- By sharing your Web map with Everyone, you can also embed the Web map into a Web page (such as your organization's Web page or on a blog).

STEP 4.8 Sharing Your Web Map as a Web Application

- For a final step, you can take your Web map and create a Web application from it (see the Related Concepts for Chapter 4 for more about Web applications). A **Web application** will take your Web map, place it into a template, and give you access to the application's source code so that you can tweak or change it, plus add additional content. For now, we'll make a very simple Web application from your map. In the Share dialog box, select **Make a Web Application**.

- In the new dialog that opens, select one of the following template options for creating a Web application: Basic Viewer, Chrome-Twitter, Edit, Legend, or Simple Map Viewer. Each of them will make your map look different. From the **Publish** pull-down menu under each one, select **Preview** and ArcGIS Online will open a new tab showing what your map will look like as a Web application. Each template will take your Web map (complete with the features, their color scheme, and the pop-up information) and place it into the pre-determined setup of the template.

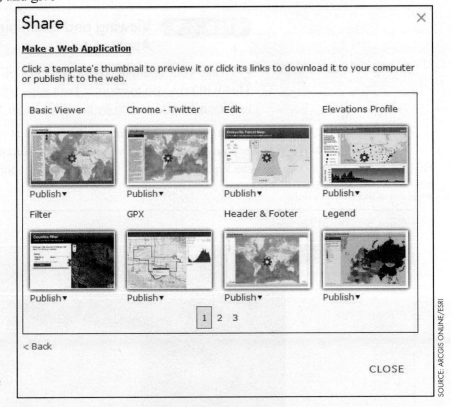

- Some of the choices are more appropriate for this particular task (creating an interactive choropleth map of seasonal home locations). Choose the one you want to use and from the **Publish** pull-down menu click **Publish**. By publishing it to ArcGIS Online, you make it available for others to use.

- In the next dialog that appears, give your Web application a descriptive title and then click **Save and Publish**. If you wanted to continue changing and editing

Web application A structure that uses shared content or data within a specific ArcGIS Online template.

your Web application you could do it here, but for now, click **Close** in the next dialog that appears.

- Back in ArcGIS Online, click on **Home** and then select **My Content**. This will show you the items you have available to you in ArcGIS Online. You should see your feature service, your Web map, and your new Web application. You will also see which items are shared and which are not shared. To share a Web map or Web application, place a checkmark in its box and click **Share**. Make sure that both your Web Map and your Web Application are shared with either your organization or everyone.

SOURCE: ARCGIS ONLINE/ESRI

STEP 4.9 **Viewing and Examining Your Web Map and Web Application**

- To see your resulting Web map, click on its name in the My Content section. This will take you to another Web page where you can choose to view your Web map (which you can do by clicking on the thumbnail image of the Web map).

- Return to My Content and take a look at the Web application you created by opening it in the same way. Note the differences in appearance and function between the Web map and the simple Web application you created from it (Figure 4.4).

FIGURE 4.4 An example Web application (using the Legend template) of this chapter's data.

SOURCE: ARCGIS ONLINE/ESRI

Checklist

Use the following checklist to make sure that you have put together your Web map (or Web application) for this chapter in a complete fashion:

_____ Publishing your ArcGIS for Desktop data as a feature service

_____ Adding the feature service to the Web map

_____ Changing the transparency level of the features

_____ Selecting a new and appropriate basemap to use for the Web map

_____ Selecting a new and appropriate color ramp to use for the Web map

_____ Configuring the pop-ups so that only the pertinent information is displayed as a result of a query

_____ Giving the Web map a descriptive title and appropriate tags

_____ Properly sharing the Web map with everyone or your organization

_____ Selecting an appropriate template (Web application only)

_____ Properly sharing the Web application with everyone or your organization

Closing Time

This chapter showed you how to take your GIS data and content and easily share them with others via the Internet by setting thcm up as an online map. Creating Web maps in ArcGIS Online is a straightforward process, but it has some limitations. Unlike the map layouts you designed and printed in Chapter 3, Web maps offer fewer design options. In a Web map, you can change the symbology and labels of layers, use different basemaps, and control which attributes are shown in a query, but the design of the map is pre-set. For instance, you can't move map elements (such as the text or zoom slider) to another location, and you can't easily add a north arrow or inset map. However, by designing a map application, you can provide much more customization to a Web map. See the _Related Concepts for Chapter 4_ for more about using Esri products for designing Web applications.

ArcGIS Online can also be used to share GIS data with others. You didn't do it in this chapter, but you could share your ArcMap work as a map package for others to use via ArcGIS Online. Similarly, rather than setting up a service, Web map, or Web application, you could have simply shared the seasonal home data through ArcGIS Online for others to download. In Chapter 5, you'll be obtaining data directly through ArcGIS Online (as well as from other free online GIS data sources).

| Related Concepts For Chapter 4 | How to Set Up GIS Web Applications |

ArcGIS Online makes it easy to create Web applications. The one you made in this chapter was quite simple; you just applied your Web map to a basic template. Several other options allow you to create fancier or more in-depth Web applications. For instance, some templates allow you to add several

story map A Web application designed to convey a specific theme or concept to the user using a special set of Web templates.

panels containing data to the map at the same time, or to add descriptive text and photos or images as well. Esri **story maps** use Web applications to inform a user on a particular topic. As its name implies, a story map is a Web application that's used to communicate a story (or theme) to its user—such as an interactive guide to San Diego tourist destinations or a guide to the cities that have hosted the Olympics since 1896 (Figure 4.5). For instance, with the Olympics story map, you can select a particular location on the map where the Olympic games have been hosted and the story map will present you with a photo and information about the event. Conversely, selecting an option from the timeline at the top of the page will bring up the photo and information about the Olympic games of that date as well as center the map at that location. Think of story maps as a way of bringing together geospatial data (such as the location of cities that hosted the Olympics) with non-spatial data (photos, attributes, and descriptive text about that city and the event) in an interactive format. Story maps allow you to place your content into a set of templates and share those data across the Internet via ArcGIS Online.

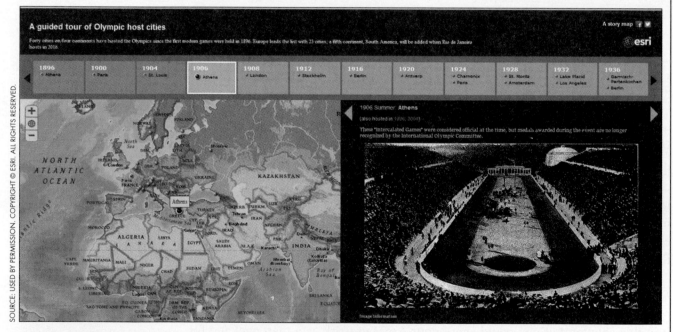

FIGURE 4.5 Esri story map showcasing the history of the Olympics.

Esri makes many types of story map templates available—the Olympics map is a custom designed story map, but other basic templates including map tours (where photos and text are linked to locations), the side accordion (where several related interactive maps with geospatial data and related attributes can be used together), or the swipe and spyglass (where two maps or images can be examined in conjunction with one another) are also available. Story maps can be created using either the free or subscription version. For more about designing and using story maps, see here: http://storymaps.esri.com.

ArcGIS for Server An Esri software product used for distributing maps, data, and services via a user's own servers via the Internet, as well as designing powerful Web applications.

Esri's other key product for developing Web applications is **ArcGIS for Server**, a separate software package designed for Web GIS. With ArcGIS for Server, users can easily distribute maps, data, and tools; publish map services; and produce detailed and in-depth Web applications via their own computer servers. For instance, the Service Definition that gets created when you publish a feature service or tiled map service can be used to take your ArcGIS Online service and set it up with ArcGIS for Server. Using Web mapping API tools such as Microsoft Silverlight, Adobe Flex, or Javascript, you can design new applications to suit the needs of a particular organization, all with a great deal of flexibility regarding the appearance of the final product.

See Figure 4.5 for an example of a Web application powered by ArcGIS for Server. The Pennsylvania Department of Conservation and Natural Resources has set up an interactive Web application to provide users with information about state parks and state forests. Users can search for parks that have certain amenities, view a variety of different basemaps, make measurements, get weather information, print maps, and extract the park data into ArcMap or Google Earth. For a look at several other Web applications created with the resources of ArcGIS Server, go to http://www.esri.com/software/arcgis/arcgisserver/showcase.

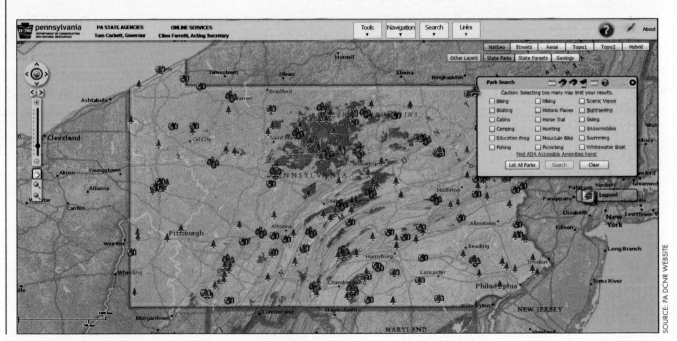

FIGURE 4.6 Pennsylvania State Parks Map Application created with ArcGIS Server.

SOURCE: PA DCNR WEBSITE

For More Information

For further in-depth information about the topics presented in this chapter, use the ArcGIS Help feature to search for the following items:

- A framework for deploying Web GIS
- About Web GIS
- About publishing services
- Hosted feature services
- Hosted tiled map services
- How to publish a service

- Publishing a hosted feature service using an ArcMap document
- Publishing a map service
- Sharing your map in a Web application
- Signing in to ArcGIS Online in ArcGIS for Desktop applications
- Types of ArcGIS maps (RC)
- Using ArcGIS Online in ArcGIS for Desktop applications
- Using Bing Maps
- Using hosted services
- Viewing subscription status (RC)
- What are ArcGIS Online hosted services?
- What is an ArcGIS Web map?
- What is a feature service?
- What is a map service?

For more information about sharing data as map packages, see: Pratt, M. 2011. "Share and Document Maps with Map Packages." *ArcUser*. Fall 2011. Online at http://www.esri.com/news/arcuser/0911/files/mapackages.pdf.

For more information about the Web Mercator projection and using different projections and basemaps in ArcGIS Online, see: Szukalski, B. 2010. "Using Your Own Basemaps (and Projection) with the ArcGIS.com Map Viewer." Online at http://blogs.esri.com/esri/arcgis/2010/12/16/using-your-own-basemaps/.

Key Terms

ArcGIS Online (p. 89)	hosted services (p. 95)	credits (p. 95)
cloud (p. 89)	feature service (p. 95)	basemap (p. 102)
Web map (p. 89)	tiled map service (p. 95)	share (p. 104)
SaaS (Software as a Service) (p. 90)	map service (p. 95)	Web application (p. 105)
publish (p. 94)	Web Mercator auxiliary sphere (p. 95)	story map (p. 108)
service (p. 94)		ArcGIS for Server (p. 108)

How to Obtain and Use Online Data with ArcGIS 10.2

▨ Introduction

For four chapters now, you've been using data from a variety of sources that has already been acquired, created, and/or processed for your use. When you're working with GIS, you'll sometimes create layers on your own (as we'll do in the next chapter), but often you'll be using layers created by someone else. For instance, if you need a layer with information on census tract boundaries, land parcels, streams, or interstates, there's no need to reinvent the wheel and create it yourself—chances are it's already been made and is available.

There's a lot of GIS data out there. While much of it is created or compiled by private companies who will sell it to you, there's much more that can be had for free. For instance, U.S. government agencies such as the USGS (United States Geologic Survey) and the U.S. Census Bureau make many GIS datasets available free via the Internet. The goal of this chapter is to introduce you to some sources of GIS data, download layers obtained from these sources into ArcGIS, and do some analysis with this data. You'll be getting data from The National Map (a key source of freely available GIS data provided by the U.S. government) as well as ArcGIS Online. In addition, you'll start using pre-made basemaps from ArcGIS Online (which you'll also use in later chapters).

▨ Chapter Scenario and Applications

This chapter puts you in the role of an urban planner compiling data that you will use to analyze a local smart-growth initiative for the city of Youngstown, Ohio. Before you can start working, you need to get your hands on some basic data layers representing city features and examine their geospatial features and attributes. In addition, to assess the quality of the data layers you're working with, you'll obtain additional data from multiple sources and compare them with actual imagery of the city.

Here are some additional examples of other real-world applications of this chapter's theory and skills:

• A realtor is setting up a presentation of new housing developments in the area for prospective buyers. She wants to see the houses' proximity to the local wastewater treatment plants. Before performing her analysis, she must first determine the locations of the plants. She can accomplish this task by downloading that geospatial information from The National Map.

• A local government is setting up an emergency response plan that will allow police, fire, and EMT units to respond quickly to situations at nearby schools.

Before the government can implement the plan, it first needs to determine the locations of the emergency response centers as well as all schools in the area. All of these items are available as point layers via The National Map.

• An environmental scientist is going to conduct an assessment of water quality for the county. However, he first needs GIS data layers showing the location and dimensions of all the county's lakes, rivers, streams, and other water bodies. He can download this dataset from The National Map prior to starting his study.

ArcGIS Skills

In this chapter, you will learn:

• How to use The National Map viewer to select downloadable geospatial data.

• How to download multiple datasets from The National Map.

• How to add data from The National Map into ArcGIS.

• How to use the Identify and Find tools to examine geospatial data.

• How to download data layers from ArcGIS Online to use with other data layers.

• How to stream basemaps from ArcGIS Online to use with other data layers.

Study Area

• For this chapter, you will be working with data from the city of Youngstown, Ohio.

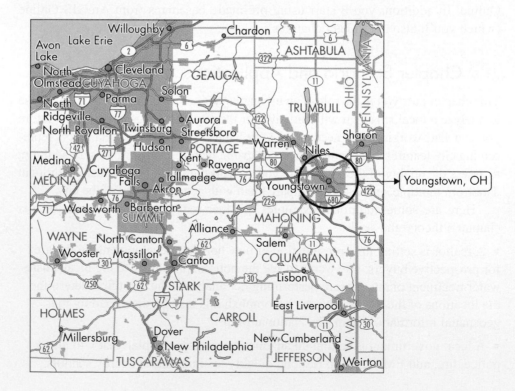

Data Sources and Localizing This Chapter

This chapter's data focus on features and locations within the city of Youngstown, Ohio. However, you can easily modify this chapter to use data from your own city or local area instead.

In Step 5.2, you will be downloading structures, boundaries, transportation, and hydrography data from The National Map for the city of Youngstown. This same data is available for all other U.S. cities. Follow the same steps to obtain the data for your own city.

In Step 5.5, you will use railroads data from the Ohio Department of Transportation that has been uploaded to ArcGIS Online. Similar data may be available for your own state via ArcGIS Online (search for "rail"); similar datasets may also be available for layers other than railroads (such as more detailed data on water bodies). Another source of national rail lines distinct from The National Map data is available on ArcGIS Online as a map service; it's entitled "USA Railroads" and is the rail data from the 2010 U.S. Census (search for "Census rail" to find it).

In addition, the basemap layers you'll use from ArcGIS Online in Step 5.6 cover the entire United States as well as other countries, ensuring that you'll have access to this type of data for your own local area (and you can then compare the roads layer with the roads around your own local campus as well).

STEP 5.1 Getting Started

• Start ArcMap and use Catalog to copy the folder called **Chapter5** from the C:\ GISBookdata\ folder to your own D:\GIS\ drive. Be sure to copy the entire directory, not just the files within it.

• Chapter5 contains the following item:

  • An empty folder called **TNMdata**

• For now, minimize ArcMap.

STEP 5.2 Downloading USGS Geospatial Data from The National Map

• Start a Web browser (the operations in this chapter were performed using the most recent version of Google Chrome).

• We'll start by getting GIS data from The National Map (and see **Smartbox 28** for more information about The National Map and its place in the U.S. National Geospatial Program). Navigate to The National Map website: http://nationalmap.gov.

Smartbox 28

What are The National Map and the National Geospatial Program?

The **National Geospatial Program (NGP)** is a federally funded initiative that provides geospatial data (in GIS-ready format) to the public. The NGP uses data from a variety of sources and works to ensure that high-quality data products are available for use in numerous settings. For instance, the Emergency Operations Office component of the NGP is part of an effort to respond rapidly to disasters and emergency situations by quickly deploying necessary geospatial data to emergency services.

National Geospatial Program (NGP) A federal U.S. initiative for managing and distributing geospatial resources.

The National Map An online resource for distributing U.S. geospatial data.

The National Map is at the core of the NGP and forms the heart of the geospatial data available through the program. The National Map data is available for free download via The National Map Viewer in geodatabase or raster format. Many different types of data are available from The National Map. In this chapter, you'll be using transportation (roads and railroads), structures, and boundaries layers from the USGS, and lakes and rivers layers from National Hydrology Data (Figure 5.1). Beyond these, you can also freely download geographic names, historical topographic maps and the new interactive US Topos (see Chapter 16), National Land Cover Data (see Chapter 12), contour lines (see Chapter 16), orthoimagery (see Chapter 13), and National Elevation Dataset files (see Chapter 15).

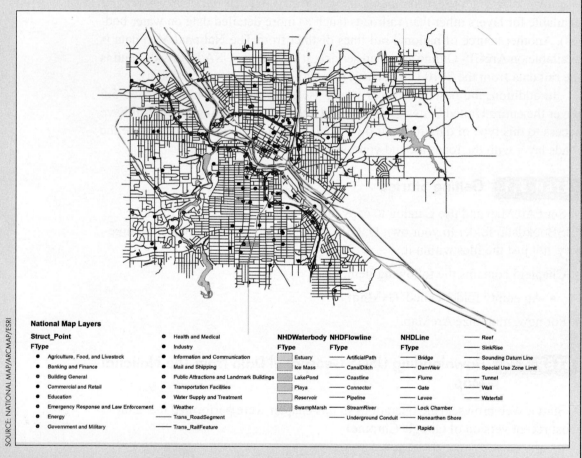

SOURCE: NATIONAL MAP/ARCMAP/ESRI

FIGURE 5.1 Various structures, transportation, and hydrography data layers of The National Map.

The USGS often makes changes and enhancements to The National Map data, the viewer, and the download methods to allow for easier access to its geospatial data. To keep abreast of changes, check out The National Map's Website at http://nationalmap.gov or follow them on Twitter at @USGSTNM.

- Choose the option for **The National Map Viewer and Download Platform**.
- On The National Map Viewer and Download Platform Website, select the link for **Click here to go to The National Map Viewer and Download Platform!**
- A new tab will open with the Viewer itself. Switch to that tab and wait for the viewer to load.

- In the Search box, type **Youngstown, OH** and click **Search**.

- From the Search Results options on the left-hand side, choose the first one (**A**).
The Viewer will shift so that you can see Youngstown and its surrounding areas.
If the view doesn't automatically center on Youngstown, click on the **Zoom To**
option.

- Next, click on the **Download Data** button in the upper right-hand corner.

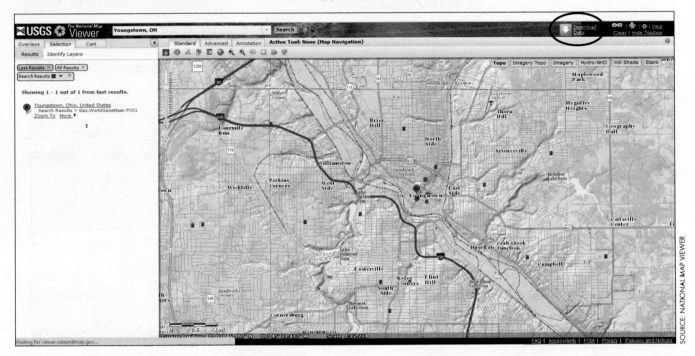

SOURCE: NATIONAL MAP VIEWER

- A new tool (the Active Data Download tool) will appear. From its pull-down
menu, select the option for **Incorporated Places**.

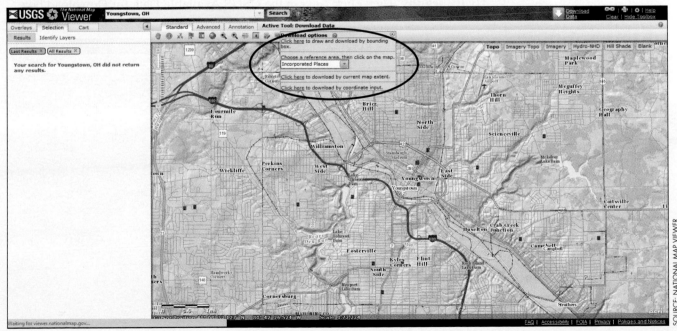

SOURCE: NATIONAL MAP VIEWER

• Next, click the mouse directly onto an area of Youngstown State University (YSU) campus or nearby downtown Youngstown. The entire boundary of Youngstown will be highlighted in a hatched green color. This is the extent of the data that you will be downloading.

• On the left side of the screen under the Selection tab, you'll see the selected boundaries of Youngstown as an option. Under this Youngstown, Ohio option, click on **Download**.

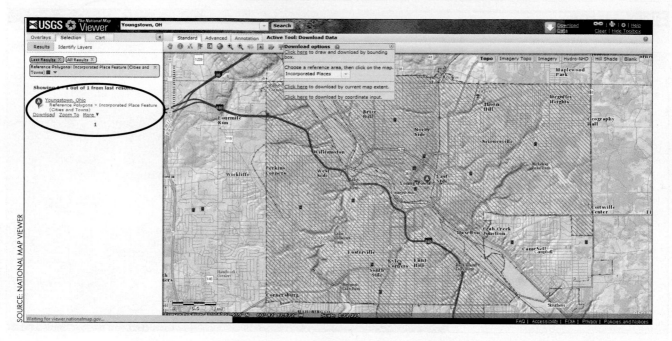

• A new pop-up will appear (called USGS Available Data for Download) that lists the data for this selected area that are available for downloading. Place checkmarks in the boxes for **Structures**, **Transportation**, **Boundaries**, and **Hydrography**. Then click **Next** in the USGS Available Data for Download box.

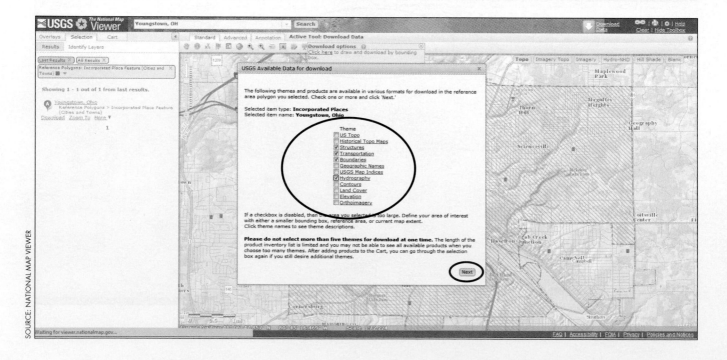

- For each of the four options (Structures, Transportation, Boundaries, and Hydrography) place a checkmark in the option related to the specific USGS Dataset for Ohio and use the **File GDB 10.1** option for each one (doing so will provide the data in a File Geodatabase format—see Chapter 6 for more information). For the Hydrography layer choose the File GDB option for "Dynamic Extract." Then click **Next** in the USGS Available Data for Download box.

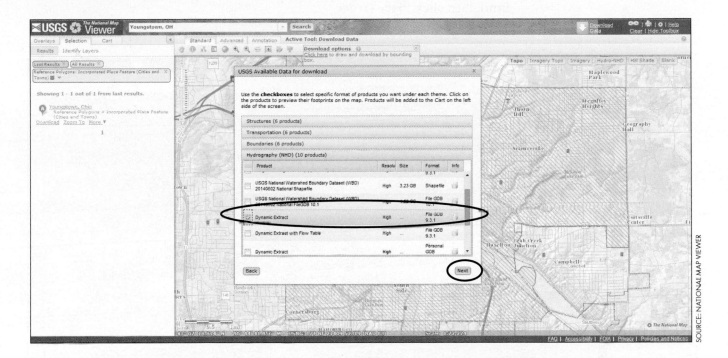

- These items are in your "Cart" for you to retrieve from The National Map. On the left-hand side of the screen, in the window called Cart, you will see your four selected layers. Make sure you have the correct four selected and all of them are in File GDB format.

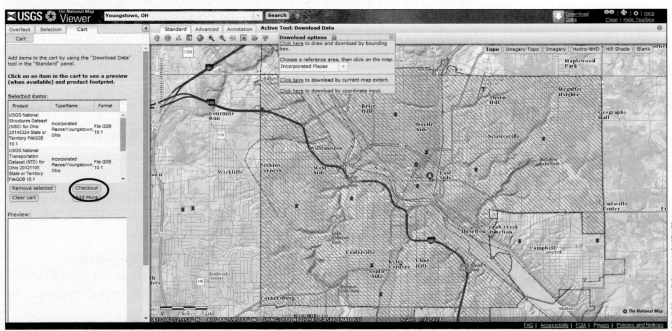

- Click the **Checkout** button to start the retrieval process for your data.

- Here's where it gets tricky. Rather than just allowing you to download your selected data, The National Map will have you input your e-mail address and will then e-mail you a confirmation of your "checkout." (Don't worry about terms like "cart" or "checkout"—all of The National Map data is free.) In the Cart box, type the e-mail address that you want the info sent to (and then type it again to confirm), then click **Place Order**.

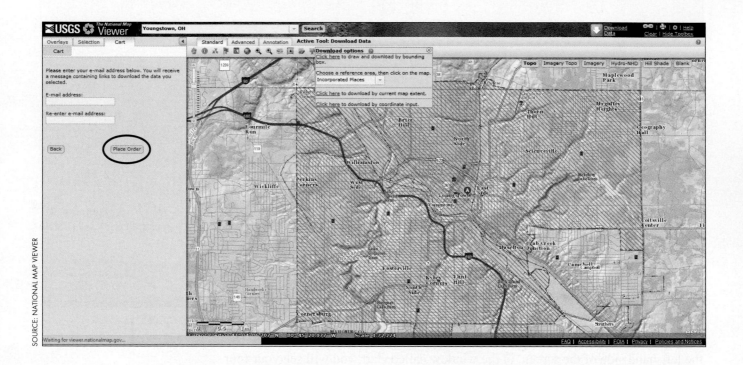

- You will receive an e-mail from the USGS indicating your "order" has been received. You will then receive one or more e-mails containing links to a USGS ftp site, from which you can retrieve your data. The e-mails indicate that this data should be "immediately" available but also indicate that the process could take up to two days.

- When you receive the e-mail, click on the Download Link URL.

- This will allow you to retrieve the data. Use the link to download the files. (***Important Note:*** Some Web browsers will automatically download the file to your computer's My Documents\Downloads folder. With other Web browsers, you may have to right-click on the file and choose an option to Save or Save As.)

- Use Windows to copy your zipped files from their downloaded location (such as the My Documents\Downloads folder) into the D:\GIS\Chapter5\TNMdata folder.

  - Note that The National Map will likely send you two separate e-mails—one containing links to the Structures, Boundaries, and Transportation data and a second e-mail containing the link to the Hydrography data.

Product	Extracted by	Format	Download Link
USGS National Structures Dataset (NSD) for Ohio 20140324 State or Territory FileGDB 10.1	Incorporated Places/Youngstown, Ohio	File GDB 10.1	Click here to download

SOURCE: NATIONAL MAP VIEWER

- The National Map server will likely place all four of the different files into their own zip files.

- Before moving on, carefully check that despite the number of e-mails or zip files, you will have access to all four of the requested items: Structures, Boundaries, Transportation, and Hydrography.

- After downloading all files, you can close the Web browser.

STEP 5.3 **Unpacking The National Map Data for Use**

- All of your files have been downloaded in **.zip** format, which is a "zipped" or compressed file format, so you will have to use Windows to unzip the files before you can get to the actual National Map layers to use in ArcGIS. First, use Windows to navigate to the D:\GIS\Chapter5\TMNData folder.

Name	Date modified	Type	Size
GOVTUNIT_39_Ohio_GU_STATEORTERRI...	6/4/2014 2:00 PM	Compressed (zipp...	31,721 KB
NHD318216	6/4/2014 1:58 PM	Compressed (zipp...	342 KB
STRUCT_39_Ohio_GU_STATEORTERRITO...	6/4/2014 1:59 PM	Compressed (zipp...	5,112 KB
TRAN_39_Ohio_GU_STATEORTERRITORY	6/4/2014 2:02 PM	Compressed (zipp...	81,495 KB

SOURCE: WINDOWS EXPLORER/MICROSOFT

- Each of the zip files contains the file geodatabases you downloaded from The National Map. To get to them, you'll have to unpack the contents of these zip files.

- We'll start with the Hydrography data. Right-click on the hydrography zip file (this will be the one that starts with the letters NHD) and select an option for **Extract All**.

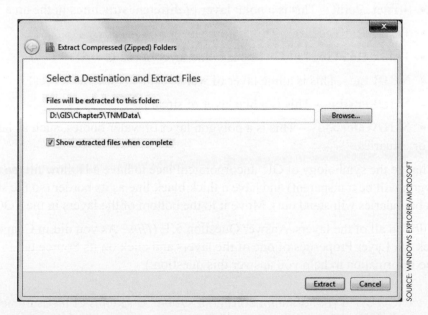

SOURCE: WINDOWS EXPLORER/MICROSOFT

- Set the folder for the files to be extracted to as D:\GIS\Chapter5\TMNData\ and click **Extract**. Windows will unpack the contents of the file and place them in the TMNData folder. The new folder labelled as the file folder is the file geodatabase you will use.

- Use the same procedure on the other zip files. You should now have a file geodatabase that corresponds to the downloaded data for each zip file.

- *Important Note:* Before proceeding, make sure that all of your files are downloaded and unzipped into the proper folder.

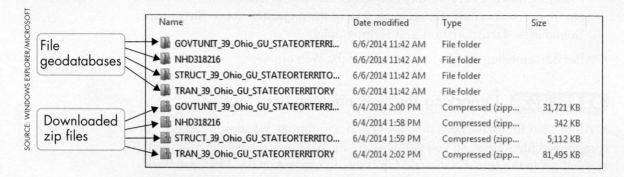

Name	Date modified	Type	Size
GOVTUNIT_39_Ohio_GU_STATEORTERRI...	6/6/2014 11:42 AM	File folder	
NHD318216	6/6/2014 11:42 AM	File folder	
STRUCT_39_Ohio_GU_STATEORTERRITO...	6/6/2014 11:42 AM	File folder	
TRAN_39_Ohio_GU_STATEORTERRITORY	6/6/2014 11:42 AM	File folder	
GOVTUNIT_39_Ohio_GU_STATEORTERRI...	6/4/2014 2:00 PM	Compressed (zipp...	31,721 KB
NHD318216	6/4/2014 1:58 PM	Compressed (zipp...	342 KB
STRUCT_39_Ohio_GU_STATEORTERRITO...	6/4/2014 1:59 PM	Compressed (zipp...	5,112 KB
TRAN_39_Ohio_GU_STATEORTERRITORY	6/4/2014 2:02 PM	Compressed (zipp...	81,495 KB

File geodatabases

Downloaded zip files

STEP 5.4   Examining National Map Datasets

- Return to ArcMap and use the Catalog to navigate to your D:\GIS\Chapter5\ TMNData\ folder. You should see four geodatabases. Expand each one of them.

- Also expand the **Government Units, Hydrography, Structures,** and **Transportation** feature datasets inside the geodatabases.

- You'll see that each geodatabase contains many feature classes and layers. At the scale we're working with (that is, a single incorporated area), many of them will be empty. In this lab, we'll be working with only the following feature classes (so drag and drop each of them into ArcMap's Table of Contents from the Catalog):

  - **GU_IncorporatedPlace** – This is a polygon boundary for the area.

  - **Struct_Point** – This is a point layer of different structures in the area.

  - **Trans_RailFeature** – This is a line layer of railroads.

  - **Trans_RoadSegment** – This is a line layer of roads.

  - **NHDLine** – This is a line layer of water features (such as levees).

  - **NHDFlowline** – This is a line layer of streams.

  - **NHDWaterbody** – This is a polygon layer of water bodies, such as lakes or estuaries.

- Change the symbology of GU_IncorporatedPlace to have a Hollow fill (so the polygon will be transparent) and have a thick black line as its border (so the study area boundaries will stand out). Move it to the bottom of the layers in the TOC.

- Turn on all of the layers. Answer Question 5.1. (*Hint:* As you did in Chapter 1, check the Layer Properties of one of the layers and click on its Source tab to get some information to help you answer this question.)

QUESTION 5.1 What datum, coordinate system, and units of measurement are used in The National Map data? (Keep in mind that all of the downloaded layers have the same spatial reference.)

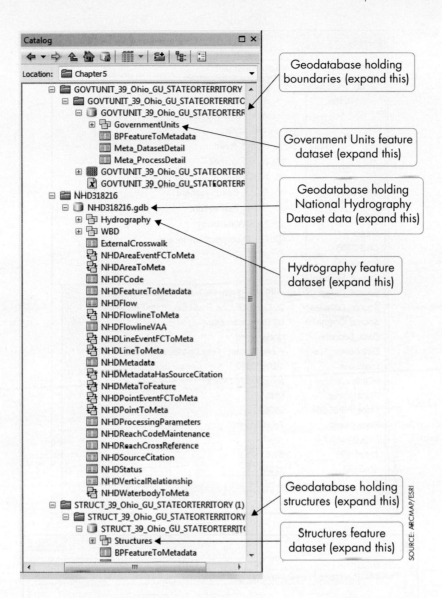

SOURCE: ARCMAP/ESRI

• For the questions in this step, you will be looking at the various features in The National Map dataset and how they relate to one another. We'll use the Identify and Find tools on the Tools Toolbar to help answer these questions.

SOURCE: ARCMAP/ESRI

• To use Identify, select the tool and click on a feature shown in the View. A new window will open to show you all of the attributes for that feature.

• To select a particular layer to get attributes for, select it from the **Identify from:** pull-down menu at the top of the Identify dialog box.

• To use Find, select the tool, then select the **Features** tab.

• In the **Find:** field, type what you are looking for.

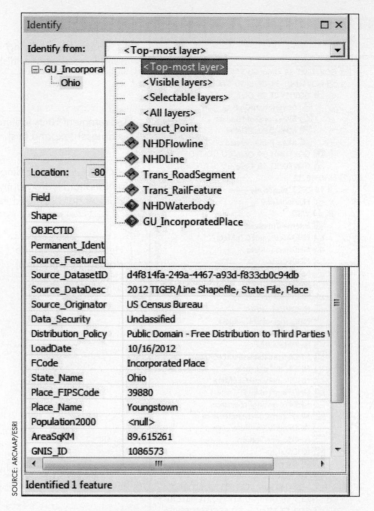

SOURCE: ARCMAP/ESRI

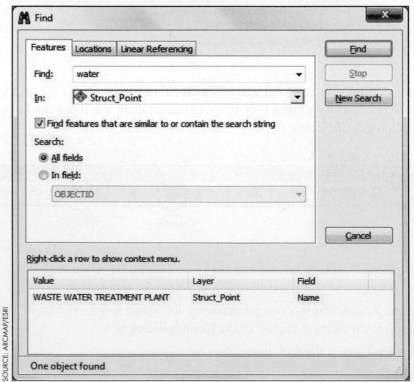

SOURCE: ARCMAP/ESRI

- In the **In:** field, choose the layer you want to search.

- Click the **Find** button. The results will be shown in the bottom dialog box.

- Click on a result and it will become briefly highlighted in the View (with lines drawn to it to help draw attention to it).

- Using these Identify and Find tools with the National Map data layers, answer Questions 5.2 to 5.7. (You may want to change the symbology of features to make them stand out better on the map.)

QUESTION 5.2 What are those "water" polygons near the City of Youngstown Wastewater Treatment Plant? (Hint: To properly locate this area with Find, you may have to search for "water" or "Youngstown City Of.")

QUESTION 5.3 Youngstown Fire Department Station 7 is at the corner of which two Youngstown roads?

QUESTION 5.4 What water feature flows between Lake Cohasset and Newport Lake? What type of water body does The National Map data classify this as?

QUESTION 5.5 Vittorio Ave is adjacent to what type of water body?

QUESTION 5.6 Coitsville Ditch connects to what body of water?

QUESTION 5.7 East High School is closest to what stream/river?

STEP 5.5 Using ArcGIS Online Data in ArcMap

- You can bring a lot of other freely available data directly into ArcGIS (that is, you don't have to download it from Websites and then import it). ArcGIS allows you to connect to the available resources of ArcGIS Online and download those maps and services directly into ArcMap. In Chapter 4, you shared your data through ArcGIS Online, but this time you'll be retrieving and using data shared there by others. These data are delivered to you via the Internet at no cost to you and no reduction of credits to your organization. (See **Smartbox 29** for more information.)

Smartbox 29

What kinds of data are available through ArcGIS Online?

Five broad types of geospatial information are available through ArcGIS Online.

- Maps: This is the term that Esri uses to describe a collection of data organized for visual representation. It includes map documents and map packages (see Chapter 1), Web maps (see Chapter 4), globe documents (see Chapter 18), and scene documents (see Chapter 15), among others. For example, you could take all of the data you currently have in ArcMap, save it as a map package, and upload that map package to ArcGIS Online. Once you've shared it, others can discover, download, and unpack your map package and have your map document (and the layers you're working with) available for use in their version of ArcMap. Suppose a team is out in the field mapping the trails of a local metropark. They can add their data to ArcMap and save it as a map package. Via ArcGIS Online, they can then share the entire package, which others can then download and use (to create a Web map of the data, for instance).

• Data Layers: This term describes hosted services on ArcGIS Online, such as feature services and tiled map services (among others). Esri makes a variety of different types of maps and services available for free download into ArcMap via ArcGIS Online. For instance, you can add tiled map services of topographic maps for your area from ArcGIS Online to your ArcMap session at the click of a mouse. Feature services and tiled map services have been shared through ArcGIS Online by many different providers and are available for your use. However, keep in mind that tiled map services are fixed layers; they can be viewed but often cannot be queried or changed. For instance, a tiled map service showing the median household income level for U.S. states, counties, and block groups can be downloaded from ArcGIS Online, but it can be used only for visual purposes. While you can identify features, you cannot query the database and extract certain counties or edit values in the attribute table.

• Data Files: This term describes data layers that you can work with. For instance, if you need a set of railroad data that you can query and edit, this dataset may be available through ArcGIS Online. Data Files are often available through ArcGIS Online as a layer package (see Chapter 18 for creating and sharing a layer package). A **layer package** consists of one or more layers stored in a file that can be unpacked for use in ArcMap. For example, a layer showing the political boundaries of all of the world's countries can be quickly downloaded as a layer package from ArcGIS Online. Keep in mind that some layer packages contain data that can be edited, while other layer packages cannot be edited (it depends on how the layer packages were created). For instance, a layer package of world time zones can be downloaded from ArcGIS Online, but it can only be overlaid and queried, or have its symbology changed. You cannot alter it or make changes to the polygons comprising the time zones.

• Tools: These files aren't tools in the sense of a tool from ArcToolbox. Rather, they are utilities used in working with GIS data. For instance, you could download a geoprocessing package (see Chapter 21) that would contain the data layers and the modeling tools used in a project.

• Applications: In Chapter 4, you created a Web application from seasonal home data and shared it on the Internet. Web applications are designed to be viewed and used online in a Web browser rather than downloaded into ArcMap.

layer package One or more ArcGIS layers combined together into a single file for portability or distribution via ArcGIS Online.

- To directly access ArcGIS Online, select the pull-down menu next to the Add Data button and choose **Add Data From ArcGIS Online…**.

- From the dialog that appears, search ArcGIS Online for the term **Ohio rail** (and select the **Data** radio button), then click on the search button (the magnifying glass) or hit **Enter** on your keyboard.

- Choose the **rail_line** dataset (a layer package), which has been made available from ODOT (Ohio Department of Transportation), and click **Add** in its box.

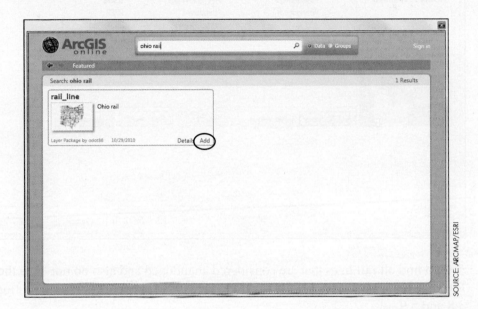

- Click **Close** if asked questions about different projections.

- The rail_line data should appear at the top of your Table of Contents, directly added from ArcGIS Online. Because it's a layer package, you'll be able to make some changes to the data. Bring up its Layer Properties and click on the **Symbology** tab.

- Under Show: select **Categories**, then select **Unique Values**.

- Choose **RR_STATUS** for the Value Field to display.

- Click **Add All Values**. This will add all of the unique data for the RR_STATUS attribute. In this case, a rail line will be classified as Abandoned or Active.

- Click **Apply**, then click **OK**. Back in the TOC, you'll see the lines in the rail_line layer are now being drawn with one line as Abandoned rail lines and another line as Active rail lines. Change the symbology of each one to distinguish them from each other (as well as to distinguish them from the separate Trans_RailFeature layer you got from The National Map).

- Right-click on the Trans_RailFeature and select **Zoom to Layer** to reset the view. Zoom in to several of the active rail lines (from ODOT via ArcGIS Online) and compare them with the rail features from The National Map.

- Zoom out a bit so that you can see the entire Youngstown area and the rail lines feeding into it from around the region. Make a select by attributes query of the

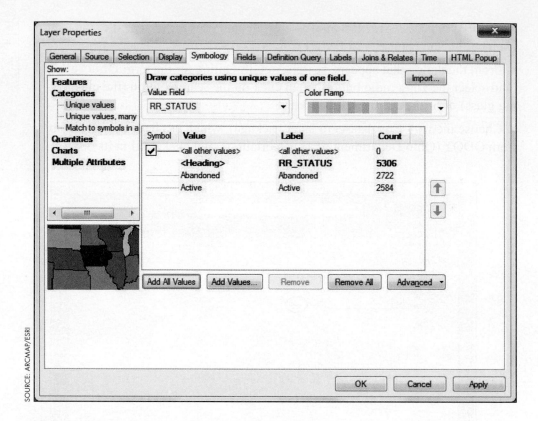

SOURCE: ARCMAP/ESRI

data to find all rail lines that are considered abandoned and also do not have the Railroad Name (the RR_NAME attribute) listed as Unknown. Answer Questions 5.8 and 5.9.

QUESTION 5.8 Which railroads have abandoned lines coming toward Youngstown from the southeast?

QUESTION 5.9 Which railroads have abandoned lines coming toward Youngstown from the northwest?

STEP 5.6 **Using Basemap Data from ArcGIS Online in ArcMap**

• By examining the different rail lines, you'll see that the two datasets don't necessarily match up. What we're going to need is another source to compare these lines; our goal is to figure out which dataset is the most accurate representation of the county railroads. Seeing how the data layers match up with the actual real-world features will give you a start on assessing the usefulness of the data for future analysis. To begin evaluating the datasets, you can access high-resolution aerial imagery as a **basemap** from the cloud resources of ArcGIS Online. This basemap layer has been spatially referenced so that your other GIS layers will align with it in ArcMap.

• There are several basemaps available to use as map services from ArcGIS Online, including topographic maps, street maps, and different types of remotely sensed imagery, all at different levels of detail, depending on the scale at which

basemap A map service available from ArcGIS Online (such as topographic maps or high resolution imagery) that can be used in ArcGIS for Desktop.

you're working. Note that the same basemaps available to you for setting up a Web map in ArcGIS Online (see Chapter 4) are also available for use here in ArcGIS for Desktop 10.2. See **Smartbox 27** on page 102 for more information.

• From the Add Data pull-down menu, select **Add Basemap**, then choose **Imagery with Labels**. The Imagery with Labels layer contains high-resolution imagery from Esri with labels added for things like locations and cities. Click **Add** to add the basemap to the View.

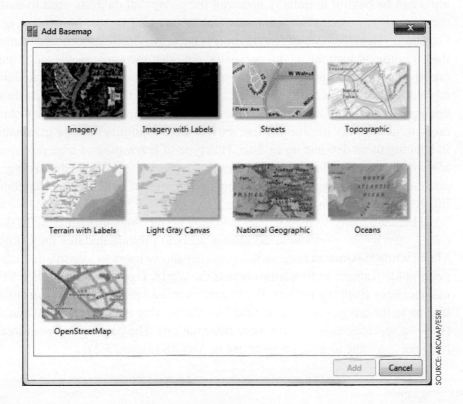

• Give ArcGIS a minute to stream the basemap via the cloud to the Desktop. The map service will be added to the bottom of the Table of Contents. You likely saw that the ODOT Active rail data from ArcGIS Online did not match up with rail line data from The National Map. With the layers on top of high-resolution imagery, you can now judge which dataset is the most accurate by examining where they match up with imagery of the landscape. Zoom in on some of the areas for the two rail lines. Answer Question 5.10.

QUESTION 5.10 By comparing the Active rail lines from the ODOT data with the rail lines from TNM and the Imagery with Labels basemap from ArcGIS Online, which of the rail lines is the most accurate? How can you tell?

• Turn off the Imagery with Labels basemap. Add a new basemap, but this time select the **OpenStreetMap** basemap. OpenStreetMap is an example of Volunteered Geographic Information (VGI) data that can be used in ArcGIS (see **Smartbox 30**).

Smartbox 30

What is VGI and how can it be used in ArcGIS 10.2?

Volunteered Geographic Information (VGI) User-generated geospatial data.

crowdsourcing Leveraging the knowledge and resources of multiple individual users for a larger project.

OpenStreetMap An open-source collaborative mapping project whose data are also available as a basemap in ArcGIS.

Volunteered Geographic Information (VGI) refers to geospatial data that's supplied by everyday persons, often people without GIS skills. VGI allows people to contribute their geographic knowledge in a way that others can use. For example, when a road is closed in your neighborhood or construction creates a new traffic pattern on your campus, your firsthand knowledge of the situation can be helpful in quickly updating the geospatial datasets used to make the maps of your area. Online sources such as Google Maps or MapQuest, or a Garmin or Magellan satellite navigation system, provide locations and routing that are only as accurate as the geospatial data that they use. However, it may take the companies who create this data a very long time to make small updates to your local area. This is where VGI comes into play. If you can provide those updates instead, your knowledge of new road patterns, out-of-business restaurants, or new library locations (these are just three examples) will be invaluable in keeping these datasets up to date. This type of leveraging of user resources toward the completion of larger projects is referred to as **crowdsourcing**, a phenomenon that's becoming increasingly common in building and maintaining geospatial datasets.

There are many VGI resources available on the Web. Google Map Maker (http://www.google.com/mapmaker) allows users to provide updates for Google Maps, while Wikimapia (http://wikimapia.org) allows users to identify and label geographic features and locations across the world. **OpenStreetMap** is a VGI collaborative mapping project. Registered users can provide updates or information to the online datasets, and the OpenStreetMap data are freely available (see http://openstreetmap.org for more information). The data from OpenStreetMap are available as a basemap to use in ArcGIS (Figure 5.2).

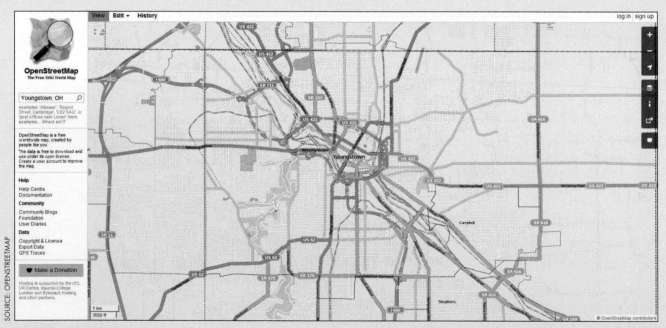

FIGURE 5.2 Youngtown, Ohio, in OpenStreetMap.

- Zoom in to the Youngstown State University (YSU) campus area (to find its general area, locate the point in the structures layer that corresponds with Youngstown State University) and examine the OpenStreetMap imagery with respect to the rail line layer from ArcGIS Online. Answer Question 5.11.

QUESTION 5.11 What major road (just south of the Youngstown State University campus) is an abandoned rail line?

- While you're looking at the campus area, examine the OpenStreetMap basemap with respect to The National Map road line data layer. Answer Question 5.12.

QUESTION 5.12 What roads on the YSU campus area (as included in The National Map Trans_RoadSegment layer) don't match up with the Open-StreetMap basemap imagery (and how don't they match)?

- A lot of freely available GIS data is available on the Internet. In this chapter you used data from only two of them (The National Map and ArcGIS Online) but there are many more sources of data. See **Smartbox 31** for information on other online data sources.

Smartbox 31

What other GIS data are available online?

Beyond The National Map and ArcGIS Online, there are a lot of GIS data available on the Internet (and a lot of them are free). As GIS usage and the importance of geospatial data have grown, so has the amount of GIS data. Many local county governments have established GIS data and resources online. For instance, Mahoning County, Ohio, maintains an extensive collection of GIS data and makes all of it available for free download. Everything from county culvert location shapefiles to high-resolution orthophotos is available for use in ArcGIS. (For more information, see http://gis.mahoningcountyoh.gov.) If you're looking for GIS data for nearby areas, check to see if your county government or county auditor makes GIS data available.

State-level GIS data is often available from state GIS portals or clearinghouses. For example, the state of Pennsylvania has the Pennsylvania Spatial Data Access (Pasda), which serves as a clearinghouse for the state's geospatial data. Pasda is a repository for many types of statewide GIS data, including road layers, watershed data, and elevation surfaces (for more information and the opportunity to browse the downloadable data, see http://www.pasda.psu.edu). Your state likely has a GIS clearinghouse online with similar GIS data for you to access.

The U.S. federal government operates a national geospatial data clearinghouse online at Data.Gov (see their Website at https://catalog.data.gov/dataset). This is the latest version of the U.S. government data access portal (which also included the former Geospatial One Stop and the former Geo.Data.Gov). Data.Gov provides access to all manner of downloadable geospatial data along with several map services and layer packages in a variety of different formats. The latest version for accessing some of this data is through the government Geospatial Platform, available here: http://www.geoplatform.gov (Figure 5.3).

SOURCE: GEOPLATFORM.GOV

FIGURE 5.3 The Website of geoplatform.gov.

Sometimes, data will be available for FTP download, and sometimes they are available by connecting to a GIS Server using the Catalog functions of ArcGIS that give you direct access to data via the cloud (you'll be using these functions in Chapter 13 to obtain data). In addition to these resources, there are numerous other online sources of free GIS data (and this list is by no means exhaustive):

• EarthExplorer: This USGS utility allows you to download GIS land cover and elevation data (including lidar data—see Chapter 17), as well as **Digital Line Graphs (DLG)** files. These DLG files are the digitized versions of features on USGS topographic maps (such as roads, railroads, hydrography, or hypsography contour lines). Because new topographic maps (see Chapter 16) are no longer being produced, these DLG features extracted from them are considered "legacy" data (that is, they represent GIS data from an older snapshot in time). However, DLGs are still useful sources of data and are available at 1:24000 and 1:100000 scales. Find EarthExplorer online at http://earthexplorer.usgs.gov (Figure 5.4).

Digital Line Graphs (DLGs) The digitized features from USGS topographic maps.

FIGURE 5.4 The USGS EarthExplorer Website.

- Landfire: This Website gives access to environmental GIS data as it relates to such attributes as wildlife habitat, vegetation cover, and forest canopy cover. Much of the data from Landfire comes in raster data format (see Chapter 12). Find Landfire online at http://www.landfire.gov (Figure 5.5).

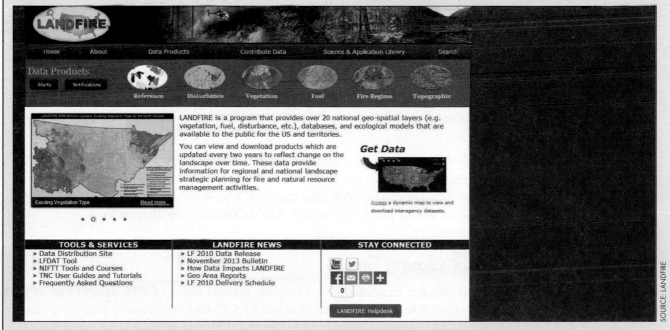

FIGURE 5.5 The Landfire Website for acquiring ecological data.

- Libremap Project: This Website features different types of GIS data for free download. It is a great source for finding scanned and georeferenced versions of USGS topographic maps (see Chapter 16) called **Digital Raster Graphics (DRGs)**. Find Libremap online at http://libremap.org.

- U.S. Census Bureau: The U.S. Census Bureau is a source of free TIGER (Topologically Integrated Geographic Encoded Referencing system)

Digital Raster Graphics (DRGs) Scanned and georeferenced versions of USGS topographic maps.

TIGER/Line Topologically Integrated Geographic Encoded Referencing boundary and road network GIS data created by the U.S. Census Bureau.

Spatial Data Transfer Standard (SDTS) A neutral file format used for the distribution of GIS data.

Data Interoperability A concept that allows GIS data of many different file types to be imported into ArcGIS or ArcGIS file formats to be converted to another file type.

products, including the **TIGER/Line** files of cartographic boundaries such as counties, census blocks, or block groups (see Chapter 2). National road data are also available as TIGER/Line files (see Chapter 10). However, as noted in Chapter 2, TIGER/Line boundary files do not contain socioeconomic attributes—for those, you can use the American FactFinder Website to download attribute tables and join them to the geospatial data. You can find TIGER products online at http://www.census.gov/geo/maps-data/data/tiger.html.

While much of the data you can download is available as a raster, shapefile, or geodatabase format, some older data will be delivered in **Spatial Data Transfer Standard (SDTS)** format. SDTS was set up as a "neutral" file format to allow geospatial data to be transferred to multiple GIS formats. ArcGIS has tools that allow for the import of SDTS files into an Esri file format. In fact, ArcGIS has a wide range of **Data Interoperability** tools (and an extension) that allow the import and translation of different types of GIS data into a class in a geodatabase.

Note that this Smartbox covers only GIS-specific data. There is plenty of remotely sensed imagery available. For instance, EarthExplorer allows you to download several different types of aerial photography or satellite imagery for use in ArcGIS, while The National Map gives you access to high-resolution orthoimagery. See Chapter 13 for much more about using remote sensing imagery in ArcGIS.

STEP 5.7 **Printing or Sharing Your Results**

- Save your work as a map document.

- Change the symbology of your layers so that the railroad features of The National Map and those you obtained from ArcGIS Online can be clearly distinguished from one another. For now, turn off the other layers from The National Map.

- Finally, either print a layout (see Chapter 3) of your final version of your Youngstown dataset (including all of the usual map elements and proper map design for a layout) or share your results as a feature service (and then create a Web map or Web application) through ArcGIS Online (see Chapter 4).

Closing Time

This chapter examined how to obtain various types of GIS data to use in your projects. While The National Map and ArcGIS Online are easily accessible sources of data, there are plenty more out there (even beyond the resources described in **Smartbox 31**). However, no matter what the source of your data, you should have information about the data you're receiving. See the *Related Concepts for Chapter 5* for more information regarding this "data about data."

Starting in the next chapter, you'll be creating your own GIS layers. While you will often use available data as a starting point for your analysis, many times you will be making your own data to use. For instance, if you're creating a map of the trails of a local metropark, you'll probably find that kind of data is unavailable for easy access. Similarly, if you have acquired the coordinates of field locations (such as culverts or water wells), you'll have to convert those numbers into a GIS data layer before you can use them.

Related Concepts for Chapter 5   How Metadata Is Used in GIS

Almost all GIS data are accompanied by some sort of descriptive information about the data (usually as a separate "readme" text file or an XML document). Useful information about the data might include a statement of how accurate the data are, what geographic or projected coordinate system is being used, and what each of the fields in the attribute table represents. Without this information, you might not be able to use the data properly. For instance, if you have to define the layer's projection, you have to know what projected coordinate system to use. Similarly, if the fields in the attribute table have strange names or if the attributes themselves use coded values, you would need something to explain what each of these items means. For instance, The National Map NHDFlowline feature class contains attribute fields entitled "ReachCode," "FlowDir," "FType," and "FCode." When you're using this hydrography data, it would help to know what these things represent. In GIS (and other fields), this descriptive information or "data about data" is called **metadata**.

With the proliferation of GIS data, standards have developed regarding what information should be included with a layer's metadata. The United States **Federal Geographic Data Committee (FGDC)** has set up a series of guidelines called the **Content Standard for Digital Geospatial Metadata (CSDGM)**. These content standards include the following:

- Identification of the dataset (a description of the data)
- Data quality (information about the accuracy of the data)
- Spatial data organization information (how the data are represented, such as vector format)
- Spatial reference information (the geographic or projected coordinate system used)
- Entity and attribute information (what each of the attributes means)
- Distribution information (how you can obtain the data)
- Metadata reference information (how current the metadata is)
- Citation information (how to cite the data)
- Time period information (what date do the data represent)
- Contact information (whom to contact for further information)

International metadata standards are also being adopted in North America as well. The **ISO 19115** metadata standards are becoming a commonly referenced format. (In the United States, this set of standards is adopted as the North American Profile—NAP—of ISO 19115.) The ISO 19115 structure contains similar information to the CSDGM, but it also contains some new elements, extended elements, and support for new geospatial applications.

In ArcGIS, a layer's metadata is accessible via the Item Description, which doesn't adhere to a particular format. However, ArcGIS can be set up to allow

metadata Descriptive information about data.

Federal Geographic Data Committee (FGDC) A committee that is responsible for setting GIS metadata content standards.

Content Standards for Digital Geospatial Metadata (CSDGM) A set format created by the FGDC for what items a metadata file should contain.

ISO 19115 An international metadata standard being adopted in the United States as the North American Profile of ISO 19115.

the use of specific metadata formats (such as CSDGM and NAP ISO 19115). In ArcMap, select the **Customize** pull-down menu, then choose **ArcMap Options**. From there, select the **Metadata** tab, which will give you several options for structuring how you want to work with metadata in ArcGIS.

For More Information

For further in-depth information about the topics presented in this chapter, use the ArcGIS Help feature to search for the following items:

- A quick tour of the ArcGIS Data Interoperability extension for Desktop
- Adding data from ArcGIS Online
- Choosing a metadata style
- Illustrated guide to complete FGDC metadata
- Redistribution rights
- Sharing your map in a Web application
- Geographic information (RC)
- Using ArcGIS Online in ArcGIS for Desktop applications
- Using Bing Maps
- What is metadata?
- Working with basemap layers

For further information about the National Geospatial Program, see:

- Carswell, W. J. 2012. *National Geospatial Program*. USGS Fact Sheet 201-3078. Available online at http://pubs.usgs.gov/fs/2011/3078/pdf/fs2011-3078.pdf.
- USGS National Geospatial Program: http://www.usgs.gov/ngpo/

For a great reference for finding data, sources of data, and working with all different types of GIS data available in the public domain, see:

- Kerski, J., and J. Clark. 2012. *The GIS Guide to Public Domain Data*. Redlands: Esri Press.

For more information about VGI, see:

- Sui, D. Z. 2008. "The Wikification of GIS and Its Consequences: Or Angelina Jolie's New Tattoo and the Future of GIS." *Computers, Environment and Urban Systems* 32 (2008): 1–5.

For more about current metadata standards, see the following FGDC Web pages:

- Content Standards for Digital Geospatial Metadata: http://www.fgdc.gov/metadata/csdgm
- Geospatial Metadata Standards: http://www.fgdc.gov/metadata/geospatial-metadata-standards

Key Terms

National Geospatial Program (NGP) (p. 113)

The National Map (p. 114)

layer package (p. 124)

basemap (p. 126)

Volunteered Geographic Information (VGI) (p. 128)

crowdsourcing (p. 128)

OpenStreetMap (p. 128)

Digital Line Graphs (DLGs) (p. 130)

Digital Raster Graphics (DRGs) (p. 131)

TIGER/Line (p. 132)

Spatial Data Transfer Standard (SDTS) (p. 132)

Data Interoperability (p. 132)

metadata (p. 133)

Federal Geographic Data Committee (FGDC) (p. 133)

Content Standards for Digital Geospatial Metadata (CSDGM) (p. 133)

ISO 19115 (p. 133)

How to Create Geospatial Data with ArcGIS 10.2

Introduction

In the first five chapters, you used geospatial data that had already been created and processed for you to use. While a huge amount of data is available, you will sometimes need to create your own datasets to use in GIS. If the data you need is unavailable, out of date, or incomplete, you have only one option: to make your own version of it.

Sometimes, you will have to create data from scratch, and there are many ways to create data to use in GIS. For example, measurements taken in the field by hand are often used to build GIS data. GPS devices or surveying can provide you with a set of latitude and longitude coordinates for a location. In turn, you can change these coordinates into a point layer to use in ArcGIS.

However, it is sometimes impossible to obtain data first-hand. In this case, you'll have to rely on a secondary source to build GIS data from. You'll begin with a pre-existing map or remotely sensed image and use that as the base to start building your data. For instance, if you're going to create a GIS layer of the walking paths of a local fairgrounds, you could get an aerial photograph that shows an overhead view of the fairgrounds (which would allow you to clearly see the paths) and use that photo to create a set of lines that represent the length and dimensions of the walking path.

The process of creating GIS data (whether points, lines, or polygons) from another secondary source is referred to as **digitizing**. Commonly, **heads-up digitizing** is performed as just described—you add a map or image (usually referred to as a *basemap*) to ArcGIS, and it becomes the basis for creating your own data. In digitizing, you're in essence drawing points, lines, or polygons on the screen, and these sketches are being converted to data layers inside the GIS. In the fairgrounds example, you could simply draw lines that follow the walking paths you could see on the aerial photo, or you could carefully draw polygons by tracing around the dimensions of the buildings.

Chapter Scenario and Applications

In this chapter, you'll be taking the role of an event planner for a university campus. Your university is developing a new campus map to be used for planning and hosting conferences, arts festivals, and other community events. The university's marketing department has a lot of ideas about visualizing and presenting this information, but first it needs a map. It's your job to create a new map of the campus showing off the roads, sidewalks, parking lots, and university buildings.

digitizing The process of creating digital GIS data from a secondary source.

heads-up digitizing Using a map or imagery as a backdrop in the digitizing process.

The following are some other examples of other real-world applications of this chapter's theory and skills.

• A geologist is drilling wells for groundwater sampling; the locations of these wells are denoted with a set of coordinates (perhaps latitude and longitude obtained through GPS). The geologist can use ArcGIS to plot the coordinates for these wells as points on a map for future analysis.

• An archeologist is recording the location of artifacts at dig sites. She can use these coordinates to measure these positions and plot them as points using ArcGIS.

• A historian is using old maps of historic buildings in an area to recreate the appearance or layout of these places. He can use digitizing to create new GIS data from these old sources.

ArcGIS Skills

In this chapter, you will learn:

• How to create a file geodatabase.

• How to convert a table of coordinates into a point layer.

• How to create feature classes within a geodatabase.

• How to obtain and use a basemap in ArcGIS as a data source for digitizing.

• How to set a bookmark in ArcMap.

• How to digitize geospatial data and save it as a geodatabase feature class.

• How to measure the real-world length of the data you're creating as a new field in an attribute table.

Study Area

For this chapter, you will be examining a portion of the Youngstown State University (YSU) campus in Youngstown, Ohio.

Data Sources and Localizing This Chapter

This chapter's data focus on features and locations within the YSU campus. However, you can easily modify this chapter to use data from your own campus or local area instead. Because you will be creating the data directly from the Imagery basemap, simply find your own campus area that you're familiar with on the basemap and digitize from there. You can quickly create the campusbound shapefile for your own area as well. You can use a source such as GPS measurements or Google Earth for obtaining coordinates of prominent points in your own local area for Step 6.3.

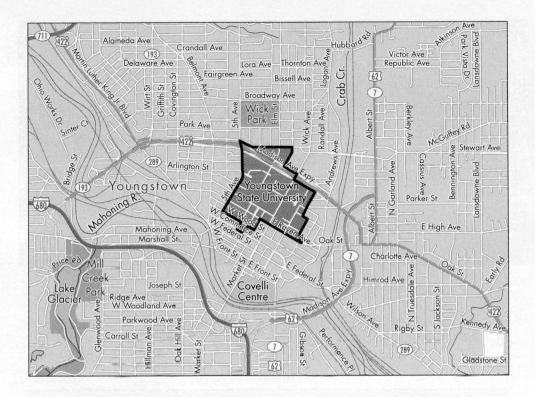

STEP 6.1 Getting Started

• Start ArcMap and use the Catalog to copy the folder called **Chapter 6** from the C:\GISBookdata\ folder to your own D:\GIS\ drive. Be sure to copy the entire directory, not just the files within it.

• Chapter 6 contains the following items:

  • **Campusbound.shp** – a simple polygon shapefile showing a simplified boundary of the YSU campus.

  • **Parkingcoords.xlsx** – a Microsoft Excel file containing the latitude and longitude of parking lift arms on campus.

  • **Parkmap2012.pdf** – a map of the YSU campus in PDF format. You won't be able to see this in Catalog because it's just a PDF file. Open this file from Windows now to use as a reference for this chapter.

• *Important Note:* Please refer to the PDF of the YSU map for the proper location and orientation of objects. Figure 6.1 shows a visualization of the section of campus used in this chapter.

• You may also want to check one of the YSU campus maps (including the interactive campus map) available here: http://web.ysu.edu/gen/ysu/Directions_and_Virtual_Tour_m162.html.

• You may find the "Birds eye view" options on Bing Maps, available at http://www.bing.com/maps, very useful in looking at oblique imagery of the YSU campus.

• Add the Editor toolbar (see Chapter 2) to ArcMap by selecting the **Customize** pull-down menu, then select **Toolbars**, then select the **Editor** toolbar. Like other toolbars, the Editor toolbar can be moved and docked in ArcMap.

SOURCE: COURTESY OF YOUNGSTOWN STATE UNIVERSITY

FIGURE 6.1 A section of YSU campus.

• You can also add the Editor toolbar by pressing the Editor Toolbar button on the Standard Toolbar:

SOURCE: ARC-MAP/ESRI

• In this chapter, you'll be using a basemap streamed from ArcGIS Online resources as the source for digitizing (in this case, a high-resolution image from Esri). ArcGIS Online contains imagery for the entire United States (and many other countries as well). To get your bearings, you'll want to look only at the portion around YSU and create data at that scale. To get this process started, add the **Campusbound** shapefile to the Table of Contents. It's a polygon showing an estimated boundary of campus.

• Next, from the pull-down menu next to the Add Data button, select **Add Basemap….** From the new options that appear, select **Imagery** and click **Add**. Give ArcGIS a minute or two to stream the imagery to you over the Internet. The high-resolution imagery should snap into place, centered around your shapefile of campus. If the imagery does not appear, right-click on Campusbound and select **Zoom to Layer**. This will refocus the View on the dimensions of the campus boundaries.

• Turn off the Campusbound shapefile but don't remove it from the Table of Contents just yet (we used it only to get the imagery to center on YSU when the basemap was added). The high-resolution imagery should now be displayed on the screen.

• Right-click on the Data Frame and select **Properties**, then select the **General** tab. Select **Feet** for Units: Display. Now you'll be able to make measurements from the imagery in feet. For more information about measurements and scale, see **Smartbox 32**.

Smartbox 32

Why is scale important when digitizing?

In Chapter 3, we discussed the importance of scale when making a map. The scale of the map that's used for the digitizing source is critical. For instance, if you're going to be digitizing features of your neighborhood, you will need a very large-scale map as a source to capture things such as the dimensions of houses, backyard sheds, or driveways because these kinds of things would not be apparent in smaller-scale maps.

The level of detail available to be digitized will change as the scale of the map source changes. For example, Figure 6.2 shows two topographic maps of McCarran International Airport, with map (a) at a 1:24000 scale and map (b) at a 1:100000 scale. If you were digitizing the airport based on each of these maps, you'd clearly end up with two completely different datasets. If you wanted to capture very detailed features of the airport, the 1:24000-scale map would be the better choice. If you just needed to mark a rough location of the airports and runways, the 1:100000-scale map could be used.

The geographic scale of the project you're working on will affect the choice of what map scale to choose for a digitizing source. For instance, using the 1:24000-scale map of McCarran to digitize the runways would provide you with a lot of detail for analysis. However, you wouldn't be able to digitize individual building or housing footprints at such a scale. Similarly, if you digitized the detail of the airport from the 1:24000-scale map and then the roads from the 1:100000-scale map, you'd have two very different datasets that may cause issues in future analysis.

FIGURE 6.2 The features of McCarran International Airport in Las Vegas as shown on (a) a 1:24000-scale topographic map, and (b) a 1:100000-scale topographic map.

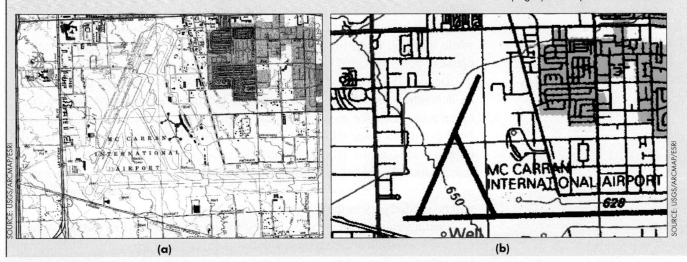

(a)　　　　　　　　　　　　　　　　(b)

SOURCE: USGS/ARCMAP/ESRI

STEP 6.2　Creating a Geodatabase to Contain GIS Data

• Because you'll be digitizing objects that represent the features of YSU campus— roads, sidewalks, campus and non-campus building footprints, and parking lots— you'll be creating separate feature classes to contain each of these types of data.

• First, we'll make a geodatabase to contain all of these files. Return to the Catalog and select your **D:\GIS\Chapter6** folder. Right-click on this folder in the Catalog and select **New**, then select **File Geodatabase**. A new geodatabase icon will appear in the folder. Rename it to **YSUDigitize**.

• For more information on geodatabases and how they're used to store GIS data, see **Smartbox 33**. Keep in mind that, when you selected New, you had a lot of other choices besides creating a new geodatabase. See **Smartbox 34** for more information on other data types available for use in ArcGIS.

Smartbox 33

What is a geodatabase, and how does it store geospatial data?

geodatabase An ArcGIS structure for a single item that contains multiple datasets, each as its own feature class.

personal geodatabase A geodatabase that can hold a maximum of 2 GB of data and has the file extension .mdb.

file geodatabase A geodatabase that can hold an unlimited amount of data and has the file extension .gdb.

feature class A GIS layer stored inside a geodatabase—it can contain multiple items of the same data type.

feature dataset A grouping of related feature classes stored within a geodatabase.

A **geodatabase** is a structure for storing GIS data in ArcGIS. In Catalog, the geodatabase symbol is a gray cylinder, which is a good way of thinking about a geodatabase—it's designed to hold several different types of geospatial (as well as non-spatial) data. In ArcGIS, you'll be using one of two types of geodatabases:

• **Personal geodatabase**: This type of geodatabase has an overall storage limit of 2 GB and uses the file extension .mdb. A personal geodatabase is stored on your computer as a single file in Microsoft Access format.

• **File geodatabase**: This type of geodatabase has no limit on the number of datasets it may contain (although each dataset in the file geodatabase is limited to 1 TB of storage space). It uses the file extension .gdb. A file geodatabase is stored on your computer as a folder of files.

A geodatabase is a single item that can contain many different types of GIS data. Geodatabases store data as a **feature class**. Each feature class holds a different type of object (such as points, lines, or polygons), so the type of feature class being created in a geodatabase depends on what you're modeling in ArcGIS. For instance, roads or streams would be stored as line feature classes, while building footprints or land parcels would be stored as polygon feature classes.

Beyond the three vector object feature classes, a geodatabase can also store an annotation feature class (text used for labeling), a dimension feature class (annotation used for measurements), a multipoint feature class (storing very large point sets—see Chapter 17), or a multipatch feature class (used for 3D objects—see Chapter 18). Geodatabases can also hold tables of non-spatial data (see Chapter 2) and raster datasets (see Chapter 12), while other data types such as shapefiles can be imported to become new feature classes. All feature classes will be shaded grey in the Catalog.

Feature classes can also be grouped together into a **feature dataset**, which can hold several feature classes with the same coordinate system information. For instance, if you have many feature classes related to transportation, you may want to create a transportation feature dataset within the geodatabase to hold them all for data-management or sharing purposes. Feature datasets are also used to collect feature classes together to create things such as network datasets (see Chapter 11), topologies (see Chapter 7), terrain datasets (see Chapter 17), geometric networks (see Chapter 11), or parcel fabrics. Like feature classes, feature datasets will be shaded gray in the Catalog. See Figure 6.3 for the organization of feature classes and feature datasets within a file geodatabase.

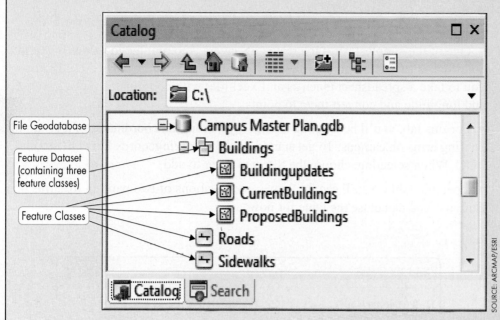

FIGURE 6.3 A file geodatabase containing a feature dataset and several feature classes.

SOURCE: ARCMAP/ESRI

Smartbox 34

What other data formats can be used in ArcGIS?

While geodatabases are very versatile for handling geospatial data in ArcGIS, they're not the only supported Esri data format. Another common format is the **shapefile**, a simple, non-topological file structure for holding GIS data. Shapefiles were introduced in the 1990s with the ArcView software program and continue to be used today. A shapefile can hold objects of a specific type. For example, you could use a polygon shapefile to store property boundaries or a point shapefile for locations of culverts. A shapefile consists of several files with different file extensions, including .shp, .shx, and .dbf. All of these many files must be present in the same folder for the shapefile to work properly in ArcGIS. Shapefiles will be shaded green in the Catalog.

Another Esri data format is the **coverage**, the file format used with the Arc/Info software program. Coverages could contain different types of objects, including points, polygons, and arcs (lines). Coverages are stored in folders called workspaces. A workspace consists of one or more folders containing the files that make up the coverages, along with a separate INFO folder that contains related files for each coverage. For transferability or downloading, coverages are frequently found as interchange files with the file extension .e00. These files are a "zipped" version of the coverage and can be imported to coverage format using a utility available in ArcToolbox. Coverages will be shaded yellow in the Catalog.

As Chapter 5 noted, Esri promotes the concept of data interoperability, whereby Esri data formats can be exported for use in other software packages. Likewise, a multitude of software-specific GIS formats can be imported into ArcGIS, included CAD datasets and files in the KML (Keyhole Markup Language) formats. The Data Interoperability extension allows for more than 100 different data formats to be used in ArcGIS.

shapefile A series of files (with extensions including .shp, .shx, and .dbf) that make up one vector data layer.

coverage A data layer represented by a group of files in a workspace consisting of a folder structure filled with files and also associated files in an INFO folder.

STEP 6.3 Converting X/Y Coordinates to a Point Feature Class

• One of the simplest ways to create GIS layers is to plot the locations of *x* and *y* coordinates as points and then save those points as a feature class. ArcGIS allows you to take a spreadsheet (such as an Excel file) with coordinates such as latitude and longitude and convert them to points.

• For this lab, you'll be creating a point layer from the coordinates of a group of parking arms on campus. To get started, add the Parkingcoords Excel file to the TOC. When selecting, choose the **Sheet1$** layer to add.

• Open the table. You'll see the names and locations of two parking lift arms on campus. You can close the table for now.

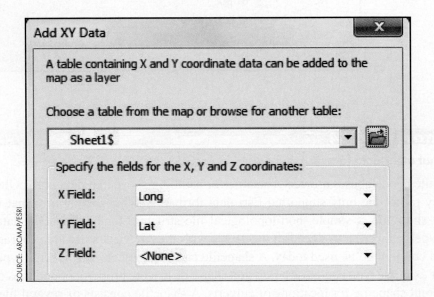

SOURCE: ARCMAP/ESRI

• To convert these coordinates to a layer, from the **File** pull-down menu choose **Add Data**, then select **Add XY Data….**

• The table to use is **Sheet1$**.

• Use **Long** for X Field.

• Use **Lat** for Y Field.

• Next, you'll set the coordinate system of the input latitude and longitude coordinates. Press the **Edit** button.

• In the Spatial Reference Properties dialog, type **WGS 1984** into the filter. ArcGIS will give you options for this datum only.

• Select **Geographic Coordinate Systems**, then select **World**, then select **WGS 1984**. It will be an option for the WKID of 4326.

• *Important Note:* If you were plotting your own coordinates that had not been measured in WGS84, you could use the Spatial Reference Properties dialog to select the proper coordinate system of the input coordinates for your own Excel file.

• Click **OK** when all settings are correct.

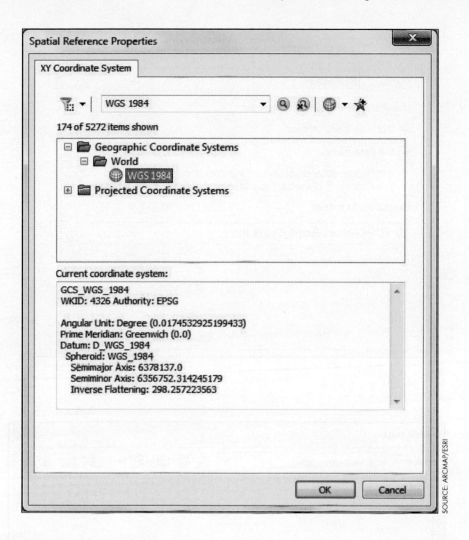

SOURCE: ARCMAP/ESRI

- Back in the Add XY Data dialog, click **OK**. You will receive a warning that the Excel table doesn't have an Object-ID field. That's okay; the Add XY Data dialog only creates an **Event Layer**. If you want to have a layer you can query or work with, you'll have to convert the Event Layer into a feature class (with just a few additional steps).

- The Event Layer with the two points will be added to the TOC (and by default will be named Sheet 1$ Events). Zoom in on the section of campus where the two points are located to see them plotted.

- You'll want to convert these points to a Point Feature Class and place it within your YSUDigitize geodatabase to complete your dataset. To do so, right-click on the **Sheet1$ Events** layer in the TOC, select **Data**, and then select **Export Data**.

- Use the same coordinate system as **this layer's source data**.

- In the Output feature class option, press the **browse** button. By default, ArcGIS will want to name the new shapefile Export_Output.shp, but you'll give it a new name (and also choose to save it as a feature class instead of a shapefile) in the next few bullet points.

Event Layer A GIS data layer created from tabular data.

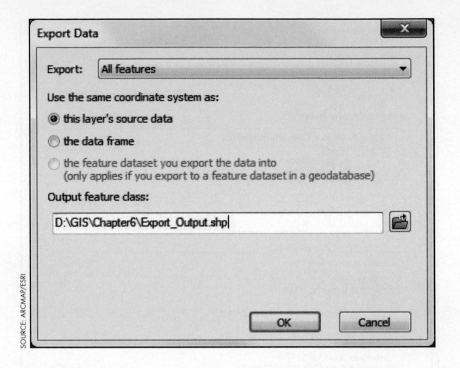

SOURCE: ARCMAP/ESRI

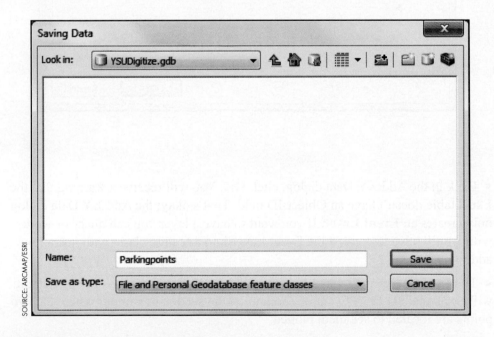

SOURCE: ARCMAP/ESRI

- In the Saving Data dialog, navigate to the D:\GIS\Chapter6 folder. Select the **YSUDigitize.gdb** geodatabase.

- Call the new layer **Parkingpoints**.

- Choose **File and Personal Geodatabase feature classes** for the Save as type option.

- When all settings are correct, click **Save**. This will return you to the Export Data dialog. Now your Output feature class will be called Parkingpoints and saved in the YSUDigitize geodatabase. Click **OK** in the Export Data dialog and then choose to add the exported data to the map.

• At this point, remove the Event Layer version of the points from the TOC and change the symbology of the Parkingpoints feature class to something more distinctive.

• Look in the Catalog at the YSUDigitize geodatabase. You'll see that it can be expanded to show the Parkingpoints feature class successfully added to it.

STEP 6.4 Creating New Feature Classes to Hold Data

• Next, because you'll be digitizing several different types of things (roads, sidewalks, parking lots, and building footprints), each of these items will be stored in its own feature class inside the YSUDigitize geodatabase. The first thing you'll be digitizing is the roads around campus, so you'll need to create a new feature class in the geodatabase in which to store them. In the Catalog, right-click on the **YSUDigitize** geodatabase and select **New**, then select **Feature Class**. The New Feature Class dialog will appear.

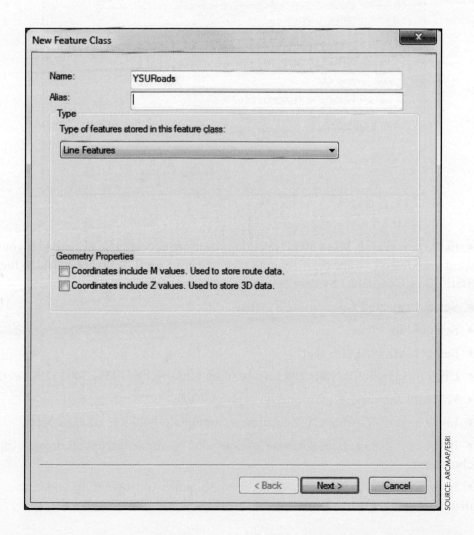

- For Name, call the feature class **YSURoads**.

- For the Type, select **Line Features** to be stored (because you will be digitizing roads as lines).

- Leave the other options alone. Click **Next**.

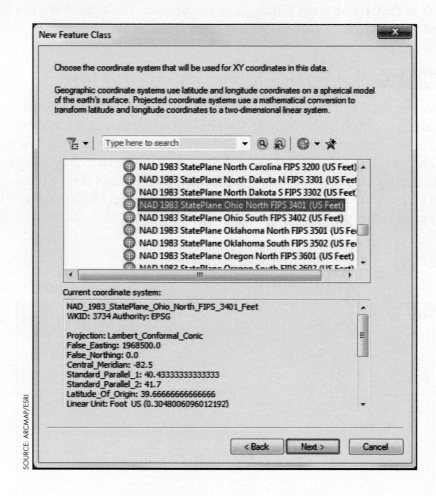

SOURCE: ARCMAP/ESRI

- In the next dialog, you will be selecting the proper coordinate system that you want your features to be digitized into. For this lab, you'll be creating data in the State Plane Coordinate System.

- Select **Projected Coordinate Systems**.

- Select **State Plane**.

- Select **NAD 1983 (US feet)**.

- Click the option for **NAD 1983 StatePlane Ohio North FIPS 3401 (US Feet)**.

- When all parameters are properly set, click **Next**.

- In the next dialog about X/Y tolerances, accept all defaults and click **Next**.

- In the next dialog about database storage configurations, accept all defaults and click **Next**.

- In the final dialog, you'll be adding room for a new attribute to be added so that when you digitize a feature, you can also add information about it.

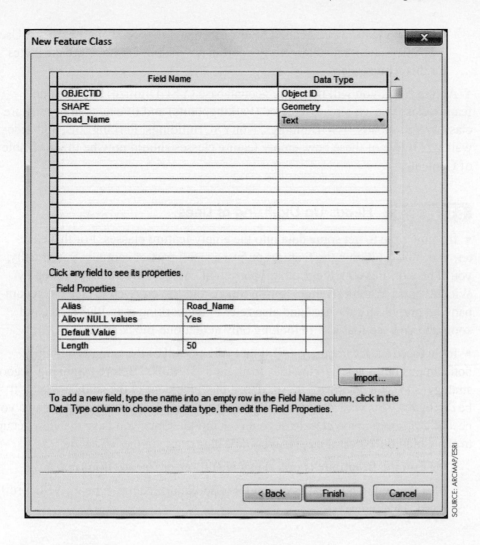

SOURCE: ARCMAP/ESRI

- Under Field Name, type **Road_Name**. (Be sure to use the underscore between the two words.)

- Under Data Type, click in the open field and a pull-down menu will appear. Select **Text** for this option (because the name of a road will be stored as Text).

- Lastly, click **Finish**.

- Return to Catalog again and you'll see that a new feature class called YSURoads has been added and it will now accept lines.

- Follow these same steps to create four other feature classes and store them in your YSUDigitize geodatabase as follows:

  - Create a feature class called **YSUBuildings** that will store **Polygon Features**. Use the same State Plane Ohio North coordinate system. In the Field Name, create a field called **Building_Name** to have Data Type **Text**.

  - Create a feature class called **NonYSUBuildings** that will store **Polygon Features**. Use the same State Plane Ohio North coordinate system. In the Field Name, create a field called **Building_Name** to have Data Type **Text**.

  - Create a feature class called **ParkingLots** that will store **Polygon Features**. Use the same State Plane Ohio North coordinate system. In the Field Name, create a field called **Lot_Name** to have Data Type **Text**.

- Create a feature class called **Sidewalks** that will store **Line Features**. Use the same State Plane Ohio North coordinate system. Do not store attributes for this feature class.

- At this point, you will have one geodatabase (YSUDigitize) that has one feature class you've already created (Parkingpoints) and five new empty feature classes (YSURoads, YSUBuildings, NonYSUBuildings, ParkingLots, and Sidewalks). All five of these new, empty feature classes should now be in your Table of Contents.

STEP 6.5 Heads-Up Digitizing of Lines

- It's now time to get some data into the empty feature classes. For this chapter, you'll be digitizing the features on just a small part of campus. Specifically, you'll be working with the southern portion of campus, the block bounded by Wick Avenue, Walnut Avenue, Spring Street, and Rayen Avenue (see the accompanying graphic for the area and also refer to the parking map PDF). Pan and zoom the view so that you're looking only at that one-block area.

- In this section of campus, you'll want to digitize all of the campus buildings, non-campus buildings, parking lots, roads, and sidewalks. Before beginning, a good strategy is to set a bookmark in ArcMap at the full extent of the study area you'll be digitizing. This way, you can easily reset the view to show you only the area you need if you zoom in too closely or zoom out too far. Once you have the view set up to show only the block of campus you'll be digitizing, do the following:

  - From the **Bookmarks** pull-down menu, select **Create Bookmark**.

  - In the dialog box that asks you for a Bookmark Name, type in: **YSU Study Area**. Then click **OK**.

- Now, under the Bookmarks menu, you have a new option called "YSU Study Area." Selecting this will reset the view to the boundaries and scale you currently see on the screen.

- We'll start by digitizing roads that will be stored in your YSURoads feature class in your Table of Contents.

  - Note that, even with this feature class turned on, nothing is displayed because it contains no data (yet).

  - Change the color and width of YSUroads to something that will distinguish it from the colors of the features on the basemap.

- From the **Editor** pull-down menu on the **Editor Toolbar**, select **Start Editing**.

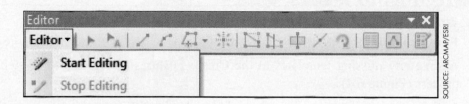

- When prompted, select the **D:\GIS\Chapter6\YSUDigitize.gdb** (your File geodatabase) as the Source you want to edit. Then click **OK**. Your cursor will change to become the Edit Tool.

- Next, click on the Create Features button on the Editor toolbar. You'll now see a new window open on the right side of the screen called **Create Features** and each of your feature classes stored there (as "templates," in ArcGIS terminology). Whenever you want to digitize a new feature, you'll select its template from this Create Features dialog. If you find that some or all of your data layers are not available as templates, see **Troublebox 5** for a solution. Also, for more information about the digitizing process itself, see **Smartbox 35**.

Troublebox 5

What if some templates are missing from the Create Features box?

Adding any missing templates to the Create Features dialog is a three-step process (Figure 6.4):

1. From the list of templates on the right-hand side of the screen, click on the **Organize Templates** button.

2. In the Organize Feature Template dialog that opens, press the **New Template** button.

3. In the Create New Templates Wizard, place checkmarks in the layers that are in the Table of Contents but not in the templates list for which you want to set up templates, then click **Finish**. The templates should then be added to the Create Features listings.

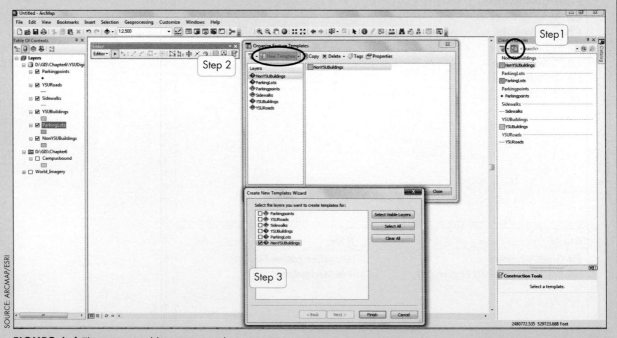

FIGURE 6.4 The steps to add missing templates.

How is digitizing performed in ArcGIS 10.2?

When you're digitizing in ArcGIS 10.2, you're basically drawing a sketch on the screen, using a source (such as a map or remotely sensed image) as your backdrop. A sketch consists of one or more **segments** that you draw. Multiple segments are connected by a **vertex** (Figure 6.5). For instance, when digitizing the outline of a parking lot as a polygon, the sketch will consist of several line segments connected by vertices and the segments will connect back to the first vertex, thus closing in the polygon.

When a sketch is complete, it will become a feature (and thus, geospatial data). Each feature will then be saved as a separate record in the layer's attribute table and will thus have a real-world length or area. For instance, if you sketch the outlines of five houses in a subdivision, the polygon layer will have five records, and you will be able to access information about the area and perimeter of each object. If you then digitize five line segments into a "driveways" feature class, it will contain five records, each measuring the real-world length of the driveways you just sketched.

Point digitizing refers to placing a vertex at each location on which you click the mouse. **Stream digitizing** refers to a mode where vertices are automatically placed at a certain distance as you move the mouse. Point mode (the default) is useful when you're controlling where vertices are placed (such as the corners of a building), while stream mode is useful when sketching long, continuous features such as coastlines. When sketching, press the F8 key on the keyboard to enter or exit streaming mode. You can change the streaming tolerance (or the distance the cursor is moved before a new vertex is created) by selecting the **Editor** pull-down menu, choosing **Options**, and then selecting the **General** tab.

segment A single digitized line.

vertex The beginning and ending points of a segment.

point digitizing The digitizing mode in which the user places a vertex by clicking the mouse button.

stream digitizing The digitizing mode in which vertices are automatically generated at a pre-determined distance of moving the mouse.

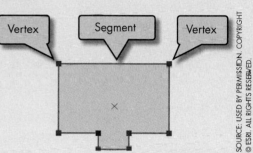

FIGURE 6.5 The segments and vertices that make up a sketch.

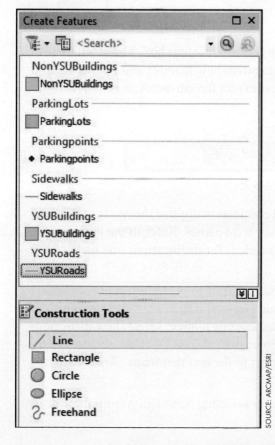

- Start by clicking on the **YSURoads** template in the Create Features box.

- You'll see a new set of options appear below the box; the new options are called Construction Tools. These are the shapes that you can begin digitizing with. Because you'll be digitizing roads as lines, select **Line** from the Construction Tools.

- Finally, you can start digitizing. Let's start at the intersection where Wick Avenue meets Rayen. Zoom closely into this area.

- You'll see that your cursor has turned into a crosshairs. Click once at the Wick and Rayen intersection and a red dot will appear; this will be the first vertex in the line. Drag the cursor up to the intersection of Wick and Spring and double click. The ending node of the line will be digitized and the line segment will turn a solid light-blue color. The line has now been digitized and a line object has been created in the YSURoads feature class.

- From the Editor toolbar, select the Edit Cursor.

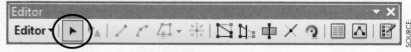

- Your cursor will turn into the Edit Cursor. Right-click on the line you just created and select **Attributes**.

- A new box will open on the right-hand side of ArcMap called Attributes. You'll see the YSURoads layer with one object in it.

- Below this will be a box that has the attribute you created (Road_Name). Click in the box next to it that says <null> and type the name of the road you just digitized: **Wick Avenue**. Then press the **Enter** key on the keyboard.

- You'll see the object name changed from the number 1 to the name of the road. What you've just done is assigned an attribute to the object you just digitized.

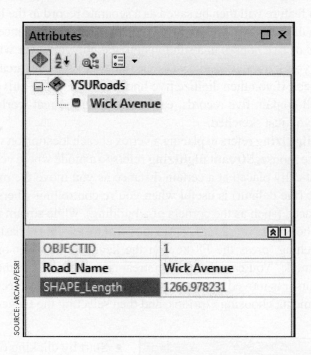

SOURCE: ARCMAP/ESRI

- You'll see a set of tabs at the bottom of the Attributes box. These will allow you to switch back and forth between creating new features and editing their attributes. You'll probably find it easiest to edit the attributes of an object right after you digitize it.

SOURCE: ARCMAP/ESRI

- Switch back to Create Features and continue to digitize the roads of this section of the YSU campus and update their road names. Refer to the Parkmap2012. pdf file (and the other online maps, if needed) for the locations and names of roads in this section of campus.

  - *Important Note:* By default, you can draw only straight-line segments. More options for drawing lines (such as curves) are available from the Feature Construction toolbar. To access this toolbar, select the **Editor** pull-down menu from the **Editor Toolbar**, then choose **Options**, then select the **General** tab. Place a checkmark next to the text that reads **Show feature construction toolbar**.

- You can save your edits at any time by selecting **Save Edits** from the **Editor** pull down menu on the **Editor Toolbar**.

- You'll notice that when digitizing, you may have some unconnected lines. Editing of vector data will be covered in Chapter 7, but for now, some simple techniques for editing your work as you digitize are as follows:

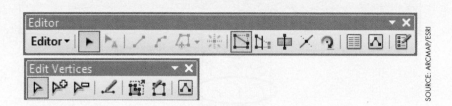

SOURCE: ARCMAP/ESRI

- Select the Edit Tool (that is, the edit cursor) from the Editor Toolbar and double-click on a line segment—you can now drag the lines to other locations on the map. The Edit Vertices toolbar will also appear.

- To edit the location of vertices, double-click on a line segment with the Edit Tool. The vertices will be highlighted in green, while the line segment is highlighted in light blue. Clicking on a vertex will allow you to move that point or create new bends in the line.

- To delete a line segment, select it with the Edit Tool (it will appear in a light-blue color) and then click the **delete** key on the keyboard.

- Once you have digitized all of the roads be sure to save your edits.

- Each line segment you digitize is stored as a separate record within the attribute table. You can open the attribute table for the YSURoads feature class to see the number of lines you have digitized and what their names are.

- Answer Question 6.1.

QUESTION 6.1 How many road line segments were digitized (and which roads do they represent)?

STEP 6.6 Heads-Up Digitizing of Polygons

- With the roads digitized, you will now digitize both the campus and non-campus buildings.

- From Create Features, choose the YSUBuildings feature class. You'll see that you have some new options in the Construction Tools, including Polygon and Rectangle. Given the different shapes of the building footprints you'll be digitizing, choose the **Polygon** option.

- When digitizing a polygon shape, click the crosshairs once at the spot you want to begin creating the starting point. As you click around the outline of building, you'll see that ArcMap will begin to show you the full shape of the polygon you're creating as a translucent colored overlay. Be sure to sketch all of the dimensions of the building. When you're done, double-click the mouse and the polygon will complete itself.

- More options for drawing the lines of polygons (such as digitizing at right angles, something that would be especially important when creating building footprints, or being able to create curved lines, which may be useful for the semicircular areas on Bliss Hall or the McDonough Museum of Art) are available from the Feature Construction toolbar. To access these choices, select the pull-down menu next to the **Trace** tool on the Editor toolbar.

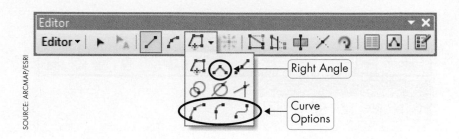

SOURCE: ARCMAP/ESRI

- You'll also want to change the colors of your polygon shapes to differentiate them from one another. This will help you to differentiate (for example) a YSU building from a non-YSU building or a parking lot.

- When digitizing a building, you'll want to zoom in very closely on the image to be able to see the building dimensions. You want to be close enough to the building so that you're not digitizing a crude blob (and doing a poor job in this chapter). The accompanying graphic shows the boundaries of Meshel Hall (a YSU building) at a good zoom level to digitize from. (You may want to zoom in even closer for some areas.)

SOURCE: ARCMAP/ESRI

- You can update the attributes with the names of the buildings just like you did with the names of the roads.

- Be careful which layer (YSUBuildings or NonYSUBuildings) you are digitizing features into.

- A few notes:
  - For our purposes, consider the M1 Parking Deck a YSU building.
  - You will not be digitizing any of the features of the Courtyard Apartments.
  - There are some non-YSU buildings shown in this area that are not listed on the parking map. Be sure to digitize them and add their names to their attributes. (If you're familiar with the area, be sure to use the proper names of the buildings, but if you're unfamiliar with the area, just use names like "non-campus building 1" or something similar to complete your digitizing.)

- Carefully check out which buildings are on the basemap to digitize. Save your edits as you go and especially when you're done. Answer Questions 6.2 and 6.3 when you've completed digitizing the buildings.

QUESTION 6.2 How many YSU campus building polygons were digitized, and which buildings are they? (Be careful to check one or more of the maps to be sure you have digitized all campus buildings.)

QUESTION 6.3 How many non-YSU campus building polygons were digitized, and which buildings do they represent? (Be sure you found and identified all non-campus buildings.)

STEP 6.7 Finishing Up Digitizing

- Now digitize the polygon shapes representing the parking lots into the Parkinglots feature class. Zoom in as you did while digitizing the buildings to make sure you accurately capture the lot dimensions. Update their attributes during the digitizing by completing the Lot_Name attribute (like M16 or V4). If the lot is not a YSU parking lot, give it a descriptive name, such as "Church Parking."

- Save your edits as you go and after you finish. Answer Question 6.4.

QUESTION 6.4 How many parking lot polygons were digitized, and which ones were they?

- Finally, digitize the lines making up the sidewalks in the area (you'll probably have to zoom in to see the walking paths for the area) and save these lines in the Sidewalks feature class. ***Important Note:*** With the sidewalks, all you have to do is digitize them, not update any attributes.

- Save your edits as you go and after you finish.

- When you are finished digitizing everything, select **Stop Editing** from the **Editor** pull-down menu on the **Editor Toolbar**. This will stop your edit session with layers and prompt you to save your work (if you haven't already done so).

- When you have completed all of your edits and digitizing, save all edits and save your ArcMap project.

STEP 6.8 **Measuring the Length of Digitized Objects**

• As noted in **Smartbox 35**, you're not simply drawing sketches. Because each object you digitize represents geospatial data, polygons measure a real-world area just as lines measure a real-world length. ArcGIS can compute the real-world measurements of your sketches through a process called "Calculate Geometry." In this chapter, you'll be computing the total length of all the sidewalk segments that you digitized.

• Right-click on the Sidewalks Feature Class in the TOC and select **Open Attribute Table**. You'll see a field called SHAPE_Length—this represents the real-world length of each digitized sidewalk segment in the feature class in the units used (i.e., feet).

   • *Important Note:* The geodatabase will store this information, but a shapefile will not. To find the length of a segment in a shapefile, you would have to use the Calculate Geometry option.

• What we'll do is compute the total distance in miles of all the digitized sidewalk segments. To begin, you'll need a new field in the attribute table to hold this data. From the **Table Options** pull-down menu in the attribute table itself, select **Add Field**.

SOURCE: ARCMAP/ESRI

• Give the new field the name of **Mileslength**.

• Select Float as the **Type**.

• Click **OK**. A new field called Mileslength will be added to the attribute table and will have null values for each entry.

• Next, start an edit session. Select **Start Editing** from the **Editor** pull-down menu on the **Editor** Toolbar.

• Back in the Sidewalks' attribute table, right-click on the name of the new Mileslength field and select **Calculate Geometry**.

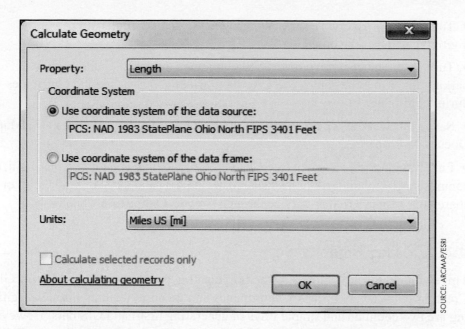

- Select **Length** for the Property for which you want to calculate geometry.

- For Coordinate System, use the **coordinate system of the data source**.

- For Units, select **Miles US [mi]**.

- Click **OK** when you're ready. If Calculate Geometry won't work properly, check Troublebox 6 for help.

Troublebox 6

Why doesn't the Calculate Geometry function work properly in ArcGIS 10.2?

Calculate Geometry will only work if the dataset you're trying to use it with is in a projected coordinate system. If your feature class (or shapefile) is still in GCS, you'll need to change it to a projected coordinate system for Calculate Geometry to work.

- Each entry in the Mileslength field will now be filled in with the length of that sidewalk segment in miles. To tally up all of the segment lengths, right-click on the name of the Mileslength field in the attribute table and select **Statistics**. In the new "Statistics of Sidewalks" dialog that appears, the Sum field will be the total value of all entries in the field. Answer Question 6.5 (and then close the "Statistics of Sidewalks" dialog box).

QUESTION 6.5 What is the total length (in miles) of all the digitized sidewalk segments?

STEP 6.9 Printing or Sharing Your Results

- Next, display your YSU buildings, non-YSU buildings, and parking lot polygons, with each category in a different color.

- Also display your roads and sidewalks, using different colors and line thicknesses.

- Turn off the imagery basemap and change the display so that all of your digitizing can be seen on the screen.

- Turn on the labels for the roads, YSU buildings, non-YSU buildings, and parking lots using the names you gave them when you assigned their attributes during digitizing.

- Save your work as a map document. Include the usual information in the Map Document Properties.

- Finally, either print a layout (see Chapter 3) of the final version of your digitized campus map (including all of the usual map elements and design for a layout) or share your results as a map service through ArcGIS Online (see Chapter 4).

Closing Time

This chapter described various methods of creating your own data to use in ArcGIS. Creating point layers from XY coordinates and creating features through digitizing are two common and simple ways of generating first-hand GIS data. However, these are not the only ways to create GIS data. Readings obtained using the Global Positioning System (GPS) can measure points and coordinates, and using a GPS receiver or a mobile version of GIS can allow you to directly measure points, lines, and polygons in the field. See the *Related Concepts for Chapter 6* for more about using mobile GIS and GPS to create geospatial data.

Even with careful digitizing or data input, some errors may creep into the process. For example, a road may be digitized at an incorrect length, a parking lot with the wrong shape, or a sidewalk incorrectly created in the roads layer. Chapter 7 will describe how to assess the quality of datasets you create (or obtain from others) as well as how to edit and fix possible errors in the data.

| **Related Concepts for Chapter 6** | Using GNSS and Mobile GIS for Data Creation |

A common method of data creation is to simply go out into the field and collect the data yourself. For example, a historian working on plotting out troop movements at Civil War battlefields would probably find little in the way of second-hand GIS data. Rather, she would likely end up on site on battlefield land trying to make measurements. Also, instead of trying to locate spots on a basemap or aerial imagery and digitize them, she would likely be identifying locations while out in the field. Much GIS data is created in this way and several tools are available to aid this process.

Many field data are collected via the **Global Positioning System (GPS)**, a series of satellites that broadcast signals to Earth. A GPS receiver can pick up these signals (from a minimum of three satellites, but commonly four or more) and use the information in them to determine the user's location on Earth's surface. These coordinates are given in GCS using the WGS84 datum, but often the device can translate the coordinates to another datum or projected coordinate system. Data collected from GPS are often used in other programs in the **GPX** data format, a standard readable by many software programs, including ArcGIS (which allows you to take data in GPX format and convert them to feature classes).

The name "GPS" refers to the Navstar GPS, which was set up by the United States Department of Defense and operated by the U.S. Air Force, but

Global Positioning System (GPS) A technology that uses signals broadcast from satellites for position determination on Earth.

GPX The standard format for data collected by a GPS receiver.

there are other satellite navigation systems available as well, including GLONASS (the Russian version of GPS), Compass (the Chinese version, currently in development), and Galileo (the European Union version, also in development). These satellite navigation systems are referred to with the blanket name of **Global Navigation Satellite Systems (GNSS)**.

GNSS receiver capability is a common function of mobile devices such as smartphones and tablets, and some GNSS receivers are capable of running GIS software. Esri products are also designed to run on mobile devices; for example, the **ArcPad** program is designed for field data collection combining GIS and GPS capabilities. Aerial imagery can be loaded onto ArcPad, and a GPS receiver can plot your location on the image. As you move, your positions can be saved directly into a shapefile, allowing you to create point, line, or polygon shapefiles based on your locations in the field. You can then edit these shapefiles or add them directly to ArcGIS (Figure 6.6) rather than having to convert or import data.

FIGURE 6.6 Using ArcPad for mobile field data collection.

For More Information

For further in-depth information about the topics presented in this chapter, use the ArcGIS Help feature to search for the following items:

- A quick tour of editing
- About creating segments
- About feature templates
- About geographic data formats
- About importing feature classes
- Adding x and y coordinate data as a layer
- An overview of working with feature datasets
- Calculating area, length, and other geometric properties
- Feature class basics
- GPX to Features (Conversion) (RC)
- Types of geodatabases
- What is a personal geodatabase?
- What is the Data Interoperability extension?

> **Global Navigation Satellite System (GNSS)** An overall term for the technologies that use signals from satellites for finding locations on Earth's surface.
>
> **ArcPad** An Esri software product for mobile devices that allows for GPS integration and field data collection.

Key Terms

digitizing (p. 137)

heads-up digitizing (p. 137)

geodatabase (p. 142)

personal geodatabase (p. 142)

file geodatabase (p. 142)

feature class (p. 142)

feature dataset (p. 142)

shapefile (p. 143)

coverage (p. 143)

Event Layer (p. 145)

segment (p. 153)

vertex (p. 153)

point digitizing (p. 153)

stream digitizing (p. 153)

Global Positioning System (GPS) (p. 160)

GPX (p. 160)

Global Navigation Satellite System (GNSS) (p. 161)

ArcPad (p. 161)

How to Edit Data with ArcGIS 10.2

⬛ Introduction

When you create GIS data, errors may creep into the data. Digitizing is often a time-consuming process that requires a lot of attention to detail, and it's easy enough to make mistakes when tracing building footprints or stream patterns from a basemap. Similarly, attribute information (such as a name or a number) can be entered incorrectly. GPS isn't perfectly accurate all the time and coordinates can sometimes be off by several meters. Incorrect projection or datum information, poor field conditions, or out-of-date basemaps can all lead to flawed final GIS data. All data-quality issues must be addressed before GIS data can be useful to you. This chapter looks at several issues of data quality in GIS; it also teaches you how to edit GIS data using the many editing tools available in ArcGIS.

The terms "accurate" and "precise" are often used interchangeably with regard to the quality of data and measurements. However, in terms of GIS data quality, they refer to two distinct things. **Accuracy** describes how closely a measurement reflects the actual value being measured. **Precision** describes how consistent or exact the measurements are, or the level of detail (or significant digits) used in making a measurement. For instance, if you're shooting arrows at the bull's-eye of a target and you strike the bull's-eye dead center, your shot was very accurate. If you put three more arrows tightly clustered around the first, your shooting was also very precise. However, if all of your shots hit far away from the bull's-eye, but they are still tightly bunched together, your shooting was not accurate but it was precise—you hit the wrong location but you did it with every shot (Figure 7.1). You can think of GIS data quality in the same way. If your digitizing of a hiking trail followed all of the curves and bends of the path, but you consistently placed the path 10 feet north of where the trail should actually be, your GIS line data would be precise but not accurate.

accuracy How closely a measurement matches up with its real-world counterpart.

precision How consistent or exact a measurement is.

FIGURE 7.1 Four "archery" results—(a) accurate and precise, (b) not accurate but precise, (c) accurate but not precise, (d) not accurate and not precise.

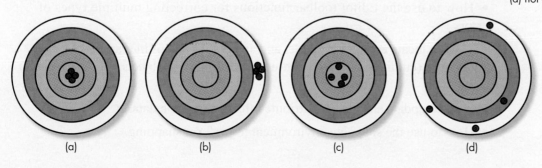

(a)   (b)   (c)   (d)

lineage The data sources and processes used for creating a dataset.

The metadata that accompanies a layer (see Chapter 5) should contain several metrics that will allow you to assess the quality of the data. One of these metrics is the data's **lineage**, which describes the original source used in creating the data, along with the steps and actions taken during the data-creation process. For instance, if streams were digitized from a 1992 USGS 1:100000 topographic map, that information would be contained in the lineage and would aid you in assessing the quality of the data layer itself. Metadata should also include five other quantitative measures of data quality: temporal accuracy, positional accuracy, attribute accuracy, completeness, and logical consistency. We'll examine each of these measures in this chapter.

Chapter Scenario and Applications

In this chapter, you'll continue in the role of an event planner for a university campus (the same role you had in Chapter 6). The new campus map has been completed (as in Chapter 6), but now it needs to be checked for accuracy; any errors must be fixed before the map can move to production. It's your job to assess the quality of the new campus map and fix any problems related to the roads, sidewalks, parking lots, and university buildings.

The following are additional examples of other real-world applications of this chapter's theory and skills.

SOURCE: MONTY RAKUSEN/CULTURA/GETTY IMAGES

- Public utilities offices need their datasets to be as up-to-date and error-free as possible. GIS analysts would use editing tools to fix any errors in infrastructure-related datasets.

- A historian receives data related to the locations of generals and troop movements during Civil War battles. She would want to edit the data to be sure that all spatial and non-spatial qualities are as accurate as possible before proceeding with her study.

- A park manager needs the online Web map of the park updated to reflect the opening of a new set of trails and the seasonal closing of others. He would use the GIS editing tools in ArcGIS to make the appropriate changes to the park data and then share the new map as a map service.

ArcGIS Skills

In this chapter, you will learn:

- How to use the Editor toolbar functions for correcting multiple types of errors.

- How to locate errors in a dataset and apply a variety of interactive editing techniques (cutting polygons, splitting lines, reshaping objects, moving objects) to fix them.

- How to update and change attribute information for a dataset.

- How to use the snapping environment to perform snapping.

Study Area

For this chapter, you will be examining a portion of the Youngstown State University (YSU) campus in Youngstown, Ohio.

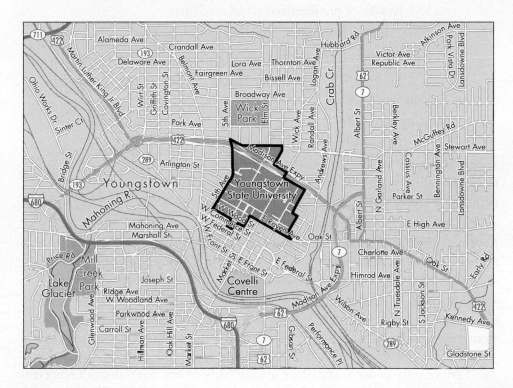

Data Sources and Localizing This Chapter

This chapter's data focus on features and locations within the YSU campus. However, you can easily modify this chapter to use data from your own campus or local area instead. Because you will be creating the data directly from the Imagery basemap, simply find your own campus area that you're familiar with on the basemap and digitize data for feature classes (roads, sidewalks, parking lots, campus buildings, and non-campus buildings) and edit these datasets.

STEP 7.1 Getting Started

- Start ArcMap and use the Catalog to copy the folder called **Chapter 7** from the C:\GISBookdata\ folder to your own D:\GIS\ drive. Be sure to copy the entire directory, not just the files within it.

- Chapter 7 contains the following items:

  - **Parkmap2012.pdf** – a map of the YSU campus in PDF format. You won't be able to see this in Catalog because it's just a PDF file. Open this file from Windows now to use as a reference for the chapter.

  - **Edittests** – a personal geodatabase containing four feature classes:

    - **Testbuilds** – a pre-digitized set of building polygons.

    - **Testcourts** – a pre-digitized set of tennis court polygons.

  - **Testlots** – a pre-digitized set of parking lot polygons.

  - **Testroads** – a pre-digitized set of road lines.

- **YSUEdits** – a personal geodatabase containing five feature classes:

  - **YSURoads** – a pre-digitized set of road lines.

  - **Sidewalks** – a pre-digitized set of sidewalk lines.

  - **Parkinglots** – a pre-digitized set of parking lot polygons.

  - **YSUBuildings** – a pre-digitized set of polygons of YSU building footprints.

  - **NonYSUbuildings** – a pre-digitized set of building footprints of buildings not owned by YSU.

- Add all four of the feature classes from the Edittests geodatabase to the Table of Contents and display them. (Leave the feature classes from the YSUEdits geodatabase alone for now—we'll get to them in Step 7.8.)

- Change colors and symbologies for the layers, so you're able to easily see them and distinguish them from one another (i.e., you should be able to easily see the differences among the three polygon layers). Some or all of these layers contain several editing errors that you will be correcting in this chapter.

- You'll also be adding a basemap to work with (see Chapter 5 or 6 for how to do this). Add the **Imagery** basemap; give ArcGIS a minute or two to stream it in from the Internet. The imagery for the campus area should load, centered around your four feature classes. Note that the imagery represents one snapshot in time. The Esri imagery of the Youngstown area is fairly recent, but it doesn't represent what the campus exactly looks like on today's date. See **Smartbox 36** for more information about how this time lag affects data-quality assessment.

Smartbox 36

What is temporal accuracy?

temporal accuracy The time period and currentness of a dataset.

The time period associated with the data you're working with is an indicator of the data's quality. **Temporal accuracy** is a measure of how current the data are, as well as the time period to which the data correspond. For instance, if you're digitizing a map of your campus with the goal of planning campus events, using current (or very recent) imagery as a source would provide very good temporal accuracy. However, if you were using an image from 1995 to digitize a campus map, it is likely that several changes have occurred since that year (such as new or expanded buildings or different road patterns) that would reduce the temporal accuracy. However, not every dataset requires present-day measurements. For example, land cover data (see Chapter 12) and digital elevation models (see Chapter 15) may have been created a few years ago; but because large-scale land cover or terrain changes so slowly, they will likely still be useful (and temporally accurate).

- *Important Note:* Please refer to the PDF of the YSU map for the proper location and orientation of objects. Figure 7.2 shows a visualization of the section of campus used in this portion of the chapter.

- You may also want to check one of the YSU campus maps (including the interactive campus map) available here: http://web.ysu.edu/gen/ysu/Directions_and_Virtual_Tour_m162.html.

- Zoom in to the area of campus that has the digitized feature classes. You'll be working with a section of the northern portion of campus, roughly bounded by Wick Avenue, Elm Street, and Service Road (see the accompanying graphic for the area and refer to the parking map PDF).

- You may also find it useful to make the polygon feature classes semi-transparent so that you can see the basemap imagery underneath the feature classes for editing purposes. To do so, right-click on the layer in the TOC and select **Properties**. In the **Layer Properties**, select the **Display** tab and type a new value (such as **50**) in the box next to **Transparent**. Click **Apply** and **OK** to see that your polygon layer is now semi-transparent (adjust the transparency value as you see fit). Do this for each polygon layer (Testlots, Testbuilds, and Testcourts).

- Finally, add the **Editor Toolbar** to ArcMap (see Chapter 6 for how to do this).

SOURCE: COURTESY OF YOUNGSTOWN STATE UNIVERSITY

FIGURE 7.2 A section of the YSU campus.

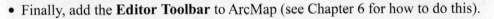

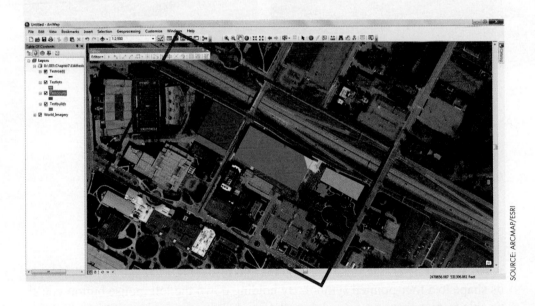

SOURCE: ARCMAP/ESRI

STEP 7.2 Starting, Stopping, and Saving the Editing Process

- To begin editing, choose **Start Editing** from the Editor Toolbar's **Editor** pull-down menu.

- You will use the Edit Tool (the edit cursor) on the Editor toolbar when performing editing:

SOURCE: ARCMAP/ESRI

- To save your edits at any time, choose **Save Edits** from the **Editor** pull-down menu.

- When you're done and want to stop editing, choose **Stop Editing** from the **Editor** pull-down menu.

STEP 7.3 Reshaping a Feature

- You can easily edit the shape and length of objects. For instance, examine the digitized version of Service Road—you'll see that its eastern edge does not extend to the intersection with Wick Avenue. To extend the length of the road, first click on it with the Edit Tool—you'll see that the line segment turns a default cyan color to indicate that it's now selected.

- Double-click on the line and you'll see the vertices that were created to give the line its shape.

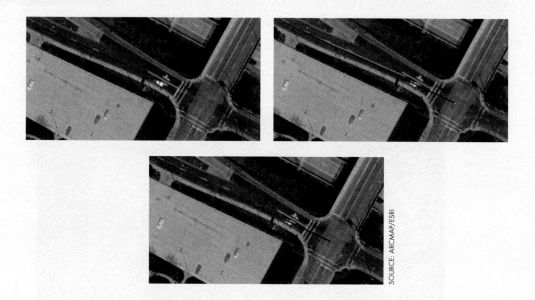

SOURCE: ARCMAP/ESRI

- Hover the cursor over the vertex on the end of the line. The cursor will change its shape to a four-pointed symbol. By holding down the left mouse button you'll be able to drag that vertex (and thus, the whole line connected to it) over to the intersection of Wick and Service. When you have the vertex positioned where you want it, release the mouse button and click somewhere else on the screen. The vertex (and thus the entire line segment) will be extended and moved to its new location. When you save your edits, the new length and location will be updated in the Testroads feature class.

- You can also use the Edit Tool to move a vertex and reshape a feature. For instance, the Watson and Tressel Training Sites (WATTS) building has not been digitized properly (one corner has not been digitized). Double-click the polygon so that the vertices of the feature are displayed.

- Place your cursor on one of the vertices at the missing corner and the cursor will change form again, allowing you to drag that vertex to a new location.

- Drag the vertex to the new location in the corner and depress the cursor button at the location to which you want to move the vertex.

- You will see the new and old locations displayed.

- Click the edit cursor somewhere else in the image. The polygon will move to the new location where the vertex has been placed and the polygon will reshape itself.

SOURCE: ARCMAP/ESRI

- Note that you can reshape a feature only by manipulating its vertices. Thus, you must sometimes add vertices to a feature before you can properly reshape it.

- To insert a vertex along a line to reshape it by adding more vertices, first double-click on the object to show its vertices. A new toolbar called Edit Vertices will appear on the screen.

SOURCE: ARCMAP/ESRI

- Select the button showing the edit cursor and the plus button, then click the location you want to add a vertex to. A new vertex will be added there.

- Alternatively, you can also right-click at the location on the object where you want to place a vertex and select **insert vertex** from the available options.

- To remove a vertex, select the button showing the edit cursor and the minus button, then click the vertex you want to remove.

- Remember: Whenever you reshape a feature, you're changing its spatial features. For more information about how reshaping affects the measurement of data quality, see **Smartbox 37**.

Smartbox 37

What is positional accuracy?

positional accuracy How closely the spatial features of a dataset match their real-world locations.

Positional accuracy is a measure of how closely the spatial features of a dataset match their real-world equivalents. For instance, if all of the digitized campus buildings are in their proper geospatial location, then the dataset has very high positional accuracy. Several factors can affect positional accuracy, including (a) the accuracy of the source map that data was derived from, or (b) the accuracy of the GPS used to obtain data from the field. If the initial source contained errors, then these errors will be carried through to the final product and affect the data's overall positional accuracy.

STEP 7.4　Cutting Polygons and Splitting Lines

- During editing, you can also split a polygon into two or more separate polygons. For instance, examine the polygon covering the tennis courts adjacent to the WATTS building. Turn the Testcourts layer on and off—you'll see that instead of two polygons (for each separate court complex) being digitized, only one polygon was created instead. Instead of reshaping the polygon down to cover one court complex and then creating a new second polygon, you could instead cut the single polygon into two smaller polygons.

- First, use the Edit Cursor to select the polygon object covering the tennis courts.

- Next, select the **Cut Polygons Tool** option from the Editor Toolbar.

SOURCE: ARCMAP/ESRI

- Select the location on one side of the polygon that you want to use as a starting point for cutting. Click the mouse at this location and you'll see a red marker appear.

- Next, use the mouse to draw a line across the length of the polygon to the other side. This line signifies where the two polygons will be cut.

- Finally, double-click to end the line on the other side of the polygon. The polygon will split into two smaller polygons along the boundary you just created. The two polygons can then be treated as separate objects. When you save your edits, you'll see that the Testcourts feature class now contains two records instead of just one.

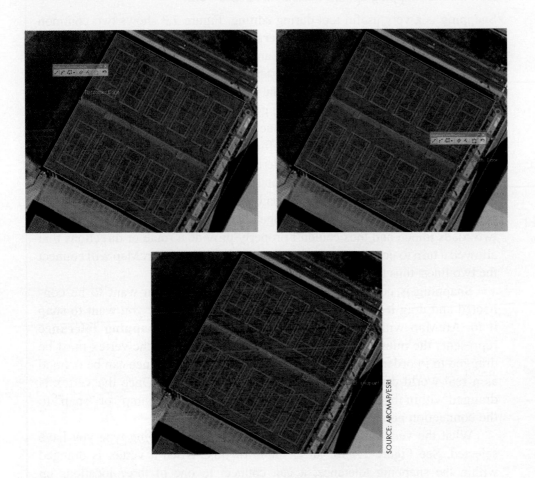

SOURCE: ARCMAP/ESRI

- You can also split lines into multiple segments. First, use the Edit Tool to select the line you want to split, and then choose the **Split** tool from the Editor Toolbar.

SOURCE: ARCMAP/ESRI

- After selecting the Split tool, click the selected line where you want to split it. The two lines can now be manipulated as separate objects.

STEP 7.5 Using Snapping for Editing

- **Snapping** (sometimes referred to as "node snapping" in GIS) is a useful editing technique for joining features together (such as connecting the ends of two lines). With snapping, you can connect lines to one another without having any dangling vertices or you can link new polygons to existing ones. See **Smartbox 38** for more information about the snapping process.

snapping When two vertices link together to become a single vertex.

Smartbox 38

How do the snapping options function in ArcGIS 10.2?

Snapping is a very useful tool during editing. Figure 7.3 shows two common types of digitizing errors: (a) an **undershoot**, in which a line doesn't extend far enough to an intersection, and (b) an **overshoot**, in which a line has been extended too far. While you can reshape the line easily enough to lengthen or shorten it, getting the line to the proper length to touch the intersection point would be very difficult. If these lines were not connected, there would be a substantial error in the dataset. For example, if these unconnected road lines were used in putting a network together, ArcGIS wouldn't see that the two roads joined and thus couldn't properly provide a route or directions that allowed a turn to occur at the intersection. With snapping, ArcMap will connect the two lines, thus leaving no gaps between them.

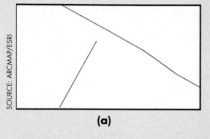

SOURCE: ARCMAP/ESRI

(a) (b)

FIGURE 7.3 Two roads, digitized with (a) an undershoot and (b) an overshoot.

Snapping is done manually. You grab the vertex you want to be connected and drag it within a certain distance of the feature you want to snap it to. ArcMap will then join the items together. The **snapping tolerance** represents the minimum distance around the feature that the vertex must be dragged to in order for snapping to occur. Snapping tolerance can be defined as a real-world distance or a certain number of pixels. Once the vertex is dragged within this distance, it will try to automatically "jump" or "snap" to the connection point.

What the vertex connects to is defined by what snapping type you have selected. See Figure 7.4 for an example of this. When a vertex is dragged within the snapping tolerance, it can connect to one of three locations on the line. It can link to (a) the endpoint of the line, (b) the edge of the line, or (c) another vertex along the line. Each of these will result in a different outcome,

undershoot When a vertex of a line falls short of its target location.

overshoot When a vertex of a line goes beyond its target location.

snapping tolerance How closely a vertex must be placed within a feature to activate snapping.

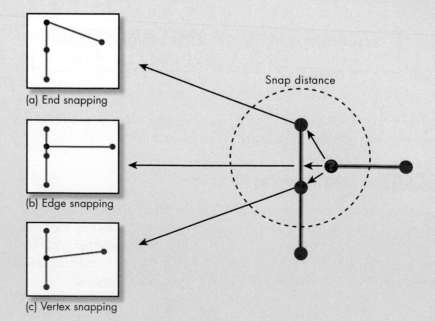

(a) End snapping

(b) Edge snapping

(c) Vertex snapping

Snap distance

FIGURE 7.4 Three different outcomes for snapping, depending on which snapping type is chosen.

and the lines will be connected differently depending on the option you have chosen. In ArcMap, you should first select the snapping type you need before dragging the vertex within the snapping distance. For instance, if the two roads were going to intersect perpendicularly, you should choose the "edge" option.

• Take a look at the intersection of Elm Street and Service Road. The line representing Elm Street should be extended to connect with the intersection at Service Road. Instead of just reshaping the Elm Street line, you want it to link together with the Service Road line. In this case, the vertex at the end of the Elm Street line can be snapped to the edge of the Service Road line.

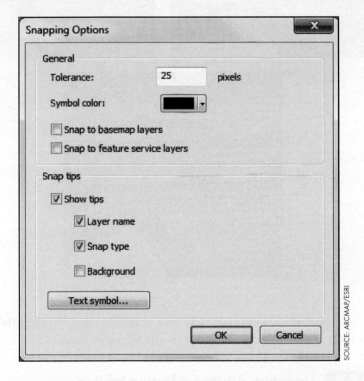

SOURCE: ARCMAP/ESRI

• First, set snapping tolerance from the **Editor** pull-down menu on the **Editor Toolbar** by selecting **Snapping**, then selecting **Options**.

• You can select the number of pixels that make up the snapping distance. To begin with, try 25 pixels—if this number turns out to be too small, you can return and change it to a larger number.

• To select which type of snapping (end, vertex, or edge) you're using, from the **Editor** pull-down menu on the **Editor Toolbar**, select **Snapping**, then select the **Snapping Toolbar**.

SOURCE: ARCMAP/ESRI

• Press the appropriate button on the toolbar for the type of snapping you want to perform (when the button has a blue box around it, that type of snapping is enabled). In this case, select the **Edge** button.

• Double-click on the Elm Street line to select it. Next, use the mouse to grab hold of the vertex at the end of the line and drag it within the distance you specified in the snapping tolerance. The vertex will "jump" to try and "snap" to the Edge option you selected. (It would "snap" to a point feature, the end of a line, another vertex, or along an edge, depending on which option you chose.)

SOURCE: ARCMAP/ESRI

• Click the mouse somewhere else and you'll see that the vertex at the end of the Elm Street line has snapped to the edge of the Service Road line.

STEP 7.6 Updating Attributes During Editing

• You can update or alter the attributes for a feature in a manner similar to how you originally created them in Chapter 6. For instance, use the Edit Tool to select the building polygon for Weller House, then select the **Attributes** button from the Editor toolbar:

SOURCE: ARCMAP/ESRI

• The Attributes box will open on the right-hand side of the screen. You'll see that the Name attribute of the building has been incorrectly called "Learning House" instead of "Weller House." Type the proper building name into the Name field and hit the Enter key on the keyboard. The building name is now correct, and when you save your edits, you'll see that the Name field in the Testbuilds' attribute table has been updated as well.

• Having correct attributes for each object is an important part of data quality. See **Smartbox 39** for more information about attribute accuracy.

Smartbox 39

What is attribute accuracy?

Just as the spatial features of a layer can contain errors, so can the non-spatial attributes. **Attribute accuracy** refers to how well the non-spatial parts of a dataset match their real-world equivalents. For instance, one of the attributes in the YSUBuildings layer is "Name." If all of the buildings in the layer have the same name as the actual campus buildings, then attribute accuracy (for that field) is very high. However, if building names are different or not up-to-date, then the attribute accuracy would be low. Attribute accuracy doesn't just apply to names, but to other non-spatial attributes as well. Land use codes, parcel ID numbers, and tax-assessment values are examples of other non-spatial data for which high attribute accuracy is very important.

attribute accuracy How closely the non-spatial features of a dataset match their real-world counterparts.

STEP 7.7 Deleting, Moving, and Creating Features

• During editing, you can easily delete or move features; you can also create new features. To delete a feature, select it with the Edit Tool, then right-click and select **delete** from the menu options. Alternately, once a feature is selected, you can press the delete key on the keyboard.

• To move a digitized point, line, or polygon, select it with the Edit Tool, drag it to a new location, and release it. The object will be moved and, when you save your edits, the object's new location will be saved in ArcGIS.

• Creating new features (to add missing points, lines, or polygons) is done in the same way as heads-up digitizing. Follow the steps from Chapter 6 to select the proper template and the correct construction tool to digitize any new features you may need.

• When editing, you may find that features that belong in one layer have been improperly created or placed in a different layer. For instance, in the feature classes you're working with now, you have two separate feature classes—one for parking lots and another for buildings. Take a closer look at the F6 parking lot—it's been incorrectly digitized as part of the Testbuilds feature class when that polygon should be in the Testlots feature class. The brute-force way of fixing this problem would be to delete the F6 parking lot polygon from the Testbuilds feature class and then create a new polygon in the Testlots feature class. However, ArcGIS provides an easier way of making this change, wherein you simply cut and paste objects from one feature class into another, as follows:

  • Use the Edit Tool to select the object you want to switch from one feature class to another. In this case, select the **F6** polygon.

  • Right-click on the **F6** polygon and select **Copy**.

  • In the Table of Contents, select the feature class you want the object to go into. In this case, select the **Testlots** feature class.

- From the **Edit** pull-down menu, select **Paste**.
- In the Paste dialog, select **Testlots** from the pull-down menu, then click **OK**.
- Now use the Edit Tool to move the F6 lot out of the way. You'll see a copy of the F6 polygon underneath it in its same place, but the new polygon will be from the Testlots feature class instead. Delete the F6 polygon that belongs to the Testbuilds feature class, and you will have an F6 polygon of the same spatial dimensions in the same location, but in the Testlots feature class.

SOURCE: ARCMAP/ESRI

- Ensuring that you haven't inadvertently added objects that shouldn't be there or left out objects that *should* be there is key to evaluating data quality. See **Smartbox 40** and **Smartbox 41** for more information related to these two types of data-quality assessment metrics.

Smartbox 40

What is completeness?

A dataset should be complete. That is, it should contain all the necessary features but no unnecessary data. In terms of data quality, **completeness** can be judged in two ways:

- **Errors of omission**: A dataset is not complete if features have been left out. For instance, if there are 55 buildings on campus, and the "buildings" feature class contains only 53 polygons, then at least two buildings on campus are not properly represented ("at least two" because some of the 55 polygons could possibly be in error) and errors of omission have occurred.

- **Errors of commission**: A dataset is not complete if additional features have been added in. For instance, if the campus suddenly has an additional building entry for "Stately Shellito Hall" (which doesn't really exist), then an error of commission has occurred. Errors of commission can also occur

completeness A measure of the wholeness of a dataset.

error of omission When items are left out of a dataset that should be there.

error of commission When extra items are added to a dataset that should not be there.

when some features are incorrectly included in the dataset. For example, while there may be only 55 official named campus buildings, there could be additional storage sheds, maintenance areas, or restroom facilities that could be captured as polygons in a "campus buildings" dataset but not really be considered "official" buildings in a campus directory or map.

Smartbox 41

What is logical consistency?

Another factor of data quality is **logical consistency**, which examines whether the same rules were followed throughout the creation of the dataset. For instance, if more than one person is mapping the trails of a metropark, is everyone following the same rules and methods of data creation and attribute coding? Logical consistency is commonly related to how topological rules are implemented throughout a dataset (see the *Related Concepts for Chapter 7* section at the end of this chapter for more information about topology).

logical consistency A measurement for examining if the same rules were used throughout a dataset.

STEP 7.8 Locating and Editing Errors in a GIS Dataset

• At this point, you can remove the Edittests geodatabase from the TOC. In its place, add the five feature classes from the YSUEdits geodatabase. Change their symbology so that each layer is easily distinguished from the others. You may find it useful to make the polygon layers semi-transparent (like you did to the Edittests data in Step 7.1). You now have the same area that you digitized in Chapter 6, except that intentional errors have been added into the various feature classes.

• Zoom in to the same area you digitized in Chapter 6—you'll be working with the southern portion of campus, the block bounded by Wick Avenue, Walnut Avenue, E Spring Street, and Rayen Avenue (see the accompanying graphic for the area and refer to the parking map PDF).

SOURCE: ARCMAP/ESRI

- Within the five feature classes, there are 11 notable errors in the digitizing. Note that these are *significant* errors with the attributes, polygons, and lines, not just minor or insubstantial errors.

- Describe what the problem was and how you fixed it for each of the errors (e.g., "Wick and Lincoln lines were overshot—fixed by snapping nodes together," or "YSU building was digitized as a parking lot—switched the building into the parking lots feature class," etc.). For the significant errors for which you're looking:

  - six are polygon errors.

  - two involve overshoots or undershoots (and will need snapping to fix).

  - one is some other type of line error.

  - two are attribute errors.

- Some hints to keep in mind when looking for errors to fix:

  - The M1 parking deck should be a YSU Building, not a parking lot.

  - The bridge going over Wick Avenue and its related features are not YSU buildings, non-YSU buildings, or parking lots.

  - The road that loops around in front of the Courtyard Apartments is not intended to be digitized because the whole road does not fit into our study area.

  - The Courtyard Apartments are not part of the study area either.

  - For our purposes, the drive-up loop in front of McDonough Museum is not a parking lot.

  - Roads will be named or digitized as roads, not entrances into parking lots.

- ***Important Note:*** Please refer to the PDF of the YSU map for the proper location and orientation of objects. Figure 7.5 shows a visualization of the section of campus used in this chapter.

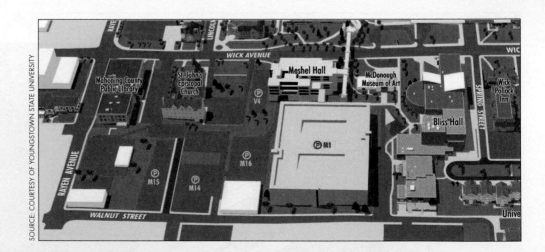

FIGURE 7.5 A section of the YSU campus.

SOURCE: COURTESY OF YOUNGSTOWN STATE UNIVERSITY

- Note that some of the features that appear on the paper map do not appear on the aerial basemap, and vice versa, so stick to what's on the basemap for the location and dimension of objects.

- You may also want to check one of the YSU campus maps (including the interactive campus map) available here: http://web.ysu.edu/gen/ysu/Directions_and_Virtual_Tour_m162.html.

- You may also find the "Birds eye view" options on Bing Maps available here: http://www.bing.com/maps. It is very useful for looking at oblique imagery (from four different angles) of the YSU campus when doing your editing.

STEP 7.9 **Printing or Sharing Your Results**

- Save your work as a map document. Add the usual information to the Map Document Properties.

- Turn off the imagery basemap. Display your final versions of the YSU buildings, non-YSU buildings, and parking lot polygons all in different colors. Also, display your roads and sidewalks in different colors and line thicknesses.

- Turn on the labels for the roads, YSU buildings, non-YSU buildings, and parking lots using the name of each for the labeling.

- Finally, either print a layout (see Chapter 3) of your final version of your edited campus map (including all of the usual map elements and design for a layout) or share your results as a map service through ArcGIS Online (see Chapter 4).

Closing Time

The chapter examined the basics of editing data in ArcGIS as well as concepts surrounding data quality. With the widespread usage of geospatial technologies and GIS, using high-quality data is essential to ensuring that the results of analyses or models are reliable. Error-filled or low-quality data will likely produce inaccurate or unreliable products. For example, county auditors use GIS to examine things like property boundaries, land valuations, and property-tax assessments. If the geospatial features (such as the dimensions of a land parcel) or the non-spatial features (such as the zoning code) are incorrect, numerous problems can arise.

How objects in a dataset connect and relate to one another is an important aspect of GIS data. When editing, rules for the types of connections that exist between features (independent of their coordinates) can be used in the editing process as well. See *Related Concepts for Chapter 7* for more about these topological relationships and how they're used in GIS.

Starting in the next chapter, we're going to be looking at how all of these geospatial objects relate to one another. For instance, once we have a well-digitized and edited map of campus, we can begin using that data for analysis, such as measuring the distances between buildings, the proximity of parking lots to buildings, and how many potential students live within a mile of campus.

Related Concepts for Chapter 7 Topology and Topological Editing in ArcGIS 10.2

When working with vector data, it's important that the GIS understands that the various pieces of data are connected to one another independent of their coordinates. For instance, when two lines that represent roads are digitized, the GIS

topology How objects relate or connect to one another independent of their coordinates.

geodatabase topology A set of rules applied to the feature classes of a feature dataset in order to remove errors.

needs to know that there's an intersection at the place where those two roads cross. Visually, we can easily see two lines crossing at a point, but the GIS needs to be told explicitly that there's an intersection there; this information is critical for applications that provide vehicle routing. Similarly, when two land parcels are digitized side by side, they share a common polygon segment between them as their boundary. Again, we can easily see this boundary, but the GIS needs explicit information telling it that these two polygons are adjacent. **Topology** refers to how objects relate or connect to one another independent of their coordinates. The three factors that topology establishes for data are:

• Adjacency: How one polygon relates to another polygon, such as a common property boundary.

• Connectivity: How two lines connect or intersect, as in the intersecting roads example.

• Containment: How all locations are situated within a polygon boundary.

In ArcGIS, a separate item can be created for a geodatabase feature dataset, simply referred to as a "topology" (sometimes called a **geodatabase topology**). In this case, a topology is a set of rules that can be applied to the feature classes within the feature dataset. This set of rules can be used to establish connectivity, adjacency, or containment for the objects in those feature classes. For example, a topology rule may state that no polygons may overlap (and an error will be generated whenever two or more polygons overlap), or that lines may not overlap themselves (and errors will be flagged wherever this occurs). See Figure 7.6 for examples of some topology

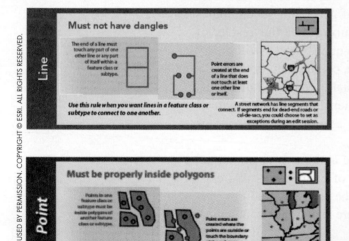

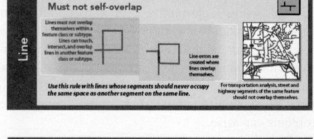

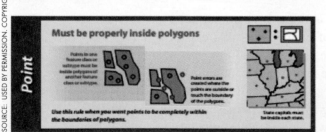

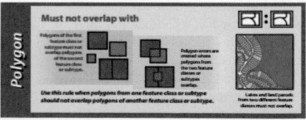

FIGURE 7.6 Examples of four geodatabase topology rules available in ArcGIS 10.2.

rules. For a complete list of the 32 available geodatabase topology rules, see the "topology rules poster" available in PDF format via the Resource Center here: http://resources.arcgis.com/en/help/main/10.2/01mm/pdf/topology_rules_poster.pdf. Different rules may be applied to different feature classes, so a line feature class representing walking paths may be allowed to have dangling nodes, while a line feature class representing paved driving paths may not.

Once one or more rules are chosen for a feature class, those rules are then applied to the objects within that feature class (in what is referred to as the *validation process*).

For instance, if the "must not have dangles" topology rule was chosen for a line feature class, all line segments within a certain distance of one another (a value referred to as the *cluster tolerance*) would be automatically edited to remove the dangling vertices. The lines will usually be extended or trimmed accordingly and the dangling vertex snapped to the nearest line or vertex.

Often, however, there will be certain vertices that are not within the cluster tolerance. These will have to be located and edited by hand. A set of tools is available within the **Topology toolbar** (available by selecting the **Customize** pull-down menu and choosing **Toolbars**) that will inspect the entire feature class. Any lines and any vertices that violate the rule will be highlighted. You can then visit each of these in turn, manually selecting how each will be edited and adjusted to comply with the rule. You can also mark a rule violation as an exception to the rule and leave it as is. See Figure 7.7 for an example of using topology rules to edit a feature class. The vertices highlighted in red are those that do not follow the "must not have dangles" rule. The vertex in black has been selected for editing to make it comply with the rule, and a set of possible actions is given to the editor. Note that these geodatabase topology functions are available only in ArcGIS 10.2 Standard and ArcGIS 10.2 Advanced, not in ArcGIS 10.2 Basic.

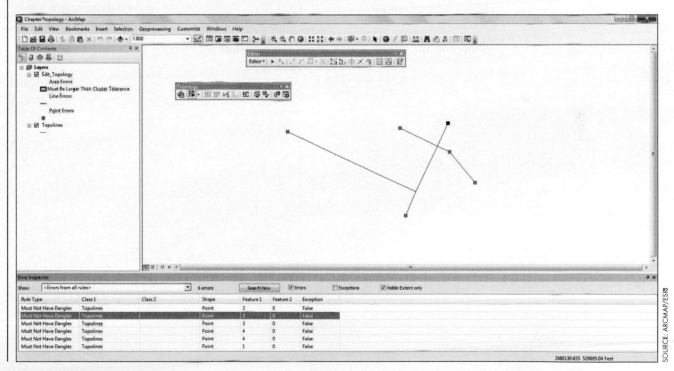

FIGURE 7.7 Using a topology rule for editing a dataset.

For More Information

For further in-depth information about the topics presented in this chapter, use the ArcGIS Help feature to search for the following items:

- About edit sessions
- About snapping
- An overview of topology in ArcGIS
- Creating a map topology

- Enabling snapping
- Exercise 4b: Using geodatabase topology to fix line errors
- Geodatabase topology rules and topology error fixes
- Illustrated guide to complete FGDC metadata (RC)
- Topology basics
- Topology in ArcGIS

Also, for more information about data quality issues and topology, see: Jensen, J., and Jensen, R. 2012. *Introductory Geographic Information Systems.* Boston: Pearson, Chapters 4 and 5.

Key Terms

accuracy (p. 163)

precision (p. 163)

lineage (p. 164)

temporal accuracy (p. 166)

positional accuracy (p. 170)

snapping (p. 171)

undershoot (p. 172)

overshoot (p. 172)

snapping tolerance (p. 172)

attribute accuracy (p. 175)

completeness (p. 176)

error of omission (p. 176)

error of commission (p. 176)

logical consistency (p. 177)

topology (p. 180)

geodatabase topology (p. 180)

How to Perform Spatial Analysis in ArcGIS 10.2

Introduction

Using GIS to work with geospatial data means more than simply making a map or performing a query to obtain attribute information about certain features. When you have two or more layers of data, you will usually use GIS to answer questions about how they relate to one another spatially. For example, these are some spatial questions that a realtor might ask:

- "What school district is this house in?"

- "How many rental properties are within one mile of this house?"

- "Which waste-water treatment plant is closest to this house?"

A law-enforcement officer might ask these spatial questions:

- "How many robberies have occurred within a one-mile radius of this location?"

- "How many registered sex offenders are living within 500 feet of this school?"

- "How many 911 calls have been made from each neighborhood within the city?"

Each professional can answer all of these questions by using GIS to examine how features relate to one another with regard to their spatial locations or their proximity.

Spatial analysis refers to examining the spatial characteristics of features and how they relate to one another across distances. In many ways, spatial analysis lies at the heart of GIS; once you have geospatial representations of the real world's features or phenomena, you can begin to examine how they interact. For example, in Chapter 2, we discussed how to join two tables together on the basis of a common attribute. In this chapter, we'll examine how to join tables not on the basis of their attributes, but rather on the basis of their features' spatial locations.

Similarly, in Chapter 2 we explained how to build a query to select records from an attribute table and how to extract their corresponding features to their own layer. This type of query isn't unique to the analysis of geospatial data—many types of database or spreadsheet programs could perform the same task. What sets GIS apart is its ability to perform a **spatial query** that selects features based not on their attributes, but rather on the basis of their location. For instance, see Figure 8.1, which shows roads and airports of northern Ohio. The records that comprise the length of Interstate 80 have been selected (as seen in cyan). A spatial query has been

spatial analysis Examining the characteristics or features of spatial data, or how features spatially relate to one another.

spatial query Selecting records or objects from a layer based on their spatial relationships with other layers rather than their attributes.

posed to ArcGIS asking "Which airports are within five miles of Interstate 80?" The airports that match this spatial query are shown in the default cyan color used by ArcGIS to denote selected features. In this chapter, we'll use ArcGIS to answer these kinds of spatial questions.

FIGURE 8.1 The results of a spatial query asking "Which airports in Ohio are within five miles of the selected features of Interstate 80?"

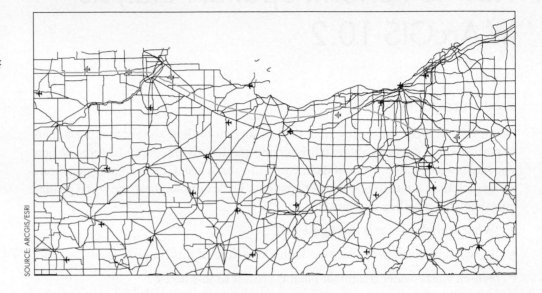

SOURCE: ARCGIS/ESRI

Chapter Scenario and Applications

In this chapter, you will take the role of a developer working for the state of Ohio. You are looking to find viable places for construction of a new airport. You'll first conduct an assessment of the locations and distributions of regional and international airport complexes in the state (for purposes of this chapter, you'll not be working with smaller airstrips or runways). Next, you'll identify the Ohio counties with populations that are sufficient to justify the expense of a new airport. You'll then look at the places in these counties (a "place" in this sense is an incorporated or non-incorporated location within the state) to find those that are near a major road (such as an interstate or U.S. highway) but are also sufficiently far enough away from an existing airport complex (so the new site won't be in competition with another airport).

The following are additional examples of other real-world applications of this chapter's skills:

• A law-enforcement officer is beginning to investigate geographic patterns of crime. As a first step, she wants to know how many houses made 911 calls in proximity to a certain location. She can quickly determine the answer to this question by performing a Select By Location query.

• An urban planner has two different geospatial layers, one of housing locations and another of U.S. Census blocks. He wants to determine the number of abandoned housing units within each census block. He can perform a spatial join of the two layers to determine this information.

SOURCE: PHOTO BY OLI SCARFF/GETTY IMAGES

• A researcher for the Centers for Disease Control and Prevention is provided with a map of reported dead bird locations within a county. She wants to locate the closest house to each of these places for potential disease monitoring. A spatial join of the two layers will allow her to conduct her analysis.

ArcGIS Skills

In this chapter, you will learn:

• How to make distance measurements with the Measure tool.

• How to perform two different types of spatial joins.

• How to use the Select by Location database query tool.

• How to perform basic spatial analysis with the different ArcGIS tools.

Study Area

For this chapter, you will be working with data for counties, places, and airport complexes for the state of Ohio.

Data Sources and Localizing This Chapter

This chapter's data focus on datasets from within the state of Ohio. However, you can easily modify this chapter to use data from your own state instead. For instance, if you were going to perform this chapter's activities in California, the same types of data would be available. The counties' dataset was downloaded (for free) and extracted from the Esri USA Counties dataset available on ArcGIS Online (and also available for download here: http://www.arcgis.com/home/item.html?id=a00d6b6149b34ed3b833e10fb72ef47b). By querying the USA Counties layer for "STATE_NAME = 'California'" you can select all California counties and export them to a new feature class.

The roads and places (incorporated and unincorporated) datasets were downloaded and extracted from The National Map and are available for all states. Use the Trans_RoadSegment feature class as the source for your state's roads, and the GU_IncorporatedPlace and GU_UnincorporatedPlace feature classes for your state's places. The airports data were also downloaded and extracted from The National Map (use the TransAirport_Point feature class). Only those airports labeled as complexes (use Ftype = 'Complex') that were international or regional airports (use Airport_Class = 'Regional Airport' or 'International Airport') were separated out for use in this chapter.

STEP 8.1 Getting Started

• Start ArcMap and use the Catalog to copy the folder called **Chapter8** from the C:\GISBookdata\ folder to your own D:\GIS\ drive. Be sure to copy the entire directory, not just the files within it.

• Chapter8 contains a file geodatabase called **Ohioselect** that contains the following feature classes:

  • Ohiocountiesproj: a polygon feature class of the 88 counties of Ohio.

  • Ohiomajrdsproj: a line feature class of the major roads in Ohio.

  • Ohioairportsproj: a point feature class of international and regional airport complexes in Ohio.

  • Ohioplacesproj: a polygon feature class of Ohio's incorporated and non-incorporated places.

• Add the Ohiocountiesproj layer to the TOC.

• The Ohiocountiesproj layer (as well as the other layers being used) was projected from its initial coordinate system to another one for use in this chapter. The Coordinate System being used in this chapter is **WGS 84 Web Mercator (auxiliary sphere)**, using Map Units of **meters**. Check the Data Frame to verify that this system is being used.

STEP 8.2 Making Basic Measurements

• Add the Ohioairportsproj and Ohioplacesproj layers to the TOC and arrange them so that they lie on top of the counties and you can clearly see all three layers (changing the symbology as necessary).

• To begin, zoom in on Fulton and Lucas counties in northwest Ohio:

• There is an airport in each county—Fulton County Airport (in Fulton) and the Toledo Express Airport (in Lucas). Use the Identify tool to determine which airport is which.

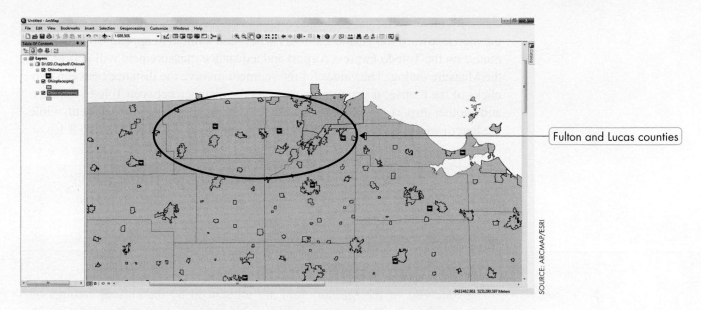

Fulton and Lucas counties

SOURCE: ARCMAP/ESRI

- The Measure tool in ArcGIS allows you to make basic measurements of features or the distance between features. To answer a simple question such as "How far is Fulton County Airport from Toledo Express Airport?" you could use the Measure tool. Choose it from the icons on the Tools toolbar:

SOURCE: ARCMAP/ESRI

- In the Measure tool dialog box, you can first select the type of measurement you'll be making. The default is Planar (which will make measurements using the projected coordinate system of the view), but we'll want to use Geodesic, which will be a "real-world" spherical distance (like how an airplane would fly).

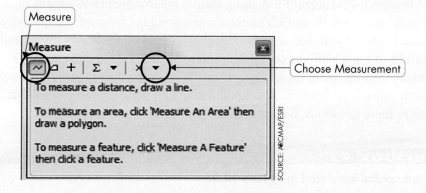

Measure

Choose Measurement

SOURCE: ARCMAP/ESRI

- From the **Choose Measurement Type** pull-down menu, select **Geodesic**.

- Next, select the **Measure Line** button. The cursor will turn into a measuring tool with a crosshairs. You will also see a concentric circle appear around the point, in essence "anchoring" the crosshairs to one point to allow you to more easily make measurements. If this circle option does not appear when you click with the Measure tool, see Troublebox 7 for how to activate it. Zoom in closely, so you can see both airports, and click the crosshairs on the Fulton County Airport. Drag the mouse toward the Toledo Express Airport and a line will appear. Click the mouse on the Toledo Express Airport and a distance measurement will appear in the Measure dialog. The values for the segment involve the distance between two clicks of the mouse; if you were to measure the distance between Toledo Express and another airport, that new distance would become the value for segment, while the length would be the sum of all segments combined. Answer Question 8.1.

QUESTION 8.1 What is the geodesic distance between Fulton County Airport and Toledo Express Airport?

Troublebox 7

Why doesn't the Measure tool snap to a point?

The functions of the Measure tool are influenced by the current settings of the snapping environment (see Chapter 7 for more about using snapping in ArcGIS). To set up the snapping options, from the **Customize** pull-down menu, select **Toolbars**, then select **Snapping**. The Snapping toolbar will appear. On the Snapping toolbar, choose the Point option (so that it has a blue square around it; see Figure 8.2). You can close the Snapping toolbar after doing this. Now, when you use the Measure tool to start measuring from a point, you'll see your measurements begin snapped to a point and then will also end snapped to a point.

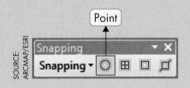

SOURCE: ARCMAP/ESRI

FIGURE 8.2 Choosing the point option of the Snapping toolbar.

• Close the Measure dialog box when you're done.

STEP 8.3 Using Spatial Joins for Analysis

• The Measure tool is useful for making simple measurements of lengths or areas, but several tools in ArcGIS perform more complex types of spatial analysis. For instance, part of your scenario for this chapter is to assess the locations and distributions of Ohio's airport complexes and to answer spatial questions like "How many airports are in each county?" or "What is the closest airport to each population center?" In ArcGIS, you could use spatial joins as a way to analyze and answer these questions. See **Smartbox 42** for more about spatial joins.

Smartbox 42

How are spatial joins used in ArcGIS 10.2?

In Chapter 2, we discussed how two attribute tables are joined together: One of the table's fields will be appended to the other table on the basis of both tables sharing a common field. This is a good method for linking non-spatial data to geospatial features, but you can perform other kinds of joins as well. In a **spatial join**, the attributes of one layer are joined to another layer not because they have a common field, but rather on the basis of their geospatial locations.

Spatial joins are very versatile techniques for spatial analysis. Each of the three vector object types (points, lines, and polygons) can have their attributes spatially joined to any of the other types (i.e., points may be joined to polygons,

spatial join Appending attributes from one layer to another based on the spatial relationship of the layers.

lines may be joined to points, and so on). Four different types of spatial joins can be performed: *is within*, *contains*, *intersects*, and *closest distance*. These can be used with any of the three vector objects, although some types of joins are just not possible (for example, you can't compute how many polygon features are within a point layer or how many lines contain polygons).

For example, as part of a study of Ohio's higher education facilities, you want to know which county each college is in. You have a point layer showing the name and location of each university and a separate polygon layer of Ohio counties that contains the name, population, and spatial dimensions of all 88 counties. These two layers don't have a common field, so we can't use the normal joining of the counties' attribute information to the records of the colleges' layer (as we did in Chapter 2). What we *can* do is perform a spatial join that will append the attributes of each county to the college that lies within the county's boundaries. The end result will be a new point layer (of colleges), where each record now has a county name and population matched to it—this matching has occurred on the basis of the spatial relationship between college point locations and county polygon boundaries (see Figure 8.3). This result acts like a many-to-one join

FIGURE 8.3 The result of performing a spatial join to append the counties' polygon layer attributes to the colleges' point layer attribute table.

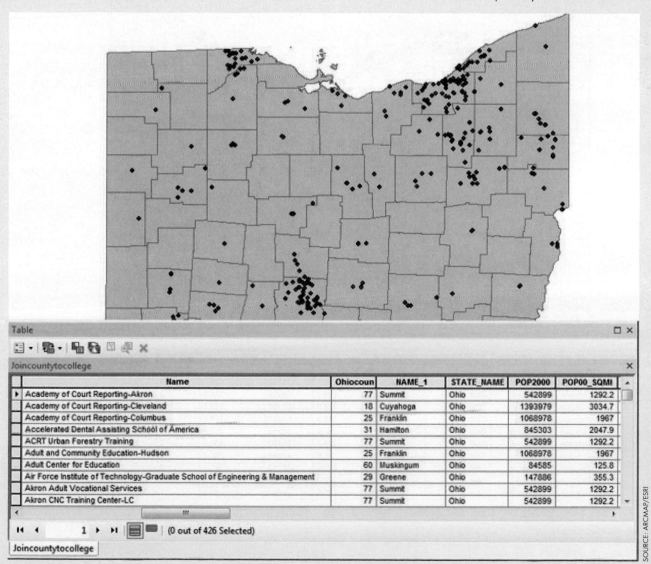

SOURCE: ARCMAP/ESRI

(see Chapter 2), in which a county's attributes could potentially be matched up with several different universities, if more than one university lies within a county.

You could also reverse the direction in which you join layers to end up with a different kind of result. For instance, if you wanted to know how many colleges are in each county, you could perform a spatial join by which a summation of the colleges' points that are contained inside the county boundaries is calculated for each county, and this value is appended to the counties' polygon layer's attribute table as a new field (which ArcGIS 10.2 will call "Count"). By performing this action, you are joining information about the points' attributes to the polygons' attributes in the form of a statistic computed about the points' attributes. The result is a new polygon layer with a new field called Count appended to it (see Figure 8.4); Count represents the sum of the points found within the polygon boundaries.

Another type of spatial join will join attributes of one layer to another based on the proximity of the closest feature in the layer to be joined. For example, if you wanted to know which road was closest to each of the colleges,

FIGURE 8.4 The result of performing a spatial join to append a summation (Count) field of the number of features from the colleges' points layer that fall within the boundaries of each polygon of the counties layer.

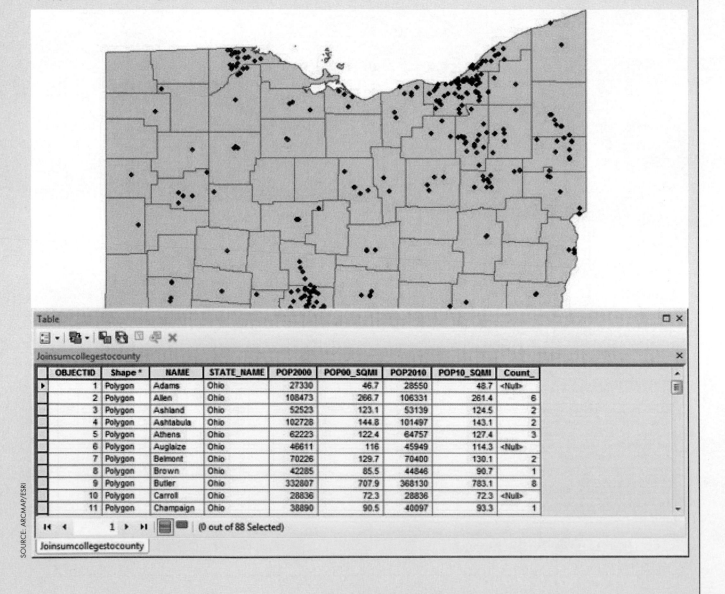

	OBJECTID	Shape*	NAME	STATE_NAME	POP2000	POP00_SQMI	POP2010	POP10_SQMI	Count_
▶	1	Polygon	Adams	Ohio	27330	46.7	28550	48.7	<Null>
	2	Polygon	Allen	Ohio	108473	266.7	106331	261.4	6
	3	Polygon	Ashland	Ohio	52523	123.1	53139	124.5	2
	4	Polygon	Ashtabula	Ohio	102728	144.8	101497	143.1	2
	5	Polygon	Athens	Ohio	62223	122.4	64757	127.4	3
	6	Polygon	Auglaize	Ohio	46611	116	45949	114.3	<Null>
	7	Polygon	Belmont	Ohio	70226	129.7	70400	130.1	2
	8	Polygon	Brown	Ohio	42285	85.5	44846	90.7	1
	9	Polygon	Butler	Ohio	332807	707.9	368130	783.1	8
	10	Polygon	Carroll	Ohio	28836	72.3	28836	72.3	<Null>
	11	Polygon	Champaign	Ohio	38890	90.5	40097	93.3	1

Table

Joinsumcollegestocounty

(0 out of 88 Selected)

Joinsumcollegestocounty

you could join the attributes of the roads' lines layer to the colleges' points layer. Performing this spatial join will determine which road line is the closest distance to a college point. ArcGIS 10.2 will then append the record (and its attributes) for that road line to the record for the point to which it is closest. It will also add a new field (called "Distance") to the table that represents the straight-line distance from each point to its closest road (using the units of the coordinate system of the features). The end result will be a new point layer with the attributes of the closest road appended to each record (see Figure 8.5).

FIGURE 8.5 The result of performing a spatial join to append the attributes of the feature of the roads layer closest to the features of the colleges' points layer, along with the computed straight-line distance of the road feature closest to each point feature (the Distance field).

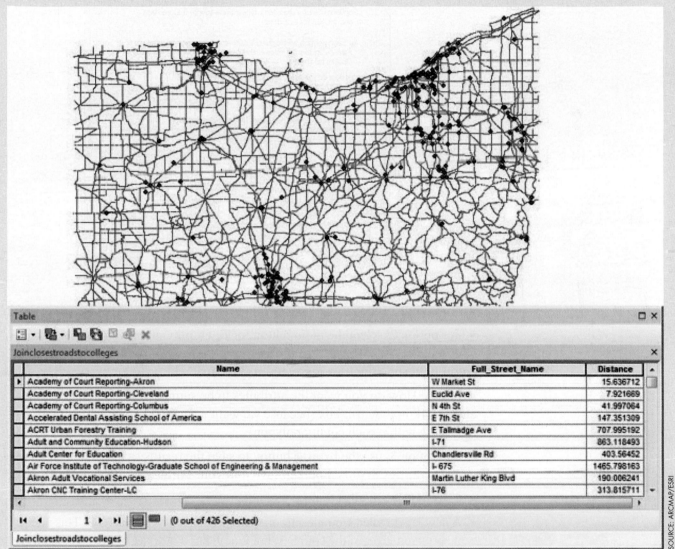

Table

Joinclosestroadstocolleges

Name	Full_Street_Name	Distance
Academy of Court Reporting-Akron	W Market St	15.636712
Academy of Court Reporting-Cleveland	Euclid Ave	7.921669
Academy of Court Reporting-Columbus	N 4th St	41.997064
Accelerated Dental Assisting School of America	E 7th St	147.351309
ACRT Urban Forestry Training	E Tallmadge Ave	707.995192
Adult and Community Education-Hudson	I-71	863.118493
Adult Center for Education	Chandlersville Rd	403.56452
Air Force Institute of Technology-Graduate School of Engineering & Management	I-675	1465.798163
Akron Adult Vocational Services	Martin Luther King Blvd	190.006241
Akron CNC Training Center-LC	I-76	313.815711

|◄ ◄   1 ► ►| (0 out of 426 Selected)

Joinclosestroadstocolleges

• The first spatial join we'll perform will examine how many airport points are within each of the polygon county boundaries. Because we want to join the airport points to the county polygons, right-click on the **Ohiocountiesproj** layer and select **Joins and Relates**, then select **Join**.

• From the **What do you want to join to this layer?** pull-down menu, select **Join data from another layer based on spatial location**.

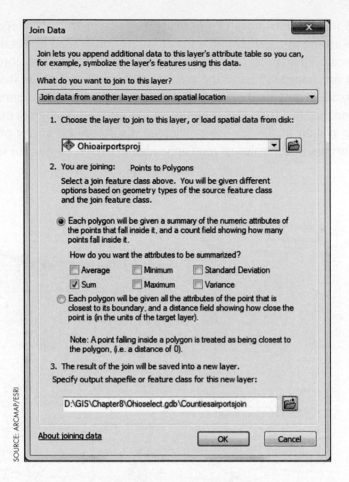

SOURCE: ARCMAP/ESRI

- For option 1 (choose the layer to join to this layer), select **Ohioairportsproj**.

- For option 2, select the radio button for **Each polygon will be given a summary of the numeric attributes of the points that fall inside it, and a count field showing how many points fall inside it.**

- For the How do you want the attributes to be summarized option, place a checkmark in the **Sum** box (because we want a sum total of how many airport points are within each county polygon).

- For option 3, use the **browse** button to save the results as a new feature class called **Countiesairportsjoin** and save it in your **Ohioselect** geodatabase.

- Click **OK** when all settings are correct.

- A new layer called Countiesairportsjoin will be added to the TOC. This is the same polygon layer of the counties, but it now has a new set of fields joined to it, summarizing the fields of the Ohioairportsproj layer based on their spatial locations.

- Open the attribute table for Countiesairportsjoin and scroll to the far-right side of the table to see these new fields. The field called count represents the sum of how many points (the airports) are within each polygon (the counties). Answer Questions 8.2 and 8.3.

QUESTION 8.2 How many airport complexes are in Cuyahoga County?

QUESTION 8.3 How many airport complexes are in Mahoning County?

- We'll now do a second spatial join to answer the question "For each populated place, what is the closest airport complex (and how far away is that complex)?"

- Since you'll be spatially joining information about the airports to the places layer, right-click on **Ohioplacesproj** and choose **Joins and Relates**, then select **Join**.

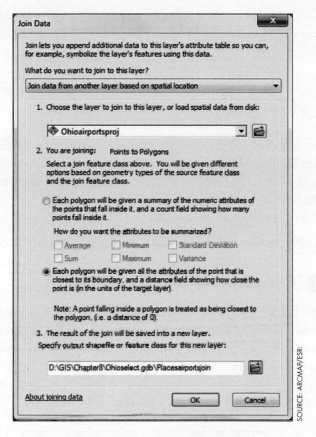

SOURCE: ARCMAP/ESRI

- From the **What do you want to join to this layer?** pull-down menu, select **Join data from another layer based on spatial location.**

- For option 1 (choose the layer to join to this layer), select **Ohioairportsproj**.

- For option 2, select the radio button for **Each polygon will be given all the attributes of the point that is closest to its boundary, and a distance field showing how close the point is (in the units of the target layer).** This will use the "closest" option for the spatial join; and because the Ohioplacesproj layer has been projected, this distance will be computed in meters.

- For option 3, use the **browse** button to save the results as a new feature class called **Placesairportsjoin** and save it in your **Ohioselect** geodatabase.

- Click **OK** when all settings are correct.

- A new layer called Placesairportsjoin will be added to the TOC. This is the same polygon layer of the places, but it now has a new set of fields joined to it; these fields deal with the distance of each place to the closest point of the Ohioairportsproj layer, based on their spatial locations.

- Open the Placesairportsjoin layer and scroll to the far-right side of the table to see these new fields. The fields from the Ohioairportsproj layer that correspond with the spatial locations of the places will be joined for each record in the places layer. A new field called Distance will be added to the end; this is the computed distance (in meters) that each populated place is away from the closest airport complex. Answer Questions 8.4, 8.5, and 8.6.

> **QUESTION 8.4** If you lived in Uniopolis, Ohio, what airport complex would be the closest to you and how far away would it be (give this value in miles)?

> **QUESTION 8.5** What populated place is the farthest away from an airport complex, and how far away is it (give this value in miles)?

> **QUESTION 8.6** Why (specifically) was a value of 0 computed for the distance from Richmond Heights to an airport complex? (*Hint:* You may want to locate Richmond Heights on the map and zoom in closely to examine it to help answer this question.)

- Turn off the Placesairportsjoin and the Countiesairportsjoin layers.

STEP 8.4 **Using Select By Location for Analysis**

- Now that we have information about the airports and their distribution, we want to determine the location for the new airport complex. Keeping in mind the criteria from the Chapter Scenario and Applications section, we want the new airport to be located in a county with a large population (more than 50,000 persons). We can find this information with a simple query. From the **Selection** pull-down menu, choose **Select By Attributes**.

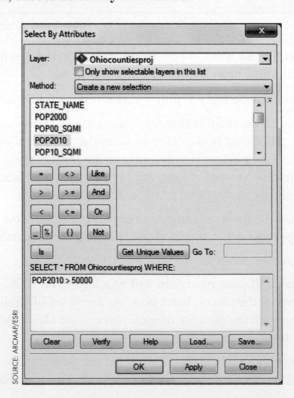

- Build a query to select all records in the Ohiocountiesproj layer that have the values greater than 50000 in their "POP2010" layer (see Chapter 2 for how to construct a simple query using Select By Attributes). Click **OK** to run the query.

- The counties meeting this criterion are now selected by ArcGIS (and outlined in the cyan color representing selected features). Answer Question 8.7.

QUESTION 8.7 How many counties in Ohio have a population of greater than 50,000 persons?

- Next, we want to select all of the populated places within the borders of these counties. We can do this by performing a Select By Location operation. (See **Smartbox 43** for more about Select By Location in ArcGIS.)

Smartbox 43

How does Select By Location work in ArcGIS 10.2?

The **Select By Location** tool in ArcGIS 10.2 allows you to perform spatial queries. With Select By Location, you can pose a spatial query and ArcGIS will return the records/features that match that query—except the query conditions are based on spatial features rather than attributes. The selection process is the same as the Select By Attributes query (see Chapter 2); selected features will be highlighted in the default cyan color along with the corresponding records in the layer's attribute table. These selected records can be extracted to their own layers or used as the basis for new queries. The same four types of selections can be performed as in Select By Attributes: *select features from, add to the currently selected features, remove from the currently selected features*, and *select from the currently selected features*.

With a Select by Location query, two distinct types of layers have to be specified: target and source. **Target** specifies the one or more layers that are to be queried through the Select By Location process, while **source** specifies the layer to be used as the layer to query from. For example, if you wanted to select all airports within 5 miles of an interstate, the airports layer would be the target (and features/records will be selected from this layer), while the interstate layer would be the source (you will be looking for features in the target layer that are 5 miles away from the features in this layer). Similarly, if you wanted to select all cities that are within a certain congressional district, the cities layer would be the target (and these would be the features that are selected), while the congressional district boundaries layer would be the source (because you will be selecting features that fall within the spatial dimensions of the district).

Figure 8.6 on the next page shows the results of a Select By Location query that searches for all airports (the target layer) that are completely within the boundaries of a populated place polygon (the source). You'll see that, of the subset of the airports and places shown in Figure 8.6, only one airport is selected because it is the only airport completely within the populated place boundaries (and thus the only feature that meets the criteria of the query).

Because many different types of spatial queries can be posed using the Select By Location tool (see Table 8.1 on pages 196–197), it's important to be sure you're choosing the right one. For example, say you're involved in a real-estate study to determine which houses in a city are prone to flooding.

Select By Location An ArcGIS tool used to select features from one or more layers based on their spatial relationship to another layer.

target (Select By Location) The layers from which features will be selected when using the Select By Location tool.

source (Select By Location) The layer containing the features for which selections will be made in relation to when using the Select By Location tool.

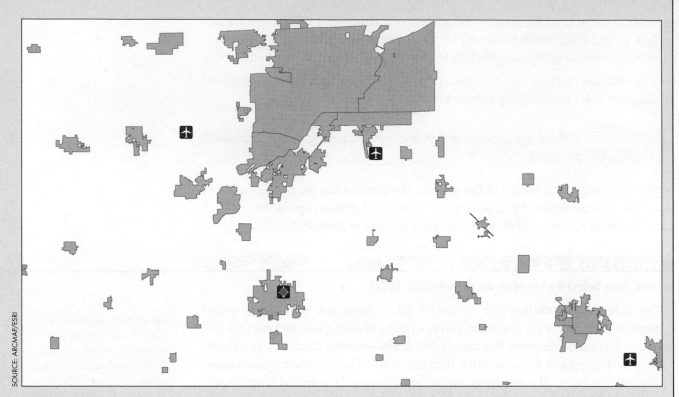

FIGURE 8.6 The results of a Select By Location query to find all features of a target layer (the airports) that are completely within the boundaries of a source layer (the polygons representing populated places).

You have two polygons—one represents the residential land parcels and the other represents the floodplain. By treating the parcels layer as the target and the floodplain layer as the source, you can use Select By Location to select the parcels on the floodplain. However, depending on the type of selection you do, you'll likely end up with different sets of selected parcels. If you're trying to select the parcels of land that are at least partially on the floodplain, you should use the "Intersect the source layer feature" option. If you're trying to select the parcels of land that are entirely inside the floodplain, you should use the "are completely within the source layer feature" option. If you're trying to select the parcels of land that are at least partly on the floodplain but at its edges, you should use the "Are crossed by the outline of the source layer feature" option.

Type of Select By Location – Target Layers:	What Gets Selected?
Intersect the source layer feature	Features from the target layers that overlap or occupy any of the same space with features from the source layer (either a complete overlap or just a partial overlap)
Intersect (3d) the source layer feature	Features from the target layers that overlap or occupy any of the same space with features from the source layer (either a complete overlap or just a partial overlap), but this is used only with three-dimensional features (see Chapter 18)
Are within a distance of the source layer feature	Features from the target layers that fall inside a user-defined buffered distance away from the features of the source layer

Table 8.1 The different actions available for the Select By Location tool and their results

Type of Select By Location – Target Layers:	What Gets Selected?
Are within a distance of (3d) the source layer feature	Features from the target layers that fall inside a user-defined buffered distance away from the features of the source layer, but used only with three-dimensional features (see Chapter 18)
Contain the source layer feature	Features from the target layers that hold some or all of the source layer's features within their boundaries
Completely contain the source layer feature	Features from the target layers that hold all of the source layer's features within their boundaries— but if the target layer's features touch the boundaries of the source layer's features, they will not be selected
Contain (Clementini) the source layer feature	Features from the target layers that hold all of the source layer's features within their boundaries— but if the target layer's features are completely on the boundaries of the source layer's features instead of holding them, they will not be selected
Are within the source layer feature	Features from the target layers that are inside the boundaries of the source layer
Are completely within the source layer feature	Features from the target layers that are entirely within the boundaries of the source layer—but if the target layer's features touch the boundaries of the source layer's features, they will not be selected
Are within (Clementini) the source layer feature	Features from the target layers that are inside the boundaries of the source layer—but if the target layer's features are completely on the boundaries of the source layer's features instead of within them, they will not be selected
Are identical to the source layer feature	Features from the target layers that have the exact same spatial features and dimensions as features in the source layer—but if the target layer is of a different vector type (point, line, or polygon) than the source layer, then nothing will be selected
Touch the boundary of the source layer feature	Features from the target layers that share some or all of the same boundaries with the features of the source layer
Share a line segment with the source layer feature	Features from the target layers that have at least one piece of line (from vertex to vertex) in common with the features of the source layer
Are crossed by the outline of the source layer feature	Features from the target layers that have their boundaries passed through by the boundaries of the source layer—but features from the target layer that have a common boundary with the source layer will not be selected
Have their centroid in the source layer feature	Features from the target layers that have their weighted center location inside the boundaries of the source layer's features

Table 8.1 Continued

• From the **Selection** pull-down menu, choose **Select By Location**.

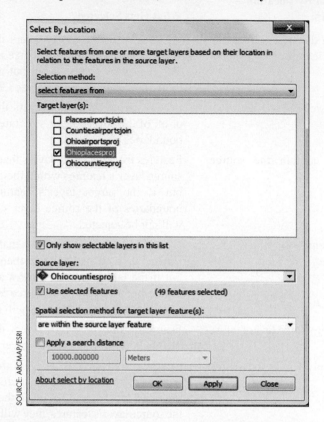

SOURCE: ARCMAP/ESRI

• For the Selection method, choose **select features from**.

• For the Target layer(s), choose **Ohioplacesproj**. This is the layer from which you will be selecting features.

• For the Source layer, choose **Ohiocountiesproj**. This is the layer you will be using as the basis of making the selection.

• Place a checkmark in the **Use selected features** box. This will limit the features used from Ohiocountiesproj to only those that you selected in the Select By Attributes query.

• For the Spatial selection method for target layer feature(s), choose **are within the source layer feature**.

• Click **Apply**. The selection process will proceed—when it's done, click **OK** to close the dialog. You'll see the populated places that were within a selected county highlighted in cyan, as they are now selected as well. Answer Question 8.8.

QUESTION 8.8 How many places are within the boundaries of a county with more than 50,000 persons?

• The new airport will have to be built in one of these selected populated places. Our second criterion is that the places must be near a major road (within 0.25 miles). Add the **Ohiomajrdsproj** layer to the TOC.

• We'll use another Select By Location criterion to choose these places. From the **Selection** pull-down menu, again choose **Select By Location**.

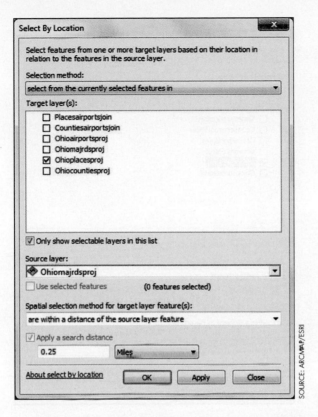

SOURCE: ARCMAP/ESRI

- For the Selection method, choose **select from the currently selected features in**. This will limit your selection, asking you to choose only from the pool of previously selected features.

- For the Target layer(s), choose **Ohioplacesproj**. This is the layer from which you will be selecting features.

- For the Source layer, choose **Ohiomajrdsproj**. This is the layer you will be using as the basis of making the selection.

- For the Spatial selection method for target layer features(s), choose **are within a distance of the source layer feature**.

- Be sure a checkmark is in the **Apply a search distance** box.

- Type **0.25** for the search distance and choose **Miles** from the pull-down menu.

- Click **Apply**. The selection process will proceed—when it's done, click **OK** to close the dialog.

- You'll see the populated places that met your criteria highlighted in cyan (and thus, selected). There should be fewer of them this time because your newly selected records were chosen from the previously selected bunch. Answer Question 8.9.

QUESTION 8.9 How many places are within an Ohio county with more than 50,000 persons and also within 0.25 miles of a major road?

- Our last criterion is that the chosen places must be far away (35 miles) from an airport complex. One last Select By Location criterion can determine this, so bring up the Select By Location dialog again.

- For the Selection method, choose **remove from the currently selected features in**. Your currently selected group of places has met the first two criteria,

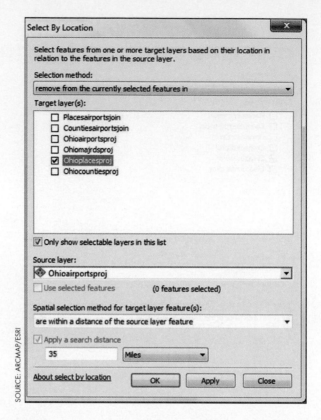

so you want to start with only these places and take out all of the places near an existing airport complex.

• For the Target layer(s), choose **Ohioplacesproj**. This is the layer from which you will be selecting features.

• For the Source layer, choose **Ohioairportsproj**. This is the layer you will be using as the basis of making the selection.

• For the Spatial selection method for target layer feature(s), choose **are within a distance of the source layer feature**.

• Be sure a checkmark is in the **Apply a search distance** box.

• Type **35** for the search distance and choose **Miles** from the pull-down menu.

• Click **Apply**. The selection process will proceed—when it's done, click **OK** to close the dialog.

• You'll see the populated places that met your criteria highlighted in cyan (and thus, selected). There should be far fewer of them this time because you just removed many of them from the previously selected bunch. Answer Question 8.10.

> **QUESTION 8.10** How many places are within an Ohio county with more than 50,000 persons, within 0.25 miles of a major road, and more than 35 miles away from an existing airport complex?

• One of these places (your answer to Question 8.10) will be the site of the new airport complex. You'll want to create a new layer of only these final places for continued analysis (as you did in Chapter 2). Right-click on the **Ohioplacesproj** layer in the TOC and select **Data**, then select **Export Data**.

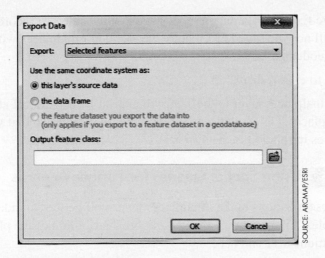

SOURCE: ARCMAP/ESRI

- For the Export option, choose **Selected features**.

- For the Use the same coordinate system as: option, choose the radio button for **this layer's source data**.

- Press the **browse** button next to the Output feature class and navigate to inside the **Ohioselect** geodatabase in your D:\GIS\Chapter8 folder.

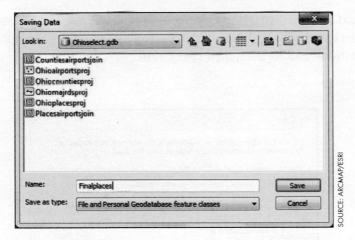

SOURCE: ARCMAP/ESRI

- Call the new feature class **Finalplaces**.

- For the Save as type, choose **File and Personal Geodatabase feature classes**.

- Click **Save**.

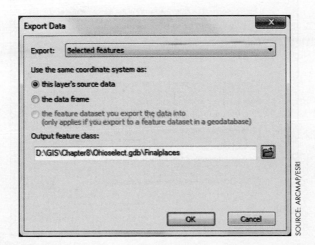

SOURCE: ARCMAP/ESRI

- Back in the Export Data dialog, you'll see that the selected records that you're exporting will now be saved in a feature class called Finalplaces within your Ohioselect geodatabase.

- Click **OK** to export them.

- Add the Finalplaces layer to the TOC when prompted. Turn off all the layers except Finalplaces, Ohiocountiesproj, and Ohioairportsproj so that you can see your final sites in relation to the airports.

STEP 8.5   **Using Spatial Queries for Further Analysis**

- One of these locations in the Finalplaces layer will be the candidate for a new airport complex. First, examine the geographic distribution of the places and answer Questions 8.11 and 8.12.

QUESTION 8.11 What are the names of these places? Which of them are classified as cities and which are classified as villages?

QUESTION 8.12 In what county is each of these places located?

- We'll next see to which of the existing airports these places are closest. As in Step 8.3, we can do this with a spatial join. Right-click on the **Finalplaces** layer, then select **Joins and Relates**, then select **Join**.

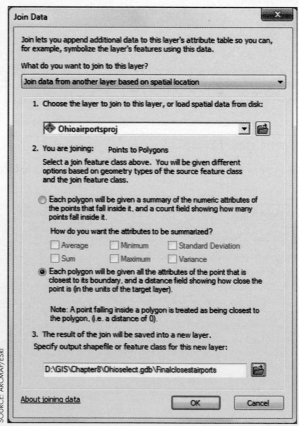

SOURCE: ARCMAP/ESRI

- From the **What do you want to join to this layer?** pull-down menu, select **Join data from another layer based on spatial location.**

- For option 1 (choose the layer to join to this layer), select **Ohioairportsproj**.

- For option 2, select the radio button for **Each polygon will be given all the attributes of the point that is closest to its boundary, and a distance field showing how close the point is (in the units of the target layer).** This will use the "closest" option for the spatial join and because the Ohioplacesproj layer has been projected, this distance will be computed in meters.

- For option 3, use the **browse** button to save the results as a new feature class called Finalclosestairports and save it in your **Ohioselect** geodatabase.

- Click **OK** when all settings are correct.

- A new layer called Finalclosestairports will be added to the TOC. This is the same polygon layer of the places, but it now has a new set of fields joined to it, specifying the distance of each place to the closest point of the Ohioairportsproj layer, based on their spatial locations. Answer Question 8.13.

QUESTION 8.13 For the potential places, what is the closest airport to each one of them and what is the distance (give this value in miles) that each airport is away from the place?

- Keep the Finalclosestairports layer open. You'll see a field indicating the population value of each of these final places. The place with the highest population value will serve as the site of the new airport complex. Select only that one record (see Step 2.4 in Chapter 2 for one way of selecting only a single record in a table).

- Next, export that one selected feature into its own feature class (as you did in Step 8.4). Call this new feature class **chosensite** and save it in your **Ohioselect** geodatabase. Chosensite should consist of a single polygon of your final site.

- With your site selected, we'll do some last analyses of the areas around this site. You'll have to do two separate Select By Location operations. First, perform a Select By Location to see how many airports (in the Ohioairportsproj layer) are within 100 miles of the chosen site and Answer Question 8.14. Then perform a second Select By Location to find out how many places (from the Ohioplacesproj layer) are within 35 miles of the chosen site (and answer Question 8.15).

- *Hint:* Be very careful with the selection methods you're using in each of these two spatial queries, the layers you're choosing as Target and Source, and the Select By Location action you are using. You may want to clear all selected features before performing each of these selections.

QUESTION 8.14 How many Ohio airport complexes are within 100 miles of the chosen site?

QUESTION 8.15 How many populated places are within 35 miles of the chosen site?

STEP 8.6 Printing or Sharing Your Results

- Save your work as a map document. Include the usual information in the Map Document Properties.

- Next, zoom in on your chosen site and show it, the surrounding counties, the nearby places, and any surrounding airports. Make sure your chosen site is labeled, along with the counties and airports.

- Finally, either print a layout (see Chapter 3) of your final version of your airport site (including all of the usual map elements and design for a layout) or share your results as a map service through ArcGIS Online (see Chapter 4).

Closing Time

This chapter introduced the two different methods of spatial analysis included with ArcGIS 10.2. Joining tables based on their spatial location and performing spatial queries are powerful tools for answering many location-based questions. Determining the closest feature or how many objects are within a certain distance from other objects allows us to delve more deeply into geospatial data. Some additional types of analysis methods related to patterns and clusters are discussed in the *Related Concepts for Chapter 8*.

We'll use many kinds of different spatial analysis methods in later chapters. For instance, two expansions of the concepts we learned in this chapter are creating a buffer zone polygon around a feature and combining two or more layers into a single layer. In the next chapter, we'll begin examining the geoprocessing techniques for doing some of these kinds of tasks.

Related Concepts for Chapter 8   Analysis of Patterns and Clusters in ArcGIS 10.2

Both spatial queries and spatial joins allow you to determine information related to distance, whether you are seeking the closest feature or selecting the objects within a certain distance of other objects. However, these types of distance measures don't tell us anything about how clustered various items might be, based not only on their distance apart, but also the values associated with the features. For instance, a police analyst could compute a summation of the number of car break-ins occurring in each neighborhood in a city, but this summation will not tell her if these break-ins are happening near one another, if high numbers are grouped together in certain locations, or if these break-ins are happening in random locations. In ArcGIS, a set of complex techniques called spatial statistics is used to measure and assess patterns and clusters of different phenomena. Many types of spatial statistical tools are available in ArcGIS 10.2; this section presents a brief overview of a few of them.

The Spatial Autocorrelation Global Moran's I tool in the Analyzing Patterns toolset of the Spatial Statistics toolbox examines the distribution of feature locations and their values (for instance, the distance between city parking lot locations and the number of car break-ins at each) to determine the extent to which items are clustered together, evenly dispersed, or simply distributed randomly (the degree of clustering is referred to as the **spatial autocorrelation** of the features). The results of Moran's I allow you to examine the spatial

spatial autocorrelation A measure of the degree of clustering of objects and their data values.

autocorrelation of the locations and their values. See Figure 8.7 for an example of different patterns of polygons and their values, which range from dispersed to clustered.

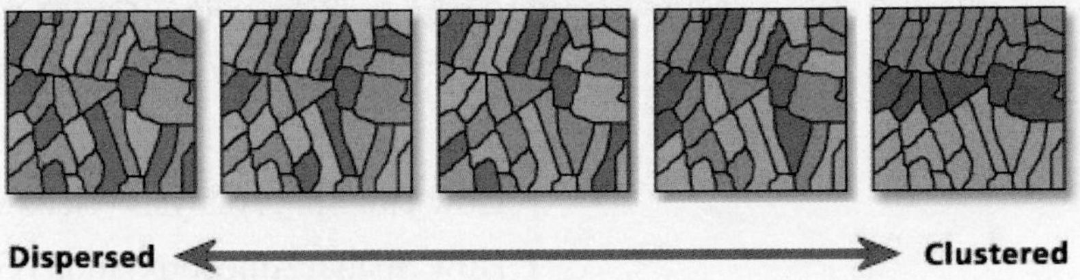

Dispersed ←————————————————→ **Clustered**

FIGURE 8.7 A set of polygons with associated values (denoted by color) that range from dispersed to clustered.

Another tool for cluster analysis is the High/Low Clustering (Getis Ord General G) tool in the Analyzing Patterns toolset of the Spatial Statistics toolbox. This tool allows you to determine the degree of clustering (for high values or low values within a dataset). ArcGIS 10.2 uses related types of techniques to examine clusters of values (for example, with the Hot Spot Analysis Getis Ord Gi* tool). Hot spots in the data refer to locations where clusters of high values are found (while cold spots refer to locations where clusters of low values are found). For instance, police analysts would use **hot spot analysis** to determine where spikes in certain types of crimes are occurring. See also Figure 8.8, which shows the mapped result of hot spot analysis to determine hot spots or high levels (in shades of red and orange) and cold spots or low levels (in shades of blue) of parking enforcement in Vancouver, British Columbia.

hot spot analysis A technique that determines spatial clusters of high values and low values.

FIGURE 8.8 A map of hot spots and cold spots of parking enforcement in Vancouver, British Columbia.

SOURCE: ARCGIS/ESRI CANADA

For More Information

For further in-depth information about the topics presented in this chapter, use the ArcGIS Help feature to search for the following items:

- About joining the attributes of features by their location
- Finding the nearest feature
- Finding what intersects a feature
- Finding what is inside a polygon

- High/Low Clustering (Getis-Ord General G) (Spatial Statistics)
- Hot Spot Analysis (Getis-Ord Gi*) (Spatial Statistics)
- How High/Low Clustering (Getis-Ord General G) works
- How Hot Spot Analysis (Getis-Ord Gi*) works
- How Spatial Autocorrelation (Global Moran's I) works
- Proximity analysis
- Select By Location: graphic examples
- Selecting features interactively
- Spatial Autocorrelation (Global Moran's I) (Spatial Statistics)
- Spatial Join (Analysis)
- Spatial joins by feature type
- Spatial relationships by feature type
- Using Select By Location

Key Terms

spatial analysis (p. 183)
spatial query (p. 183)
spatial join (p. 188)
Select By Location (p. 195)

target (Select By Location) (p. 195)
source (Select By Location) (p. 195)

spatial autocorrelation (p. 204)
hot spot analysis (p. 205)

How to Perform Geoprocessing in ArcGIS 10.2

Introduction

Many types of GIS analysis involve taking a layer, performing an action on it, and then getting a new output layer in return. **Geoprocessing** is the general term used for this type of task. For example, you could take a point layer representing culvert locations and create a new polygon layer that represents a one-mile area around each point, which you could then use for examining the land-cover characteristics in each of these one-mile areas. In ArcGIS, geoprocessing covers a wide range of tasks that are used for multiple applications. In fact, the tools in ArcToolbox are referred to as *geoprocessing tools*. You already used some of these tools in Chapter 1; we'll examine more of them in this chapter.

Geoprocessing is often used to create zones of spatial proximity around features or to combine similar polygons. Another common form of geoprocessing is an **overlay**, which involves combining two (or more) layers that have some of the same spatial boundaries but different properties or attributes to create a new layer. Think of creating an overlay as the process of placing one layer over top of another layer, determining where these layers match up and combine, and then producing a new layer as a result (Figure 9.1). For instance, to determine what sites are available for developing new coastal properties, a polygon representing a buffer around a lake could be overlaid with a second polygon layer representing the parcel boundaries that are zoned for commercial use. The resultant overlay would show only those polygons within the buffer zone that are also available for commercial usage.

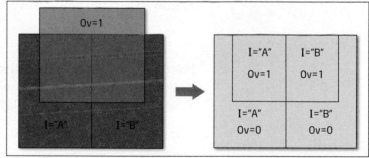

FIGURE 9.1 An example of overlaying two layers and the resultant third layer that combines the spatial and attribute characteristics of the two inputs.

Chapter Scenario and Applications

This chapter puts you in the role of an Ohio teacher examining the distribution of schools in one of Ohio's regions as well as access to those schools. As part of your study, you will be trying to determine which schools are located within place boundaries (whether an incorporated or an unincorporated place), as well as which schools are near major roads. You want to figure out which schools are in which places, as well as which schools are not in a place and not near a major road.

The following are additional examples of other real-world applications of this chapter's theory and skills:

geoprocessing When an action is taken to a dataset that results in a new dataset being created.

overlay The combining of two or more layers together in GIS.

• An urban planner wants to determine the distribution of residential land parcels on a floodplain. He can use polygon overlay to find out which homes are in the floodplain.

• A county engineer needs to examine which areas of the county cannot hear tornado sirens. He can create a buffer around each siren's location to denote the extent of the siren, then determine which populated areas are not reached by the sirens.

• A developer wants to see the potential effect of the placement of a new shopping center and its parking area. She can overlay the digitized footprint of the proposed development onto the existing land use to gauge the amount of wetlands and forested areas that would be removed by such a project.

ArcGIS Skills

In this chapter, you will learn:

• How to merge multiple feature classes together into a single feature class.

• How to interactively select and extract objects to their own layer.

• How to choose which layers can have objects selected from them.

• How to dissolve the boundaries of a set of polygons and combine them.

• How to extract a subset of the features in a layer by using the shape of another layer.

• How to build a buffer around features.

• How to perform polygon overlay as well as point-in-polygon overlay.

Study Area

For this chapter, you will be starting with data for the entire state of Ohio, then working only with a subset of four counties representing Ohio Educational Regional Service System #5: Ashtabula, Trumbull, Mahoning, and Columbiana counties.

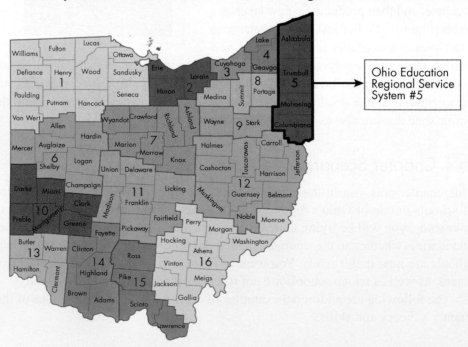

⬛ Data Sources and Localizing This Chapter

This chapter's data focus on datasets from within the state of Ohio. However, you can easily modify this lab to use data from your own state instead. For instance, if you were going to perform this lab's activities in Missouri, the same sets of data would be available. The Ohio counties' dataset was downloaded (free) and extracted from the USA Counties dataset available on ArcGIS Online (and also available for download here: http://www.arcgis.com/home/item.html?id=a00d6b6149b34ed3b833e10fb72ef47b). By querying the USA Counties layer for "STATE_NAME = 'Missouri'" you can select all Missouri counties and export them to a new feature class. From there, you can choose the counties within Missouri that correspond to the educational region you wish to examine.

The roads and places (incorporated and unincorporated) datasets were downloaded and extracted from The National Map and are available for all states. Use the Trans_RoadSegment feature class as the source for your state's roads, and the GU_IncorporatedPlace and GU_UnincorporatedPlace feature classes for your state's places. The schools data were also downloaded and extracted from The National Map (use the Struct_Point feature class and use Ftype = "730 - Education").

STEP 9.1 Getting Started

• Start ArcMap and use the Catalog to copy the folder called **Chapter 9** from the C:\GISBookdata\folder to your own D:\GIS\ drive. Be sure to copy the entire directory, not just the files within it.

• Chapter 9 contains a file geodatabase called **Ohiogeo** that contains the following feature classes:

   • Ohiocountiesproj: a polygon feature class of the 88 counties of Ohio.

   • Ohiomajrds: a line feature class of the major roads in Ohio.

   • Ohioschools: a point feature class of elementary and secondary schools in Ohio.

   • Ohioincplaces: a polygon feature class of Ohio's incorporated places.

   • Ohiononincplaces: a polygon feature class of Ohio's unincorporated places.

• Add the **Ohiocountiesproj** layer to the TOC.

• Before proceeding, you'll want to make sure your Geoprocessing **environment settings** are correct for the work you'll be doing in this chapter. When performing geoprocessing in ArcGIS, several settings will affect all the actions that you take during a project. For example, you can specify the extent or coordinate system that all results should have. Several upcoming chapters will instruct you to fix these settings in place before starting the chapter. To set up the environment settings for this chapter, from the **Geoprocessing** pull-down menu, select **Environments...** to open the Environment Settings dialog box.

environment settings
Geoprocessing options that control activities in a map document, such as the extent of outputs, the coordinate system of outputs, or the cell size of outputs.

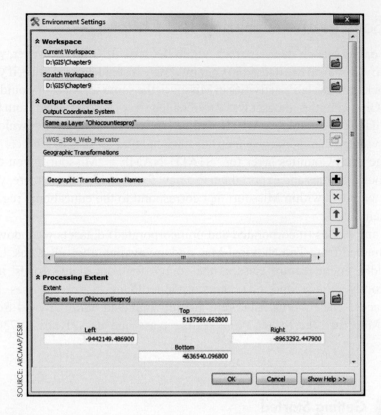

SOURCE: ARCMAP/ESRI

- Under the **Workspace** options, for **Current Workspace**, use the browse button to navigate to: **D:\GIS\Chapter9**.

- Under the **Workspace** options, for **Scratch Workspace**, use the browse button to navigate to: **D:\GIS\Chapter9**.

- Under the **Output Coordinates** options, for **Output Coordinate System**, select **Same as Layer "Ohiocountiesproj"**. This selection tells ArcGIS to use the same coordinate system as the Ohiocountiesproj layer for all of its output.

- Under the **Processing Extent** options, for **Extent**, select **Same as Layer Ohiocountiesproj**. This selection tells ArcGIS to use same extent as the Ohiocountiesproj layer for all output.

- Click **OK** when all the environment settings are in place.

- The Ohiocountiesproj layer was projected from its initial coordinate system to another one for use in this chapter. The coordinate system being used in this chapter is **Web Mercator (WGS 84)**, using Map Units of **Meters**. Check the Data Frame to verify that this coordinate system is being used.

STEP 9.2 Merging Multiple Layers

- Add the Ohioincplaces and Ohiononincplaces layers to the TOC and position them so that you can see both of them on top of the Ohiocountiesproj layer.

- The National Map separates areas into two categories: incorporated places and unincorporated places. For purposes of this chapter, we'll not be making a distinction between these two. Thus, we want to combine these two layers into a single layer, rather than work with two different layers. In ArcGIS, we can do this by using the merge process (see **Smartbox 44** for more information).

Smartbox 44

How can multiple layers be merged in ArcGIS 10.2?

A common geoprocessing operation combines several different layers into a single layer. For instance, if a team of three people is mapping the trails of a large metropark, each person will be producing feature classes of a third of the park. Ultimately, you do not want three separate layers of trails, but rather a single layer of trails that combines the geography of all three. You can accomplish this with the Merge tool in ArcGIS. **Merge** combines the geospatial features and attributes of two or more vector layers into a single new vector layer.

See Figure 9.2 for an example of a merge. Two different polygon layers (with features in different locations) can be combined into a single polygon layer that will contain all features (and attributes) from both layers. If attributes are present in one layer and not another, a value of "null" will be assigned to the missing values. Keep in mind that only features of the same type can be merged (for instance, two line feature classes can be merged into a single line feature class, but a point feature class and a line feature class cannot be merged into a single feature class). A related operation in ArcGIS is **append**, which combines one or more layers with an already existing feature class.

merge A geoprocessing operation that combines two or more layers of data together into a single layer.

append A geoprocessing operation that combines one or more layers of data together with an already existing layer.

FIGURE 9.2 Using the Merge tool to combine two layers into one.

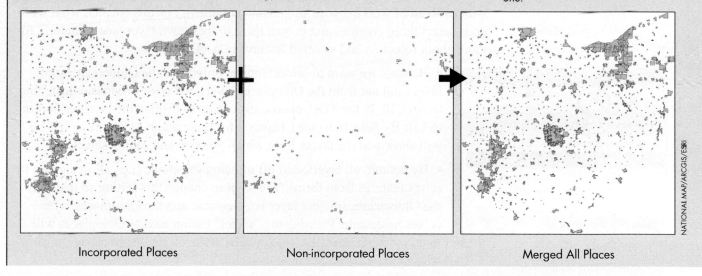

Incorporated Places   Non-incorporated Places   Merged All Places

NATIONAL MAP/ARCGIS/ESRI

- From the **Geoprocessing** pull-down menu, select **Merge**.

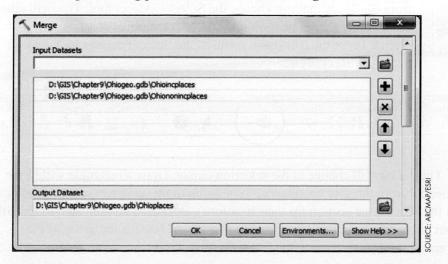

SOURCE: ARCMAP/ESRI

- Press the **browse** button next to Input Datasets and navigate to your D:\GIS\ Chapter9 folder and look inside the Ohiogeo geodatabase. Select the **Ohioinc-places** layer. Back in the Merge dialog box, you'll see it at the top of the list of Input Datasets (this is the list of layers to be merged together).

- Press the **browse** button a second time and follow the same procedure to add the **Ohiononincplaces** layer to the Input Datasets list.

- For the Output Dataset, call it **Ohioplaces** and save it in your Ohiogeo geodatabase.

- Click **OK**. A new layer called Ohioplaces will be added to the TOC and will have both of the layers combined in it. At this point, you can remove both the Ohioincplaces and Ohiononincplaces layers from the TOC (since you now have a new merged layer to work with).

STEP 9.3 Interactively Selecting and Extracting Data Layers

- The focus of this chapter is on the four counties in northeast Ohio that make up Educational Region #5—Mahoning, Trumbull, Columbiana, and Ashtabula. Thus, the first thing to do is to select and extract only those four counties to work with (instead of working with all 88 Ohio counties). For this chapter, what we'll do is select those counties and extract them to their own layer (see Chapter 2 for more about selection and selected features in ArcGIS).

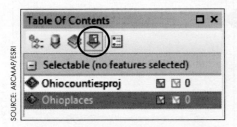

- Because we want to select features only from the Ohiocountiesproj layer (and not from the Ohioplaces layer), we want to communicate this to ArcGIS. In the TOC, choose the option to display the layers according to List By Selection (see Chapter 1 for more about TOC settings). This will show you the layers from which you can and cannot select.

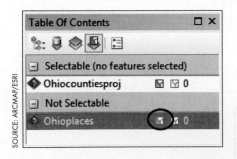

- By default, all layers start off as being selectable (i.e., allowing you to select features from them). You want to change this setting so that only the Ohiocountiesproject layer is Selectable and the Ohioplaces layer is Not Selectable. Pressing the "select" button next to Ohioplaces will toggle it from Selectable to Not Selectable. (Note that pressing it again will toggle it back to being Selectable.)

- Now, only Ohiocountiesproj should be in the Selectable category. Next, from the **Tools toolbar**, choose the Select Features tool. This will allow you to draw shapes on the View and select features that fall within those shapes. From the pull-down menu of the Select Features tool, choose **Select By Rectangle**.

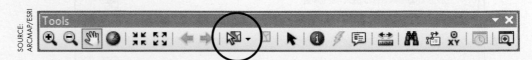

- The cursor will change to the selection cursor. Draw a rectangle within the boundaries of the four counties we'll be using in this chapter (see the graphic in the Study Area section for their locations). Because you indicated to ArcGIS that only features within Ohiocountiesproj would be selected, none of the corresponding features in Ohioplaces that fell within the rectangle were selected.

• *Important Note:* Select only the four counties. If you find that you've se-lected other counties, you can de-select them by holding down the Shift key on the keyboard and clicking on the counties you want to de-select to remove them from the selection. If you want to clear all selected counties and start over, click on the Clear Selected Features button on the Tools toolbar:

SOURCE: ARCMAP/ESRI

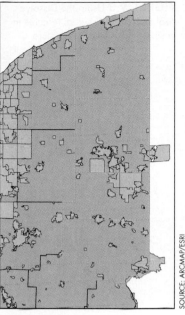

SOURCE: ARCMAP/ESRI

• The next step is to create a new feature class containing just these four selected counties (see Chapters 2 and 8 for how to export data into a new feature class in a geodatabase). First, change how you're viewing items in the TOC to **List By Drawing Order**. Next, in the TOC, right-click on the Ohiocountiesproj layer, choose **Data**, then choose **Export Data**. Export your **selected features** to create a new feature class called **fourcounties** and place this in your **Ohiogeo** geodata-base. Add the new exported data to the map as a layer when prompted.

• At this point, turn off the Ohiocountiesproj layer and zoom in on the four counties that we will be using in this lab.

• Switch the fourcounties polygons color to hollow (see Chapter 1 for how to change the color or symbology of a layer) so that you can see through it (that is, make it transparent). Give it a thicker outline as well, so it stands out.

STEP 9.4 Dissolving Polygon Boundaries

• The next step to defining our four-county region is to dissolve the borders between the counties so that we can use the entire region as a boundary, not just four separate counties (see **Smartbox 45** for more about using dissolve for geoprocessing). We will use this new "region" in further analysis.

Smartbox 45

How does dissolve operate in ArcGIS 10.2?

The **dissolve** function in ArcGIS combines multiple objects in a feature class into a single object when these objects have the same attributes. For example, a polygon feature class of U.S. counties may have an attribute indicating the state to which they belong (such as a FIPS code or state name). When dissolve is used, the boundaries of the counties' polygons are removed and they are com-bined into a single polygon, as long as those polygons have the same attribute (for example, all counties that have the same state FIPS code or state name). Figure 9.3 shows how dissolve can be used to remove county boundaries and create a larger object of the geography of the combined area. By dissolving polygon boundaries, you can create "regions" of data that have the same at-tributes. For instance, a polygon feature of parcels may contain an attribute indicating how that parcel is zoned. You can dissolve the polygon boundaries according to their zoning type so that you can create larger polygons of areas zoned residential, commercial, or industrial.

dissolve A geoprocessing operation that combines polygons with similar attributes together.

FIGURE 9.3 Using the Dissolve tool to combine several features in a single layer into one feature.

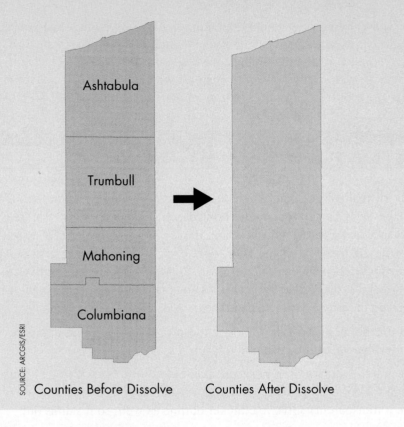

SOURCE: ARCGIS/ESRI

Counties Before Dissolve   Counties After Dissolve

- From the **Geoprocessing** pull-down menu, select **Dissolve**.

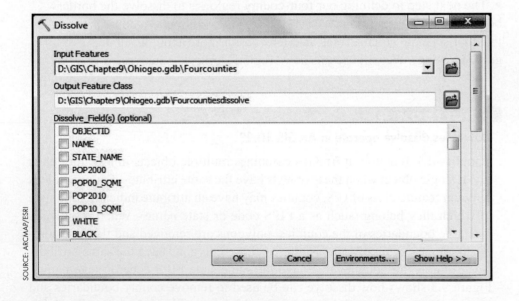

SOURCE: ARCMAP/ESRI

- For the Input Features use **fourcounties** (from your Ohiogeo geodatabase). This is the feature class you wish to dissolve.

- For the Output Feature Class, use the browse button to save this newly dissolved feature class in your **Ohiogeo** geodatabase and name it **fourcountiesdissolve**.

- Don't choose a Dissolve Field. When all options are set, click **OK**.

- You will now have a "borderless" four-county region layer called fourcounties-dissolve added to your table of contents. Change its symbology to Hollow and turn off the fourcounties layer.

STEP 9.5 Clipping Feature Classes

- Add the Ohioschools and Ohiomajrds layers to the TOC. Right now, these two layers and the Ohioplaces feature class contain information about all of Ohio. We want to focus only on those schools and major roads that are within the boundaries of the four-county region. To select only those areas, you can use the Clip operation in ArcGIS. (For more information about Clip, see **Smartbox 46**.)

Smartbox 46

How does clip operate in ArcGIS 10.2?

The **clip** operation in ArcGIS uses the shape and geometry of one feature class to extract objects out of another feature class. In Figure 9.4, the roads layer covers the entire state, but we want to use only the roads that fall within the multi-county polygon boundaries. The roads can be clipped so that a new dataset consisting of only the roads within the geographic area of the polygon can be extracted for use. Using clip is like using a feature class (such as a polygon) as a "cookie cutter"; all features from another layer that fit within this "cookie cutter" are removed to a new layer for use. Polygons, lines, and points can all be clipped using this operation. For instance, if you have a land-cover map of your entire state but will be focusing your environmental analysis on a single county, you could use the polygon boundaries of the county to clip out the land-cover polygons that are within that single county.

clip A geoprocessing operation that extracts objects from one layer based on the geometry of a second layer.

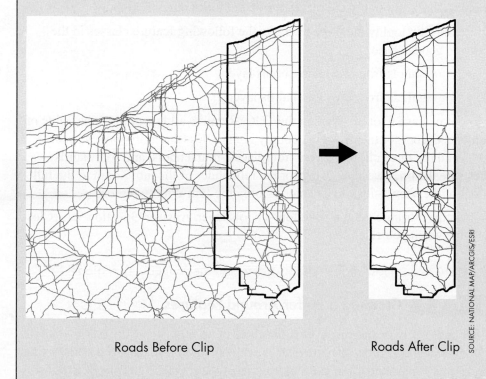

FIGURE 9.4 Using the Clip tool to create a new feature class.

Roads Before Clip   Roads After Clip

SOURCE: NATIONAL MAP/ARCGIS/ESRI

- From the **Geoprocessing** pull-down menu, select **Clip**.

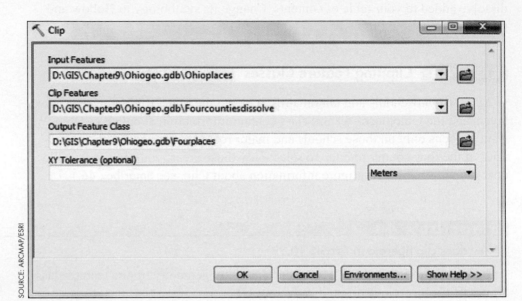

- For the Input Features, select **Ohioplaces** from the Ohiogeo geodatabase. This is the feature class that you want to clip.

- For the Clip Features, select **fourcountiesdissolve**. This is the feature class that will define the "cookie cutter" shape used for clipping.

- For Output Feature Class, call the layer **fourplaces** and save it in the Ohiogeo geodatabase.

- Click **OK** when done.

- You will now have a new feature class called fourplaces added to the Table of Contents that shows only the places inside the four-county region.

- Use the Clip tool twice more to create the following feature classes in the Ohiogeo geodatabase:

  - **fourschools** from the Ohioschools layer

  - **fourmajrds** from the Ohiomajrds layer

- You now have a lot of layers in the TOC. Remove all layers from the Table of Contents except for the dissolved county boundaries (the fourcountiesdissolve layer) and the three clipped layers (fourplaces, fourschools, and fourmajrds). These are the ones we'll be using for the remainder of the lab. Zoom to the full extent of the View—it should center on only the four-county region.

- Open the attribute table for the fourschools layer. Answer Question 9.1. (Close the table when you're done.)

QUESTION 9.1 How many schools are within the four-county region?

STEP 9.6 Creating a Buffer Around Features

- Because we're examining access to schools, our next step is to examine information concerning areas that are near the region's major roads. To do this, we can create a buffer around all of the major roads in the region. See **Smartbox 47** for more information on buffers.

Smartbox 47

How do buffers operate in ArcGIS 10.2?

A **buffer** is a region of spatial proximity created around a set of objects. In ArcGIS, a buffer is a new feature class created by calculating a distance around a set of objects in another layer (whether points, lines, or polygons). See Figure 9.5 for an example—a feature class contains a set of lines representing roads, and a half-mile buffer is created around these roads and saved as a polygon feature class. Distances and polygons are calculated around each road, and in this example, the borders between those polygons have been dissolved in order to create a single new object that is the buffer. Creating buffers is a common geoprocessing operation. For instance, engineers may create a buffer around a road that needs widening to see which sections of neighboring properties will be affected, or emergency planners may create buffers around tornado sirens that represent the distance that the sirens' sound reaches.

Note that a buffer is different than a Select By Location query (see Chapter 8). In Select By Location, one of the options allows you to select how many objects are within a certain distance of other objects. The Buffer tool creates a new polygon object that represents this distance, allowing you to work with the actual dimensions of the buffer zone as a separate layer.

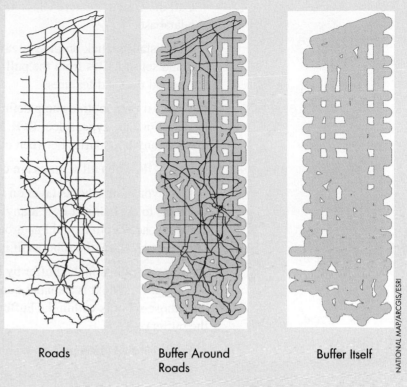

Roads   Buffer Around Roads   Buffer Itself

NATIONAL MAP/ARCGIS/ESRI

FIGURE 9.5 Creating a half-mile buffer around the roads of a feature class.

buffer A zone of spatial proximity around a feature or set of features.

- We'll start by creating a half-mile buffer around the major roads. From the **Geoprocessing** pull-down menu, select **Buffer**.

Buffer

Input Features
D:\GIS\Chapter9\Ohiogeo.gdb\Fourmajrds

Output Feature Class
D:\GIS\Chapter9\Ohiogeo.gdb\halfmilebuffer

Distance [value or field]
⦿ Linear unit
                                        0.5   Miles

○ Field

Side Type (optional)
FULL

End Type (optional)
ROUND

Dissolve Type (optional)
ALL

OK   Cancel   Environments...   Show Help >>

SOURCE: ARCMAP/ESRI

- For Input Features, select **fourmajrds**. This is the layer whose features you want to create a buffer around.

- Call the Output Feature Class **halfmilebuffer** and save this in your Ohiogeo geodatabase.

- For Distance, click on the **Linear unit** radio button. Type **0.5** for the Linear unit and select **Miles** from the pull-down menu. This will create a half-mile buffer around the road features.

- The next three options help define the shape of the buffer. For Side Type, select **FULL**. For End Type, select **ROUND**. For Dissolve Type, select **ALL** (this will dissolve any boundaries between overlapping buffers).

- Click **OK** when all settings are correct.

- Place your fourschools layer on top of the half-mile buffer layer in the Table of Contents to eyeball just how many schools are (and are not) within a half mile of a major road. Zoom in and Pan around the map to see what the dimensions of the buffer are and how the schools fit in the buffer.

- Counting points in the buffer by hand will be inefficient and likely inaccurate. What you'll do next is a spatial query operation (as in Chapter 8) that will select only those schools within the buffer (and will thus not select schools not inside the buffer).

- From the **Selection** pull-down menu, choose **Select By Location**.

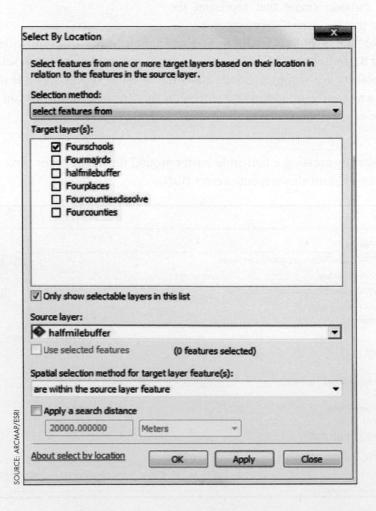

• Build a Select By Location query (see Chapter 8 for instructions on how to use the Select By Location tool) to select features from the fourschools layer that are within the source layer (the halfmilebuffer layer). Answer Questions 9.2 and 9.3.

QUESTION 9.2 How many schools in the four-county region are within half a mile of a major road?

QUESTION 9.3 How many schools in the four-county region are not within half a mile of a major road?

• Clear your selected features.

STEP 9.7 Performing Polygon Overlay

• Knowing about the proximity of schools to roads is only part of our analysis. We also want to know which schools are within the boundaries of an incorporated or unincorporated place and also near a major road. We want to identify those areas on the ground that are in the place boundaries and also near a road, something that a spatial query could not do for us. What we can do is create a new feature class that shows us the boundaries of the buffer and the boundaries of the places together—but there are several different ways we can overlay these two types of polygons. See **Smartbox 48** for more information about overlaying polygons.

Smartbox 48

What are the different types of polygon overlays in ArcGIS 10.2?

When two polygons are overlaid, the two layers are combined to create a new, third, layer that has qualities retained from the two polygons used to create it. This new layer combines not only the geospatial qualities and geometry of the two polygons but also their attributes. For an example, see Figure 9.6—two polygon layers (a set of land parcels and the floodplain) are overlaid to create a

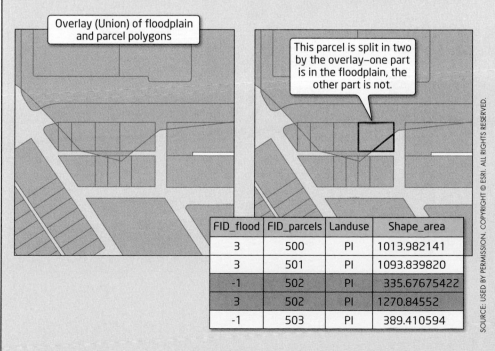

Overlay (Union) of floodplain and parcel polygons

This parcel is split in two by the overlay—one part is in the floodplain, the other part is not.

FID_flood	FID_parcels	Landuse	Shape_area
3	500	PI	1013.982141
3	501	PI	1093.839820
-1	502	PI	335.67675422
3	502	PI	1270.84552
-1	503	PI	389.410594

FIGURE 9.6 An overlay of two polygon layers (land parcels and the floodplain) to create a new third layer and how the attribute table of the new layer combines both layers.

new, third layer that has the geographic characteristics of both inputs. However, where a land parcel polygon is split into two polygons (with one on the floodplain and one not on the floodplain), the attribute table is updated to include two separate records: One polygon has the attributes of being on the floodplain (FID_flood = 3), and another polygon has the attributes of not being on the floodplain (FID_flood = −1).

Overlay operations are used for multiple types of applications. Figure 9.7 shows some of the overlay methods available in ArcGIS 10.2:

FIGURE 9.7 Six different types of polygon overlay operations—Intersect, Identity, Symmetrical Difference, Union, Erase, and Update.

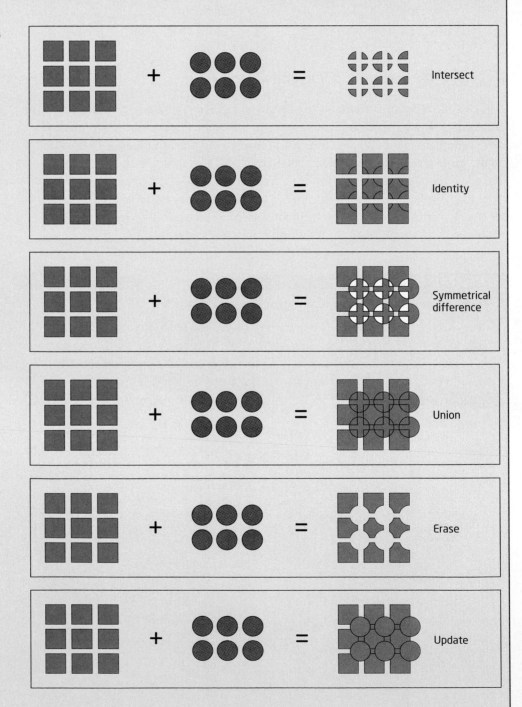

- **Intersect**: In this type of overlay, only the features that both input layers have in common are retained in the output layer. If the land parcels and the floodplain are overlaid using an Intersect operation, the new layer would show only areas that have land parcels on the floodplain, not any other parcels or any other sections of the floodplain.

- **Identity**: In this type of overlay, all of the features of the first input layer are retained in the output layer, along with all of the intersecting features of the second layer. If the land parcels and floodplain are overlaid using an Identity operation, the new layer would show all available land parcels plus the area of the floodplain covering these parcels (but no other parts of the floodplain).

- **Symmetrical Difference**: In this type of overlay, the output layer retains all features of the two input layers except for those areas they have in common. If the land parcels and floodplain are overlaid using a Symmetrical Difference operation, the new layer would show all land parcels and all areas of the floodplain except for those parcels that were on the floodplain.

- **Union**: In this type of overlay, the output layer retains all features of the two layers. If the land parcels and floodplain are overlaid using a Union operation, the new layer would show all land parcels as well as all areas of the floodplain combined together.

- **Erase**: In this type of overlay, the output layer retains all features of the first input layer except for those areas that the first layer has in common with the second input layer. If the land parcels and the floodplain are overlaid using an Erase operation, the new layer would show all land parcels except those on the floodplain.

- **Update**: In this type of overlay, the output layer retains all features of the first input layer with all features of the second input layer placed on top of them, sort of like a "cut and paste" operation. All of the features of the first layer that are in common with the second layer would be covered over by the second layer. If the land parcels and the floodplain are overlaid using an Update operation, the new layer would show all of the floodplain and any parts of the land parcels not covered by the floodplain.

Intersect A type of GIS overlay that retains the features that are common to both layers.

Identity A type of GIS overlay that retains all the features from the first layer along with the features it has in common from the second layer.

Symmetrical Difference A type of GIS overlay that retains all of the features from both layers except for the features that they have in common.

Union A type of GIS overlay that retains all of the features from both layers.

Erase A type of GIS overlay that retains all of the features from the first layer except for what they have in common with the second layer.

Update A type of GIS overlay that retains all features from the second layer as well as those features from the first layer that are not in common with the second.

- To show the areas that are both near a road and also within a place boundary, we'll use the Intersect option. Answer Questions 9.4 and 9.5.

QUESTION 9.4 Why did we use the Intersect operation and not the Union operation? What would the resulting overlay have been if Union were used?

QUESTION 9.5 Why did we use the Intersect operation and not the Symmetrical Difference operation? What would the resulting overlay have been if Symmetrical Difference were used?

- From the **Geoprocessing** pull-down menu, select **Intersect**.

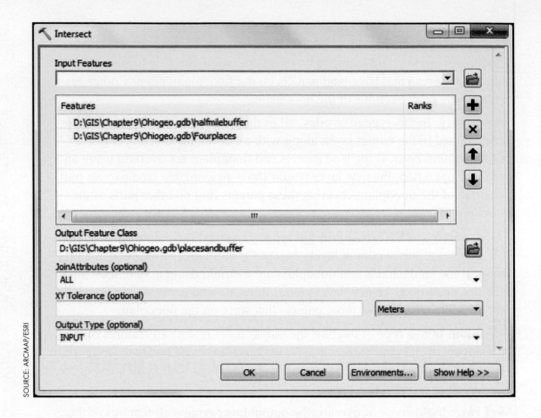

• For Input Features, add the **halfmilebuffer** and the **fourplaces** layers from the Ohiogeo geodatabase (either open both at the same time or open them one at a time). These are the features that you want to overlay using an Intersect operation.

• For Output Feature Class, call it **placesandbuffer** and save it in the Ohiogeo geodatabase. This is the name of the new layer created from the overlay.

• Use **ALL** for the JoinAttributes option.

• Click **OK** when you're done. The placesandbuffer layer will be added to the TOC. Turn off the fourplaces and halfmilebuffer layers to see only the new overlay.

• Use Select By Location to determine which of the schools in the fourschools layer are within the source layer feature (use placesandbuffer for the source layer). Answer Question 9.6.

QUESTION 9.6 How many schools are within half a mile of a major road and also within the boundaries of a place?

• In the fourschools attribute table, you can switch the selection to select the records (see Chapter 2) that did not meet the selection criteria. Do so and answer Question 9.7.

QUESTION 9.7 How many schools are not within half a mile of a major road and also not within the boundaries of a place?

• Export these points (the schools that do not fall within the placesandbuffer layer) to a new feature class in your Ohiogeo geodatabase called **fourschoolsoutside**.

- Examine the attribute table of the fourschoolsoutside layer. Answer Question 9.8.

> **QUESTION 9.8** Of the schools not near a road and not within a place, how many of them are considered to be affiliated with the city of Youngstown? How many of them are considered to be affiliated with the city of Warren? (*Hint*: You may want to use Select By Attributes queries on the fourschoolsoutside layer to help determine the answers to these questions.)

- Clear your selected features.

STEP 9.8 Performing Point-in-Polygon Overlay

- For the last step of analysis, for those schools that are near a major road and within the boundary of a place, we want to identify which place they are in. We can do this by overlaying the points of the schools with the polygon boundaries of the placesandbuffer layer as a point-in-polygon overlay (see **Smartbox 49** for more information).

Smartbox 49

How is a point-in-polygon overlay performed?

Overlay operations can be performed with more than just polygons. As its name implies, a **point-in-polygon overlay** operation will overlay points onto polygons. The end result is a new point layer, in which each point will contain the attributes of the polygon that it was overlaid in (a process similar to a spatial join; see Chapter 8). Figure 9.8 shows an example of a point-in-polygon overlay. The figure shows a set of polygons representing land parcels in one

point-in-polygon overlay A type of GIS overlay that results in a new point layer where the points have the attributes of the polygons within which they are.

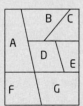 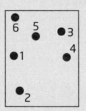 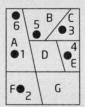

FIGURE 9.8 The layers and output of a point-in-polygon overlay operation.

Land Parcel	Owner
A	Smith
B	Jones
C	Roberts
D	Hill
E	James
F	Young
G	Cross
Parcel polygons layer	

Wells
1
2
3
4
5
6
Wells points layer

Wells	Owner
1	Smith
2	Young
3	Roberts
4	James
5	Jones
6	Smith
Points with polygon attributes layer	

line-on-polygon overlay A type of GIS overlay that results in a new line layer where the lines have the attributes of the polygons within which they are.

layer and a set of points representing water wells in a second layer. You want to determine which wells are on which parcels of land. Because you want to see where these two layers intersect, you can perform the point-in-polygon overlay using the Intersect tool. Like the Intersect overlay operation, it will determine which points and which polygons are in common with one another.

The output layer will be a new point layer in which each point now also has the attributes of its corresponding polygon parcel. For instance, in Figure 9.8, point 2 was in polygon F. In the resultant new point layer, point 2 will have the attributes of polygon F (that is, the record for point 2 will now contain information that it is in a parcel owned by Young).

A similar operation is a **line-on-polygon overlay**, which overlays lines onto polygons. The result of this operation is a new line layer in which each line is assigned the attributes of the polygon through which it passes. If a single line were to pass through three polygons, that line would be split into three separate lines (and thus, the attribute table would have three records instead of one) and each of these segmented lines would have the attributes of the corresponding polygon through which it passed.

- Out of our available geoprocessing operations, an Intersect operation will do the job, so from the **Geoprocessing** pull-down menu, select **Intersect**.

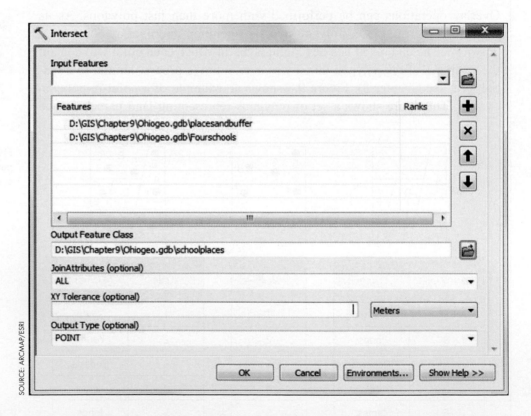

SOURCE: ARCMAP/ESRI

- For Input Features, add the **fourschools** and the **placesandbuffer** layers from the Ohiogeo geodatabase (either open both at the same time or open them one at a time). These are the features that you want to overlay using an Intersect operation.

- For Output Feature Class, call it **schoolsplaces** and save it in the Ohiogeo geodatabase. This is the name of the new layer created from the overlay.

- For the JoinAttributes option select **ALL**.

- For Output Type, select **POINT** (because you are performing a point-in-polygon overlay, the output will be a point feature class with polygon attributes attached to it).

- Click **OK** when you're done. The schoolsplaces layer will be added to the TOC. Turn off the fourschools and fourschoolsoutside layers to see only the new overlay. Using the information from the schoolsplaces layer, answer Questions 9.9, 9.10, 9.11, and 9.12. (*Hint:* You may want to use Select By Attributes queries to help find the answers.)

QUESTION 9.9 In which incorporated or unincorporated place is Cardinal Mooney High School located?

QUESTION 9.10 In which incorporated or unincorporated place is John F. Kennedy High School located?

QUESTION 9.11 In which incorporated or unincorporated place is Saints John and Paul Elementary located?

QUESTION 9.12 Which three schools are considered to be in Edgewood?

STEP 9.9 Printing or Sharing Your Results

- Save your work as a map document. Include the usual information in the Map Document Properties.

- Next, turn off all layers except the placesandbuffer and the schoolsplaces layers and change the symbology so that the overlaid areas and the points can be clearly seen.

- Finally, either print a layout (see Chapter 3) of your final version of your sites (including all of the usual map elements and design for a layout) or share your results as a map service through ArcGIS Online (see Chapter 4).

Closing Time

Geoprocessing operations are common and versatile methods for performing spatial analysis in GIS. Buffers, overlay operations, merge, and dissolve operations are all highly useful. In ArcGIS, geoprocessing tools can be incorporated into more complex projects such as chaining several commands together or executing iterative procedures. See the *Related Concepts for Chapter 9* for information on how scripts using the Python programming language can be used to work with geoprocessing within ArcGIS.

We'll continue using geoprocessing tools throughout future chapters. In the next chapter, we'll continue working with other GIS applications, such as taking a table of addresses and turning that information into a layer of geospatial data. Once we've created GIS data through this geocoding process, we can use spatial analysis techniques to examine the data.

Related Concepts for Chapter 9   Geoprocessing and Python

Python An object-oriented, open-source programming language that is used for developing scripts for ArcGIS.

script A short section of computer code.

ArcPy The Python add-in that allows for ArcGIS tools and settings to be used in Python scripts.

The **Python** scripting language is used for programming in ArcGIS, especially for creating **scripts** (sections of computer code) to run tools, create apps, and perform geoprocessing operations. By using Python, you can execute command-line operations rather than using the tools from ArcToolbox; you can also create more in-depth operations. For instance, if you need to build ten different buffers of different sizes, you can write a Python script with a few lines of code to automate those tasks for you.

Python (created by Guido van Rossum and named for the Monty Python comedy team) is an object-oriented, open-source programming language that is very versatile and used in several different fields. In ArcGIS, the **ArcPy** add-in allows you to gain access to the tools of ArcGIS and use them when writing scripts or executing code. When using Python in ArcGIS, the first command you should execute is "import ArcPy," which gives you access to the ArcPy add-in (and thus, the ArcGIS tools and Environment Settings).

There are two ways to work with Python in ArcGIS. The first of these is through the Python window accessible from the Standard toolbar (see Chapter 1). The Python window is a command line that allows you to type and execute Python code directly; it's best used for short statements, such as running a tool. See Figure 9.9 for an example of using the Python window to run the Merge tool as used in this chapter. The left-hand pane is where Python code is typed at the command line; the right-hand pane shows the results of the code's implementation. In the left pane, the first line ("import arcpy") gets the ArcPy add-in so that you can work with ArcGIS functions in Python. The second line sets one of the Geoprocessing Environment Settings for the workspace (i.e., the location of the data that you'll be working with). The third line runs the Merge tool by taking two layers in the Ohiogeo geodatabase (Ohioincplaces and Ohiononincplaces), merging them to become a third layer (Ohioallplaces), and writing that output to the Ohiogeo geodatabase.

FIGURE 9.9 An example of running the Merge tool using the Python window.

```
Python                                                    □ x
>>> import arcpy                              Executing: Merge
arcpy.env.workspace = "D:/GIS/Chapter9"       D:/GIS/Chapter9
arcpy.Merge_management(["Ohiogeo.gdb/Ohioincplaces",   \Ohiogeo.gdb/Ohioincplac
 "Ohiogeo.gdb/Ohiononincplaces"],             es;D:/GIS/Chapter9
 "Ohiogeo.gdb/Ohioallplaces")                 \Ohiogeo.gdb/Ohiononincp
>>>                                           laces D:/GIS/Chapter9
                                              \Ohiogeo.gdb
                                              \Ohioallplaces
                                              "Permanent_Identifier
                                              "Permanent_Identifier"
                                              true true false 40 Text
                                              0 0 ,First,
```

A second way of working with Python in ArcGIS is to use a separate program to write lengthier or more complex scripts. Keep in mind that Python scripts are simply text files, so you could use a text-editing utility like Notepad to write the script. However, a more effective means of preparing scripts would be to use a program specifically designed to aid you in writing Python scripts, such as the PythonWin program (available for free download from http://python.org—note that you should use version 2.7 because that's the version supported by ArcGIS). Using a program like PythonWin as an environment for writing scripts will help you keep your coding, commenting, and syntax clearer when writing lengthy scripts. A Python script is saved with the extension .py and can be run through a program like PythonWin.

Python is commonly used with ArcGIS for writing scripts related to geoprocessing. In the ArcGIS Help utility (see Chapter 1), the help page for each tool also lists examples of Python code that can be used in the Python window and as a stand-alone script. For instance, if you want to use Python to overlay two layers through the Intersect operation, look up the Intersect tool in ArcGIS Help and you'll see examples of Python code showing how Intersect can be implemented in Python scripting.

For More Information

For further in-depth information about the topics presented in this chapter, use the ArcGIS Help feature to search for the following items:

- A quick tour of ArcPy
- A quick tour of geoprocessing
- Buffer (Analysis)
- Clip (Analysis)
- Creating a new Python script
- Dissolve (Data Management)
- Erase (Analysis)
- Essential geoprocessing vocabulary
- Essential Python vocabulary
- How Dissolve (Data Management) works
- Importing ArcPy
- Merge (Data Management)
- Overlay analysis
- Using the Python window
- What is ArcPy?
- What is Python?
- What is the Python window?
- Writing Python scripts

Key Terms

geoprocessing (p. 207)
overlay (p. 207)
environment settings (p. 210)
merge (p. 211)
append (p. 211)
dissolve (p. 213)
clip (p. 215)

buffer (p. 217)
Intersect (p. 221)
Identity (p. 221)
Symmetrical Difference (p. 221)
Union (p. 221)
Erase (p. 221)
Update (p. 221)

point-in-polygon overlay (p. 223)
line-on-polygon overlay (p. 224)
Python (p. 226)
script (p. 226)
ArcPy (p. 226)

How to Perform Geocoding in ArcGIS 10.2

Introduction

A common application of geospatial technologies is locating an address on a map. When you're using a service like MapQuest or Google Maps, you type in an address. The service takes that string of letters and numbers and converts that information into a location plotted on a map. This matching of an address with its corresponding geospatial location is called **geocoding**. In ArcGIS 10.2, you can easily plot a single address on a map, much as you would in MapQuest. ArcGIS 10.2 is more versatile, however, in that it has the ability to perform **batch geocoding**, in which multiple addresses are plotted at the same time and then mapped as a point feature class (which can then be used for further analysis).

The key to geocoding is matching addresses to good reference data. Because we're dealing with street addresses, the **reference data** will take the form of a geospatial data layer representing roads or streets. In GIS, a road is not represented as one long line; for example, a road like Fifth Avenue in New York City does not consist of one line object and a single record in an attribute table. Instead, it is split into multiple line **segments** (likely at each intersection with another street or road) and each segment has different attributes. Thus, a road like Fifth Avenue might be represented with hundreds of different records in the line layer. See Figure 10.1 on page 230 for another example: In Mahoning County, Ohio, Mahoning Avenue is a major road that runs across the county. In the roads layer, however, Mahoning Avenue is represented not as one line, but rather as 191 segments.

When a road is represented as a series of segments, each segment can have different attributes assigned to it, such as the name of the road, the range of addresses on the left and right sides of the road, the zip code on the left and right sides of the road, the type of road, and the speed limit for each stretch or the measured length of the road. By separating sections of a road into smaller bits, each with a narrow range of addresses assigned to it, the GIS can use this information when attempting to match an address.

The result of a geocoding process is a new set of points representing the matched locations of each address found using this kind of reference data. Because geocoding has converted a set of addresses in a table into geospatial data, each point can also be assigned coordinates (such as the latitude and longitude) that represent that location and the point layer can be used alongside your other GIS layers. Several steps occur in the geocoding process, and in this chapter, we'll examine how they work in ArcGIS 10.2.

geocoding The process of using the numbers and letters of an address to plot a point at that location.

batch geocoding Matching a group of addresses together at once.

reference data The base street layer used as a source for geocoding.

segment A single part of a line that corresponds to one portion of a street.

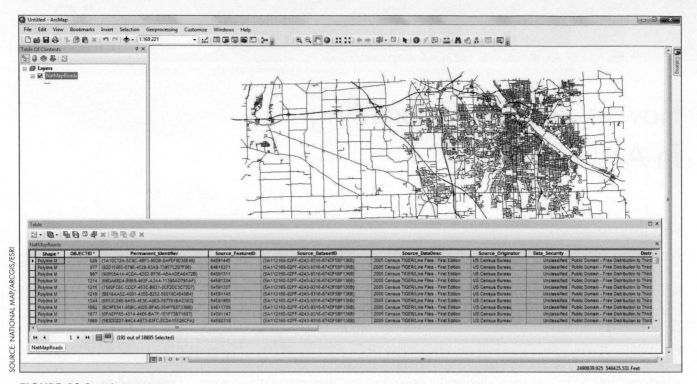

FIGURE 10.1 Mahoning Avenue shown selected in ArcGIS—one road, but 191 segments.

Chapter Scenario and Applications

This chapter puts you in the role of a county librarian working to coordinate several library services, such as deliveries and community outreach. Before you can begin coordinating these services, you need to have a GIS layer of the libraries' locations. Starting with an Excel file of library names and addresses, you'll create a new GIS layer of the libraries' locations and perform some initial spatial analysis of the libraries in relation to the county's main roads.

The following are additional examples of other real-world applications of this chapter's theory and skills:

- An operator at a 911 call center needs to quickly determine a caller's address so that she can convey this information to emergency services. She would use geocoding to determine the location from which the call was placed.

- A marketing department for a retail store is putting together a targeted advertising campaign for certain customers. The marketing department would use geocoding to create a map of where these customers live and can then examine the demographics of these neighborhoods.

- A police detective is investigating a series of break-ins at houses. He can use geocoding to plot the locations of the break-ins as a starting point for analyzing their locations and determining possible patterns.

ArcGIS Skills

In this chapter, you will learn:

• How to construct and use an Address Locator.

• How to perform geocoding by matching street addresses from a spreadsheet to their locations on a road network.

• How to perform interactive rematching and how to fix address information in the geocoding process.

• How to use geocoded points in basic spatial analysis.

Study Area

For this chapter, you will be working with data from Mahoning County, Ohio.

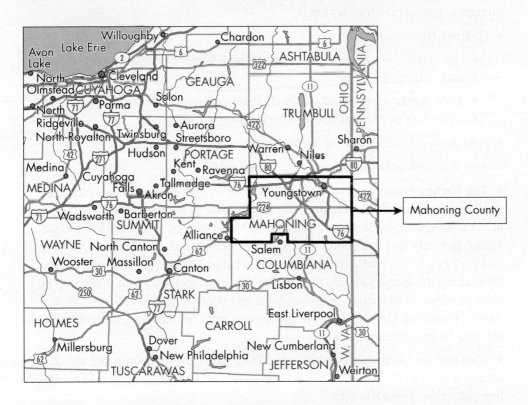

Data Sources and Localizing This Chapter

This chapter's data examine addresses and roads in Mahoning County, Ohio. However, you can easily modify this chapter to use the streets of your own county. The roads layer used in this chapter was downloaded from The National Map, and the same kind of data are available for all U.S. counties. For example, if you were examining Johnson County, Iowa, you would download the Transportation data from The National Map and use the Trans_RoadSegment feature class.

For purposes of this chapter, the roads file was converted from NAD 83 GCS to NAD 83 Ohio North State Plane (to keep network measurements in a projected coordinate system). A new attribute called NEWLENGTH was added to the attribute table, and Calculate Geometry (see Chapter 6) was used to compute the length of each road segment in feet. You can also do the same kind of projection (for instance, Johnson County is in the Iowa North State Plane zone).

This chapter also examines the location of public libraries in the county. The address information for the county libraries comes from the Mahoning County Public Library's Website at http://libraryvisit.org. Using your own county's library Website (or the phone book) you can put together a Microsoft Excel file of library names and addresses (for instance, Johnson County, Iowa's library addresses are online here: http://www.icpl.org/pljc). If you do not have a cluster of libraries in your area, use the addresses of something else like high schools, gas stations, coffee shops, or drugstores.

STEP 10.1 Getting Started

• Start ArcMap and use the Catalog to copy the folder called **Chapter10** from the C:\GISBookdata\ folder to your own D:\GIS\ drive. Be sure to copy the entire directory, not just the files within it.

• Chapter 10 contains:

  • **libraries.xlsx**—an Excel file containing address information for Mahoning County libraries.

  • a file geodatabase called **Geocode** that contains **NatMapRoads**, a line feature class of Mahoning County's road.

• Add the **Geocoding toolbar** to ArcMap.

• Add the NatMapRoads feature class to the map.

• The NatMapRoads layer was projected from its initial coordinate system to another one for use in this chapter. The Coordinate System used in this chapter is **NAD 83 Ohio North State Plane**, using Map Units of **feet**. Check the Data Frame to verify that this coordinate system is being used. Also, in the **Data Frame Properties**, under the **General** tab, change the Display Units to **Miles**. By making these changes, you ensure that any measurements you make will be done in miles (even though the roads layer has been projected to feet). Click **Apply**, then click **OK**. Note that in the bottom right of the View, measurements are now being made in miles.

• Also add the libraries.xlsx table to the TOC (see Chapter 2 for more about adding a table to the TOC). When prompted to add a "Sheet" when adding the libraries table, select **Sheet1$**.

STEP 10.2 Using a Streets or Roads Layer for Geocoding

• Open the attribute table for NatMapRoads and answer Question 10.1. Each of the records represents a segment of a road, not a complete road. Scroll across the attribute table; you'll see several attributes assigned to each of the segments that will be used in the geocoding process. This roads layer will serve as your reference data for the geocoding process in this chapter. See **Smartbox 50** for more about using road layers in the geocoding process.

QUESTION 10.1 How many records (segments) are in the NatMapRoads layer?

Smartbox 50

How is a streets layer used as a reference database for geocoding in ArcGIS 10.2?

A geospatial layer that represents streets or roads will separate a street into a series of segments, each containing attributes that help in identifying the address-related information for a particular segment. The exact attributes and their names will vary depending on the source of the streets layer, but common attributes include:

- the name of the street,
- the prefix of the street (such as North or West),
- the suffix of the street (such as NW or East),
- the type of street (such as Avenue, Blvd, or Road),
- the range of address numbers on each side of the street (usually separated into four categories—the starting and ending addresses on the left side and the starting and ending addresses on the right side),
- the state and city of the segment, and
- the zip code of the segment (sometimes the zip code on the left side of the street and zip code on the right side of the street).

With this level of detail assigned to each segment, the streets layer can serve as reference data for the geocoding process, which will search for a particular segment to which to match an address.

These types of street layers are usually referred to as **street centerline** files. They are line layers used in GIS, created to model each street as a series of line segments. Street centerline files are common forms of GIS data and are available from a variety of sources. Because they are used for geocoding, companies may have their own proprietary street centerline files commercially available, but there are several free alternatives as well. For instance, the roads file from The National Map you're using in this chapter is derived from the U.S. Census **TIGER/Line** files (see Chapter 2 for more about U.S. Census GIS data). TIGER (Topologically Integrated Geographic Encoded Referencing) files for streets are line layers that model streets as segments with attributes similar to those described above and are sources of reference data for geocoding. TIGER/Line files are created by the U.S. Census Bureau and are freely available.

Some sources of streets data for use in ArcGIS 10.2 include:

- TIGER/Line files of roads (at the county level) can be downloaded for free from the U.S. Census Bureau at http://www.census.gov/geo/maps-data/data/tiger-line.html.
- Also, your local county or state GIS division may have street centerline files available for you to download or use.

street centerline A file containing line segments representing roads.

TIGER/Line A file produced by the U.S. Census Bureau that contains (among other things) the line segments that correspond with roads in the United States.

- In ArcMap, open the libraries.xlsx table (use **$Sheet1**). You'll see several fields containing information about the addresses of each library. In the next step, we'll create an Address Locator to help match up the data in this table with their proper locations using the reference data.

STEP 10.3 **Using an Address Locator**

- In the libraries table, you'll see 16 addresses, with their information broken into separate fields called Name, Address, State, City, and Zip (you may have to expand the far-right side of the table to see all attributes and the full five-digit zip code). Close the table when you're done.

- The first step in geocoding is to set up an Address Locator to use for matching. See **Smartbox 51** for more about what an Address Locator is and how it operates.

Smartbox 51

Address Locator An ArcGIS utility that specifies the reference data, geocoding style, and necessary attributes for geocoding.

address locator style The format that will be used for geocoding addresses.

What is an Address Locator and how does it work in ArcGIS 10.2?

An **Address Locator** is the key component of geocoding in ArcGIS 10.2; it establishes the parameters for how geocoding will be performed. The Address Locator will have information on what layer is used as the reference data, what address locator style is used, and which specific attributes from the reference data layer are used (and how they will be used) during the geocoding process. In other words, creating an Address Locator is a "pre-processing" step you must undertake before you can perform the geocoding. Once you have created a single Address Locator, it will be established and you will not need to re-create it if you perform further address matching operations.

When creating an Address Locator, you will have to select which address locator style to use for matching addresses. An **address locator style** determines the criteria that are used for matching an address to a segment in the reference data (and thus also determines what attributes need to be present in the reference data for use in geocoding). An example of an address locator style is the *Dual Ranges* style (which you will be using in this chapter). Dual Ranges is used to find the location of a house on a specific side of a street. To function properly, it needs the following attributes to be present for the segments in the reference data: the name of the street, the low value for the addresses on the left side of the street, the high value for the addresses on the left side of the street, the low value for the addresses on the right side of the street, the high value for the addresses on the right side of the street, the zip code on the left side of the street, and the zip code on the right side of the street. If your reference data don't have these attributes, then Dual Ranges could not be used.

Several other address locator styles are available, as there may be times when you don't need to match an address to a particular house on a street. Retail sales analysts, for example, may target marketing or mailings to customers in a particular zip code; thus, they may need to determine the location of customers only by zip code instead of house address. The *US Address - Street Name* style requires only the name of the street, the city, and the state—but the geocoding will match only to a particular street, not to a particular address on that street. Similarly, the *US Addresses - Zip 5-Digit* style requires only the zip code of an address—but it will geocode points only to a zip code, not to a street address. In short, your choice of address locator style determines not only the kind of output data that will be generated by the geocoding process, but also the kind of attributes needed in the reference data.

- To create your own Address Locator to use for geocoding, return to Catalog and navigate to your Chapter10 folder. Right-click on the **Chapter10** folder, then select **New**, then select **Address Locator**. The Create Address Locator tool will open.

- Press the **Browse** button next to Address Locator Style.

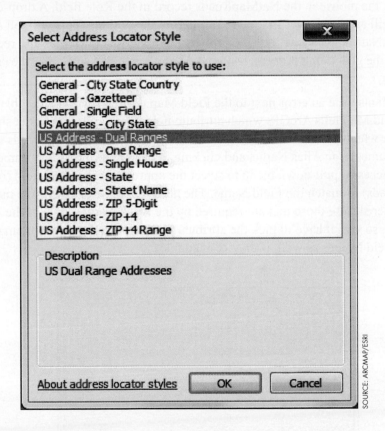

- In the Select Address Locator Style dialog box, choose **US Address – Dual Ranges**, then click **OK**.

- Press the **Browse** button next to Reference Data and select the **NatMapRoads** feature class.

- You'll see you now have an error symbol next to the Reference Data option. Hover the cursor over the error symbol and you'll be told not enough reference data tables are available for this style. You're getting this error because of the role that's been assigned to the NatMapRoads layer.

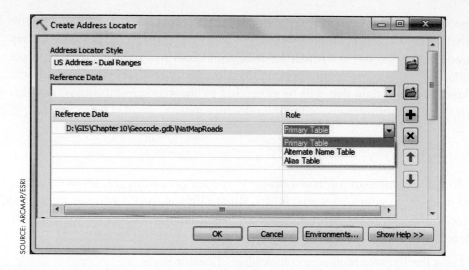

SOURCE: ARCMAP/ESRI

• Click the mouse in the NatMapRoads record in the Role field. A drop-down menu will appear. Select **Primary Table** from the options. This will tell ArcGIS that the NatMapRoads will be the primary source of the attributes and records used in the geocoding process. (After you have made this selection, the error will go away.)

• You'll also see an error next to the Field Map (which is completely blank). The Field Map tells ArcGIS which attribute fields in the reference data should be used for which parts of the Address Locator. Next to each of the records in the Field Names is an Alias Name, and clicking in each record of Alias Name allows you to access a pull-down menu to select the appropriate attribute field (of Nat-MapRoads) to match the Field Name. The headings under Field Name marked with asterisks are those that are required by the Dual Ranges style of the Address Locator, so you'll have to pick the attribute fields to match up with them (and other Field Names as well, if they're available).

SOURCE: ARCMAP/ESRI

- Pick the attribute from Alias Name to match up with Field Name as shown in the table on the right (see **Smartbox 51** on page 234 for more about what each of these attributes represents):

- Under Output Address Locator, name your address locator **NatMapRoadsLocator** and save it as type **Locators** in your D:\GIS\Chapter10 folder.

- Click **OK**. Give the tool time to run, then check Catalog again. A new item, called NatMapRoadsLocator (with the Address Locator icon of a red and white house) will be added to the Chapter10 folder in the Catalog.

Field Name	Alias Name
From Left	Low_Address_Left
To Left	High_Address_Left
From Right	Low_Address_Right
To Right	High_Address_Right
Street Name	Full_Street_Name
Left Zip Code	Zip_Left
Right Zip Code	Zip_Right

STEP 10.4 The Geocoding Process

- Now we can begin geocoding. On the **Geocoding toolbar**, press the **Geocode Addresses** button.

SOURCE: ARCMAP/ESRI

- The Choose an Address Locator to use... dialog will open. Select the **NatMapRoadsLocator** Address Locator. If it's not on the list of options, press the **Add** button, navigate to your D:\GIS\Chapter10 folder, and select it—it will then be added to the list and available for you to select. After you've selected it, click **OK**. The Geocode Addresses dialog will appear.

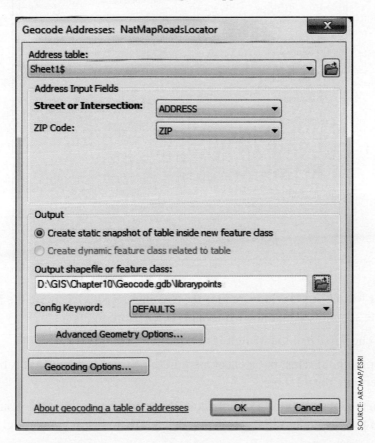

SOURCE: ARCMAP/ESRI

- For Address Table, select the **Sheet1$** from the libraries Excel table as the Address table (this is the table that contains the address information you want to geocode).

- For the Address Input Fields, for Street or Intersection select **Address** and for Zip Code select **Zip**. These reflect the names of the fields in the Excel table to match the items used by the Address Locator. (If they were called something different in the Excel table, you'd have to choose different field names at this point.)

- For Output, select the **Create static snapshot of table inside new feature class** option.

- For the resulting output shapefile or feature class, navigate to your **D:\GIS\ Chapter10** folder and call the output **librarypoints** and save it in your **Geocode** geodatabase. Make sure to save it as the file type **File and Personal geodatabase feature classes**.

- Next, click the **Geocoding Options…** button.

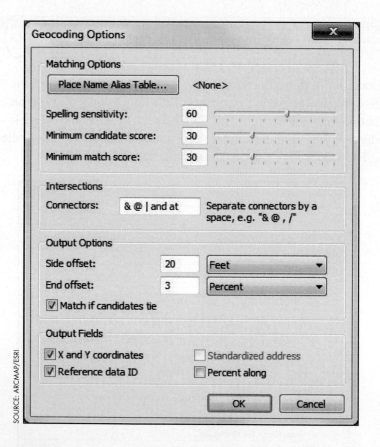

SOURCE: ARCMAP/ESRI

- Under **Matching Options**, reduce the Spelling sensitivity to **60**, the Minimum candidate score to **30**, and the Minimum match score to **30**.

- Put a checkmark in the **Match if candidates tie** box.

- Under Output Fields, put a checkmark in the **X and Y coordinates** box and the **Reference data ID** box.

- Leave the other options as the default. When finished, click **OK**. You will return to the Geocode Addresses dialog. Click **OK** to begin geocoding.

- You will receive a report indicating to you how many of the 16 library addresses were matched, how many were tied, and how many were unmatched. You'll see that 14 of the libraries were matched but two were not. A lot happened "behind the scenes" to geocode these addresses according to the settings you gave ArcGIS 10.2. For more information on what happens during the geocoding process, see **Smartbox 52**.

Smartbox 52

How does the geocoding process work in ArcGIS 10.2?

The geocoding process converts the string of letters and numbers in an address into a point matching a location referenced by a line segment. How? First, the addresses are parsed into the various components that make up the address. **Parsing** breaks a string of numbers and characters (such as an address) into smaller components, separating the street number, the street prefix, and the street name into individual items. Once an address is parsed, its pieces are matched with the segments in the reference data, as indicated by the address locator style. See Table 10.1 for an example of parsing—the second column of the table shows the street address, while the columns to its right show how pieces of the address are parsed into separate headings.

parsing Breaking an address into its component parts.

Location	Address	Prefix Direction	Number	Street Name	Street Type	Suffix Direction
White House	1600 Pennsylvania Avenue NW		1600	Pennsylvania	Avenue	NW
Smithsonian Air and Space Museum	600 Independence Ave SW		600	Independence	Ave	SW
U.S. Government Printing Office	732 North Capitol St NW	N	732	Capitol	St	NW

Table 10.1 Three Washington, D.C., addresses parsed into their component pieces for geocoding

Once an address is parsed, multiple representations of the components of the address are assigned. This is necessary because there are often several ways of representing some of the pieces of an address. For instance, in a street address, "avenue" could be written out as "avenue" or shortened to "ave" or "av," but the geocoding should be able to recognize all three of these representations. As Table 10.1 shows, the White House is on Pennsylvania Avenue, while the Smithsonian Air and Space Museum is on Independence Ave. Both "Avenue" and "Ave" are used in writing these addresses, and the geocoding should recognize that both of these strings of characters mean the same thing. Similarly, the address of the U.S. Government Printing Office could be written as "North Capitol" street or "N Capitol" street, indicating the north section of Capitol Street, or the street itself could simply be called "North Capitol" and not indicate a prefix direction. The geocoding process will thus set up multiple representations of this address to cover these conditions.

Next, the segments are searched according to the address locator style and the various attributes are used to find the best options for segments to match

linear interpolation A method used in geocoding to place an address location among a range of addresses.

the address. For instance (using the Dual Ranges address locator style), if the reference data used to match the address for the White House contain a segment with its name equal to "Pennsylvania Ave," its address range on one side of the street from 1522 to 1608, and its address range on the other side of the street from 1523 to 1609, that segment would likely be one of the best matches, but there will likely be some other potential candidates as well. ArcGIS will assign a ranking of 0 to 100 to each potential candidate segment; higher numbers indicate a greater likelihood that the particular segment is the correct match. ArcGIS will then select the best candidate and choose that segment for the matching. Before beginning geocoding, you can adjust the settings for the minimum score that a candidate should have to be considered a match, as well as the minimum acceptable score that a candidate should have for an address to be matched to it.

Finally, ArcGIS will plot a point corresponding to the address somewhere along the selected segment. If using address ranges (like the Dual Ranges style), ArcGIS will use the ranges of the addresses to determine the side of the street (right or left) on which the geocoded point ends up, as well as how far along the segment the point should be placed. For instance, if an address is at 150 Shady Street and the range of addresses for a Shady Street segment extends from 100 to 198, then the point for 150 will be placed at about the halfway point of the segment. If the address was for 120 Shady Street, it would be placed about one-fifth of the way from the low end of the segment. This method used for placing a point at an approximate distance along a segment is referred to as **linear interpolation**. See Figure 10.2 for an example of plotting a geocoded point in ArcGIS using linear interpolation.

FIGURE 10.2 Plotting a geocoded point for an address in ArcGIS 10.2.

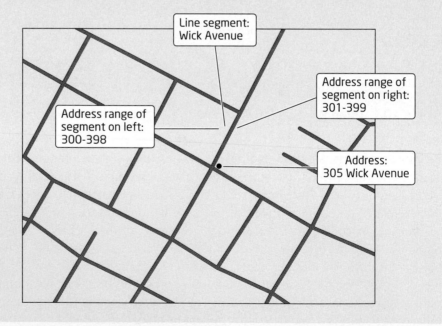

Line segment:
Wick Avenue

Address range of segment on right:
301-399

Address range of segment on left:
300-398

Address:
305 Wick Avenue

STEP 10.5 Rematching Addresses by Editing

• Two of your library addresses did not match. To attempt to rematch them (and to see what was wrong and figure out why they did not match the first time around) click on **Rematch**. This will bring up the Interactive Rematch dialog.

- You may want to expand the Interactive Rematch dialog by grabbing and dragging one of its corners to better work with it on the screen.

- The addresses will be listed at the top of the screen (you may need to scroll across the table to see all of the information). The Status column will show the results of the geocoding—"M" means matched, "T" means tied, and "U" means unmatched.

- We'll examine the first library that was unmatched (Greenford). Locate this record in the Show results box and select it (so the record is highlighted in cyan).

- This address just has a minor problem that can be fixed through editing.

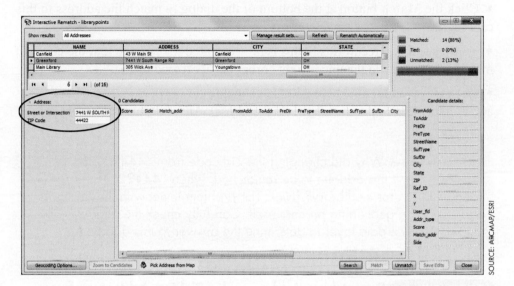

- In the Address options on the left-hand side of the dialog box, delete the ZIP Code value of 44422 and instead type in **44460** and click the **Enter** key on the keyboard.

- Several new options will appear as potential candidates for this address to get matched to. Answer Question 10.2.

QUESTION 10.2 How many candidates are listed as a potential match for the Greenford address?

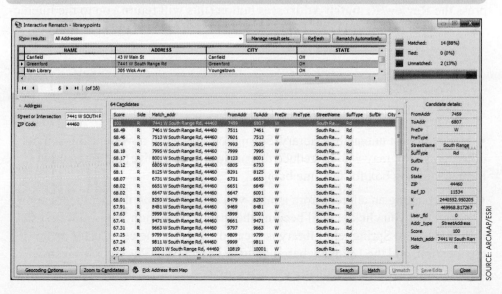

- Select the candidate with the highest score (it should be the candidate at the top of the list) and it will be highlighted in blue. Information about this candidate will be displayed at the right of the Interactive Rematch dialog. Answer Question 10.3.

> **QUESTION 10.3** Which side of the street will ArcGIS place the geocoded point on? Why?

- Click the **Match** button at the bottom of the dialog to match the address to this candidate.

- You'll see that Greenford's Status is now changed to "M" and that the count of matched addresses in the upper-right corner has increased to 15.

- Click Close on the Interactive Rematch dialog for now (we'll come back to it again in the next step to match the final address). Answer Question 10.4.

> **QUESTION 10.4** Why did changing the Zip code from "44422" to "44460" allow this address to be rematched, when "44422" is the correct, listed, address for the library? (*Hint*: The problem is not with the original address or the geocoding process itself. Carefully check the attribute table of the reference data layer to determine the answer to this question.)

STEP 10.6   Rematching Addresses by Picking Addresses from the Map

- Sometimes, you can correct geocoding errors by editing the address information to get things to match up, but other times you may find yourself simply selecting the proper location interactively from the map, in essence telling ArcGIS "put the point at this location." ArcGIS refers to this process as *Pick address from map* and it is how we will geocode the second unmatched library.

- In the TOC, select the **librarypoints** layer. Then, on the Geocoding toolbar, press the **Review/Rematch Addresses** button (this will return you to the Interactive Rematch dialog box).

SOURCE: ARCMAP/ESRI

- Select the last unmatched library, Springfield. As we'll now be using some other tools, you may want to reduce the size of (or minimize) the Interactive Rematch dialog box for the time being.

- When picking an address from a map, you'll need to find its specific location. The spot that you click on will become the matched location. Thus, our first step is to find where Springfield Library is located. To get some context for picking the location, add the **Imagery** basemap to the TOC (see Chapter 5 for instructions on how to add a basemap).

- Next, from the Tools toolbar, select the **Go To XY** icon.

SOURCE: ARCMAP/ESRI

- Go To XY allows you to type in an x and y coordinate (such as decimal degrees or UTM coordinates). ArcGIS will shift the View to those coordinates.

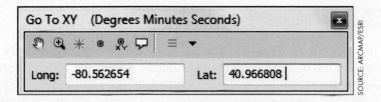

SOURCE: ARCMAP/ESRI

- For an approximate set of coordinates to use for Springfield Library, type **-80.562654** for Long and **40.966808** for Lat and then press the **Enter** key on the keyboard. ArcGIS will use four lines and a dot to show you the location of these latitude and longitude coordinates.

- Carefully look at the location of the X/Y coordinates (which, remember, are an approximation) and the imagery to locate the library. Locate a place in the view where you wish to geocode Springfield Library (keep in the mind that the geocoded location of your other points will be slightly offset from the road on the proper side of the street). When you've identified a good location on the map to place the point, answer Question 10.5.

QUESTION 10.5 Where will you place the geocoded point for Springfield library? Why?

- Return to the Interactive Rematch dialog and make sure the unmatched Springfield library record is selected. Press the **Pick Address from Map** button at the bottom of the dialog.

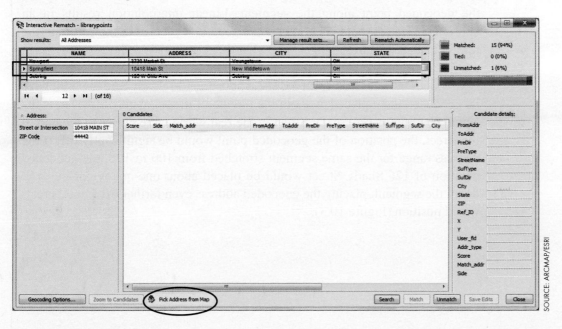

SOURCE: ARCMAP/ESRI

• You'll see that your cursor has turned into a crosshairs. Return to the view and click the left button of the mouse on your selected location for Springfield Library. Before doing anything else, right-click on the mouse and choose **Pick Address**. You'll see a point now placed at that location.

• Return to the Interactive Rematch dialog; You'll see that you're up to a full 16 matches. In the Interactive Matching dialog, click **Close**.

STEP 10.7 Examining Geocoded Results

• The fact that a result is returned to you as "Matched" doesn't necessarily mean that its point has been placed in the absolutely correct geocoded location. Points are sometimes placed in a location that's not correct, yet may be flagged as "matched" by a geocoding process (and not just in ArcGIS 10.2). For more information about why geocoding results may contain inaccuracies, see **Smartbox 53**.

Smartbox 53

Why are geocoded results sometimes inaccurate in ArcGIS 10.2?

Geocoded results are only as good as the reference data used for performing the geocoding. If the reference data are out of date (for instance, if they don't contain streets that correspond to the new subdivision to which you're trying to match an address), then you may end up with strange results as a point is placed on a different street with a similar name. Similarly, if the reference data contain errors or missing data in address ranges, zip codes, or other essential attributes, the accuracy of the geocoded results may be compromised. Sometimes, when a street address cannot be located, the center of its zip code is used as the next best source for matching; in this case, the point is "matched" not to a house address, but to the zip code instead.

Another source of inaccuracy in geocoded results is linked to linear interpolation, which uses the range of addresses as a guide for where to place the geocoded point along the segment. Linear interpolation results in an approximation of where the point should go, and it adheres to the values in the reference data's address ranges, not what the distribution of houses may be in real life. For instance, in the reference data, the segment for Shady Street may have values between 100 and 148, and a house at 126 Shady Street would have its geocoded point placed at about the middle of the segment. However, if the actual position of 126 Shady Street in the real world were closer to the end of the street, the position of the geocoded point would be significantly off. If the address range for the same segment stretched from 100 to 198, the geocoded position of 126 Shady Street would be placed about one-quarter of the way down the segment, placing the geocoded address even farther off from its real-world position (Figure 10.3).

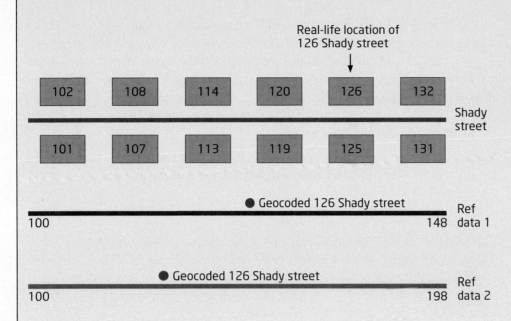

Real-life location of
126 Shady street

FIGURE 10.3 The location of 126 Shady Street in the real world and its geocoded position (via linear interpolation) using two different sets of reference data.

- As a way of checking your geocoded results, you can compare the points you geocoded using the Dual Ranges address locator style and the free road layer from The National Map against the results generated by a Geocoding Service (see the *Related Concepts for Chapter 10* for more about using Geocoding Services in ArcGIS 10.2). Through the Geocoding toolbar, you can use free ArcGIS Online resources to find the location of one address at a time.

- To start, open the **libraries.xlsx** table and position it somewhere on the screen so that you can read the full address information and still see most of the view.

- Next, on the Geocoding toolbar, from the pull-down menu, select **World Geocode Service (ArcGIS Online)** as the Address Locator to use. This is a Geocoding Service available through ArcGIS Online.

- Type the address of the first library in the list (Austintown Library) in the Find Address box (**600 S Raccoon Rd, Youngstown, OH 44515**). Be sure to type the address information just as it is in the libraries.xlsx file and include all parts—street address, city, state, and zip code. The Geocoding Service will use all of these items to match the address (not just the address ranges and zip code information you used in your NatMapRoadsLocator Address Locator).

- After you've typed the address, press the **Enter** key on the keyboard.

- A pop-up saying the address was Found will appear. After this appears, right-click on the address in the Find Address box and select **Zoom To**.

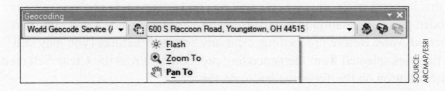

• The view will shift to the location of the Austintown Library and four lines and a large dot will show its location. Change the symbology of your librarypoints layer to really stand out. You should see the geocoded Austintown Library from the librarypoints layer on top of the Imagery basemap (if you turned that off, turn it back on again). To see the results from the Geocoding Service again, right-click on the address on the Geocoding toolbar and choose **Zoom To** (or select **Flash** to simply show the dot and lines again).

• Zoom in on the view so that you can clearly see the location of your geocoded results as well as the location returned from the Geocoding Service. Answer Question 10.6.

QUESTION 10.6 Where are your geocoded results of the Austintown Library located in relation to the location provided by the ArcGIS Online Geocoding Service? You should also be able to see the library on the Imagery basemap. Where are the two locations positioned in relation to the actual library?

Some hints to aid you with this question:

• If necessary, turn on the NatMapRoads layer and use the Identify tool to get information about the names of roads.

• You may want to use a utility like the Birds Eye View from Bing Maps (see Chapters 6 and 7) to aid in identifying which building on the basemap is the library in question.

• Continue through the other 15 library addresses in the list, comparing the results of your geocoding with that of the Geocoding Service locations. As you complete each one, answer Question 10.7. *Important Note:* When checking the Greenford Library, remember that you properly geocoded it by changing the zip code to 44460. Use that as the zip code here as well.

QUESTION 10.7 For each of the libraries, describe the position of both results (your geocoded points and the results provided by the Geocoding Service) in relation to each other and the library building as you can observe it in the Imagery basemap. Be specific in your answers.

• After you have reviewed all 16 points, you can remove the Excel version of the libraries table from the TOC (because you've already converted its data into a set of points).

STEP 10.8 Basic Spatial Analysis of Geocoded Points

• In this last part of the chapter, you will be conducting some analysis of your geocoded points in relation to the roads layer to answer some spatial questions. *Important Note:* Before proceeding, clear any selected features (you may still have libraries selected from the geocoding procedure). Press the **Clear Selected Features** button on the Tools toolbar to do this.

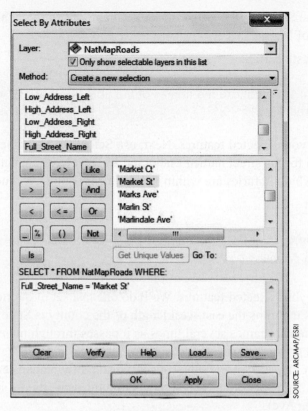

First, we will examine how many libraries are near a major road (Market Street) in Mahoning County. To begin, use the Select By Attributes tool (see Chapter 2) and build a query to find all of the road segments that comprise Market Street. Build your query to find all of the road segments (records) in the NatMapRoads layer that have the attribute of **Full_Street_Name** equal to **'Market St'**.

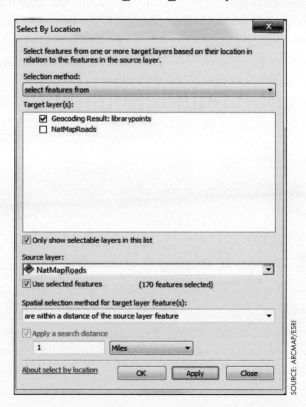

- Next, use the Select By Location tool (see Chapter 8) to select all libraries within 1 mile of Market Street.

- Answer Question 10.8.

QUESTION 10.8 What are the names of the libraries that are within 1 mile of Market Street?

- Clear all of your selected features. Next, use Select By Attributes to find all the road segments for the street named Mahoning Ave, then use Select By Location to determine which libraries are within 1 mile of Mahoning Avenue. Answer Question 10.9.

QUESTION 10.9 What are the names of the libraries that are within 1 mile of Mahoning Avenue?

- Clear all of your selected features. We'll do one last set of queries. One of the main roads that runs the east-west length of the county is State Route 224. However, it changes names several times as it passes through many different townships and populated places. We want to select all the road segments of State Route 224, but this will mean finding several differently named segments (as opposed to all segments of the same road having the same name, such as Market St or Mahoning Ave).

    - Use Select By Attributes to find all segments with the following names: US Hwy 224, Akron Canfield Rd, W Main St, Boardman Canfield Rd, Boardman Poland Rd, W McKinley Way, E McKinley Way, and Center Rd. State Route 224 is also composed of road segments with the name Main St; however, there are two Main St's in the county, and the ones that correspond with State Route 224 have the Zip code 44406 (so be sure to select only these segments, not any other segments with the name Main St).

    - When these are all selected, you'll see the whole of State Route 224 stretched across the county with all road segments selected.

    - Next, use Select By Location to find all libraries within 1 mile of State Route 224.

- Answer Question 10.10.

QUESTION 10.10 What are the names of the libraries that are within 1 mile of State Route 224?

STEP 10.9 Sharing or Printing Your Results

- Save your work as a map document. Include the usual information in the Map Document Properties.

- Next, display only the NatMapRoads layer and the geocoded librarypoints. Turn on the labels for the librarypoints, using the Name field. You may have to adjust the font or symbol to make the names stand out in relation to the lines of the roads layer.

- Finally, either print a layout (see Chapter 3) of your final version of your geocoded libraries (including all of the usual map elements and design for a layout) or share your results as a map service through ArcGIS Online (see Chapter 4).

- If you are sharing your data, you may also want to share your Address Locator via ArcGIS Online so that others can use the locator you created. You can share your locator by creating a **locator package** (.gcpk file)—this will contain the Address Locator. In Catalog, right-click on your **NatMapRoadsLocator**, select **Share As**, and then select **Locator Package**. You can then either save the locator package to a file or upload it to your ArcGIS Online account. You'll need to add information about the locator under the Item Description entry in the Locator Package dialog before you can share or save the package.

locator package A single file created in ArcGIS that allows for the sharing of Address Locators.

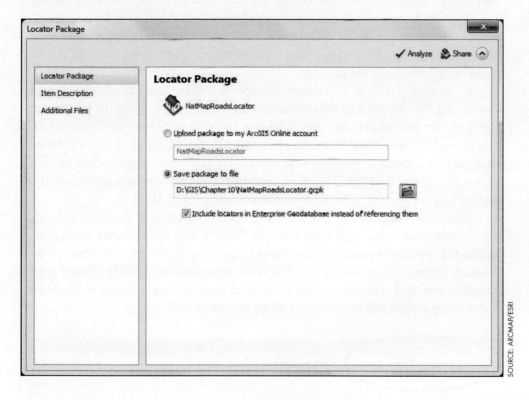

SOURCE: ARCMAP/ESRI

Closing Time

This chapter demonstrated how to take a set of addresses and match them to their geographic locations using a street layer and geocoding techniques. Geocoding is a powerful tool that you likely will use often, whether you're finding a destination with an online mapping Website, a mapping app on your phone or tablet, or a vehicle navigation system in your car. ArcGIS moves beyond simple address location by providing the tools necessary to geocode multiple addresses and then use the plotted data for spatial analysis. Through ArcGIS resources, Esri provides other types of geocoding options; see the *Related Concepts for Chapter 10* for more information about cloud-based Geocoding Services.

There's a lot more that can be done with street data and geocoded results. For instance, a service like Google Maps will allow you to geocode an origin and a destination and then show you what it computes as the shortest driving route between those two points. ArcGIS allows you to compute shortest paths between multiple points on a street network, and we'll use the same data from this chapter to do just that in Chapter 11.

Related Concepts for Chapter 10   Geocoding Services and ArcGIS 10.2

geocoding service An online utility that allows for geocoding one or more addresses.

One thing you've probably noticed is that, when you use an online system for geocoding a single address (such as Google Maps—https://maps.google.com) or batch geocoding multiple addresses (such as BatchGeo—http://batchgeo.com), you don't have to go through many of the steps outlined in this chapter, such as setting up an address locator or examining ranking results. These activities are happening "behind the scenes" on the Website. The advantage to using one of these systems (referred to as **geocoding services**) is that the steps in the geocoding process (from setting up the reference database, to selecting the type of address locators, to the matching itself) have already been implemented, leaving you to focus on the results of the process.

In ArcGIS 10.2, a geocoding service is available through the World Geocoding Service via ArcGIS Online. On the Geocoding toolbar, the pull-down menu allows you to choose from available Geocoding Services (Figure 10.4). You can select these cloud-based services to act as the Address Locator; the reference data are located within the cloud as well. By selecting the geocoding service shown in Figure 10.4 (World Geocode Service), you can type an address into the toolbar and see the geocoded location (as you did in Step 10.6).

You can also use the World Geocoding Service as a source for batch geocoding. Rather than using your own Address Locator and downloaded street centerline file, you can instead use the World Geocoding Service when selecting an Address Locator. However, use of this service requires an ArcGIS Online subscription and will consume credits from your account (see Chapter 4 for more about using credits and the ArcGIS Online subscriptions).

FIGURE 10.4 Geocoding Services on the Geocoding toolbar in ArcGIS 10.2.

SOURCE: ARCGIS/ESRI

For More Information

For further in-depth information about the topics presented in this chapter, use the ArcGIS Help feature to search for the following items:

- Commonly used address locator styles
- Creating a locator package
- Creating an address locator
- The geocoding process
- The geocoding workflow
- Understanding address locator styles
- Working with the ArcGIS Online World Geocoding Service (RC)

For more information about using the World Geocoding Service with ArcGIS 10.2, see: http://www.arcgis.com/home/item.html?id=305f2e55e67f4389bef269669fc2e284.

Key Terms

geocoding (p. 229)

batch geocoding (p. 229)

reference data (p. 229)

segment (p. 229)

street centerline (p. 233)

TIGER/Line (p. 233)

Address Locator (p. 234)

address locator style (p. 234)

parsing (p. 239)

linear interpolation (p. 240)

locator package (p. 249)

geocoding service (p. 250)

Networks aren't limited to roads. Other types of transportation networks can also be modeled using ArcGIS, such as railroads or walkways. The tools for creating and working with transportation networks are in a separate extension for ArcGIS, called the Network Analyst, which we'll be using in this chapter. While we'll be working only with a road network in this chapter, other types of non-transportation networks can be modeled in ArcGIS, such as utilities or power lines, but these are better handled with a different set of network tools (see the *Related Concepts for Chapter 11* for more information).

Chapter Scenario and Applications

This chapter puts you in the role of a local librarian who is coordinating several tasks for a county library system. Your jobs include overseeing book deliveries to libraries as well as identifying the high schools most likely to make use of a library's resources. You'll be creating a network in ArcGIS, determining the shortest driving distance between libraries to cut down on delivery times, and using the network applications to connect schools and libraries.

The following are additional examples of other real-world applications of this chapter's skills:

SOURCE: © KEITH ERSKINE / ALAMY

- A dispatcher for a messenger service needs to route a courier on the quickest path to a destination through an area where several roads are heavily congested. He can use the routing tools (explained in this chapter) to determine the shortest route to take.

- A 911 operator needs to determine the closest hospital to an accident location so that EMTs can reach the hospital as quickly as possible. She can use the closest facility tools in ArcGIS to do so.

- A realtor showing potential homes to clients wants to find the optimal driving route between houses so that she can show as many homes as possible within a certain amount of time. The route tools of Network Analyst can set up her best driving routes.

ArcGIS Skills

In this chapter, you will learn:

- How to activate ArcGIS extensions.
- How to use the features of the ArcGIS Network Analyst.
- How to create different types of "shortest distance" routes between a set of stops.
- How to incorporate barriers into route analysis.
- How to create Service Areas using Network Analyst.
- How to use the Closest Facility tools of Network Analyst.

Study Area

- For this chapter, you will be working with data from Mahoning County, Ohio.

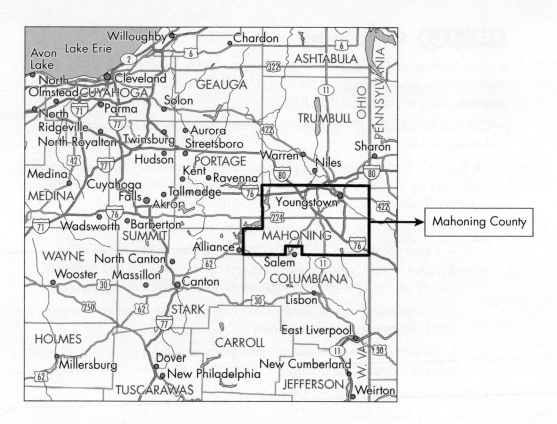

Data Sources and Localizing This Chapter

This chapter's data examines addresses and roads in Mahoning County, Ohio. However, you can easily modify this chapter to use the streets of your own county. The roads layer used in this chapter was downloaded from the National Map and the same kind of data is available for all U.S. counties. For instance, if you were examining Johnson County, Iowa, you would download the Transportation data from the National Map and use the Trans_RoadSegment feature class. TIGER/ Line files or (if available) local county or state street centerline files (see Chapter 10) may also be used as a source for building a network dataset.

For purposes of this chapter, the roads file was converted from NAD 83 GCS to NAD 83 Ohio North State Plane (to keep network measurements in a projected coordinate system). A new attribute called NEWLENGTH was added to the attribute table, and Calculate Geometry (see Chapter 6) was used to compute the length of each road segment in feet. You can do the same kind of projection (for instance, Johnson County is in the Iowa North State Plane zone).

This chapter also focuses on the location of public libraries (and high schools) in the county. The address information for the Mahoning County high schools comes from http://www.publicschoolreview.com/county_public_schools/stateid/OH/ county/39099. Similar information for other U.S. counties is available from this Website (for instance, Johnson County, Iowa, school information is available at http://www.publicschoolreview.com/county_schools/stateid/IA/county/19103). Using your own county's library Website (or the phone book), put together a Microsoft Excel file of library and school names and addresses and geocode them into a point layer (see Chapter 10 for more about geocoding).

STEP 11.1 **Getting Started**

- Start ArcMap and use the Catalog to copy the folder called **Chapter11** from the C:\GISBookdata\ folder to your own D:\GIS\ drive. Be sure to copy the entire directory, not just the files within it.

- Chapter11 contains a file geodatabase called **MahoningNet** that contains the following items:

  - Roads: a feature dataset that contains four feature classes. A feature dataset is necessary here to contain the Network Dataset that we will create in this chapter. This feature dataset contains the following feature classes:

    – NatMapRoads: a line feature class of Mahoning County roads.

    – Librarypoints: a point feature class containing the locations of 16 libraries in Mahoning County. This represents geocoded and corrected library addresses from the Mahoning County Public Library Website.

    – Construction: a point feature class containing the locations of simulated construction blocking a road.

    – Schoolpoints: a point feature class containing the locations of 30 high schools in Mahoning County.

- Add the **Network Analyst** toolbar to ArcMap.

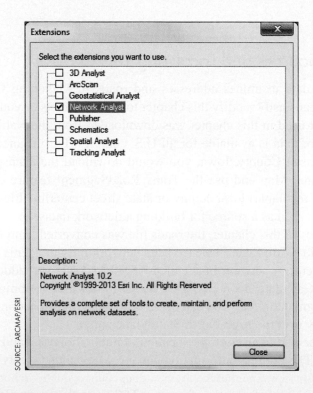

SOURCE: ARCMAP/ESRI

- Several ArcGIS network functions and tools are part of an **extension** that must first be activated before it can be used. Extensions are separate ArcGIS components that enable a wide range of features, but they do not automatically come included with the core system. You'll use different extensions in several upcoming chapters, but the one that will allow you to use different network

extension An add-on set of functions for ArcGIS.

functions is the **Network Analyst**. To activate it, from the **Customize** pull-down menu, select **Extensions** and place a checkmark in the **Network Analyst** box.

- Click **Close** in the Extensions dialog. The Network Analyst extension is now activated. When you need other extensions, you can activate them by simply placing a checkmark in their corresponding box.

> **Network Analyst** The ArcGIS extension used with transportation networks.

STEP 11.2 Creating a Network Dataset

- In the Catalog, navigate to your D:\GIS\Chapter11 folder and expand it so you can view its contents. Expand the MahoningNet file geodatabase and also the Roads feature dataset it contains.

- Right-click on the **Roads** feature dataset and select **New** and then select **Network Dataset**. A wizard (i.e., a series of menus) will walk you through the process of creating a network dataset from the road segments in the NatMapRoads feature class that the Roads feature dataset contains. If the Network Dataset option is unavailable, see Troublebox 8 for assistance. For more information about the elements of a network dataset in ArcGIS, see Smartbox 54.

Troublebox 8

Why can't I create a new Network Dataset in ArcGIS 10.2?

The Network operations (such as creating a new Network Dataset, as well as the other functions described in this chapter) are available only when the Network Analyst is activated. If these functions are unavailable, be sure that a checkmark is placed next to Network Analyst in the Extensions dialog box.

Smartbox 54

What are the elements of a Network Dataset in ArcGIS 10.2?

In ArcGIS, a network will be stored in a **Network Dataset**, which will use the files stored as a feature dataset in a geodatabase for its construction. The initial streets layer that contains the segments (which will be used as edges in the network) must be one of the files in the feature dataset. If other files are used, they will be stored in the feature dataset as well.

A Network Dataset is not simply a layer of lines that represent road segments, as a network allows you to design routes and do analysis with them. For this reason, a network has to contain much more information than just a layer of line segments. First, a network can contain information about **turns**. That is, parts of the network can have information about items such as where left turns cannot be made, where U-turns can be made, or what streets are one-way. This kind of information is important because you don't want to create routes that allow vehicles to travel the wrong way on a one-way street or allow them to make turns where (in the real world) these turns would be illegal. Turn information can be set up at a junction and is stored as a separate turn feature class.

Another key component of a network is its **connectivity**, or how all of the edges link up with one another. Proper connectivity in a network ensures that items that should link together do, while others that shouldn't, don't. For instance,

> **Network Dataset** An ArcGIS structure of a series of connected junctions and edges, commonly used for transportation-related problems.
>
> **turns** Information used by ArcGIS to determine information about valid flows along a network.
>
> **connectivity** The linkages between edges and junctions of a network.

impedance A value that represents how many units (of time or distance) are used in moving across a network edge. Also referred to as *transit cost.*

ArcGIS needs to understand that the junction where four line road segments come together allows travel to continue from one segment to another. However, the segments of a road and a freeway overpass should not connect in a way that would allow a vehicle to turn from the road onto the freeway at that point.

Similarly, if a roads layer has lines that represent both roads and railroad tracks, the network connectivity should not be set up in a way that allows drivers to continue from a road onto a railroad track. It may sound silly to think that a GIS would instruct a driver to drive on a railroad track, but keep in mind that a roads layer consists only of line segments. Without something to explicitly tell the GIS that the segments of roads and the segments of railroads should not connect, the GIS simply sees edges and junctions that fit together and does not automatically distinguish a road from a railroad. If different parts of a city transportation network are modeled separately (such as roads, railroads, sidewalks, and subways), ArcGIS uses special connectivity groups that allow the user to specify which of these groups connect. (For example, sidewalks and metro stations connect to each other, and metro stations and subways connect to each other, but sidewalks and subways do not connect.)

A network also needs a means of measuring the travel distance between destinations on a route. A route is made up of a series of edges and there is a cost associated with traveling across that edge. This travel cost (referred to as **impedance** in ArcGIS, and sometimes also referred to as *transit cost*) is used to determine the overall shortest route. For example, the cost of traveling the length of an edge could be the physical distance of that edge (for instance, a quarter of a mile). This cost could also be expressed in the time it takes to drive that distance (for instance, about 20 seconds, if you were driving at 40 mph). Depending on the type of transportation network being used, this cost could reflect the length of the segment and its assigned speed limit; it might also have additional impedance due to heavy traffic conditions, or it could reflect the amount of walking time it takes to travel the segment.

If a route consists of 10 segments, the total cost of that route will be the sum total of the travel costs of all the segments. If the impedance value being used is the actual distance you had to travel each segment, the overall cost of the route would be the 10 distance values added together. If the impedance value is the time it takes to drive each segment, the overall cost of the route would be the 10 drive time values added together. Thus, the overall cost of taking a route reflects whatever impedance values are used in the network.

• In the first menu you can give your network a name. Accept the default name for your Network Dataset (**Roads_ND**). Also select **10.1** for the version of your Network Dataset. Click **Next**.

• In the second menu, you will select which of the feature classes in the Roads feature dataset will participate in the network. Place a checkmark next to only the **NatMapRoads** line feature class—this will be the only feature class used in building the network (we'll use the other three later). If you wanted to include other layers in the network, here is where you would specify them.

- In the third menu, you can choose to model turns for the network. For this network, you want to model turns, so select **Yes** and accept the default for the use of Global Turns. Click **Next**.

- In the fourth menu, you can set connectivity options for the network. If you wanted to alter things, you could click on the Connectivity button, but for this chapter, we'll use the default connectivity settings, so just click **Next**.

- In the fifth menu, you can add elevation data to the features of the network. If you had elevation data to combine with your network, you would set it up here. However, this chapter will not be using elevation data, so select the radio button for **None** and click **Next**.

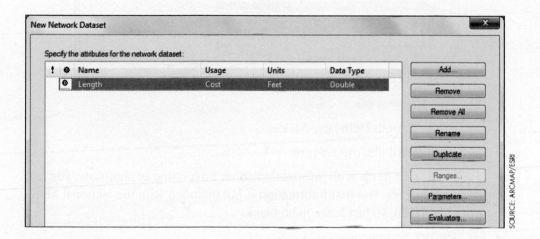

- In the sixth menu, you'll define the attributes (such as the cost of traversing an edge) of the network. Your dataset has one attribute to use as the cost field: the length of the edge itself. This is measured by the values in the Length attribute (which has already been added to the Network by default). If the NatMapRoads layer had other fields with information about other factors (such as the time equivalent of traversing an edge), you would add it here (by pressing the Add button). With the single cost field (Length) already added, just click **Next** to continue.

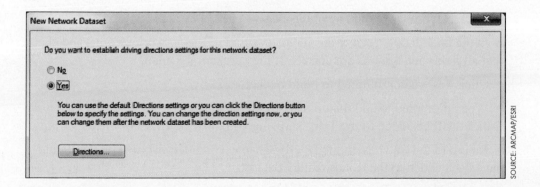

- In the seventh menu, select the radio button for **Yes** to create driving directions for your network. Click on the **Directions** button to properly specify the fields used in creating driving directions.

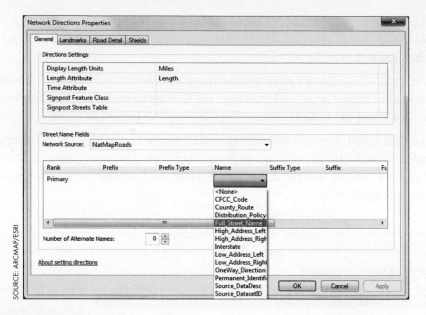

- Select the **General** tab.

- For Display Length Units, use **Miles**.

- For Length Attribute, use **Length**.

- If you had other fields with information about travel time or signposts, you could use them here. But that information is not included with the National Map transportation data, so just leave them blank.

- For Network Source, use **NatMapRoads**.

- For the other options for Street Name Fields, you'll see that the field for "Name" is currently empty. ArcGIS will not be able to properly generate directions if you do not supply the name of the road to be driven on in the network. In the row marked Primary, click under the Field heading for Name and a series of options will appear. Choose **Full_Street_Name** as the field that contains names of roads to use. If the NatMapRoads layer had contained other information about each network edge (such as the prefix or suffix), you would select those fields here.

- Click **Apply** when you're done, then click **OK** to return to the Network dialog.

- Click **Next** to leave the Directions page.

- In the final dialog, review the summary of all of the network settings and check that all fields and options are correct. If they are, click **Finish**.

- Click **Yes** when prompted to build the network.

- When prompted to add feature classes to the map, click **No**. The new Network Dataset will be added to the TOC, showing you only the edges of the network.

- Return to the Catalog. In the D:\GIS\Chapter11 folder, two new items should have been added to the Roads feature dataset:
  - the Network Dataset file called **Roads_ND**
  - a point feature class of junctions for the network called **Roads_ND_Junctions**

- Right click on **Roads_ND** and select **Item Description**. In the box that appears, select **Preview** (if prompted about adding feature classes, again select **No**). Do the same for the **Roads_ND_Junctions** file. You should now see the proper edges and junctions for the Mahoning County network.

STEP 11.3 **Creating a Route by Visiting Stops in Order**

- Next, you'll want to find the shortest driving distance route among all of the libraries. To begin, add the librarypoints feature class to the TOC.

- In Data Frame Properties, change the Display Units to **Miles** (the Map Units will remain as feet).

- On the **Network Analyst toolbar**, select the **Show/Hide Network Analyst Window** icon to open the Network Analyst Window. A new dockable and pinnable window called Network Analyst will open on the left side of the screen.

SOURCE: ARCMAP/ESRI

- From the **Network Analyst** pull-down menu, select **New Route**. A new geodatabase called "Route" will be added to the table of contents, consisting of information about stops, barriers, and so on. A new item called Route will be added to the Network Analyst Window that corresponds with this new Route in the TOC.

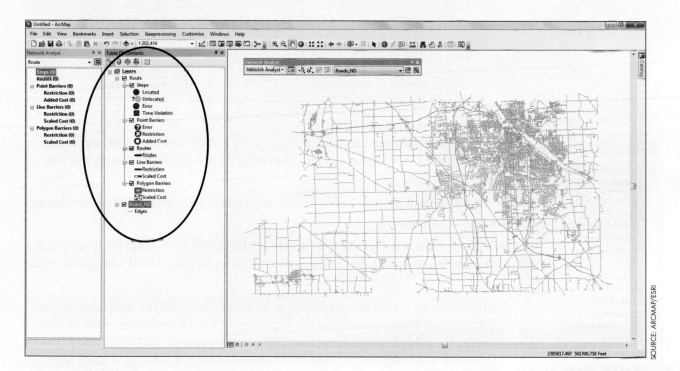

SOURCE: ARCMAP/ESRI

- In the **Network Analyst Window**, select **Route** from the pull-down menu and press the **Route Properties** button.

- In the Layer Properties dialog box, select the **General** tab and rename this route **libroute1** and click **OK**. In the Network Analyst Window, you'll see your new libroute1 and its number of Stops, Routes, and Barriers, all of which should be 0.

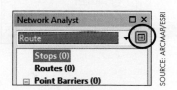

SOURCE: ARCMAP/ESRI

- In your role in this chapter as a local librarian, you are helping a delivery person who has to make a set of deliveries to the libraries. To help the delivery person, you have to locate the shortest driving distance route between all libraries. In this case, the library locations will be considered **stops** along the network. Your next step will be to add the points representing the libraries as stops to be visited.

- In the **Network Analyst Window**, select **libroute1** from the pull-down menu.

- In the **Network Analyst Window**, right-click on **Stops** and select **Load Locations**.

stops Destinations to visit on a network.

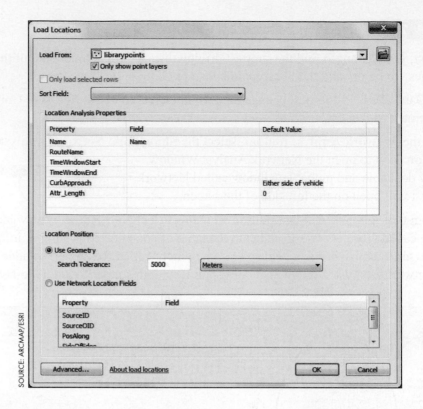

SOURCE: ARCMAP/ESRI

SOURCE: ARCMAP/ESRI

- Choose the librarypoints feature class as the source you want to **Load From** and select the **Use Geometry** radio button for the Location Position. Click **OK** when you're done.

- You'll see that libroute1 now has 16 stops added to it. Expand the Stops list to locate the 16 libraries.

- Answer Question 11.1. Keep in mind that the libraries are listed in alphabetical order and thus that will also be the numerical order method ArcGIS assigns to them.

QUESTION 11.1 Use the Identify Tool to familiarize yourself with each of the 16 libraries. Briefly sketch out the locations of the 16 points, as well as their names (i.e., Austintown, Boardman, Brownlee Woods, and so on) and the initial number assigned to each one. You are doing this for your own reference in future questions.

- You'll now create the best (shortest) route among these libraries.

- Click the **Solve** button on the Network Analyst toolbar.

SOURCE: ARCMAP/ESRI

Solve Button

- Your "best route" will appear (if you receive a message about M-values, just close it, as we are not using M-values in this chapter). A new item will be added to your Roads feature dataset in the TOC called libroute1. For further information on how these shortest routes are calculated, see **Smartbox 55**.

Smartbox 55

How is a shortest route calculated in ArcGIS 10.2?

As described in **Smartbox 54**, the cost of traveling a route is determined by summing up the impedance costs for each segment of the route. However, when you're traveling between an origin and a destination on a road network, there are multiple different routes you can take to get there. For instance, if you're driving from your house to another friend's house on the other side of town, there are likely numerous routes you could take. Some of them are likely very direct while others would take you well out of your way but would eventually get you there. A system (such as an online Website or a vehicle navigation device) that's finding your shortest route needs to have a way of determining which of these routes is the shortest (based on the impedance value being used, such as the shortest overall driving time or the shortest overall driving distance). ArcGIS uses an **algorithm**, or a set of steps used in a process, to figure out which route, of all the possible ways of traveling between two points, is the shortest one.

ArcGIS utilizes a shortest path algorithm called **Dijkstra's algorithm**, with some modifications to it based on the setup of the network and destinations. Dijkstra's algorithm is designed to determine the shortest path between an origin point and other destinations on a network. Figures 11.2 through 11.9 provide simple examples of how Dijkstra's algorithm finds the shortest path from an origin point to locations on a network.

Figure 11.2 shows an initial network of four locations (nodes or junctions) that are connected by edges. Node *a* is the origin node; we want to find the shortest path from node *a* to node *d*. Each of the edges has a cost assigned to it for traveling across that edge (the units of the impedance value), and you'll see that some of the edges have different costs depending on the direction being traveled. Note that only the origin node (node *a*) has a value assigned to it (a value of 0 because it's our starting point); right now, we have no idea how much the cost will be to travel to any of the other nodes.

In the first step of the algorithm (Figure 11.3), we will travel each of the edges from the starting node *a* to the nodes to which it directly connects (node *b* and node *c*) and we will assign a value to those two nodes for the total cost of travel to that node. For instance, the travel cost of moving from node *a* to node *b* was 5 units, so node *b* is assigned a value of 5, while the cost of moving from node *a* to node *c* was 9 units, so node *c* is assigned a value of 9. Node *a* is now colored red to signify that we've already evaluated the costs for traveling away from that node (and we don't have to do that again).

Since we've already searched from node *a* to all directly connected nodes, we need to select another node from which to search. Figure 11.4 shows this step. The algorithm will look at the values of all nodes (except for the ones from which we've already searched) and try to decide which node we should use to begin searching from again. The algorithm will choose the node with the lowest value. In this case, the value of 5 in node *b* is lower than the value of 9 in node *c*. Thus, node *b* will become our new node from which to begin searching.

algorithm A set of steps used in a process.

Dijkstra's algorithm A mathematical process that will determine the shortest path along a network from an origin node to the other nodes on the network.

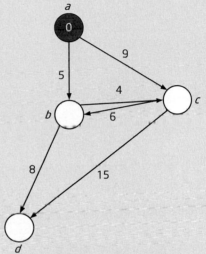

FIGURE 11.2 A sample network of four junctions and six edges.

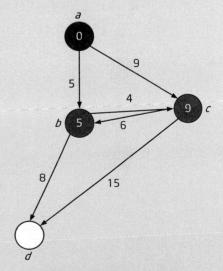

FIGURE 11.3 Searching from the origin node (node *a*) to the two nodes directly connected to it (node *b* and node *c*).

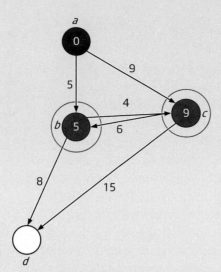

FIGURE 11.4 Selecting the new node to begin searching from again (the node with the lowest value—either node *b* or node *c*).

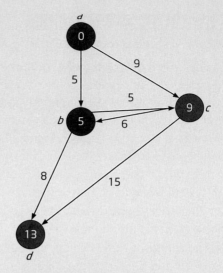

FIGURE 11.5 Searching from node *b* and updating the values for the nodes it directly connects to (node *c* and node *d*).

We will now travel from node *b* to all directly connecting nodes and update the values of those nodes (Figure 11.5). Node *b* connects directly to node *c* and node *d*. It costs 8 units to travel from node *b* to node *d*, so node *d*'s value is updated to 13 (we take node *b*'s value of 5 and add the additional 8 units it would take to travel from node *b* to node *d*). The cost of traveling from node *b* to node *c* is 5 units, so under this logic, node *c* would be updated to a value of 10 (the value of 5 in node *b* is added to the other 5 units it would take to travel from node *b* to node *c*). However, node *c* has already been assigned a cost to travel to it (a value of 9). Because node *c*'s value of 9 is lower than the travel cost from node *b* to node *c* (a value of 10), it is not updated to the higher value. A node's value is updated only if the new value is lower than the existing value (indicating that a shorter path to that node has been found). Node *b* is also now colored red, indicating that we can no longer search for paths from it.

With node *b* now exhausted, we need to find a new node from which to search. As before, the algorithm will look at the values of the nodes that haven't been searched from yet and select the lowest one (Figure 11.6). Node *c* has a value of 9 and node *d* has a value of 13, so node *c* becomes our new node from which to search.

Node *c* connects directly to node *b* and node *d*. The total cost of traveling from node *c* to node *b* would be 15 (the value of 9 for node *c* added to the cost of 6 for traveling from node *c* to node *b*). The value of 15 is higher than node *b*'s value of 5, so node *b* is not updated. Similarly, the cost of traveling from node *c* to node *d* would be 24 (the value of 9 for node *c* added to the cost of 15 for traveling from node *c* to node *d*). Because this value is higher than the current value of node *d* (13), node *d* is not updated (Figure 11.7). Node *c* is now exhausted, so it is colored red (and we can no longer search from it).

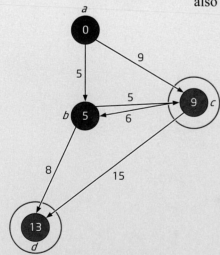

FIGURE 11.6 Determining the new node to begin searching from (the node with the lowest value, either node *c* or node *d*).

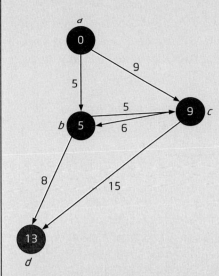

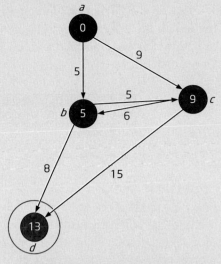

FIGURE 11.7 Updating the values for the nodes directly connected to node *c* (node *b* and node *d*). Both nodes already have lower values, so they are not updated.

FIGURE 11.8 Determining the new node to begin searching from (only node *d* remains as a choice).

The algorithm now needs to select a new node from which to begin searching. There is only one node left that is not exhausted (node *d*), so this one is selected as the new search node (Figure 11.8).

However, there are no nodes for it to directly connect to, so no comparisons of values are made and no updating is done. Node *d* is colored red because its searching is exhausted (Figure 11.9). Because there are no more nodes from which to search, the algorithm is completed and all shortest paths have now been found. The shortest path to travel from node *a* to node *d* will cost a total of 13 units and the shortest route is to travel from node *a* to node *b* and then to node *d*. Any other possible routes will be longer than this.

ArcGIS uses the same type of algorithm to find the shortest distance between an origin and a destination, with some adjustments to allow locations to be found at locations other than junctions (nodes), and also to allow for settings in the network, such as turns.

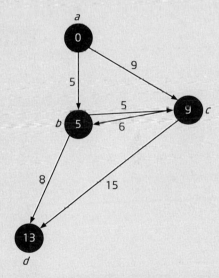

FIGURE 11.9 Node *d* does not directly connect to any other nodes, so searching is complete and the algorithm has determined all shortest paths.

- In the Network Analyst Window, you'll see that your count of Routes has increased by 1 (this is your new libroute1 route). Expand this **libroute1** (it will be called Austintown-West because the route started at Austintown Library and ended at West Library). Right-click on this route (by right-clicking on it, you'll select it) and choose **Properties**. Answer Question 11.2.

QUESTION 11.2 How long is the best route between the stops in miles? (***Important Note:*** The length being reported back from Network Analyst is in the units of the network, which are feet.)

- Click on the **Directions** button on the Network Analyst toolbar to gain information on the driving directions along the route (similar to a Google Maps set of directions). Also, closely examine the route that ArcGIS plotted for you. Answer Question 11.3.

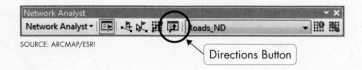

SOURCE: ARCMAP/ESRI

Directions Button

QUESTION 11.3 In examining the driving directions, do you see any particular directions that seem non-intuitive to a driver? (*Hint:* Chances are you're unfamiliar with driving throughout Mahoning County, Ohio. However, look for any directions that drivers may be unlikely to take, such as switching to an interstate for very short distances, or similar kinds of activities.) Why would ArcGIS provide these kinds of directions?

- Clear selected features when you're done.

STEP 11.4 Creating a Route by Rearranging Stops

- The libroute1 calculated the shortest path on the network among the 16 stops, but if you look at it, the route does a lot of criss-crossing and backtracking in order to visit the 16 stops in the numeric order that was specified. Some of the libraries (such as Lake Milton or Sebring) are in more remote areas of the county, while others (such as Brownlee Woods, Struthers, and Poland) are relatively close together. To save time, it would make more sense for a driver to visit the closer libraries one after the other instead of having to drive far away to a remote location and then back again. What we will do next is compute a new delivery route that rearranges the order of the stops. For more information about reordering stops to compute best routes, see **Smartbox 56**.

Smartbox 56

How does reordering stops affect shortest routes in ArcGIS 10.2?

By changing the order of the stops you visit, you can adjust the overall cost of traversing a route. For instance, see Figure 11.10, which shows two different routes used to visit five different stops (public libraries in Mahoning County, Ohio). Figure 11.10a shows the route generated by visiting the stops in a predetermined order; Dijkstra's algorithm is used to find the shortest route between library number 1 and library number 2, then the shortest route between library number 2 and library number 3, and so on. You'll see that there's an awful lot of backtracking being done in order to find the shortest path between stops in this particular order. In contrast, Figure 11.10b shows another shortest route between the five stops, but this time the order of the stops is more optimal. The route begins at one end, visits the stops nearest one another, then ends at the furthest stop.

TSP The Traveling Salesman Problem, a mathematical process involving determining the optimal configuration of rearranging a series of stops on a network.

Reordering stops gives you more flexibility when trying to determine the best route for visiting all destinations. In this case, the Traveling Salesman Problem (**TSP**) can help. The TSP is a mathematical process that determines the

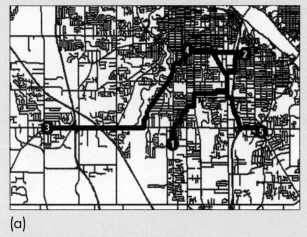

(a)

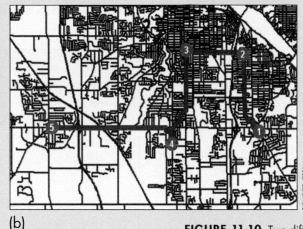

(b)

FIGURE 11.10 Two different shortest routes between stops (public libraries in Mahoning County, Ohio)—visiting stops in a predetermined order (11.10a) and visiting stops in a rearranged, more optimal order (11.10b).

SOURCE: ESRI

optimal configuration of stops on a network. Think of it this way: A traveling salesman must visit several destinations to make sales and needs to find the shortest (or least-cost) route among all of these destinations. The TSP will examine the different combinations of how to reorder all the destinations to find the best route. (For instance, is the lowest overall cost incurred by visiting the five stops as 1-2-4-5-3, or by visiting them as 2-3-4-5-1, and so on?) While algorithms can provide decent solutions to the problem (and ArcGIS uses heuristics to come up with good solutions), the only way to really know which of the different combinations of ordering the stops is the best is to work through all of the combinations, something that would be impossible with a large number of stops.

You can place additional constraints on the reordering of stops. You may want to keep the first stop and the last stop fixed and shuffle only the order of stops in between (for instance, if you have errands to run, you will leave work—a fixed stop—and then find the best order to run your errands, then arrive home at the end—a second fixed stop). You may also want to keep the first stop the same as the last stop (for example, if you were planning a new bus route, the initial stop and the final stop must both be the location of the bus depot in order to simulate the bus leaving and returning to its point of origin each day). There may be times when you are not able to shuffle the order of the stops and will have to visit locations in a predetermined order (these constraints are faced by a shipping company vehicle that has to make time-sensitive deliveries).

- From the **Network Analyst** pull-down menu, select **New Route** (this will be your second route).

- In the Table of Contents, a new Route geodatabase will appear. Minimize this (to save on space).

- In the Network Analyst Window's pull-down menu, you'll see the new Route as an option. Select **Route** from the pull-down menu.

- Load the locations of the stops like you did with libroute1, so you have the 16 libraries loaded as stops.

- Press the **Route Properties button** in the Network Analyst Window.

- Under the **General** tab, rename the Route **libroute2**.

• To make changes to libroute2, choose the **Analysis Settings** tab.

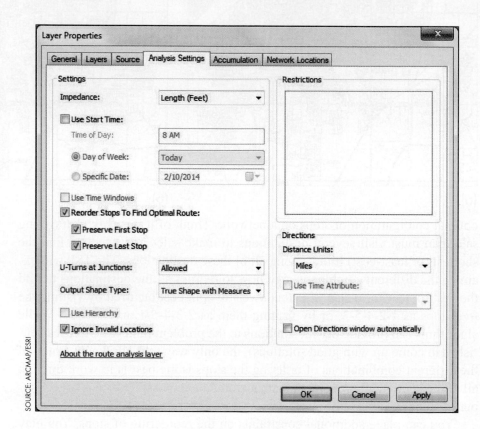

SOURCE: ARCMAP/ESRI

• Select **Length (feet)** for your Impedance.

• Click on **Reorder Stops to Find Optimal Route**.

• Also click on the **Preserve First Stop** and **Preserve Last Stop** options. This means that the first and the last stop are both preserved, or fixed (i.e., you must start at stop #1 and must end at stop #16).

• Use the other defaults. Click **OK** when done.

• Click the Solve button on the Network Analyst toolbar and libroute2 will be computed. A new item will be added to your Roads feature dataset in the TOC called libroute2. Answer Questions 11.4 and 11.5. You may find it helpful to Clear Selected Features for both libroute1 and libroute2 when examining both of them.

QUESTION 11.4 How long is this new best route (libroute2) between the libraries (in miles)?

QUESTION 11.5 How does libroute2 differ from libroute1 (and not just "it's shorter" or "it's longer"—how specifically has the route itself changed in relation to the stops and the computed path)?

- Make a third new route and name it **libroute3**. Load the librarypoints as stops for libroute3. In the Analysis Settings for this third route:

  - Select **Length (feet)** for your Impedance.

  - Click on **Reorder Stops to Find Optimal Route**.

  - Do NOT click on either the **Preserve First Stop** or **Preserve Last Stop** options. This means that the first and the last stop are no longer fixed (i.e., you no longer need to start at stop #1 and end at stop #16).

- Solve this libroute3. A new item will be added to your Roads feature dataset in the TOC called libroute3. Answer Questions 11.6, 11.7, and 11.8.

QUESTION 11.6 How long is this new best route (libroute3) between the stops (in miles)?

QUESTION 11.7 How is the arrangement of stops different among the three routes? (That is, which stop is visited first, which is visited second, and so forth in each of the three routes? List them as #1 through #16 for each route.)

QUESTION 11.8 You have three shortest paths among the libraries. What is so special about the specific route taken by libroute3 to make it the shortest of the three?

- Clear selected features when you're done.

STEP 11.5 **Creating Routes with Barriers on the Network**

- Turn off libroute1 and libroute2 but leave libroute3 on. In this next step, we'll focus only on libroute3 (the shortest or least-cost of the three paths).

- Add the construction feature class to the map. Construction consists of seven points representing places where the roads have been blocked due to construction. Network Analyst will treat these seven points as absolute barriers, not letting you cross the road at these spots. You're now going to re-examine your libroute3 with this new constraint added to them.

- Make a fourth New Route and call it libroute4. In the Analysis Settings for this fourth route, use the same settings as libroute3:

  - Select **Length (feet)** for your Impedance.

  - Click on **Reorder Stops to Find Optimal Route**.

  - Do NOT click on either the **Preserve First Stop** or **Preserve Last Stop** options. This means that the first and the last stop are not fixed (i.e., you do not need to start at stop #1 and end at stop #16),

- Load the librarypoints as stops for this libroute4. Do NOT solve just yet—you have to add the construction barriers as an added constraint to the network.

- In the Network Analyst Window, be sure that libroute4 is selected from the pull-down menu.

- In the Network Analyst Window, right-click on **Point Barriers** and choose **Load Locations**.

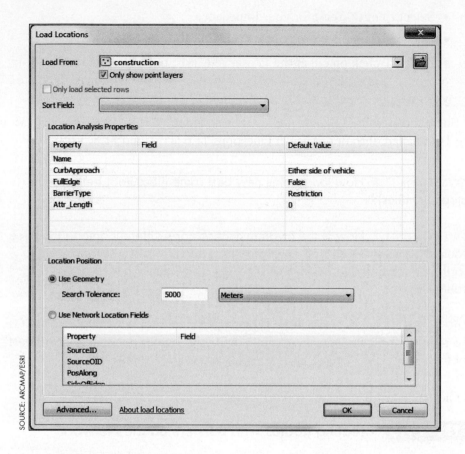

SOURCE: ARCMAP/ESRI

- Under Load Locations, choose the **construction** feature class.

- For BarrierType, choose **Restriction**. The other option here would be Added Cost. In our case, these barriers are completely blocking the path, not just increasing the transit cost, so Restriction should be used.

- Use the other defaults and click **OK**. You should now have seven barriers added to your route.

- Expand the Barriers option to see the seven barrier locations (identified as Location 1, Location 2, and so on). They will show up in the View as large red (x) symbols.

- With regard to library locations, you'll see that with the barriers:
  - Four of them are between Main and West.
  - Three of them are near Springfield (on different nearby roads).

- Solve libroute4. Keep in mind that libroute 4 has the same problem constraints as libroute3, except it now contains these construction barriers.

- Answer Questions 11.9 though 11.13. Turn off all four of the routes when you're done.

QUESTION 11.9 What is the Length (in miles) of libroute4 with these barriers?

QUESTION 11.10 How did the ordering of the stops change because the barriers were added (i.e., what is the new ordering of the stops—which is first, which is second, and so on)?

QUESTION 11.11 Why did ArcGIS compute this new order of the stops (in Question 11.10) because of the barriers being added?

QUESTION 11.12 How (specifically) did the path between West and Main change?

QUESTION 11.13 How (specifically) did the path going and coming from Springfield change?

STEP 11.6 Working with Service Areas

• Routes are not the only tools that can be created with networks. Another network function is a Service Area that will show those parts of the network near (or "served by") a particular point. See **Smartbox 57** for more information about Service Areas.

Smartbox 57

What are service areas and how are they used in ArcGIS 10.2?

A **service area** is used in network analysis to determine which parts of a network are near a particular location (referred to as a **facility**) and are thus "being served" by that location. The service area is a polygon (or series of polygons) defined by the impedance factor being used (such as the actual driving distance or the driving time) for the network. Figure 11.11 depicts two service areas for fire stations based on driving time. The service areas in red show which parts of the network can be reached from the fire station within a certain amount of time. (Note that this value can also include the *turnout*, or amount of time needed to scramble and deploy vehicles from a facility, not simply the time or distance of the network.)

Service areas help determine the accessibility of a facility and can be thought of as a buffer around the facility point (see Chapter 9), but this buffer is constrained by the properties of the network rather than just a large circle. For instance, a circular 1-mile buffer around a fire station is likely significantly different than the places on a network that can be reached within one mile of driving along city streets. Service areas are used for finding which parts of a network can be reached by emergency vehicles within a certain amount of time, which parts of a network are within a certain distance of a store, or which sections of a network are within a certain walking time of a school (among many other examples).

service area A polygon boundary created around sections of a network to determine which areas of the network are within a certain distance of a location.

facility Locations from which service areas are created.

FIGURE 11.11 Service areas for fire stations based on their driving response times.

• From the Network Analyst toolbar, select the **Network Analyst** pull-down menu, and then select **New Service Area**. A new Service Area geodatabase will be added to the TOC and a new item called Service Area will be added to the Network Analyst window. In the Network Analyst window, press the **Service Area Properties button** and under the **General** tab change its name to **libservice**.

• Back in the Network Analyst Window, select Libservice from the pull-down menu. Right-click on **Facilities** and choose **Load Locations**. Just as you loaded librarypoints before as stops, now do the same thing and load them as facilities. Libraries will now be considered the facilities you're trying to identify a service area for.

• In the Network Analyst Window, select **libservice** from the pull-down menu and press the **Service Area Properties button**. Select the **Analysis Settings** tab.

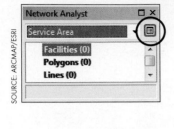

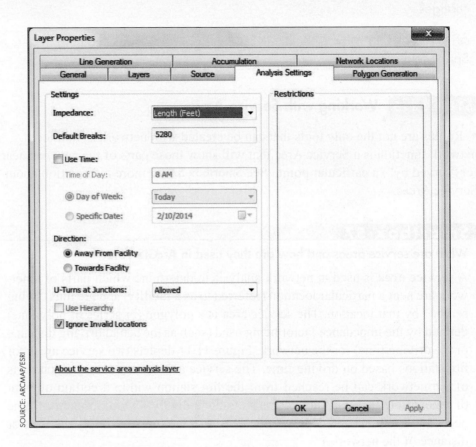

• Choose **Length (Feet)** as your impedance value.

• In Default Breaks use **5280** (because there are 5280 feet in a mile, you will be looking at 1-mile service areas).

• Select **Away From Facility** as the Direction of service.

• Click **OK** when finished.

• Press the **Solve** button on the Network Analyst toolbar to solve for the locations of your service areas. A new set of polygon areas (the service areas) will be created at 1 mile around each library.

• Answer Question 11.14.

QUESTION 11.14 Are there any overlapping service areas? If so, which libraries have overlapping areas within the 1-mile regions? (Your answer may be something like "Austintown Library and Main Library overlap service areas within 1 mile.")

Working with Closest Facilities

• The closest facility function can be used to determine which facilities are closest (in terms of network distance) to a set of incidents (see **Smartbox 58** for more information about the closest facility function). In this step, you will be determining which libraries are closest to which high schools in terms of their network distance (rather than creating a buffer or using a search radius from Select By Location). This step will help determine which libraries a high school student is most likely to use.

Smartbox 58

What is the closest facility function and how does it work in ArcGIS 10.2?

The **closest facility** function of ArcGIS determines (using a network and impedance values) which one of several locations on a network (referred to as facilities) is the nearest one via the network to a separate location (referred to as an **incident**). For instance, the closest facility function will find which hospitals (facilities) are closest in terms of network distance to an accident location (an incident).

Figure 11.12 provides an example of the closest facility function. The goal is to determine which public libraries are nearest to each high school, via travel along a network, while the impedance value is the actual physical driving distance along the network. For each high school, ArcGIS will identify the library that is closest to it via the driving distance (using Dijkstra's algorithm) and plot out that route. Note that each high school can only be closest to one library, but a single library may often be the closest facility for many high schools.

closest facility A process used to determine which locations on a network are nearest (via network distances) to a different set of locations.

incident The locations for which the closest facility function tries to determine the nearest facility.

FIGURE 11.12 The output of the closest facility function, showing the shortest network distance between high schools (the incidents, shown with flags) and the library nearest to the school (the facilities, shown with blue dots).

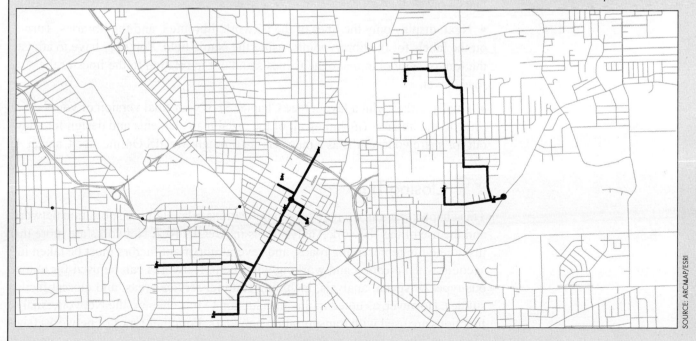

SOURCE: ARCMAP/ESRI

• Add the schoolpoints layer to the Table of Contents.

• From the **Network Analyst toolbar's Network Analyst** pull-down menu, select **New Closest Facility**. Change its name to **Closestlib**.

- Back in the Network Analyst Window, select **Closestlib** from the pull-down menu. Right-click on **Facilities** and choose **Load Locations**. The same way you loaded librarypoints before as stops, now do the same thing and load them as facilities. Libraries will now be considered the facilities for which you're trying to identify the closest facility.

- Right-click on **Incidents** and choose **Load Locations**. Use the schoolpoints as the layer to load for the incidents. These are the points for which you want to find the closest facility.

- Press the **Solve** button on the **Network Analyst toolbar** to solve for the closest library to each school (in terms of the network). A new set of lines will be created showing which schools are assigned to each library. Look at which schools connect to which libraries (and also expand the "Routes" option in the Network Analyst Window). Answer Questions 11.15 and 11.16.

> **QUESTION 11.15** Which library is considered the closest facility for Cardinal Mooney High School?

> **QUESTION 11.16** Which library is considered the closest facility for Ursuline High School?

STEP 11.8 **Saving, Printing, or Sharing Your Results**

- Save your work as a map document. Include the usual information in the Map Document Properties.

- Next, display only the lines of the network, libroute3, and the libraries. Turn on the labels for the librarypoints, using the Name field. You may have to adjust the font or symbol to make the names stand out in relation to the lines of the roads layer.

- Finally, either print a layout (see Chapter 3) of your final version of your best route to the libraries (including all of the usual map elements and design for a layout) or share your results as a map service through ArcGIS Online (see Chapter 4).

Closing Time

This chapter examined a number of applications related to transportation networks and the use of the Network Analyst tools in ArcGIS 10.2. Networks are more than just a line file representing roads, and several important factors must be taken into account when creating and working with them. Networks can be used for numerous types of spatial analysis, including routing, service areas, and closest facility studies. However, as noted in the introduction to this chapter, the Network Analyst functions are used with transportation networks. For other types of networks (such as power lines or telephone lines), it is more common to use a geometric network (see the *Related Concepts for Chapter 11* for further information).

Starting with the next chapter, we'll be switching gears from the vector data of points, lines, and polygons and looking at a new kind of GIS data representation called raster data. There are many features in the real world that vector data may not be suited for modeling in GIS, and raster data give us a new way of representing these concepts.

Related Concepts for Chapter 11 — Geometric Networks in ArcGIS 10.2

Geometric networks are used in ArcGIS 10.2 to model and analyze non-transportation related features. There are many things that move along networks that are not vehicles—for example, gas pipelines, sewer lines, and electrical mains. All of these can be modeled with GIS. In these cases, functions such as shortest routes will be less applicable than techniques such as tracing the flow of water upstream or determining the faulty electrical transformer that's causing a power outage in a subdivision.

Like a transportation network, a geometric network consists of a series of edges and junctions with proper connectivity. Also like a transportation network, a geometric network is set up with the proper feature classes present in a feature dataset, but instead of creating a new Network Dataset (as you did back in Step 11.2), you would create a new Geometric Network. A separate toolbar (the Utility Network Analyst toolbar) is used for working with the functions of a geometric network. These tools allow you to set up barriers on a geometric network as well as place flags on junctions or edges that represent key points in trying to solve problems (for example, placing a flag at the location of a broken transformer or water main).

The techniques used for solving problems in a geometric network are called **traces**. ArcGIS comes with a variety of traces. Analysts can trace how the flow along a utility network upstream or downstream will be affected when the power is shut off, or they can find which parts of a network are connected or not connected to a particular point. For instance, a trace may be used to determine the areas on a network that will be affected by a power failure (and this information can be combined with other data to figure out which homes will not have power).

geometric network An ArcGIS structure of a series of connected junctions and edges, commonly used for utility-related problems.

trace The problem solving techniques used with a geometric network.

For More Information

For further in-depth information about the topics presented in this chapter, use the ArcGIS Help feature to search for the following items:

- A quick tour of geometric networks
- About tracing on geometric networks
- Algorithms used by the ArcGIS Network Analyst extension
- An overview of the Geometric Network toolset
- Closest facility analysis
- Designing the network dataset
- Network elements
- Route analysis
- Service area analysis

- The network dataset layer
- Turns in the network dataset
- What are geometric networks?
- What is a network dataset?
- What is the ArcGIS Network Analyst extension?

Key Terms

How to Use Raster Data in ArcGIS 10.2

Introduction

The last eleven chapters have used the points, lines, and polygons of the vector data model to represent real-world items. This system is intuitive; after all, building footprints have definite boundaries that are modeled with polygons, and sidewalks have definite starting and stopping points that are modeled with lines. However, many types of phenomena aren't easily modeled with a set of objects. For instance, every location on Earth has a value for the measured temperature or elevation at that spot. If you were to represent these measurements in GIS using the vector data model, you would need a near-infinite set of points, each with an assigned value.

A different way of viewing these types of continuously varying phenomena is to think of them as a surface stretching across Earth, with every location on the surface having a value for temperature or elevation at that spot. This way of looking at the world is called the **continuous field view**. In this model, phenomena are modeled as a surface (or field), and every location on the surface can hold a value for each phenomenon and there will be no gaps in the measurements.

GIS usually uses the **raster data model** to represent these continuous surfaces. Raster data doesn't use points, lines, or polygons. Rather, it uses a set of evenly distributed square **grid cells**. While the cells themselves are square, the entire raster can be either square or rectangular and it contains a certain number of cells. For example, a raster with seven rows and five columns would contain 35 grid cells (Figure 12.1). Each cell represents the same sized area on Earth's surface. For instance, each grid cell could represent 9 square meters, or 900 square meters, or 9 square kilometers, depending on the grid cell size being used in the model. In our 7 × 5 raster grid example, if each cell was 900 square meters in size (that is, measuring 30 meters on one side and 30 meters on the other), the entire raster surface would cover a geographic area of 31,500 square meters (35 cells × 900 square meters). All grid cells represent the same size—a single raster cannot have mixed sized cells within it.

Each grid cell contains a single value representing the phenomenon being measured in that area. For example, if our sample raster represented elevation, each cell would have one value representing the height above sea level for the area of the ground covered by the cell. With raster data, it's assumed that the value for the cell is applied to the entire area of the cell. Thus, for our sample grid, all

continuous field view The conceptualization of the world that all items vary across Earth's surface as constant fields, and values are available at all locations along the field.

raster data model A conceptualization of representing geospatial data with a series of equally spaced and sized grid cells.

grid cell A single square unit of a raster.

FIGURE 12.1 A sample raster and its grid cells.

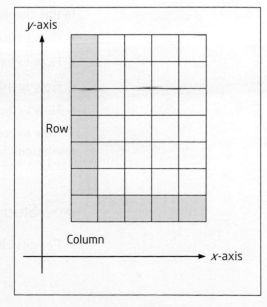

places within the 30 × 30 meter area of a grid cell would be assumed to have the same elevation value. Thus, every location within the raster surface contains an elevation value. In addition to elevation, raster data is often used to represent phenomena such as soil type or land cover and is frequently used in delineating watersheds, modeling groundwater, and measuring solar radiation levels on Earth's surface.

Chapter Scenario and Applications

This chapter puts you in the role of a county environmental planner tasked with putting together a map of the county's land cover resources (forests, water bodies, and agricultural areas). Before you can begin the mapping process, you'll need to acquire and examine land cover data in ArcGIS.

The following are additional examples of other real-world applications of this chapter's skills:

• As part of a research project about land management policies, a forest manager needs to compile information on the amount of deciduous and coniferous forests in an area. A surface of raster land cover provides the basis for him to do so.

• An urban planner needs to assess the amount of impervious surface within a suburban area as part of an analysis of water runoff. Raster data provides an excellent basis for modeling and measuring this area.

• A geologist is undertaking a large-scale study of environmental conditions across a state as these conditions relate to hydraulic fracturing. Utilizing data about elevation, soil conditions, and land cover in a raster format allows her to develop a large-scale dataset within GIS.

ArcGIS Skills

In this chapter, you will learn:

• How to use Geoprocessing Settings for raster data.

• How to handle raster data in ArcMap and the Catalog.

• How to use raster zones, regions, and raster attribute tables.

• How to the use the National Land Cover Dataset (NLCD) as a layer in ArcMap.

• How to convert vector data to raster format at a variety of raster grid cell resolutions.

Study Area

• For this chapter, you will be working with data from Mahoning County, Ohio.

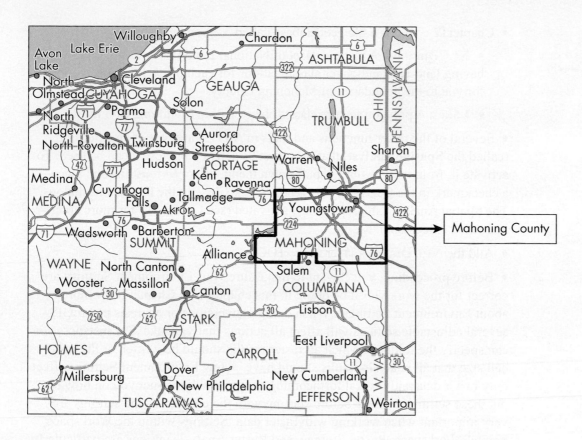

Mahoning County

Data Sources and Localizing This Chapter

This chapter's data focus on features and locations within Mahoning County, Ohio. However, you can easily modify this chapter to use data from your own county or local area instead. The lakes data were extracted from the county-level hydrography dataset available for download from The National Map (see Chapter 5). Use the NHDWaterbody feature class for the county as the lakes layer.

The NLCD 2006 dataset was downloaded (for free) from the Multi-Resolution Land Characteristics Consortium (MRLC), which is online here: http://www.mrlc.gov. NLCD 2006 data for the entire United States can be downloaded from that site, and the data for your individual county (for example, Delaware County, Indiana) can be clipped out using ArcGIS. Alternatively, you can download NLCD 2006 data for a county directly via The National Map.

STEP 12.1 Getting Started

• Start ArcMap and use the Catalog to copy the folder called **Chapter12** from the C:\GISBookdata\ folder to your own D:\GIS\ drive. Be sure to copy the entire directory, not just the files within it.

- Chapter12 contains a file geodatabase called **Mahonraster** with two items:

  - **NLCDmahon**: A raster dataset of National Land Cover Database (NLCD) having land-use/land-cover classifications for each pixel. It has been clipped for you to the boundaries of Mahoning County.

  - **Lakes**: A polygon feature class of lakes within Mahoning County.

- Several of the raster functions and tools in ArcGIS are part of an extension (called the **Spatial Analyst**) that must first be activated before it can be used. To activate it, from the **Customize** pull-down menu, select **Extensions** and place a checkmark in the **Spatial Analyst** box. Click **Close** in the Extensions dialog. The Spatial Analyst extension is now activated (see Chapter 11 for more about activating extensions).

- Add the **NLCDmahon** layer to the TOC.

- Before proceeding, you'll want to make sure your Environment Settings are correct for the work you'll be doing in this chapter (see Chapter 9 for more about Environment Settings). When doing conversions or analysis in ArcGIS, several adjustable settings will affect all actions that you take. For instance, you can specify the folder to which all raster results should be written or the raster cell size that all analysis results should have. A few Environment Settings affect only raster data and several forthcoming chapters will instruct you on how to put these settings in place before starting with the chapter, as the settings are very important when working with raster data. Settings within the Workspace, Scratch Workspace, Raster Storage, and Raster Analysis options are particularly important.

- In this chapter, you'll be doing several vector-to-raster conversions, so we want to specify the output coordinate system to be used for all of them at once. To begin, choose the **Geoprocessing** pull-down menu and select **Environments…**.

- Expand the option for **Workspace**. In the Current Workspace box, browse to **D:\GIS\Chapter12**. Also in the **Scratch Workspace box**, browse to **D:\GIS\Chapter12**.

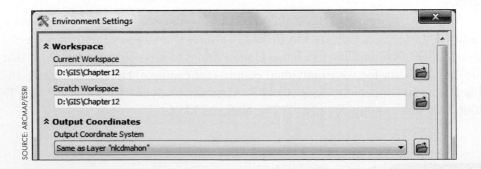

- Expand the option for **Output Coordinates**. In the **Output Coordinate System** pull-down menu, select **Same as Layer "nlcdmahon"**. The projected coordinate system of the NLCD layer (NAD 83, UTM Zone 17) will appear in the box below it. By selecting this option, you are instructing ArcGIS to create all output with the same coordinate system as the raster NLCD layer.

- Click **OK** to put these Geoprocessing Environment Settings into place.

STEP 12.2   Raster Data Basics

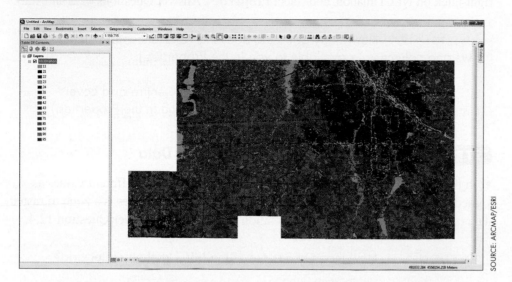

- The raster data of the Mahoning County land use should be in the View (because you added the layer in the previous step). Note that ArcMap assigns a random color to each different cell value when the layer is added, so your color scheme may be different from the one in the above graphic. Use the zoom tools to zoom in and out of the preview image to see how some features are represented through the use of grid cells. Answer Question 12.1. For more information on what a raster data set consists of, see **Smartbox 59**.

QUESTION 12.1 How are the major roads in the county being modeled with raster data?

Smartbox 59

What does a raster dataset consist of?

A raster dataset in ArcGIS is a rectangular grid consisting of values assigned to each cell. ArcGIS needs to know the geographic coordinates of the entire grid (usually the upper-left or lower-left corner of the grid) in order to properly align it with other datasets. The grid consists of rows and columns, and each grid cell is the same size. Thus, in its most basic form, a raster consists of a set of coordinates for the grid, the dimensions (rows and columns) of the grid, the size of each cell, and a list of values to populate the grid with (Figure 12.2).

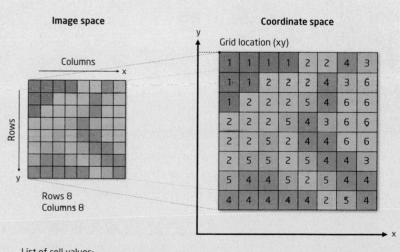

List of cell values:

[111122431122224361222546622254366225244662552544354452544444444254]

FIGURE 12.2 The basic setup of a raster grid, in which each cell receives a single value and the coordinates of the grid are known.

• Back in Catalog, open the Chapter12 folder, expand the Mahonraster geodatabase, right-click on NLCDmahon, and select **Properties**. Answer Questions 12.2 and 12.3.

> **QUESTION 12.2** How many grid cells does this dataset contain? (*Hint:* The rows and columns are available to you in the Properties.)

> **QUESTION 12.3** How many square meters does the entire grid cover? (*Hint:* The grid cell size, in meters, is available to you in the Properties.)

STEP 12.3 Working with Zones of Raster Data

• In the Mahoning County land-use raster, you'll see several different values assigned to this land-cover raster. Each of these different categories is a zone of raster data (see **Smartbox 60** for more information about zones). Answer Question 12.4.

> **QUESTION 12.4** How many zones are in the NLCDmahon raster grid?

Smartbox 60

zone One of several different values assigned to a raster grid cell.

NoData A raster data cell that contains a null value.

FIGURE 12.3 An example raster containing four zones and two NoData grid cells.

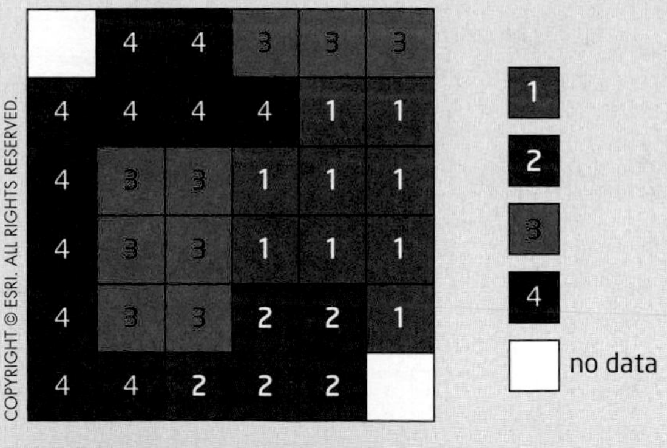

SOURCE: USED BY PERMISSION. COPYRIGHT © ESRI. ALL RIGHTS RESERVED.

What are zones of raster data in ArcGIS 10.2?

In the raster data model, a **zone** refers to each different value assigned to a grid cell. For instance, if a grid cell could contain a value of 1, 2, or 3, then that raster would have three different zones. There can be (and probably will be) multiple grid cells in each zone. A special value that is separate from the zones (but not considered its own zone) is the **NoData** value. As its name implies, NoData contains no value at all (not a value of zero, or a value of infinity; it is simply a null value). NoData values are treated as "holes" in the dataset—sometimes they are errors, but other times they are intentionally placed in the dataset to indicate grid cells that should not be included in the analysis of a raster.

The raster in Figure 12.3 contains four different zones, each assigned a value of 1, 2, 3, or 4. Note that multiple grid cells are assigned to each zone. Two grid cells are assigned a value of NoData; Neither of these cells is included in any of the four zones.

• The values assigned to each grid cell are specific values used as land-cover codes in the National Land Cover Dataset. See **Smartbox 61** for more information about the NLCD and what each grid cell value represents.

Smartbox 61

National Land Cover Dataset (NLCD) A 30 meter raster dataset of land cover for the entire United States.

What is the National Land Cover Dataset?

The **National Land Cover Dataset (NLCD)** is a raster data layer of the various types of land cover of the entire United States (Figure 12.4). Each cell of the raster covers 30 meters by 30 meters and the value of the cell represents a specific land-cover code. Each code stands for a specific land-cover type, such as open water, deciduous forest, or pasture. See Table 12.1 for a list of the land-cover types represented by each grid cell value.

FIGURE 12.4 The National Land Cover Dataset (NLCD) representing land cover for the United States circa 2006.

Value	NLCD 2006 Classification	Value	NLCD 2006 Classification
11	Open Water	51	Dwarf Scrub
12	Perennial Ice and Snow	52	Shrub Scrub
21	Developed, Open Space	71	Grassland/Herbaceous
22	Developed, Low Intensity	72	Sedge/Herbaceous
23	Developed, Medium Intensity	73	Lichens
24	Developed, High Intensity	74	Moss
31	Barren Land	81	Pasture/Hay
41	Deciduous Forest	82	Cultivated Crops
42	Evergreen Forest	90	Woody Wetlands
43	Mixed Forest	95	Emergent Herbaceous Wetlands

Table 12.1 The assigned value and the corresponding classification for the NLCD 2006

NLCD data are compiled by the Multi-Resolution Land Characteristics (MRLC) Consortium, a group of several U.S. federal agencies led by the USGS. NLCD data are available in raster layers showing the land cover of the United States circa 1992, 2001, and 2006 (also, a 2011 dataset was just recently released in 2014). Other datasets are also available that show the changes in land cover between time periods. NLCD data are distributed free either directly through the MRLC's Website (http://www.mrlc.gov) or The National Map. In addition, products related to various NLCD data are available for free download, including raster layers of impervious surface and tree canopies.

- Right-click on the NLCDmahon layer and open its layer **Properties**. Click on the **Symbology** tab. Change the color scheme of each zone to something better than the random colors that ArcGIS assigns when you add the layer. Use whatever color scheme you want, but some suggestions are: Water areas (blue), Developed areas (shades of red), Barren (brown), Forest areas (shades of green), Shrub Scrub and Grasslands (shades of orange), Pasture and Crops (shades of yellow), and Wetlands (shades of purple).

- Also change the headings under the "Label" column to text that describes what each zone is really displaying (Open Water, Grasslands, and so on. See **Smartbox 61** on page 283 for the meaning of each land-cover code value.

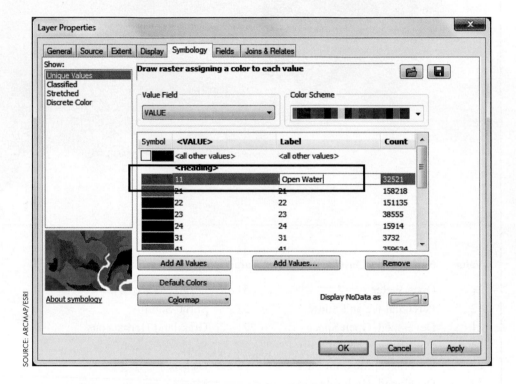

SOURCE: ARCMAP/ESRI

- When the labeling and color schemes are complete, click **Apply**, then **OK** to dismiss the dialog box. Your changes will be updated on the map and in the Table of Contents.

- Next, right-click on the NLCDMahon layer in the Table of Contents and select **Open Attribute Table**. The Value of each zone and the Count of cells in each zone will be displayed (see **Smartbox 62** for further information about raster attribute tables). Answer Questions 12.5 and 12.6.

Smartbox 62

How is a raster's attribute table set up in ArcGIS 10.2?

In a vector layer's attribute table, each object is stored as its own record with attributes assigned to that record. Thus, for a polygon layer that represents 200 different lakes in a county, there would be 200 records. However, raster data will often use a huge number of grid cells to represent similar data types. (For example, the NLCD layer you're using in this chapter for land cover of a single

Ohio county contains well over a million grid cells.) It would be impossible to navigate an attribute table in which each grid cell was an individual record and the attribute tables for raster layers are set up differently than vector layers.

A raster attribute table is referred to as a **Value Attribute Table (VAT)** because it consists of two primary columns: a field called "Value" and another field called "Count." The Value field lists each of the raster zones by their designation. For instance, in Figure 12.5, the raster grid cells have numerical values of 1, 2, 3, and 4. Thus, the Value field of the raster's VAT will contain four entries, one for each number. The Count field is a summation of how many cells comprise each zone. For instance, there are 9 cells in zone 1, so the Count for zone 1 will be 9. Additional attributes can then be created in GIS and appended to the VAT for each of the zones (such as the attributes of Type, Area, and Code shown in Figure 12.5).

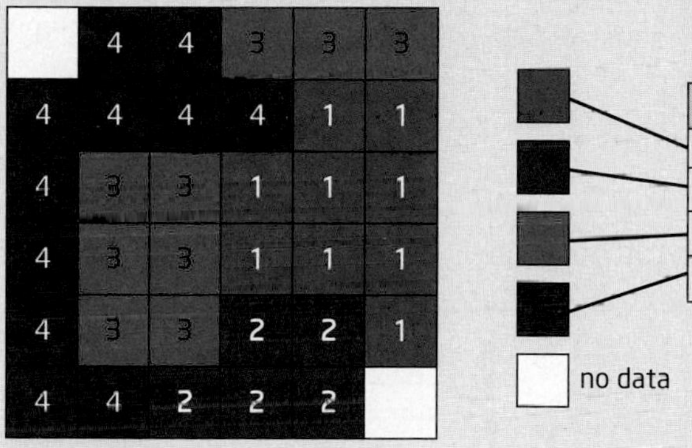

Value Attribute Table (VAT) The attribute table of a raster dataset.

Value	Count	Type	Area	Code
1	9	Forest land	8100	FL010
2	5	Wetland	4500	WL001
3	9	Crop land	8100	CL301
4	11	Urban	9900	VL040

no data

FIGURE 12.5 A raster and its corresponding Value Attribute Table (VAT).

QUESTION 12.5 How many cells are in each zone? (That is, how many pixels are in the first zone (11), how many are in the second zone (21), and so on?)

QUESTION 12.6 Why is your answer from Question 12.5 so much lower than the sum total of your answers to Question 12.2? What accounts for this discrepancy?

STEP 12.4 Regions of Raster Data

• You'll now start using some functions available via the Spatial Analyst extension. In this step, you'll be grouping your cells together to form regions (see **Smartbox 63** for more information).

Smartbox 63

What is a region in raster data?

In ArcGIS, a raster **region** is created by contiguous (touching) cells that have the same value. Each of these groups forms its own region, and a raster may consist of multiple regions. In ArcGIS, the regiongroup process is used to form regions from cells. See Figure 12.6 for an example of regiongrouping. The grid

region A set of contiguous grid cells that have the same value.

to be regiongrouped is labeled "INGRID1." The three cells in the upper-left corner are contiguous and all have the same value. Thus, they form region 1 in the regiongroup output (the grid labeled "OUTGRID"). In the OUTGRID, all cells in a region retain the same value—because these cells represent region 1, in OUTGRID they all have a value of "1." Back in INGRID1, the two cells in the upper-right corner are contiguous and have the same value; thus, they form region 2 in the OUTGRID. Each group of contiguous cells in INGRID1 is assigned to a different region in the OUTGRID.

The OUTGRID's Value Attribute Table consists of the usual fields of VALUE and COUNT, where the VALUE field represents the number assigned to each region—in Figure 12.6, the first region created (value of 1), the second region created (value of 2), and so forth. COUNT represents how many cells are in each region.

Regiongrouping creates a special field in the output grid called LINK, which represents the original grid cell value that each region had in INGRID1. For instance, region 2 originally consisted of grid cells that had a value of "0," so region 2's LINK attribute is 0. Similarly, region 6 originally consisted of grid cells that had a value of "1" so region 6's LINK attribute is 1.

Outgrid Attribute Table

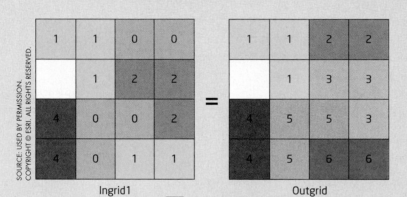

Value	Count	Link
1	3	1
2	2	0
3	3	2
4	2	4
5	3	0
6	2	1

Ingrid1   Outgrid

☐ Value = No data

FIGURE 12.6 An example of the output raster of a regiongroup procedure and its attribute table.

Regiongrouping is useful in analyzing the size or extent of large contiguous blocks of grid cells that have the same properties. For example, all of the contiguous grid cells in an area that have a land-use code representing a specific type of forested land can be grouped together to create a region representing the mapped boundaries of that type of tree stand. Similarly, several contiguous cells that all have a value representing open water can be grouped together to form a region that represents the extent of a lake.

- Open **ArcToolbox**.

- Open the **Spatial Analyst Tools** toolbox, then select the **Generalization** toolset, then select the **Region Group** tool.

  - Remember: You'll need to have the Spatial Analyst extension activated to use the tools in the Spatial Analyst Tools toolbox (see Step 12.1). Otherwise, ArcMap will give you an error message.

- This will bring up the Region Group dialog.

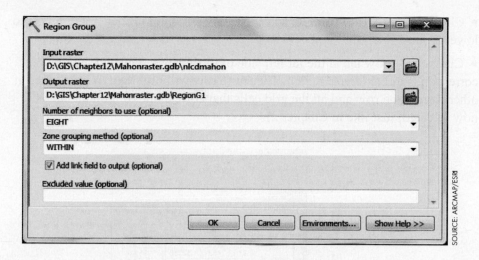

- Select your **NLCDmahon** raster as the Input Raster.

- Call the Output raster **regionG1** and place it in your Mahonraster geodatabase.

- Use **EIGHT** for number of neighbors.

- Use **WITHIN** for zone grouping method.

- Make sure the **Add link field to output** field has a checkmark in it.

- When all variables are set, click **OK**. When the "completed" message is returned, close the dialog box.

- A new grid (regionG1) will be added to the Table of Contents. Open this new Region Group grid's attribute table. Answer Question 12.7.

QUESTION 12.7 How many regions have been created?

- In regionG1's attribute table, right-click on the **Count** field (click on the word itself). A new set of options will appear:

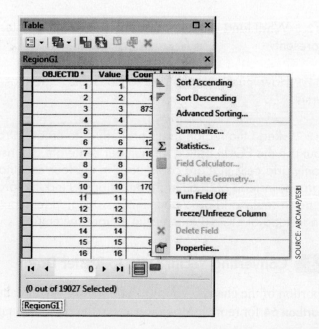

- Click on **Sort Descending** to sort the Count column in order from highest to lowest count.

- Click on the row tab at the far left side of the attribute table on the row that corresponds to the largest region. The entire row will highlight in light blue (in other words, the row and all the grid cells that correspond to it on the map are now selected, just like in Chapter 2).

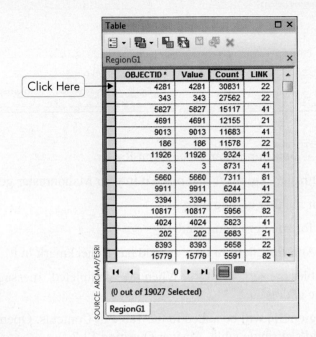

SOURCE: ARCMAP/ESRI

- Minimize (don't close) the attribute table. Examine the map to see the new selected region.

- After you've seen the region that was created, expand the attribute table from where you minimized it. Answer Questions 12.8, 12.9, and 12.10. Close the attribute table when you're done.

QUESTION 12.8 What land use does the region with the greatest number of grid cells represent?

QUESTION 12.9 How much real-world area does this region take up (in square meters)?

QUESTION 12.10 Redo the regiongroup procedure to create a raster called regionG2, but use the FOUR option for number of neighbors instead. (This means that only grid cells that are contiguous in the four cardinal directions will form a group, instead of in all eight directions.) Find the largest region again. How much real-world area does this new region (computed using the FOUR option instead of EIGHT) take up (in square meters)?

STEP 12.5 **Converting Vector Data to Raster Data**

- In the last portion of the chapter, you'll be comparing vector data to the raster data. (See **Smartbox 64** for more information about the vector data model vs. the raster data model.) You can turn off the Region Grouped raster(s) at this point, leaving the NLCDmahon raster turned on.

- Add the lakes feature class from your Mahonraster geodatabase folder.
- Zoom in on Evans Lake (in southern Mahoning County).

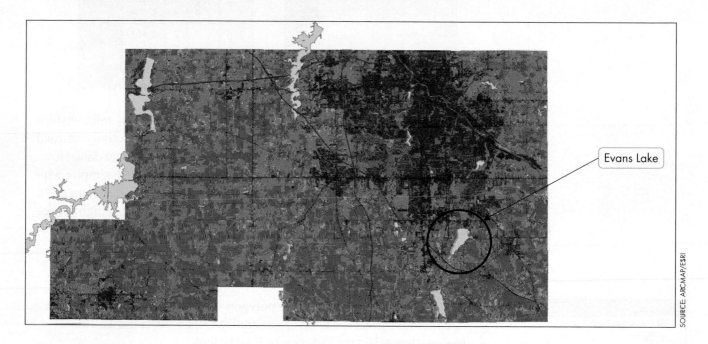

- Answer Question 12.11. When finished, turn off the NLCDmahon raster.

QUESTION 12.11 How closely (in terms of the spatial arrangement) does the raster version of Evans Lake match the vector feature class version?

Smartbox 64

How do raster data compare with vector data?

The choice between using vector or raster data to represent geospatial data often depends on what you want to represent in ArcGIS, and there are several advantages and disadvantages to each. For instance, land cover could be mapped with grid cells (like the NLCD) or by a series of polygons. Figure 12.7 shows a hypothetical landscape composed of four different land-cover types (1 = urban/built, 2 = agricultural, 4 = forested, 5 = water) modeled with polygons (Figure 12.7a) and grid cells (Figure 12.7b).

The raster data model is simpler than the vector data model; a grid of numbers is a simpler structure than several polygons of various shapes and dimensions. However, the vector data model is more compact than the raster data model; the landscape is modeled with only six objects in the vector model but with 64 grid cells in the raster model. The vector model can provide more aesthetically appealing output than the raster model. For example, the border of the "body of water" on the right side of the landscape in Figure 12.7 has naturally curved edges that are more easily represented by the polygons of the vector model than by the square blocks of the raster model.

FIGURE 12.7 A hypothetical landscape modeled using (a) vector data and (b) raster data.

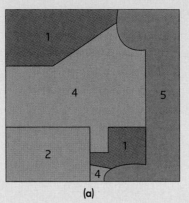

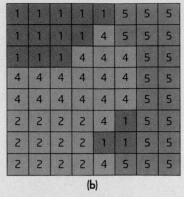

(a)   (b)

Vector data is also better for representing topological relationships (see Chapter 7). It's difficult to represent things like the connectivity of road intersections or the adjaceny of shared parcel boundaries with raster data. However, when overlaying two or more data layers, the process is much simpler with raster data (see Chapter 20) and more difficult with vector data (see Chapter 9). Last, raster data is better at representing surfaces or datasets with a high degree of spatial variability. For instance, areas with changes in elevation or temperature are more efficiently represented with raster data than with vector data.

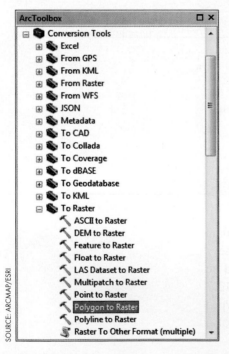

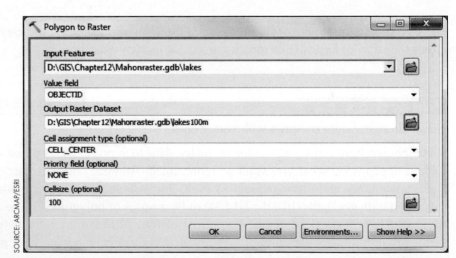

- You'll now convert the vector polygon shapefile into a raster layer. In ArcToolbox, select the **Conversion** toolbox, then select the **To Raster** toolset, then select the **Polygon to Raster** tool.

- In the Polygon to Raster dialog, choose the **lakes** feature class for the Input features.

- Use **OBJECTID** for the Value field.

- Call the Output Raster Dataset **lakes100m** and save it in your Mahonraster geodatabase.

- Use the defaults for Cell assignment type (**CELL_CENTER**) and Priority Field (**NONE**).

- Type **100** for the Cellsize (this will mean cells of 100 meter resolution).
- Click **OK**. Close the dialog when the conversion has completed.
- Open the lakes100 meter attribute table. Answer Question 12.12.

QUESTION 12.12 How many grid cells does Evans Lake consist of when modeled with 100 meter grid cells?

- Examine the vector shapefile of Evans Lake and your new lakes 100 meter grid (turn each on and off and compare the two). Answer Question 12.13.

QUESTION 12.13 How does the spatial arrangement of the lakes 100 meter grid compare to the vector feature class?

- Repeat the Polygon to Raster conversion process, but this time create a grid called lakes 10 meter and use **10** for your output Cellsize (thus creating a 10 meter grid). For more about the effect of raster resolution on data representation, see **Smartbox 65** .

Smartbox 65

How does resolution affect the raster dataset?

The **resolution** of a raster dataset affects not only the size of the raster but also the amount of detail that can be captured in that raster. Remember, each grid cell represents a fixed amount of area on the ground. The larger (or coarser) the raster resolution, the fewer grid cells it will take to cover the surface. However, due to the larger size of the cells, features will appear "blockier" and be less detailed. At smaller (or finer) raster resolutions, items will appear more detailed, but it will take more grid cells to represent items on the surface.

resolution The ground area represented by a single grid cell in a raster.

Figure 12.8 shows three different representations of a lake using raster grid cells. You can see the polygon boundaries of the lake along with how each of the three raster resolutions match up with them. The finer-resolution cells (of 1 meter) can capture more detail of the lake, but the grid requires 256 cells to do so. As resolution becomes coarser, fewer cells are used in the datasets, but the amount of detail declines as well. With 4 meter resolution, only 16 cells are used to cover the same area, but the lake is modeled in a very general way. This trade-off in the size of the raster (and the resulting number of cells being used) and the amount of detail that can be captured in the raster is inherent in the resolution used by the raster.

1m cell
16 x16 cells

2m cell
8 x 8 cells

4m cell
4 x 4 cells

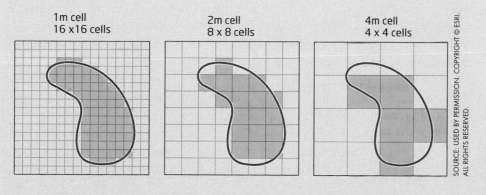

FIGURE 12.8 The effect of three different raster resolutions on the representation of data.

• Answer Questions 12.14, 12.15, and 12.16.

> **QUESTION 12.14** How many grid cells does Evans Lake consist of when modeled with 10 meter grid cells?

> **QUESTION 12.15** How does the spatial arrangement of the lakes 10 meter grid compare to the vector feature class?

> **QUESTION 12.16** What is the advantage of using the lakes 100 meter raster to represent Evans Lake and what advantage is there to using the lakes 10 meter raster to represent Evans Lake?

STEP 12.6 **Printing or Sharing Your Results**

• Save your work as a map document. Add the usual information to the Map Document Properties.

• Finally, either print a layout (see Chapter 3) showing the NLCD data for the county (along with the 10 meter lakes grid) or share your results as a tiled map service through ArcGIS Online (see Chapter 4).

• Raster data are also commonly shared by creating a **mosaic dataset** and adding the rasters to it. A mosaic dataset is a special file structure that can be created in Catalog (or through the Create Mosaic Dataset tool in ArcToolbox); it can be used to work with one or more rasters by pointing to their locations and managing the rules for visualizing and mosaicking the rasters together. (*Important Note:* Despite its name, a mosaic dataset does not actually contain the raster layers themselves—a mosaic dataset is neither a mosaic nor a dataset.) Mosaic datasets are frequently used with very large raster datasets or large numbers of rasters, or with several imagery rasters (see Chapter 13).

Closing Time

This chapter examined the basics behind the raster data model and compared raster data with vector data. Neither format is inherently superior, but certain problem-solving tasks or models lend themselves more to one format or the other. Raster data are often more associated with representing phenomena that vary continuously, such as precipitation patterns, land-cover types, or elevation, while vector data are used to represent items as a series of objects. The raster data model is simple to use and very versatile—see the *Related Concepts for Chapter 12* for more ways to use raster datasets in GIS. We'll work with raster data several more times throughout the book.

One specific use of rasters we didn't discuss in this chapter is remotely sensed imagery taken from a satellite or aircraft. Remote sensing imagery is a common data source used with GIS, and ArcGIS treats it in much the same way as the raster data you worked with in this chapter. For instance, Landsat satellite imagery (which we'll discuss in Chapter 13) was used as the initial basis for creating NLCD. In Chapter 13, we'll do much more with satellite imagery in raster format.

mosaic dataset An ArcGIS file structure designed for managing one or more rasters or collections of raster data.

Related Concepts for Chapter 12   Using Mosaics and Subsets of Rasters

When you're using raster data, it will always be in the form of a rectangular (or square) grid, but this shape may not represent the size or dimensions of the area you're studying. For example, the raster you used in this chapter of Mahoning County, Ohio, was a full rectangle, even though you were working only with the grid cells within the oddly shaped county boundary. A dataset such as NLCD covers the entire country, but you were working with only a small portion (or **subset**) of the data. The rectangular area of the raster didn't include cells with land-cover values from surrounding counties within the raster's area. Instead, ArcGIS was able to extract only the grid cells that fit within the county boundary.

ArcGIS provides a variety of Extraction tools in ArcToolbox that allow you to specify the subset of the data you want to work with. For example, you can specify the dimensions of a rectangle or polygon and extract all grid cells within those boundaries to a new raster dataset to create a subset. Another approach would be to use the area covered by another raster and remove all the cells except those that match the boundaries of this second raster to create a subset (you can do this with the Clip tool in ArcToolbox). You could also digitize a polygon boundary and use it as a "cookie cutter" shape to extract all of the cells under the digitized shape (we will do this in Chapter 15). In essence, you are creating a "mask" and extracting only the cells under that "masked" region to work with (this is the Extract By Mask tool in ArcToolbox; see Figure 12.9). All of these Extraction options allow you to create a subset of only the grid cells needed for analysis.

subset A raster created by removing a set of grid cells from a larger raster.

mosaic A raster created by joining several smaller rasters.

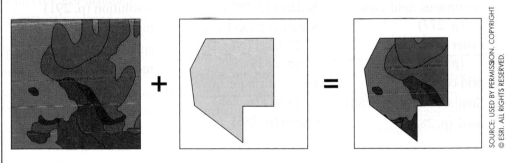

FIGURE 12.9 Extracting a subset of grid cells using the boundaries of a pre-defined "mask" polygon.

Conversely, you may have multiple rasters that need to be fitted together so that you can manipulate them as a single layer. These rasters may be adjacent to one another (like tiles) or overlap in certain ways. What you need to do is "knit" the various rasters together to form a single raster **mosaic**. ArcGIS provides several options for doing so through the use of the Mosaic tool in ArcToolbox (Figure 12.10).

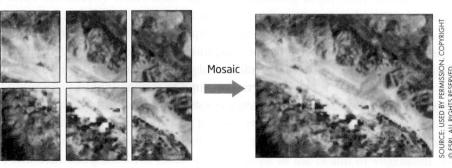

Mosaic

FIGURE 12.10 Six different raster datasets combined into a single mosaicked raster.

FIGURE 13.1 The Esri Change-Matters online tools, showing satellite images from two time periods and a measure of land-scape change between them.

Chapter Scenario and Applications

This chapter puts you in the role of an urban planner for the Cleveland, Ohio, area. You're collecting information regarding the city's waterfront area for potential redevelopment or expansion. Before you begin doing a study in GIS, you'll need current information about many of the properties and their usages (such as sports stadiums, a nature preserve, and a local airport). You'll examine these through different types of satellite imagery and aerial photos.

The following are additional examples of other real-world applications of this chapter's skills.

• A forest ranger is measuring the extent of the burn scar of a wildfire at a national forest. Recent satellite imagery can provide a basis for GIS analysis to reveal not only the spatial dimensions of the burn, but also information about the health of the trees and vegetation near the burn area.

• A public utilities worker is in the process of updating a GIS database with information about the placement of new traffic and safety infrastructure items. Recent aerial photographs of the city will aid her in plotting their locations.

• A county auditor is preparing to compute tax assessments on the privately owned properties in the county. As part of this GIS analysis, he needs current high-resolution aerial photographs to help assess if new additions (such as swimming pools or decks) are now part of the properties.

ArcGIS Skills

In this chapter, you will learn:

• How to load and examine remotely sensed imagery from Landsat in ArcGIS.

• How to examine different band combinations in a multispectral image.

- How to locate and use remotely sensed imagery from ArcGIS Online.

- How to load and examine orthoimagery in ArcGIS.

- How to access and examine imagery data from an ArcGIS Server.

- How to use the Image Analysis Window tools to examine different images together.

Study Area

For this chapter, you will be working with data of a portion of Cleveland, Ohio.

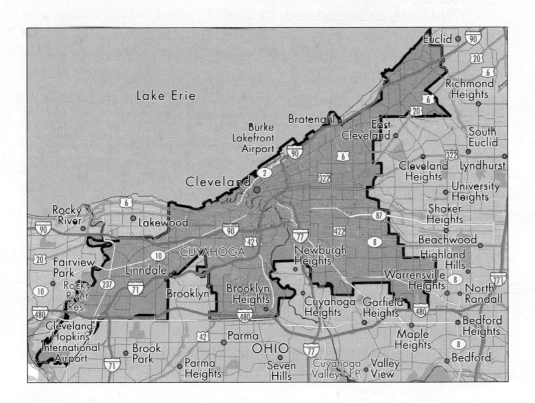

Data Sources and Localizing This Chapter

This chapter's data focus on features and locations within Cleveland, Ohio. However, you can easily modify this chapter to use data from your own city or local area instead. For example, if you were performing this chapter's activities in Norfolk, Virginia, you could easily obtain the same kinds of imagery. The Landsat data were downloaded for free from GloVis (http://glovis.usgs.gov), processed into an image, and subsetted to show only a section of Cleveland. Similar data are available for Landsat scenes containing Norfolk from GloVis and also LandsatLook (http://landsatlook.usgs.gov). In addition, the Landsat imagery from ArcGIS Online used in Step 13.4 is also available for Norfolk (and other areas around the United States and the globe).

The orthoimages used in Step 13.5 were downloaded (for free) from The National Map and multiple options for orthoimagery for your own area (such as Norfolk) within the United States are also available. Lastly, the NAIP imagery used in Step 13.6 is available free for all states within the United States. If you were examining Norfolk, you would access the Virginia NAIP imagery.

STEP 13.1 Getting Started

- Start ArcMap and use the Catalog to copy the folder called **Chapter13** from the C:\GISBookdata\ folder to your own D:\GIS\ drive. Be sure to copy the entire directory, not just the files within it.

- Chapter13 contains the following items:

  - **Cle.img**: a subset of a Landsat 5 TM satellite image from 9/11/05.

  - 17TMF395945_200909_0x3000m_CL_1.tif
    17TMF410945_200909_0x3000m_CL_1.tif
    17TMF425960_200909_0x3000m_CL_1.tif
    17TMF425945_200909_0x3000m_CL_1.tif
    These four files are orthoimages of sections of Cleveland from September 2009 and are at 0.3m resolution.

  - There are some related files for the Landsat TM image and the orthoimages. These must be kept in the folder because they are used for placing the images at the proper coordinates and locations.

- Activate the **Spatial Analyst** extension (see Chapter 12 for how to do this).

- *Important Note:* If at any time when adding a raster layer during this chapter you are prompted to build pyramids, click **Yes** and allow ArcMap to do so. A **pyramid** is a lower-resolution version of an image that is used when displaying the image at different scales. Pyramids allow images to be quickly rendered on the screen.

> **pyramids** Lower-resolution versions of an image that are used when displaying the image at different scales in order for images to be quickly rendered on the screen.

STEP 13.2 Adding Multispectral Imagery to ArcMap

- To begin, add the **cle.img** layer to ArcMap.

  - *Important Note:* When adding the satellite image, select the name of the file and click **Add** rather than double-clicking. If you click twice, ArcMap will prompt you to add different bands of imagery, which you don't want to do (at least right now).

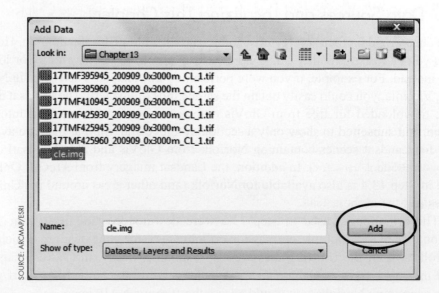

• This is a remotely sensed satellite image of the waterfront area of Cleveland, Ohio. By zooming in, you should be able to see the distinction between the built areas on the waterfront and Lake Erie. The image should be in color. If it's in black and white, see **Troublebox 9** for why this may be so and how to fix it.

Troublebox 9

Why is the imagery in grayscale (black and white) instead of color?

Unless the image is intended to be in black and white, a single remotely sensed image is often made up of several layers (or bands).

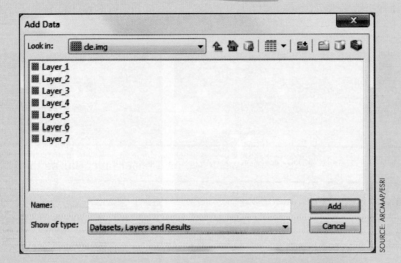

SOURCE: ARCMAP/ESRI

When adding the image in ArcMap, if you double-click the file name as usual, you'll end up in an Add Data dialog box asking you to select one of the image's bands. If you select one band it will be shown in shades of gray from black to white instead of a color composite image (see **Smartbox 67** on page 301 for more information about what these various bands represent).

• By zooming in closely, you'll see that the image becomes very pixelated and it's difficult to make out fine details in the image. This is because the spatial resolution is 30 meters—that is, each pixel (or cell) of the raster (see Chapter 12) represents an area on the ground 30 meters × 30 meters square. For more information about spatial resolution, see **Smartbox 66**.

Smartbox 66

What is spatial resolution?

Each pixel in the image represents a certain amount of size on the ground. The area of a single pixel is the **spatial resolution** of the image. When remote sensing data are collected, the spatial resolution of the sensor is fixed. That is, images with 30 meter resolution can't be "fine-tuned" to obtain 3-meter-resolution images instead.

The spatial resolution helps in determining what kinds of detail can be observed in an image. For instance, in a 30-meter-resolution image, each pixel represents a 30 meter × 30 meter area (900 square meters) on the ground; individual trees and cars are much smaller than that and would not be able to be

spatial resolution The size of the area on the ground represented by one pixel's worth of energy measurement.

resolved (seen). So, if you were digitizing tree stands, you would need a much finer spatial resolution in order to see the necessary details for your project.

Remotely sensed imagery is acquired at a variety of different spatial resolutions (see Figure 13.2 for examples of the same area on the ground sensed at differing spatial resolutions). For example, most imagery from the Landsat series of satellites is 30 meter spatial resolution. This imagery is often used for larger-scale environmental applications, where being able to examine fine detail is unnecessary. Many commercial satellite sensors and aerial photography can obtain imagery at sub-meter spatial resolutions that provide crisp detail. For instance, the Imagery basemap available in ArcGIS contains 30 centimeter resolution for the continental United States and 60 centimeter resolution for locations in Europe.

FIGURE 13.2 The same area on the ground sensed at three different resolutions.

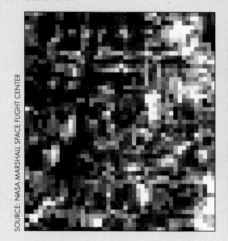

SOURCE: NASA MARSHALL SPACE FLIGHT CENTER

• Examine the areas around the waterfront and answer Questions 13.1 and 13.2.

QUESTION 13.1 What features around the Cleveland waterfront can you visibly determine from the 30 meter resolution of the Landsat image?

QUESTION 13.2 What kinds of features around the Cleveland waterfront can you not determine from the 30 meter resolution of the Landsat image?

STEP 13.3 Working with Multispectral Imagery in ArcGIS

• Right-click in the **cle.img** layer in the TOC and select **Zoom to Layer** to return to the full view of the Landsat image. You'll see in the TOC that the single raster layer is composed of three layers, each placed into a red, green, or blue channel. During the remote sensing process, multiple kinds of data are collected of the same geographic area and are placed together into a single image. For instance, this Landsat image actually consists of seven different layers, each representing a different type of remotely sensed information taken of the same section of Cleveland. ArcGIS can then display three of these seven layers simultaneously, using the red, green, and blue channels. For more information about what remote sensing is actually measuring and what all of these various image layers represent, see **Smartbox 67**. For further information about the Landsat program and the kinds of satellite images it can acquire, see **Smartbox 68**.

Smartbox 67

What is a remotely sensed image actually showing?

In the remote sensing process, a sensor on a satellite or aircraft measures the reflection of the sun's electromagnetic radiation from objects on Earth. Two exceptions to this are (a) *thermal sensors*, which measure the amount of radiated heat from objects on Earth (they absorb energy from the sun and re-emit it as heat), and (b) *active sensors*, which generate their own energy, throw it at a target, and record the reflection or backscatter of that energy pulse (radar waves are an example of this, as is lidar—see Chapter 17). However, sensors typically measure the reflection of energy.

Electromagnetic energy radiates from the sun through space as waves, and this energy has different properties depending on the energy's wavelength. For instance, energy with very short wavelengths has the properties associated with x-rays or gamma rays, while slightly longer wavelengths have the properties of ultraviolet or infrared light. Much of this energy gets trapped in Earth's atmosphere by greenhouse gasses like ozone or carbon dioxide, but some of it reaches and penetrates objects on Earth's surface and is reflected. The reflection of energy makes its way upward to the sensor on an aircraft or satellite, and the sensor records the radiance (Figure 13.3).

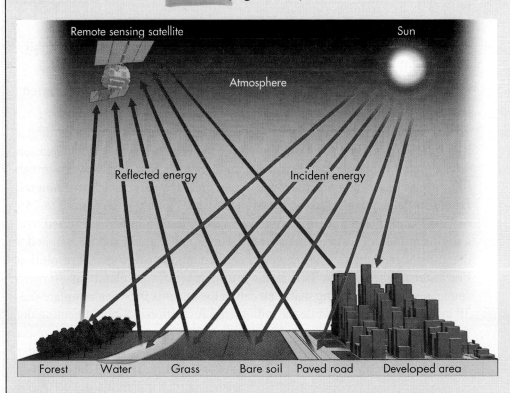

FIGURE 13.3 The remote sensing process.

The sensor often records the radiance of multiple types of energy over the same area at once. For instance, an aircraft sensor could record the three main types of visible light (blue, green, and red) and also a small portion of the wavelengths in the infrared portion of the spectrum. A satellite sensor may be tuned to record these four wavelengths and several others, including different, longer wavelengths of infrared energy. These narrow wavelengths of energy are called **bands** of energy. A sensor that records several bands at the same time is called a

band A narrow range of wavelengths being measured by a remote sensing device.

multispectral Sensing several bands of energy at once.

hyperspectral Sensing hundreds of bands of energy at once.

panchromatic Black-and-white imagery acquired by sensing the entire visible portion of the spectrum at once.

visible light Electromagnetic energy with wavelengths between 0.4 and 0.7 micrometers.

blue Electromagnetic energy with wavelengths between 0.4 and 0.5 micrometers.

green Electromagnetic energy with wavelengths between 0.5 and 0.6 micrometers.

red Electromagnetic energy with wavelengths between 0.6 and 0.7 micrometers.

near infrared (NIR) Electromagnetic energy with wavelengths between 0.7 and 1.3 micrometers.

middle infrared (MIR) Electromagnetic energy with wavelengths between 1.3 and 3.0 micrometers.

shortwave infrared (SWIR) Another term used for electromagnetic energy with wavelengths between 1.3 and 3.0 micrometers.

thermal infrared (TIR) Electromagnetic energy with wavelengths between 3.0 and 14.0 micrometers.

radiometric resolution A sensor's ability to determine fine differences in a band of energy measurements.

digital number The energy measured at a single pixel according to a predetermined scale.

brightness value Another term used synonymously with digital number.

channel The term used for the display of imagery in shades of red, green, or blue

multispectral sensor, while a sensor that can record hundreds of bands at once is called a **hyperspectral** sensor. If a sensor is recording only the range of the entire visible light spectrum at once, it's referred to as a **panchromatic** sensor. Thus, the kind of imagery (multispectral, hyperspectral, or panchromatic) being produced refers to the kind of sensor used to acquire it. The Landsat image you're using in this chapter consists of seven bands and is a multispectral image.

Bands of electromagnetic energy are commonly recorded by sensors from the following wavelength ranges:

- **Visible light** (0.4 to 0.7 micrometers): This is the only range of reflected energy that the human eye can see.

- **Blue** (0.4 to 0.5 micrometers): This is the portion of reflected energy of visible light that the human eye perceives as shades of blue.

- **Green** (0.5 to 0.6 micrometers): This is the portion of reflected energy of visible light that the human eye perceives as shades of green.

- **Red** (0.6 to 0.7 micrometers): This is the portion of reflected energy of visible light that the human eye perceives as shades of red.

- **Near infrared (NIR)** (0.7 to 1.3 micrometers): This is reflected energy in the shorter range of infrared wavelengths and is commonly used by both satellite imagery and aerial photography.

- **Middle infrared (MIR)** or **shortwave infrared (SWIR)** (1.3 to 3 micrometers): This is reflected energy in the medium range of infrared wavelengths.

- **Thermal infrared (TIR)** (3.0 to 14.0 micrometers): This is the emitted radiant heat energy from longer portions of the infrared wavelengths.

A sensor records the radiance of several bands and scales them according to a specific range as a record of the amount of radiance measured at that location, with higher numbers indicating a greater amount of radiance and lower numbers indicating a lesser amount. This range is determined by the sensor's **radiometric resolution** and common scales are 8-bit (in which numbers go from 0 to 255) or 11-bit (in which numbers go from 0 to 2047). Each number is referred to as a **digital number** or a **brightness value**. The digital number is then assigned to the grid cell, and the size of that cell is based on the sensor's spatial resolution. For instance, the Landsat TM image you're using of Cleveland has 30 meter spatial resolution, meaning that each cell in the raster is 900 meters square (i.e., 30 meters × 30 meters), while the radiometric resolution is 8-bit, indicating that the digital number values in each cell range from 0 through 255. The Landsat TM image consists of seven different bands of recorded energy, so each one of these seven layers consists of a raster with these properties.

A single band will be displayed in shades of gray, where the lowest digital numbers (such as 0) are shown in black, the highest digital numbers (such as 255) are shown in white, and all values in between are dark or light gray (Figure 13.4). In order to show the bands of imagery in color, ArcGIS has three **channels** into which bands can be placed for display: a red channel, a green channel, and a blue channel. When displaying these bands of energy on the screen in color, you're able to display only three of them simultaneously. For instance, you could show the digital numbers from the green band in

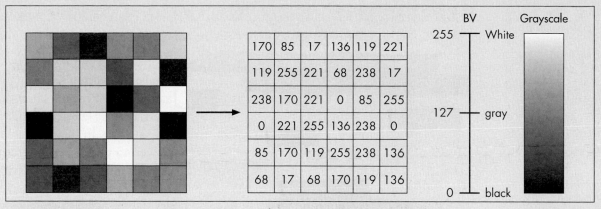

170	85	17	136	119	221
119	255	221	68	238	17
238	170	221	0	85	255
0	221	255	136	238	0
85	170	119	255	238	136
68	17	68	170	119	136

BV   Grayscale

255 — White

127 — gray

0 — black

FIGURE 13.4 A single raster band with its digital numbers and corresponding grayscale.

the blue channel, the near-infrared band in the green channel, and the middle-infrared band in the red channel, and the resulting composite would show the three bands combined, in color, on the screen. The green digital numbers would be shown in shades of blue, the near-infrared digital numbers would be shown in shades of green, and the middle-infrared digital numbers would be shown in shades of red (and all objects would then be displayed in a color composed of whatever shades of blue, green, and red the various types of digital numbers combined to be). Figure 13.5 shows a sample three-band composite.

You can show any combination of the seven bands through the three channels at any time and generate a different **color composite**. When the display is showing a blue band in the blue channel, a green band in the green channel, and a red band in the red channel, the composite is considered a **true color composite** because it represents the image in the way that the human eye sees things (that is, we see reflection of blue light in shades of blue, green light in shades of green, and red light in shades of red). A **false color composite** will be generated when the pattern of channels and bands deviates from a true color composite. False color composites are frequently used for analysis because they enable analysts to view types of energy that are typically invisible to the human eye (such as near infrared or middle infrared) displayed on the screen in shades of blue, green, or red.

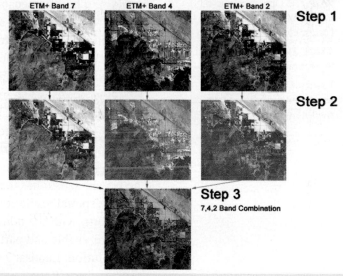

ETM+ Band 7   ETM+ Band 4   ETM+ Band 2   Step 1

Step 2

Step 3
7,4,2 Band Combination

SOURCE: NASA IMAGES ACQUIRED BY LANDSAT 5

FIGURE 13.5 A sample three-band composite.

color composite An image formed by placing a band of imagery into each of the three channels (red, green, and blue) to view a color image instead of a grayscale one.

true color composite An image arranged by placing the red band in the red channel, the green band in the green channel, and the blue band in the blue channel.

false color composite An image arranged by not placing the red band in the red channel, the green band in the green channel, and the blue band in the blue channel.

Smartbox 68

What is the Landsat program and what are its capabilities?

Landsat A long-running U.S. remote sensing program that had its first satellite launched in 1972 and continues today.

MSS The Multispectral Scanner onboard Landsat 1 through 5.

TM The Thematic Mapper instrument onboard Landsats 4 and 5.

ETM+ The Enhanced Thematic Mapper instrument onboard Landsat 7.

OLI The Operational Land Imager instrument onboard Landsat 8.

TIRS The Thermal Infrared Sensor onboard Landsat 8.

Landsat scene A single image obtained by a Landsat satellite sensor.

The Landsat series of satellites is the longest continually running satellite remote sensing program in existence. The first Landsat satellite (Landsat 1) was launched by the U.S. government in 1972, and the most recent satellite (Landsat 8, also formerly referred to as LDCM, the Landsat Data Continuity Mission) was launched in 2013. Through the seven satellites in the program, Landsat has provided global imagery coverage for over 40 years. Even better, the U.S. government has recently opened this archive of imagery and made it available to the public for free.

The first five Landsat satellites contained a sensor called the MSS (the Multispectral Scanner), which had a spatial resolution of 79 meters and recorded four bands. MSS data over an area was collected every 18 days. In 1982, Landsat 4 was launched containing a new instrument called the TM (Thematic Mapper). The TM sensor was also onboard Landsat 5 when it launched in 1984. TM imagery had a spatial resolution of 30 meters (the TIR band had a resolution of 120 meters) and sensed in seven multispectral bands (blue, green, red, NIR, two different MIR bands, and a TIR band—see **Smartbox 67** for a review of these bands). Landsat 5's mission continued until 2012, thus providing nearly 28 years of TM data, which was collected of an area every 16 days. The image of Cleveland you're using in this chapter is a subset of a Landsat TM image obtained on September 11, 2005.

Landsat 7 was launched in 1999 (Landsat 6 did not achieve orbit and was lost in 1993) carrying a new sensor, ETM+ (the Enhanced Thematic Mapper). The ETM+ sensor has the same capabilities of TM (all multispectral bands had 30 meter spatial resolution, except for the TIR band, which had 60 meter spatial resolution), with the addition of an eighth panchromatic band (which sensed the entire visible and part of the NIR region as a single band) at 15 meter spatial resolution. Landsat 7 collects data of an area every 16 days. Unfortunately, the Scan Line Corrector aboard Landsat 7 developed an uncorrectable error in 2003, which meant that imagery acquired by ETM+ contained only about 75 percent of the total pixels that should be there. Thus, if you're using Landsat 7 imagery from 2003 onward, you may see diagonal sections of pixels with no data recorded in them.

Landsat 8 is the latest mission and contains two different sensors. The first is OLI (Operational Land Imager). The OLI sensor monitors bands similar to the six multispectral and the single panchromatic on ETM+ with the addition of two other bands: an "ultra-blue" band for examining coastal areas and another IR band aimed at use with cirrus clouds. The OLI bands are sensed at 30 meters, with the panchromatic at 15 meters. The second instrument onboard Landsat 8 is TIRS (Thermal Infrared Sensor), which senses two different thermal bands at 100 meters (but are resampled to a 30 meter resolution for delivery to the end user). Table 13.1 summarizes all the Landsat sensor capabilities.

The various Landsat sensors have a swath width of 185 kilometers, which is the extent of how much of the ground a sensor can "see" during one pass. An entire Landsat scene measures a ground area about 170 kilometers long by 183 kilometers wide. Thus, when you're obtaining Landsat imagery to use in ArcGIS, you'll frequently be downloading files for each of the bands that

Satellite/ Sensors	Bands Sensed (by band number)	Spatial Resolution
Landsat 1-5 MSS	1: R, 2: G, 3: NIR, 4: NIR	79 m
Landsat 5 TM	1: B, 2: G, 3: R, 4: NIR, 5: MIR, 6: TIR, 7: MIR	Multi: 30 m, TIR: 120 m
Landsat 7 ETM+	1: B, 2: G, 3: R, 4: NIR, 5: MIR, 6: TIR, 7: MIR, 8: Pan	Multi: 30 m, TIR: 60 m, Pan: 15 m
Landsat 8 OLI/ TIRS	1: Ultra-Blue (Coastal), 2: B, 3: G, 4: R, 5: NIR, 6: MIR, 7: MIR, 8: Pan, 9: IR (Cirrus), 10: TIR, 11: TIR	Multi: 30 m, Pan: 15 m, TIR: 100 m

Table 13.1 The sensors used by Landsat missions, their specific bands and their band numbers, and their spatial resolutions

cover the area defined by a single Landsat scene. If the area of analysis isn't contained in a single scene, you'll often have to mosaic the images together into a larger image raster (see Chapter 12).

Imagery from the Landsat archives are free for downloading. The USGS has three different utilities online to help you obtain Landsat imagery. The first of these is EarthExplorer (http://earthexplorer.usgs.gov), which allows you to access Landsat imagery along with many other types of satellite imagery, aerial photography, and GIS data (see **Smartbox 31** on page 129 for more about EarthExplorer). The second is GloVis (http://glovis.usgs.gov), the Global Visualization Viewer, which allows you to search for Landsat scenes (and other satellite imagery), then download the bands from a scene. The third Website is LandsatLook (http://landsatlook.usgs.gov), which allows you to search by location, then view available Landsat scenes from multiple dates for the area of interest. LandsatLook also allows you to download the imagery bands for use in GIS (Figure 13.6).

FIGURE 13.6 Using the LandsatLook Viewer to download Landsat scenes.

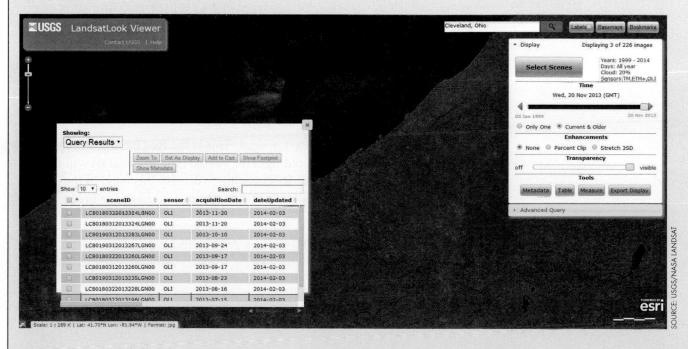

- From the **Windows** pull-down menu, select **Image Analysis**. A new dock-able window will open in ArcMap. The Image Analysis tools give you a wide variety of utilities for working with imagery in ArcGIS. The box at the top shows you which images are available for use. Right-click on the **cle.img** option in the **Image Analysis** window and select **Properties**.

- In the Layer Properties dialog, select the **Symbology** tab.

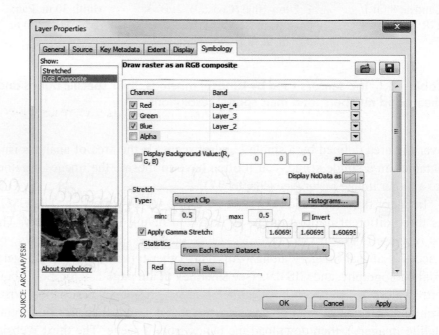

SOURCE: ARCMAP/ESRI

- Under Show: select **RGB Composite**. With the Draw raster as an RGB composite option, you can choose which of the multispectral bands to display in each of the three channels. This is a Landsat TM image, so one of the seven bands can be displayed in the red channel, a second band can be displayed in the green channel, and a third band can be displayed in the blue channel. The rest of the bands will not be used when drawing the composite on the screen.

- For the Red channel, choose **Layer_4**. This is the NIR band from TM. Healthy, natural features (such as healthy trees, grassy fields, or vegetation) reflect a lot of near-infrared light. By choosing to display this band in the red channel, you will have ArcGIS display those features in shades of red. The more infrared reflectance in a pixel, the brighter red that pixel will be.

- For the Green channel, choose **Layer_3**. This is the Red band from TM. Features that would reflect a lot of red light (or visible light in general) will have a brighter reflectance in this band. By choosing to display this band in the green channel, you will have ArcGIS display those features in shades of green. The more red reflectance in a pixel, the brighter green that pixel will be.

- For the Blue channel, choose **Layer_2**. This is the Green band from TM. Features that would reflect a lot of green light (or visible light in general) will have a brighter reflectance in this band. By choosing to display this band in the blue channel, you will have ArcGIS display those features in shades of blue. The more green reflectance in a pixel, the brighter blue that pixel will be.

• Click Apply to make the changes and then OK to close the dialog. You'll see that the Landsat image looks much different when shown in this composite. Answer Questions 13.3 and 13.4.

> **QUESTION 13.3** Why does the deep water of Lake Erie appear black in this false color composite image? (*Hint:* Water strongly absorbs all types of energy except for blue light.)

> **QUESTION 13.4** Why do all of the urban areas appear in a cyan color in this false color composite image?

• Pan and zoom the image into the Cleveland Lakefront Nature Preserve on the shores of Lake Erie:

Cleveland Lakefront Nature Preserve

SOURCE: ARCMAP/ESRI

• Keeping in mind the characteristics of near-infrared light as described above, answer Question 13.5.

> **QUESTION 13.5** What does the appearance of this area in the false color composite image say regarding the relative health of the trees and greenery at the Cleveland Lakefront Nature Preserve? Explain your answer.

• Zoom back to the extent of the full image layer.

STEP 13.4 Accessing Satellite Imagery Through ArcGIS Online

- Esri has added Landsat imagery to ArcGIS Online that you can access for free. To do so, click on the **drop-down arrow** next to the **Add Data** button on the **Standard Toolbar** and select **Add Data From ArcGIS Online** (as you did in Chapter 5).

- Select the **Data** radio button and type **Landsat** into the Search box. A number of options will be returned to you.

- Select the option for **Vegetation Analysis (543) 1990-2010** and click **Add**. ArcMap will stream the image from ArcGIS Online to the View.

- Note that the imagery from ArcGIS Online is in the **image service** format. Image services are used to serve raster data via the cloud. This image service you've just added is not the actual Landsat data, but instead is a service created from processing several bands into a new layer.

- This layer is a mosaic of processed imagery derived from several time periods over a 20-year stretch. It shows a false color composite of TM band 5 (MIR) in the red channel, band 4 (NIR) in the green channel, and band 3 (red) in the blue channel. Thus, the healthy vegetation that reflects a lot of NIR light and absorbs more red light will appear in shades of green, with very healthy areas appearing in bright green.

image service An ArcGIS Online format used for the distribution of raster data.

• Zoom out from the area to examine the greater Cleveland region. Focus on the areas shown in the next graphic and answer Question 13.6.

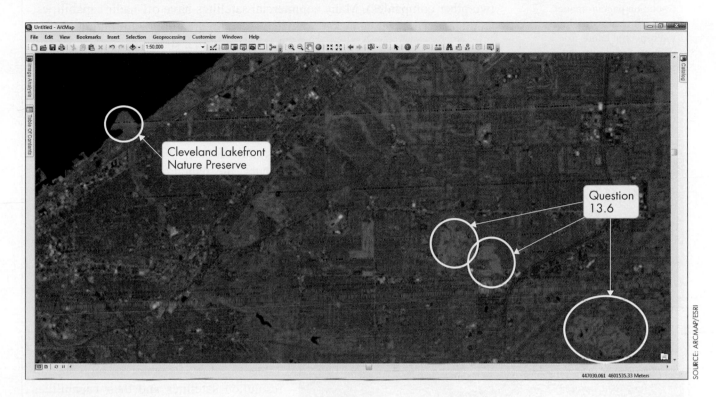

SOURCE: ARCMAP/ESRI

QUESTION 13.6 What are these circled areas on the landscape and how can you tell? *Hint:* Look at the size and pattern of the areas and think about what kinds of energy reflectance you're examining with this Landsat image service.

• There is a wide range of mosaicked Landsat image services available through ArcGIS Online, useful for studying vegetation change (NDVI), agriculture, healthy vegetation, shortwave infrared energy reflection, and land/water boundaries. You'll note, however, that while Landsat 30 meter spatial resolution is useful for examining broad landscape and environmental features, that kind of spatial resolution would be far too coarse for picking out individual houses or for doing the kind of digitizing and editing at the scale you worked at in Chapters 6 and 7. Aerial photography is a good source of high-resolution imagery, and there are several commercially available satellite imagery sources as well (see **Smartbox 69**). For now, turn off the Vegetation Analysis (543) 1990–2010 layer but don't delete it. Zoom to the extent of the cle.img layer again.

Smartbox 69

What satellites have high-resolution capabilities?

Much of the available high spatial resolution satellite data come from private companies rather than government sources. DigitalGlobe and Airbus Defence and Space are among the companies who operate commercial satellites and make the imagery available for sale (note also that, in 2013, DigitalGlobe and

pan-sharpening The technique of fusing a higher-resolution panchromatic band with lower-resolution multispectral bands to improve the clarity and detail seen in an image.

another leading commercial imagery provider, GeoEye, merged into a single company, while Airbus was formed from the reorganization of Astrium and two other companies). Many commercial satellites have off-nadir capabilities, so their sensors can be tasked to image a certain area on the ground during an orbital pass, and can then re-image that area within a short time frame, usually one to three days.

If a satellite has a multispectral sensor along with a panchromatic sensor of finer spatial resolution, the panchromatic band is usually used to pan-sharpen the multispectral bands. When **pan-sharpening** occurs, the panchromatic and multispectral bands are fused together in order to refine the image (so that the multispectral bands can be viewed at the resolution of the panchromatic). Note, however, that U.S. government regulations indicate that non-government purchasers of commercial imagery will have it resampled to a spatial resolution of 0.5 meters if the sensor resolution is finer than that (although this restriction was recently lifted in 2014). See Table 13.2 for a selection of high-spatial-resolution satellites and their capabilities and Figure 13.7 for an example of high spatial resolution satellite imagery.

SOURCE: PHOTO DIGITALGLOBE VIA GETTY IMAGES

FIGURE 13.7 A satellite image of the Magic Kingdom in Walt Disney World in Orlando, Florida.

Satellite/Sensors	Source	Launch Date	Bands Sensed	Spatial Resolution (at Nadir)
GeoEye-1	DigitalGlobe	2008	Pan, B, G, R, NIR	Pan = 0.41 m, Multi = 1.65 m
IKONOS	DigitalGlobe	1999	Pan, B, G, R, NIR	Pan = 0.82 m, Multi = 3.2 m
Pleiades-1A	Airbus	2011	Pan, B, G, R, NIR	Pan = 0.5 m, Multi = 2 m
Pleiades-1B	Airbus	2012	Pan, B, G, R, NIR	Pan = 0.5 m, Multi = 2 m
QuickBird	DigitalGlobe	2001	Pan, B, G, R, NIR	Pan = 0.61 m, Multi = 2.44 m
SPOT 5 HRG	Airbus	2002	Pan, G, R, NIR, MIR	Pan = 2.5 m and 5 m, Multi = 10 m and 20 m
SPOT 6	Airbus	2012	Pan, B, G, R, NIR	Pan = 1.5 m, Multi = 6 m
SPOT 7	Airbus	2014	Pan, B, G, R, NIR	Pan = 1.5 m, Multi = 6 m
WorldView-1	DigitalGlobe	2007	Pan	Pan = 0.5 m

Table 13.2 A selection of high spatial resolution satellites and their capabilities

Satellite/Sensors	Source	Launch Date	Bands Sensed	Spatial Resolution (at Nadir)
WorldView-2	DigitalGlobe	2009	Pan, B, G, R, NIR, Coastal, Yellow, Red Edge, NIR-2	Pan = 0.46 m, Multi = 1.85 m
WorldView-3	DigitalGlobe	2014	Pan, B, G, R, NIR, Coastal, Yellow, Red Edge, NIR-2, 8 SWIR bands, 12 CAVIS bands	Pan = 0.31 m, Multi = 1.24 m, SWIR = 3.7 m, CAVIS = 30m

Table 13.2 Continued

STEP 13.5 Adding Orthoimagery to ArcMap

• Aerial photography is an excellent source of remotely sensed imagery for GIS, especially of higher spatial resolution. Aerial imagery is used in ArcGIS as a raster and consists of several bands. A color image consists of the three visible bands (blue, green, and red), but false color aerial imagery is also used (red, green, and a NIR band), while hyperspectral sensors are also carried by aircraft. When using aerial photography in GIS, it's common to use it as orthoimagery (see **Smartbox 70**).

Smartbox 70

What is orthoimagery?

When an aerial photo is acquired, it unfortunately has a number of distortions involved in it. For one, the photo doesn't have the same scale at all locations covered in the photo. Photo scale is a function of the distance to the camera, and as the plane flies over the landscape below, some items are closer to or farther from the camera, and thus the photo won't have the same scale everywhere. In addition, the farther away tall objects are from the center of the photo (called the *principal point)*, the more they tend to "lean" away from the center, an effect referred to as *relief displacement.*

A regular aerial photo is affected by these issues, but it can be placed through a process called *orthorectification,* in which the effect of relief displacement is removed, the photo is given uniform scale (i.e., the same scale at all locations in the photo), and the photo is given real-world coordinates and georeferenced (see the *Related Concepts for Chapter 13* for further information). An aerial image that has gone through orthorectification is referred to as an **orthoimage** (or often, an **orthophoto**). Orthoimagery is often used in GIS because the images will align with other geospatial data and have a coordinate system assigned, and thus can be used for measurements or a digitizing source because they have uniform scale.

orthoimage (orthophoto) A spatially referenced aerial photo with uniform scale.

• A single aerial image will cover a much smaller area than a satellite image. You'll be using four orthoimages of the waterfront area of Cleveland. Add the first of these (**17TMF395945_200909_0x3000m_CL_1.tif**) the same way you added the Landsat image in Step 13.2 (don't add individual bands, just add the entire image). The color orthoimage should appear. If a grayscale image appears of a single band, refer back to Troublebox 9 on page 299 for instructions on how to add the entire three-band color image.

• Zoom to the extent of the orthoimage layer and answer Question 13.7.

QUESTION 13.7 What is this orthoimage showing?

• Add the other three orthoimages (**17TMF410945_200909_0x3000m_ CL_1.tif, 17TMF425960_200909_0x3000m_CL_1.tif, and 17TMF425945_200909_0x3000m_CL_1.tif**). They should appear adjacent to one another. Readjust the view so that you can see the area covered by all four images.

• Each of these orthoimages is from September 2009 and has a spatial resolution of 0.3m (approximately one foot). To get a sense of the high-resolution imagery, right-click on the **17TMF425945_200909_0x3000m_CL_1.tif** image and select **Zoom to Raster Resolution**. This will zoom the view down to the level of the spatial resolution of the image. Pan around the scene of urban Cleveland and answer Question 13.8.

QUESTION 13.8 Even at 0.3 meter spatial resolution, what kinds of features or details of the city cannot be clearly resolved? (Be specific, using examples from this particular image.)

• Zoom back out so you can see the extent of the four orthoimages on the screen.

• Turn on the Landsat image as well. The orthoimages should be displayed on top of the Landsat image. To reveal the Landsat imagery under an orthoimage, you could turn the ortho layer on and off, but there's a better way to switch between layers. ArcGIS gives you the capability to "swipe" one layer away from another to view what's underneath. Specifically, you'll be looking at the Cleveland Browns' stadium in both the high-resolution orthoimage and the Landsat image. Zoom in so that the View is centered on the stadium.

- Next, return to the Image Analysis window (re-open it if it's been closed).

- In the Image Analysis window, first select the **17TMF410945_200909_0x3000m_CL_1.tif** image in the box at the top of the dialog. This will be the layer you want to look underneath.

- Next, click on the **Swipe** tool in the Image Analysis window. You'll see the cursor has turned into a sold black triangle. Click on the image below the stadium. Hold down the left mouse button and drag the triangle up past the stadium—you'll see the Landsat image underneath be revealed and you can drag the orthoimage up and down over it, like you were opening and closing window blinds. Look carefully at the stadium and field in both images and answer Question 13.9.

QUESTION 13.9 Is the field of the Cleveland Brown's stadium real grass or artificial turf? How can you tell from using the orthoimage and the Landsat image?

- To release the swipe control, click on the cursor tool on the Tools toolbar. Zoom back out to see the extent of the four orthophotos.

STEP 13.6 Working with NAIP Imagery in ArcGIS

- Another source of remotely sensed imagery is the 1 meter aerial orthoimages from the National Agriculture Imagery Program (NAIP) administered by the United States Department of Agriculture's Farm Service Agency. For more information about NAIP, see **Smartbox 71**.

Swipe tool

SOURCE: ARCMAP/ESRI

Smartbox 71

What is NAIP imagery?

NAIP (National Agriculture Imagery Program) imagery has been collected since 2003, sponsored by the Farm Service Agency, a branch of the USDA. Originally on a 5-year cycle of image collection, in 2009 NAIP began a 3-year cycle. NAIP orthoimagery is 1 meter spatial resolution and is distributed free by the USDA. Most NAIP imagery is the usual three-band (red, green, and blue) color imagery, but some imagery is available as color-infrared (which uses the NIR, red, and green bands instead). In this chapter, you'll be downloading the imagery from a USDA server in the cloud for use in ArcGIS.

Instead of obtaining an individual orthoimage, you can also obtain NAIP imagery in a special format called a **DOQQ** (Digital Orthophoto Quarter Quad, also often just referred to as a DOQ or Digital Orthophoto Quad). A DOQQ is a special product in which the orthoimage is compiled into a single image that covers one-quarter of a 1:24000 USGS topographic map quadrangle (see Chapter 16 for more information about topographic maps). The orthoimagery in a DOQQ covers 3.75 degrees of latitude by 3.75 degrees of longitude in area. NAIP imagery is also available as a mosaicked image of an entire county.

NAIP The National Agriculture Imagery Program maintained by the USDA.

DOQQ A Digital Orthophoto Quarter Quad showing orthoimagery covering 3.75 degrees of latitude by 3.75 degrees of longitude.

- NAIP imagery is available free from the USDA, which makes the imagery available from a variety of sources, including an ArcGIS Server site in the cloud (see Chapter 4 for more information about ArcGIS Server). You can connect to this USDA service through the cloud via ArcGIS and utilize its resources. To connect to this ArcGIS Server site for obtaining data, begin by opening the **Catalog** and expanding the option for **GIS Servers**.

- Next, select the option for **Add ArcGIS Server**.

- In the next dialog box, select the radio button for **Use GIS Services** and click **Next**.

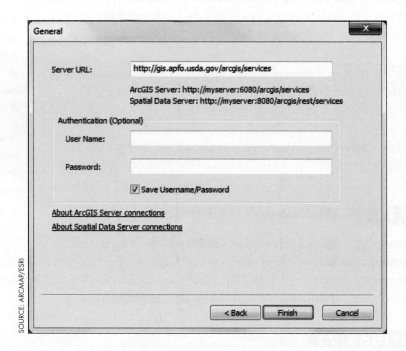

- In the next dialog box, for the Server URL, enter http://gis.apfo.usda.gov/arcgis/services.

- Click **Finish**.

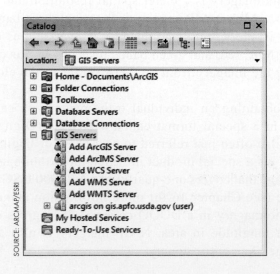

• A new GIS Service called **arcgis on gis.apfo.usda.gov (user)** has been added to the Catalog.

• Expand that option to see the folders of data you now have access to via the cloud. Expand the **NAIP** folder, then select and drag the **Ohio_2013_1m_NC** raster into the Table of Contents.

• The 1 meter NAIP imagery (of the state of Ohio) will now be in the Table of Contents with the other layers. Focus on the Burke Lakefront Airport (to the immediate northeast of the Cleveland Brown's stadium) and pan the View so that the airport (and its landing strips) fill the screen.

• In the Image Analysis window, turn on all of the orthoimages as well as the NAIP image. Select the NAIP image so it's highlighted, then swipe back and forth between the NAIP and the orthoimages underneath it.

• You should be able to view the 0.3 meter orthoimages and the 1 meter NAIP images by swiping back and forth. Answer Question 13.10.

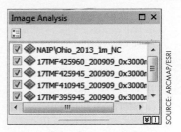

SOURCE: ARCMAP/ESRI

> **QUESTION 13.10** How many airplanes are on the tarmac in the September 2009 orthoimages? How many airplanes are present in the 2013 NAIP imagery?

Closing Time

This chapter examined the use of remotely sensed imagery in ArcGIS. Imagery is becoming increasingly integrated in GIS analysis, especially with the growing ease of access to remotely sensed data through resources like ArcGIS Online, GIS Services, GloVis, and LandsatLook. You'll be using remotely sensed imagery in several other chapters, integrating it with other layers (such as the landscape) in Chapters 16 and 18.

One thing to note is that all of the imagery you used in this chapter was already pre-processed to have real-world coordinates and be aligned with each other. However, there are many types of imagery that haven't already been processed in such a way and exist only as scanned digital images. For instance, items that you may find useful in analysis, such as historic ownership maps or original aerial photo prints, can't simply be added to ArcGIS and used as a layer because they don't have any spatial reference assigned to them. They're simply image graphics and ArcGIS doesn't know what to do with them or how they match up with other geospatial data. These types of items can still be used in ArcGIS; they just need to be properly processed in order to assign some sort of real-world coordinate system to them. See the *Related Concepts for Chapter 13* for more about the process of georeferencing raster images.

Related Concepts for Chapter 13   Georeferencing an Image

In ArcGIS 10.2, **georeferencing** is the process of taking a layer without spatial reference and aligning it with other layers that already have a coordinate system (projected or geographic) assigned to them. Georeferencing is commonly performed on raster images, such as scanned historical maps or scanned aerial photos, so that they can be used as geospatial data along with

georeferencing A process whereby spatial referencing is given to data without it.

control points Point locations where the coordinates are known—these points are used in aligning the unreferenced image to the source.

transformation A process in which data are altered from unreferenced to having spatial reference.

RMSE The Root Mean Square Error, an error measure used in determining the accuracy of the overall transformation of the unreferenced data.

other layers in GIS. For instance, if you're digitizing old home locations from a 50-year-old aerial photograph, you would want to be sure that the photo has a proper coordinate system assigned and matched to your other layers so that the digitizing results and measurements will be correct. To georeference a raster image, you'll also need an already spatially referenced source with which to match it, such as an orthoimage or satellite image. (Vector data layers, such as transportation layers, can also be used as a source.)

Georeferencing is performed by selecting locations that can be clearly identified in both images and placing a **control point** at those spots. At least three control points are necessary to tie the two images together, and often several more will be used to get a better fit of the two images. Control points are normally selected at well-delineated locations in both images, such as the intersections of roads or the corners of buildings. Once the two images are well aligned, the unreferenced image can undergo a **transformation** to convert it to the new coordinate system. For instance, if the referenced image is in the UTM Zone 17 projected coordinate system, the unreferenced image can be transformed so that its coordinates will be assigned the same coordinate system and spatial reference.

An error term, called the Root Mean Square Error (or **RMSE**) is used to compare the locations on the newly transformed image with where they match up to the image used as the reference. The lower the value for RMSE, the better the alignment between the images. Because the transformation is based on the control point locations, control points are often deleted or moved, or new points are added, to lower RMSE and achieve better alignment. When this process is complete, the image can be permanently transformed to a new georeferenced dataset in ArcGIS 10.2 for use in future projects.

A separate Georeferencing toolbar to access the georeferencing tools in ArcGIS 10.2 is available from the Toolbars option under the Customize pull-down menu. Control points can be selected, various transformation options are available, and the Rectify command can be used to permanently transform the new georeferenced image.

ArcGIS 10.2 allows for automatic georeferencing, in which the user does not have to select control points. After placing the unreferenced raster in the general location of where it should align with the referenced image, you can then use the Auto Registration command on the Georeferencing toolbar to attempt to link the two images together. Note, however, that this automatic process uses the spectral properties of the unreferenced image and the source image for georeferencing, so it can be performed only with aerial or satellite images. As a result, the unreferenced and source images should have similar properties, such as their location or the time the images were taken.

For further information about georeferencing in action, see the video in the Resource Center at http://video.arcgis.com/watch/376/georeferencing-rasters-in-arcgis and the video on automatic georeferencing available in the Resource Center at http://video.esri.com/watch/1553/automatic-georeferencing-in-arcgis-101.

For More Information

For further in-depth information about the topics presented in this chapter, use the ArcGIS Help feature to search for the following items:

- A quick tour of displaying image and raster data in ArcMap
- Adding raster data to a map
- Fundamentals of georeferencing a raster dataset
- Fundamentals of panchromatic sharpening
- Georeferencing a raster automatically
- Georeferencing toolbar tools
- Image Analysis window: Display section
- Imagery: Data management patterns and recommendations
- Key concepts for image services
- Making measurements from imagery
- Raster bands
- Raster dataset properties
- Raster pyramids
- Renderers used to display raster data
- Satellite sensor raster types

For further information about satellite capabilities, see here:

- Landsat series: http://landsat.usgs.gov/band_designations_landsat_satellites.php
- High-resolution satellite sensors: http://www.satimagingcorp.com/satellite-sensors.html

For more information on obtaining Landsat imagery and using it in ArcGIS, see the AmericaView piece by Jarlath O'Neill-Dunne, available here: http://blog.americaview.org/2011/03/accessing-landsat-data-and-using-it-in.html.

For more information about incorporating Landsat imagery into ArcGIS, see the great set of tutorials put together by members of VirginiaView here: http://virginiaview.cnre.vt.edu/education.html#RSinArcGIS10.

The Esri ChangeMatters Viewer can be accessed here: http://www.esri.com/software/landsat-imagery/viewer.

For further information about accessing NAIP data, see the USDA Website here: http://www.fsa.usda.gov/FSA/apfoapp?area=home&subject=prog&topic=nai.

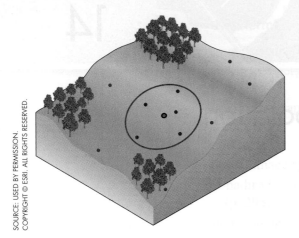

FIGURE 14.2 An example of using a subset of points for a local interpolation method.

global interpolation A method that uses all known values in the dataset when approximating an unknown value.

trend Very general patterns or overriding processes that affect measurements.

local interpolation A method that uses a subset of known values when approximating an unknown value.

method or local method is used for interpolation. A **global interpolation** method attempts to use every known value in the dataset when approximating an unknown value. For instance, if you live in northeast Ohio and you are trying to interpolate a value for precipitation at your location, a global interpolation method would use the values from all available weather stations in the state (including points in the far south and west) to estimate the precipitation level at your position. As such, global interpolation methods are inexact and produce extremely general results that are often used for determining **trends** (very general patterns or overriding processes that affect the measurements) in a dataset. For instance, a trend may show a greater amount of precipitation as one moves south to north.

For more exact results, a **local interpolation** method is used. In local interpolation, only a small subset of point measurements near the location are used to approximate the unknown value at that location. See Figure 14.2 for an example: When trying to interpolate the value at the unknown location (shown in yellow) only the values of the closest five points (those shown in red within the radius around the unknown value point) are used. The values at all the other known points (shown in blue) are not used because they are too distant to have a realistic impact on the value at the chosen location. This chapter introduces you to two different local interpolation methods in ArcGIS for creating raster surfaces from an initial set of points.

Chapter Scenario and Applications

This chapter places you in the role of a climatologist trying to create a statewide map of annual precipitation for Ohio. You also want to be able to determine the average annual precipitation for entire counties as part of a larger climate-related study. You have a set of points representing precipitation values taken at multiple places across the state, but your goal is to create a continuous surface representation that covers all locations.

The following are additional examples of other real-world applications of this chapter's skills.

• An environmental scientist is attempting to create a map of pollution levels for her state. She has samples of CO_2 (carbon dioxide) levels from a number of different measurement areas around the state and she needs to use these point samples to create a statewide pollution index using spatial interpolation techniques.

• An engineer is attempting to create a model of the bottom of a section of the Chesapeake Bay to help identify the nesting sites of endangered species of sea turtles. He has a set of sample bathymetric points obtained by dredges as they move through the bay. The engineer can use spatial interpolation methods to create a continuous bathymetric surface from these individual points.

• A geologist has measured a sampling of groundwater points based on drilling wells. She wants to create a map showing the depth to groundwater for the entire area and can use spatial interpolation methods to create a surface showing this information based on the samples from the wells.

ArcGIS Skills

In this chapter, you will learn:

- How to interpolate a surface from a set of points using the IDW method.
- How to interpolate a surface from a set of points using the Spline method.
- How to use the Zonal Statistics tool to examine an overall average of cell values within a polygon boundary.
- How to clip the data frame to conform to the boundaries of a polygon layer.
- How to use the Swipe tool on the Effects toolbar to compare multiple datasets.

Study Area

For this chapter, you will be working with data from the entire state of Ohio, along with areas near the Ohio borders.

Data Sources and Localizing This Chapter

This chapter's data focus on datasets from within the state of Ohio. However, you can easily modify this to use data from your own state instead. For instance, if you were doing this analysis in Wisconsin, the same kind of data are available. The Ohio county dataset was downloaded (for free) and extracted from the USA Counties dataset available on ArcGIS Online (and also available here: http://www .arcgis.com/home/item.html?id=a00d6b6149b34ed3b833e10fb72ef47b); you can do the same for Wisconsin counties.

The airports data were downloaded and extracted from The National Map (the trans_airportpoint layer from the Structures geodatabase) and only those airports labeled as complexes that were international or regional airports were separated out for use in this chapter.

The Ohioprecippoints layer was downloaded from the National Weather Service Advanced Hydrologic Prediction Service (located online here: http://water.weather.gov/precip/download.php), which contains precipitation data that can be downloaded (for free) for the entire United States. The points were originally derived from gridded datasets, and each point has an attribute that represents the total observed precipitation amount for the year 2011. Keep in mind that, although the dataset used for this chapter is a gridded set, the interpolation techniques used here can be used (and commonly are used) with non-gridded data.

The values used for the airport precipitation readings in Step 14.5 were obtained from information culled from values on the Wunderground (Weather Underground) Website at http://www.wunderground.com. Observed precipitation amounts for airport weather stations for 2011 (for various airports throughout the United States) are available from the "History and Almanac" information for individual stations at airports. A caveat: Some observed values being used to verify the interpolated results may have been used in creating the original Ohioprecippoints data.

STEP 14.1 Getting Started

• Start ArcMap and use the Catalog to copy the folder called **Chapter14** from the C:\GISBookdata\ folder to your own D:\GIS\ drive. Be sure to copy the entire directory, not just the files within it.

• The Chapter14 folder contains a file geodatabase called **Ohioprecipdata** that contains the following feature classes:

  • Ohioairports: A point feature class of the locations of Ohio airport complexes.

  • Ohiocountiesalbers: A polygon feature class representing the Ohio county borders (projected to the Albers Equal Area Conic projection).

  • Ohioprecippoints: A point feature class representing total observed precipitation amounts for Ohio and nearby surrounding areas in 2011.

• Activate the **Spatial Analyst** extension.

• Add the **Effects toolbar** to ArcMap.

• Add the Ohiocountiesalbers feature class to the TOC.

• Add the Ohioprecippoints feature class to the TOC.

• Use the following Geoprocessing Settings for this chapter (from the **Geoprocessing** pull-down menu, select **Environments...**):

  • Under the **Workplace** options, for **Current Workspace**, navigate to **D:\GIS\Chapter14**.

  • Under the **Workspace** options, for **Scratch Workspace**, navigate to **D:\GIS\Chapter14**.

  • Under the **Output Coordinates** options, for **Output Coordinate System**, select **Same as Layer "Ohiocountiesalbers"**.

- Under the **Processing Extent** options, for **Extent**, select **Same as layer Ohioprecippoints**.

  - Under the **Raster Analysis** options, for **Cell Size**, select the option for **As specified below**, then type the value of **100** (this will use 100-meter-resolution grid cells).

- The Ohiocountiesalbers layer was projected from its initial coordinate system to another one for use in this chapter. The coordinate system being used in this chapter is **North America Albers Equal Area Conic (NAD 83)**, using Map Units of **Meters**. Check the Data Frame to verify that this projection system is being used.

STEP 14.2 Examining the Data Points

- Rearrange the layers so that you can see all of the available precipitation points placed on top of the Ohio counties (you may want to make the counties fill color hollow so that you can see only the outlines of the counties). This will give you an indication of the distribution of points relative to the borders of Ohio.

- Answer Question 14.1.

> **QUESTION 14.1** How many points are in the Ohioprecippoints layer? (That is, how many precipitation sample points are you dealing with in this exercise?)

- Next, you'll change the symbology of the points to visually examine the distribution of precipitation levels in (and around) Ohio in 2011. Bring up the layer properties of Ohioprecippoints. In the Show: box, select **Quantities**, then select **Graduated colors**. Under the Fields options, select **Globvalue** as the attribute you want to display. Also select an appropriate Color Ramp.

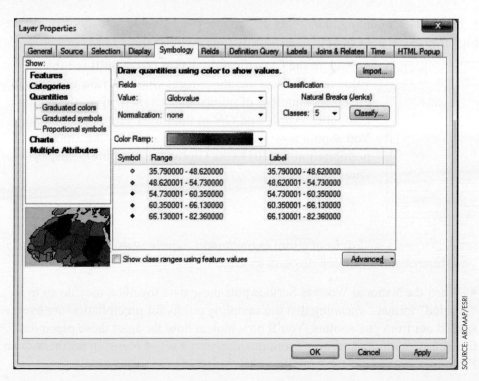

• *Important Note:* The default sampling size in ArcGIS 10.2 is 10,000 records. From your answer to Question 14.1, you'll note that you have more points than this, so ArcGIS will cut some of them off from the sample (and you should receive a pop-up box to this effect). To adjust the sample size, in the Ohiopre-cippoints Layer Properties, under the Symbology tab, first click on the **Classify** button. In the Classification dialog that opens, press the **Sampling...** button.

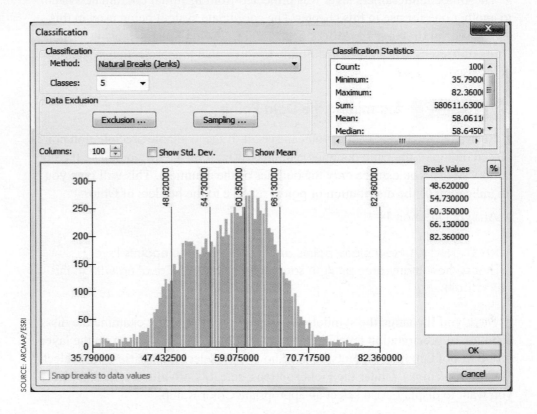

• In the Data Sampling dialog that will appear, type a value that is one number higher than the answer from Question 14.1, then click **OK**.

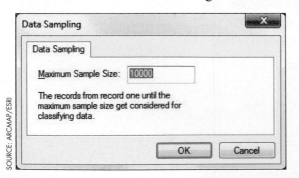

• Back in the Classification dialog box, you'll see that the value for Count in the upper-right corner is now changed to the correct number of points. Click **OK** to close the dialog. Last, click **Apply** and **OK** in the Layer Properties dialog. You should now see the points classified by the amount of precipitation for 2011 (the Globvalue attribute). Answer Question 14.2.

QUESTION 14.2 Just from visual examination, where were the greatest concentrations of rainfall amounts in Ohio in 2011?

• When the National Weather Service puts these data together, they do so in "gridded" format—meaning that the sampling points for precipitation are evenly spaced out from one another. You'll now look at how far apart these precipitation point measurements are. Zoom in closely on a set of points in northern Ohio. From the Tools toolbar in ArcGIS, select the **Measure** tool.

SOURCE: ARCMAP/ESRI

- Your cursor will change to a ruler shape (see Chapter 8 for more about using the Measure tool). Use the Measure tool to find the planar distance between two points in a row (you will be measuring slightly diagonally to follow a row). In the Measure dialog that will appear, the (planar) distance between the two points will be calculated. Answer Question 14.3.

QUESTION 14.3 How far apart are the precipitation points (in meters)?

STEP 14.3 Spatial Interpolation with IDW

- With these sample points being so far apart (see your answer to Question 14.3), we want to know how much precipitation occurred at the places between them, not just at these points. To generate this information, you'll have to interpolate a surface of total rainfall.

- The first method you'll use is inverse distance weighted (IDW). See **Smartbox 72** for more information about IDW and how it operates. From the **Spatial Analyst Tools** toolbox, select the **Interpolation** toolset, then select the **IDW** tool.

Smartbox 72

How does inverse distance weighted (IDW) interpolation function?

Inverse distance weighted (IDW) is a local interpolation method with one premise: The closer a known value point is to an unknown value location, the more important that known point is to determining the unknown value. Conversely, the farther away a known value point is from an unknown value location, the less that known point contributes to determining the unknown value. In IDW, the values of the points within the local search radius are averaged to determine the value at the unknown location. However, the value of each known point is weighted according to how far away that point is (and closer points carry a higher weight than points that are farther away). For instance, the amount of snowfall in areas closer to you is going to have a stronger influence on the amount of snow you get at your location than the amount of snowfall occurring at areas farther away. While IDW is a good and fast interpolation method, it has a tendency to create "bull's-eye" effects around known value points, especially if there are no other points nearby.

> **inverse distance weighted (IDW)** An interpolation process that assigns a higher weight to the values of known points closer to the location being interpolated and lower weights to the values of known points that are farther away.

Mathematically, the formula for determining the interpolated value at the unknown point ($F(x,y)$) is the sum of all the values of the known points (Fi) multiplied by a weight (Wi) for each known point value:

$$F(x,y) = \sum_{i=1}^{N} WiFi$$

The weights of each point are calculated by taking each known point and measuring the distance (d) from the known point to the unknown point and dividing by the sum total of all distance values. A separate value for power (p) is used in the weighting formula as well:

$$Wi = \frac{d_{i0}^{-p}}{\sum\limits_{i=1}^{N} d_{i0}^{-p}}$$

The value used for power directly affects the equation. If a high value is used for power, the points at closer distances will be weighted even more heavily. The default value for power used in ArcGIS is 2.

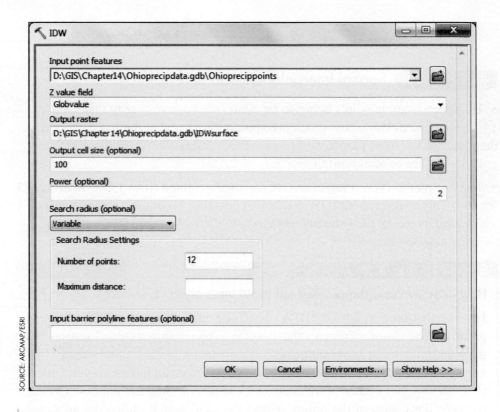

SOURCE: ARCMAP/ESRI

- For Input point features, select the **Ohioprecippoints** feature class from your Ohioprecipdata geodatabase.

- For the Z value field, select **Globvalue** (this is the attribute field that will be the source of the values you will be interpolating).

- Call the Output raster **IDWsurface** and save it in your Ohioprecipdata geodatabase.

- The Output cell size should be **100** (100 meters, which you previously set in the Geoprocessing Environments).

- Use the default Power value of **2**.

- Use **Variable** for the Search radius.

- Use the default value of **12** for Number of Points under Search Radius Settings.

- Click **OK** when ready.

- In the lower-right-hand corner of the ArcGIS window, you'll see the progress of the IDW calculation (it may take a couple of minutes to run).

- The new IDWsurface will be added to the TOC—it will be a new grid with 100 meter resolution, with the cell values reflecting the interpolated precipitation at that location.

- Change the symbology of the IDWsurface grid to show in Classified mode rather than Stretched, using 10 classes, the Natural Breaks classification method, and a suitable and intuitive color ramp.

- At this point, you have an interpolated surface for all of the data points, but we're only interested at looking at those within Ohio. ArcGIS has the capability to clip what's being shown in the view to only the boundaries of a polygon (such as the Ohio borders). Right-click on the **Layers** data frame in the TOC and select **Properties**.

- In the Data Frame Properties dialog, select the **Data Frame** tab. Under Clip Options, select **Clip to Shape** from the pull-down menu options.

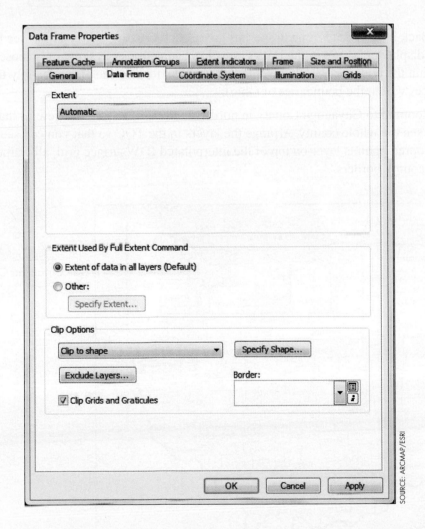

SOURCE: ARCMAP/ESRI

- Click on the **Specify Shape...** button.

- In the Data Frame Clipping dialog that will open, select the radio button for **Outline of Features**, and from the **Layer:** pull-down menu, select the **Ohiocountiesalbers** layer. Click **OK**.

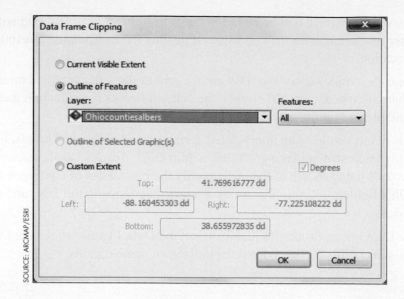

SOURCE: ARCMAP/ESRI

• Back in the Data Frame Properties layer, click **OK** to close it. You'll see that the display of what's in the Data Frame has been clipped to show only those areas within the boundaries of the Ohiocountiesalbers feature class (that is, only those places within the boundaries of Ohio).

• Zoom in to Cuyahoga County in northern Ohio and adjust the view so that you can see the whole county. Arrange the layers in the TOC so that you can see the Ohioprecippoints layer on top of the interpolated IDWsurface grid, all framed by the county borders.

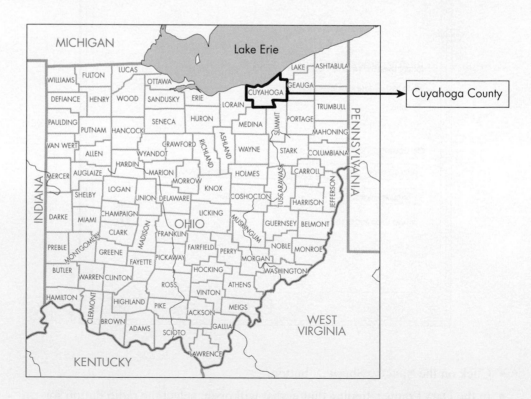

• Examine the precipitation points and their rainfall values in relation to the interpolated surface. Answer Question 14.4.

QUESTION 14.4 Why do some points have a circular set of grid cells centered around that point while others do not?

- Next, add the Ohioairports feature class to the TOC so that it sits on top of the other layers. Zoom out so that you can see the entire state again.

- At this point, you'll locate six airports across the state to determine the interpolated precipitation value at each location. The airports you'll be examining are:

Airport Name	Text in Attribute: Name
Akron Canton KCAK	Akron-Canton Regional Airport
Cleveland Hopkins KCLE	Cleveland-Hopkins International Airport
James M. Cox Dayton KDAY	James M. Cox Dayton International Airport
Mansfield-Lahm Regional KMFD	Mansfield Lahm Regional Airport
Port Columbus International KCMH	Port Columbus International Aiport
Youngstown-Warren Regional KYNG	Youngstown-Warren Regional Airport

- To find the correct airport, select the **Find** tool from the Tools toolbar (see Chapter 5 for more about the Find tool).

SOURCE: ARCMAP/ESRI

- In the Find dialog, select the **Features** tab.

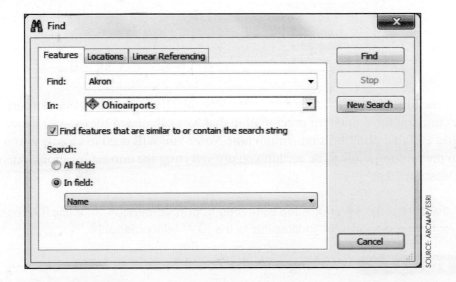

SOURCE: ARCMAP/ESRI

- In the Find: box, type part of the name of the first airport on the list: **Akron**.

- In the In: box, select **Ohioairports** as the layer you want to find information from.

- Place a checkmark in the **Find features that are similar to or contain the search string** box.

- Select the **In field:** radio button.

- From the pull-down menu, select **Name** as the field to search (this is the attribute representing the name of the airport).

- Click the **Find** button.

- For this first airport, select the option that corresponds with Akron-Canton Regional Airport (see the table above for the names of the airports and their corresponding value in the Name field). Double-clicking on **Akron-Canton Regional Airport** will shift the view to that airport and briefly highlight it with a dot and crosshairs.

- Zoom in very closely to the first airport and use the **Identify** tool on the Tools toolbar (see Chapter 5 for more about the Identify tool) to click on the Airport point and get information about the name of the airport.

- You'll notice that the results of Identify give you information about one layer at a time. Within the Identify dialog box, select **<All layers>** from the **Identify from:** pull-down menu. This will identify all layers at that location, including the value for precipitation interpolated into the IDWsurface layer.

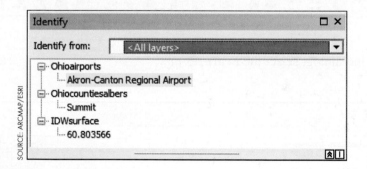

- Use Identify on all six airports, carefully selecting the point of the airport and examining the amount of precipitation that was calculated for the 100 meter grid cell that point falls on. *Important Note:* You will need to closely zoom in to the airport points to be certain you are selecting the correct locations. Answer Question 14.5.

> **QUESTION 14.5** What is the estimated total precipitation (in inches) at each of the six airport sites (according to the IDW interpolation)?

STEP 14.4 Working with the Zonal Tools

- Next, rather than examining the interpolated precipitation value at a single point, we will be looking at an average of many precipitation values for an entire area. The Zonal tools in ArcGIS allow you to do this. For more information about Zonal tools, see **Smartbox 73**.

Smartbox 73

How do the Zonal tools operate in ArcGIS 10.2?

The Zonal tools in ArcGIS allow you to obtain information about a set of raster grid cells that are spatially within a boundary (which can be defined by a polygon feature class or another raster layer). In a Zonal function, the raster for which you want to calculate information serves as the input, and the boundary serves as the "zones." If a polygon feature class is used as the source of the zones, ArcGIS will internally convert it to a raster in performing a Zonal calculation. An example of a Zonal function is the **Zonal Statistics** tool, which allows you to calculate a particular statistic (for example, the mean, the sum, the standard deviation, the median, the value that occurs most often, the value that occurs least often, the number of unique values, the highest value, the lowest value, or the range between the highest and lowest values) for the grid cells of the input raster that fall within the boundaries of the zones.

Figure 14.3 shows how the Zonal Statistics function can be used to find the average precipitation for each Ohio county. A raster surface contains interpolated values of precipitation for each grid cell. The polygon feature class (Ohiocountyalbers) will designate the boundaries of each zone (each of the 88 Ohio counties will be a separate zone). The Zonal Statistics tool will compute the mean of all grid cell values that fall within the boundaries of each of the counties. A new output raster will be created, and all grid cells that were within each zone will be assigned the same value (specifically, the mean of all interpolated values that were in that zone).

Zonal Statistics A method of calculating statistical values based on the raster grid-cell values that fall within a boundary.

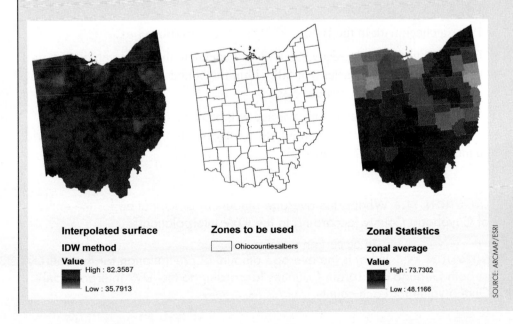

SOURCE: ARCMAP/ESRI

FIGURE 14.3 An interpolated raster surface of precipitation values, a polygon feature class defining zones as county boundaries, and a Zonal Statistics output layer showing the average precipitation value calculated for each zone.

- For this chapter, you'll be calculating the average amount of precipitation for each county in Ohio. Thus, ArcGIS will compute the average of all the interpolated grid cell values that are within the boundaries of each county. From the **Spatial Analyst Tools** toolbox, select the **Zonal** toolset, then select the **Zonal Statistics** tool.

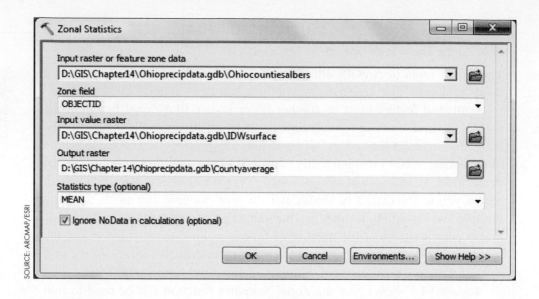

- For the Input raster or feature zone data, select the **Ohiocountiesalbers** feature class from the Ohioprecipdata geodatabase.

- For the Zone field, use the **OBJECTID** attribute.

- For the Input value raster, use the **IDWsurface** raster.

- Call the Output raster **countyaverage** and save it in your Ohioprecipdatageo-database.

- Use **MEAN** as the Statistics type.

- Place a checkmark in the **Ignore NoData in calculations** box.

- Click **OK** when you're ready. ArcGIS will calculate the mean value of all grid cells in the IDWsurface raster that fall within the boundaries of the Ohio counties as modeled in the County feature class.

- A new raster grid called countyaverage will be created in which each cell shows the average amount of rainfall per county in Ohio. Examine this new grid and answer Questions 14.6 and 14.7.

> **QUESTION 14.6** What is the average amount of precipitation for the entirety of Cuyahoga County (according to the IDW interpolation)?

> **QUESTION 14.7** What is the average amount of precipitation for the entirety of both Geauga and Lorain Counties (according to the IDW interpolation)? Give a value for each county.

STEP 14.5 Spatial Interpolation with Splines

- IDW isn't the only deterministic method of interpolating a surface from a set of points. In this portion of the lab, you'll use another method called Spline to create a new interpolated surface. See **Smartbox 74** for more information about using Spline for spatial interpolation.

Smartbox 74

How do Spline functions operate in spatial interpolation?

The **Spline** method of interpolation uses a mathematical process to fit the surface exactly through the known points and also fit a set of input points in the local search area. Here is one way to conceptualize Splines: You have a set of points with known values for some phenomenon (such as precipitation). Instead of thinking of them as dots on a map, think of them as poles that are being raised above a flat surface to a height represented by their value. For instance, if one point has a value of 50 and another has a value of 100, think of these as two poles, one that is 50 units high and another that is 100 units high. Once you have all your points set up as poles, you have to take a rubber sheet (that can be stretched in various directions) and adjust it so that it touches the top of each pole. In the end, you'd have a smooth surface that's been stretched this way and that.

The actual process is a mathematical one that tries to fit the surface to the known points as well as minimize the overall curvature. Two types of Spline methods are available in ArcGIS 10.2. The first of these is **regularized**, which results in an overall smoother surface but may also result in values that are beyond the range of the known point values. The second method, called **tension**, will generate a less smooth surface than the regularized method, but it will contain values closer to the range of the known point values.

The end result of Spline interpolation is a smooth surface and is best used for interpolating surfaces that do not have abrupt variations in their values (such as elevation measurements or temperature readings). See Figure 14.4 for an example of a regularized Spline interpolated surface compared to an IDW-generated surface. The two surfaces appear visually similar, but the Spline method produces a smoother overall surface. However, the range of interpolated values is very different between the two methods.

Spline An interpolation method that uses a mathematical process to fit the surface exactly through the known points.

regularized A Spline method that results in an overall smoother surface but may also result in values that are beyond the range of the known point values.

tension A Spline method that results in a less smooth surface than a regularized Spline but that will contain values closer to the range of the known point values.

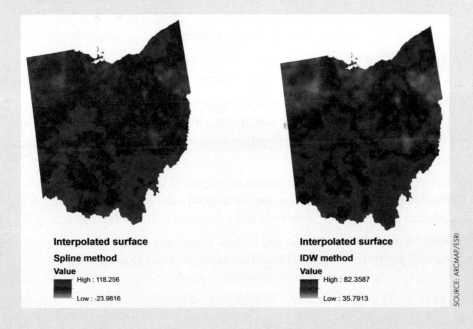

Interpolated surface
Spline method
Value
High : 118.256
Low : -23.9816

Interpolated surface
IDW method
Value
High : 82.3587
Low : 35.7913

SOURCE: ARCMAP/ESRI

FIGURE 14.4 A comparison of a Spline-generated surface and an IDW-generated surface.

- From the **Spatial Analyst Tools** toolbox, select the **Interpolation** toolset, then select the **Spline** tool.

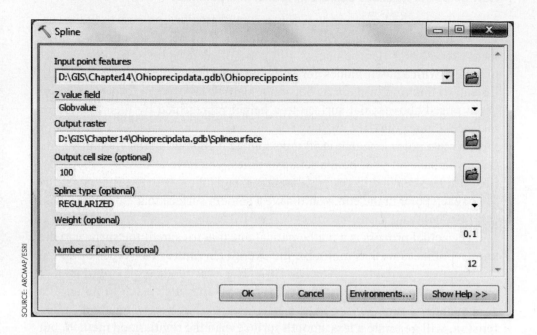

SOURCE: ARCMAP/ESRI

- For Input point features, select the **Ohioprecippoints** feature class from your Ohioprecipdata geodatabase.

- For the Z value field, select **Globvalue**.

- Call the Output raster **Splinesurface** and save it in your Ohioprecipdata geodatabase.

- The Output cell size should be **100** (100 meters, which you previously set in the Geoprocessing Environments).

- Use **REGULARIZED** for the Spline type.

- Use **0.1** for the Weight (this is the default).

- Use **12** for the Number of points (this is the default).

- Click **OK** when ready.

- In the lower-right-hand corner of the ArcGIS window, you'll see the progress of the Spline calculation (it may take a couple of minutes to run).

- The new Splinesurface will be added to the TOC. It will be a new grid with 100 meter resolution, with the cell values reflecting the new interpolated precipitation at that location.

- Change the symbology of the Splinesurface grid to show in **Classified** mode rather than Stretched, using **10** classes, the **Natural Breaks** classification method, and a suitable and intuitive color ramp (one that's different from the IDWsurface grid).

- Examine the Splinesurface in and around Cuyahoga County, comparing it to the surface generated for the IDWsurface layer. Answer Question 14.8.

QUESTION 14.8 Why does the Spline surface appear "smoother" (i.e., has less of the circular effect) than the IDW surface?

• Use the Identify tool to determine the amount of precipitation interpolated for each of the six airports used in Step 14.4, but this time examine the Splinesurface layer. Answer Question 14.9.

> **QUESTION 14.9** What is the estimated total precipitation (in inches) at each of the six airport sites (according to the Spline interpolation)?

• Redo the Zonal Statistics operation, but this time calculate the mean value for all of the Splinesurface grid cells that fall within an Ohio county boundary. Answer Questions 14.10 and 14.11.

> **QUESTION 14.10** What is the average amount of precipitation for the entirety of Cuyahoga County (according to the Spline interpolation)?

> **QUESTION 14.11** What is the average amount of precipitation for the entirety of both Geauga and Lorain Counties (according to the Spline interpolation)? Give a value for each county.

STEP 14.6 Evaluation of Interpolated Surfaces

• At this point, you have two interpolated surfaces of total precipitation for the state of Ohio. What you want to do is examine both surfaces to see which of them comes closer to predicting the actual amounts of precipitation in the state of Ohio. First, we'll visually compare the two surfaces. Turn off all of your layers except for Splinesurface and IDWsurface and place the Splinesurface at the top of the TOC. From the **Effects toolbar**'s pull-down menu, choose **Splinesurface**, then select the **Swipe** tool from the Effects toolbar.

• The Effects toolbar's Swipe tool works like its counterpart on the Image Analysis Window (see Chapter 13). After selecting the Swipe tool, your cursor will become a black triangle. Click the mouse on the image, then hold and drag the triangle from left to right or up and down; the Splinesurface layer will be replaced with the IDWsurface layer underneath, like drawing the blinds to see a window. Use Swipe to compare the two layers and zoom in on areas where the layers appear similar and areas where the layers appear different. Also pay close attention to the values of the classifications as seen for both layers in the TOC (because they will likely have different values displayed differently for each surface). Answer Question 14.12.

SOURCE: ARCMAP/ESRI

> **QUESTION 14.12** How do the two surfaces visually compare (in terms of patterns and smoothness of the surfaces)?

• By examining both surfaces (as well as your answers to Questions 14.5 and 14.9), you'll see that, while the two surfaces have some similarities, they also contain different interpolated values for the same locations. By adjusting the parameters of both the IDW and the Spline processes, you could generate other similar-but-different interpolated surfaces of precipitation. With different surfaces, you'll need a way to determine which method provides the most accurate result of interpolation.

• The table below lists the observed precipitation values for the six Ohio airports that you've looked at in this lab. This information was culled from values on the Wunderground (Weather Underground) Website at http://www.wunderground.com.

Airport Name	Name in Feature Class	Observed Precipitation
Akron Canton KCAK	Akron-Canton Regional Airport	58.38 inches
Cleveland Hopkins KCLE	Cleveland-Hopkins International Airport	65.32 inches
James M. Cox Dayton KDAY	James M. Cox Dayton International Airport	56.72 inches
Mansfield-Lahm Regional KMFD	Mansfield Lahm Regional Airport	56.68 inches
Port Columbus International KCMH	Port Columbus International Aiport	54.96 inches
Youngstown-Warren Regional KYNG	Youngstown-Warren Regional Airport	54.01 inches

• Compare the Wunderground Observed Precipitation values to those that you interpolated from each surface, and use Swipe to examine both surfaces around these six locations. Answer Question 14.13.

QUESTION 14.13 How do the interpolated values for the IDW surface and the Spline surface compare to the six actual values? Which values are too high, too low, or nearly match in each data set? Based solely on these six points, which surface is more representative of the actual observed values and why?

STEP 14.7 **Printing or Sharing Your Results**

• Save your work as a map document. Add the usual information to the Map Document Properties.

• Finally, either print a layout (see Chapter 3) with two Data Frames, with the IDW results in one Data Frame and the Spline results in the other (see Chapter 1 for how to use multiple Data Frames and Chapter 3 for how to place multiple Data Frames in a layout), or share your results as a tiled map service through ArcGIS Online (see Chapter 4). If you're sharing your data, you may want to create a Web application using a template that has the Swipe tool available to allow others to interactively examine both surfaces (like you've been doing in this chapter).

Closing Time

This chapter looked at how to take an initial set of point measurements and create a raster surface from them to "fill in the gaps" where point measurements were unavailable. Spatial interpolation methods are used to create surfaces to model continuous phenomena, including precipitation, groundwater, and pollution levels. While commonly used, IDW and Splines are not the only spatial interpolation methods available within ArcGIS. The *Related Concepts for Chapter 14* examines another powerful set of interpolation tools.

Elevation is a phenomenon that varies continuously across Earth and is commonly modeled as a raster surface. An elevation surface acts like the interpolated surfaces we've examined in this chapter (i.e., at each x and y location, a z-measurement for elevation can be determined). We will work with these kinds of "functional" surfaces in the next chapter.

Related Concepts for Chapter 14   Geostatistical Methods for Spatial Interpolation

Both the IDW and Spline techniques are deterministic interpolation methods, in that they rely solely on the location of the known points and mathematically fit the surface through these points. As such, they do not rely on information regarding how the values of the known points relate to one another. For instance, if you have several known points with similar values located close together, how will this similarity of values influence how the values for nearby unknown points are calculated? Likewise, if the values for the points are very different from one another but very far apart, how will these conditions affect the new values to be interpolated for other locations? These kinds of spatial variations in the values of the points or the similarities of values of nearby points are not considered in deterministic methods like IDW (which deals only with the distances between points, not their values) or Splines (which seeks to fit the surface through points and minimize the curvatures of the surface). However, a separate set of geostatistical interpolation methods take spatial autocorrelation (see Chapter 8) and similarities over space between points into account when calculating interpolated values for a surface.

Kriging is a commonly used geostatistical interpolation method that takes into account the variation or similarity of the known values as well as their spatial arrangement when predicting what the new interpolated values should be. To model these factors, we first need to know something about the variance between the values of the known points. In theory, points with similar values (i.e., a low variance) should be found closer together, while points with dissimilar values (i.e., a high variance) should be found farther away from one another. In reality, though, these conditions are not always met, so a model of the variance of the points' values as it relates to their distance apart can be generated. This model can then be used to determine the variance associated with points to be interpolated that are a certain distance apart. This model is called a **semivariogram**, and properly setting it up in relation to the values of the known points is key to its use in the prediction process (Figure 14.5).

Kriging is a local interpolation method that uses (a) the variation between the subset of known points being used to determine the unknown value, and (b) the variation between the known points and the unknown location. Each point is weighted through a mathematical process that takes the variance into account as well as a computed margin of error. Weights are assigned to each of the points being used in the calculation of the location's unknown value (Figure 14.6).

deterministic An interpolation method that does not take spatial variation into account.

geostatistical An interpolation method that accounts for spatial variation.

Kriging A local geostatistical interpolation method.

semivariogram A model of the variance between points and their distance apart.

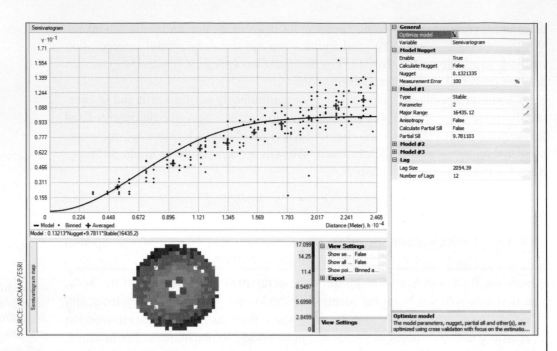

FIGURE 14.5 Fitting a semivariogram for use in the Kriging process.

SOURCE: ARCMAP/ESRI

Ordinary Kriging A Kriging method that assumes an unknown mean and a lack of trend in the data.

Universal Kriging A Kriging method that assumes a trend within the data (and removes the trend in order to perform interpolation).

Geostatistical Analyst The ArcGIS extension that contains several functions related to examining and working with spatial statistics.

Several types of Kriging are used in interpolation. A common method is **Ordinary Kriging**, which assumes an unknown mean for the known point values as well as the absence of a trend in the data. Another is **Universal Kriging**, which assumes the presence of a trend and will remove that trend from the data prior to performing interpolation. A Kriging tool is available in the Interpolation toolset of the Spatial Analyst toolbox, but further in-depth analysis with Kriging (such as adjusting the fit of the semivariogram or examining the weights assigned to points) can be performed with the functions of ArcGIS 10.2's **Geostatistical Analyst** (and its corresponding Geostatistical Analyst toolbar). This extension provides modeling techniques for examining the statistical nature of spatial data, as well as statistical tools for examining the overall accuracy and fit of an interpolated surface. Figures 14.5 and 14.6 both show the use of Geostatistical Analyst functions.

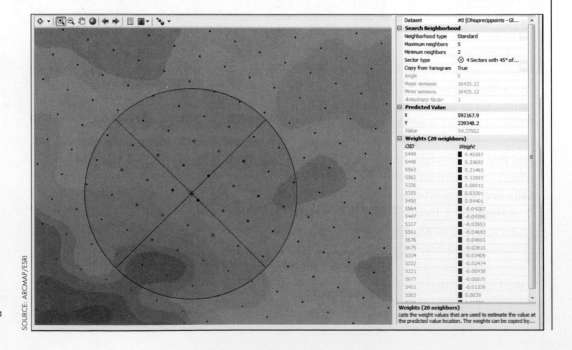

FIGURE 14.6 The assigned weights for the 20 points being used to determine the value for a location using Kriging.

SOURCE: ARCMAP/ESRI

For More Information

For further in-depth information about the topics presented in this chapter, use the ArcGIS Help feature to search for the following items:

- An overview of the Zonal tools
- Comparing interpolation methods
- Deterministic methods for spatial interpolation
- How IDW works
- How inverse distance weighted interpolation works
- How Kriging works
- How radial basis functions work
- How Spline works
- How Trend works
- Kriging (Spatial Analyst)
- Spline (Spatial Analyst)
- Understanding interpolation analysis
- Zonal Statistics (Spatial Analyst)

For further information and comparisons of different spatial interpolation methods, see Childs, Colin. 2004. "Interpolating Surfaces in ArcGIS Spatial Analyst." *ArcUser*, July–September 2004, pp. 32–35 (available online at http://www.esri.com/news/arcuser/0704/files/interpolating.pdf).

Key Terms

spatial interpolation (p. 319)
surface (p. 319)
global interpolation (p. 320)
trend (p. 320)
local interpolation (p. 320)

inverse distance weighted (IDW) (p. 325)
Zonal Statistics (p. 331)
Spline (p. 333)
regularized (p. 333)
tension (p. 333)
deterministic (p. 337)
geostatistical (p. 337)

Kriging (p. 337)
semivariogram (p. 337)
Ordinary Kriging (p. 338)
Universal Kriging (p. 338)
Geostatistical Analyst (p. 338)

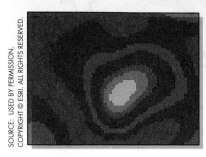 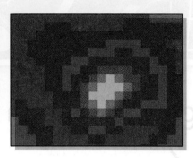

FIGURE 15.1 A finer-resolution raster elevation surface versus a lower-resolution raster elevation surface.

Figure 15.1 for an example. Numerous DEMs are available for use in GIS at a variety of resolutions, including USGS DEMs, as well as Esri's World Elevation service available through ArcGIS Online.

Chapter Scenario and Applications

This chapter puts you in the role of a planner in Hocking County, Ohio. Your job is to assess the landscape features of the county as part of a study involving the building of a new observation platform in the county. You will be examining the terrain and land cover characteristics of the county, particularly at the building site for the platform. You want to determine the minimum height at which the platform can be constructed in order to have an unobscured view of the city of Logan's fairgrounds over the terrain. From there, you'll then determine what areas of the entire city can be seen or not seen from this projected minimum platform height.

The following are additional examples of other real-world applications of this chapter's skills.

- A developer is designing the layout of a new ski resort. She could use a digital elevation model as the basis for slope and aspect calculations for planning ski runs.

- A forest manager is examining several possible sites for the location of a new lookout tower. He can perform viewshed analysis from each site to determine what areas can and cannot be seen.

- A county engineer in a coastal region wants to examine the visual impact of the development of new high-rise hotels being built on the coast. She can build viewsheds and lines of sight from a variety of key sites to determine how the addition of tall buildings will impact tourists' views of the beaches and shores.

ArcGIS Skills

In this chapter, you will learn:

- How to use the graph functions to create a histogram.
- How to work with NED data in ArcGIS.
- How to derive a hillshade.
- How to calculate slope and aspect raster layers.
- How to create a viewshed.
- How to use Line of Sight, viewsheds, and profile graphs for visibility analysis in ArcMap.
- How to use a vector layer to extract a subset of a raster.
- How to add data layers to ArcScene and work in the ArcScene environment.

- How to drape both raster and vector layers over an elevation surface.

- How to use the flying and navigation tools of ArcScene to view 3D raster surfaces.

- How to export a scene to a graphical file format.

 Study Area

The chapter data is of Hocking County, located in southeastern Ohio. The county seat (and main urban area) is the town of Logan.

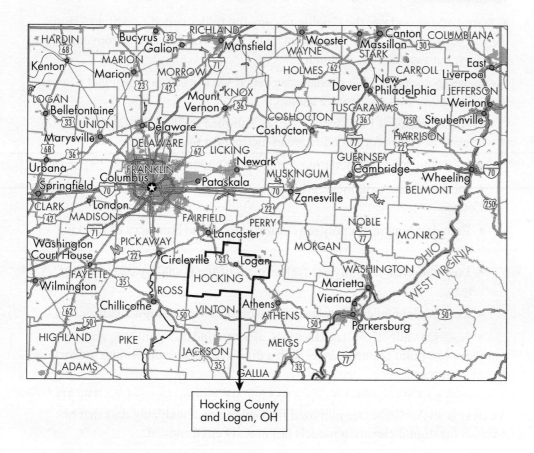

Hocking County and Logan, OH

 Data Sources and Localizing This Chapter

This chapter's data focus on Hocking County in the state of Ohio. However, you can easily modify this to use data from your own local area—for instance, Marquette County, Michigan. The NED, county boundary, and roads layers were all downloaded and extracted for free from The National Map. Use GU_county-orequivalent for the county boundary and trans_roadsegment for the roads layer. The boundary for a city within the county (such as the city of Marquette) can also be downloaded from The National Map (use GU_Incorporated_Place). The four point layers of locations in the county were digitized into their own feature classes (see Chapter 6 for how to digitize points).

The NLCD data was downloaded for free from the MRLC here: http://www.mrlc. gov. NLCD 2006 data for the entire United States can be downloaded from that Website and the data for Marquette County (or your individual county) can be extracted out using ArcGIS (through the Extract By Mask tool—see Chapter 12—using the county boundary layer). Alternatively, NLCD 2006 data can also be downloaded directly via The National Map. The Mask layer was created from processing the NLCD layer; Map Algebra was used to select all grid cells with a value of greater than 0, which resulted in a new grid in which all cells in the county had a value of 1 and those cells outside the county had a value of 0 (see Chapter 20 for more about Map Algebra). Then the grid was reclassified so that all cells with a value of 0 were given a value of NoData (see Chapter 19 for more about reclassifying cells).

STEP 15.1 Getting Started

• Start ArcMap and use the Catalog to copy the folder called **Chapter15** from the C:\GISBookdata\ folder to your own D:\GIS\ drive. Be sure to copy the entire directory, not just the files within it.

• Chapter15 contains a file geodatabase called **hockingcounty.gdb** that contains the following rasters and feature classes:

  • NEDhocking: a 1 arc-second NED resampled to 30 meter resolution (elevation values for the NED are in meters) for this chapter.

  • NLCDhocking: 30 meter resolution NLCD 2006 raster layer.

  • Maskhocking: a 30 meter grid that conforms to the boundaries of Hocking County.

  • Boundaryhocking: a polygon feature class of the boundary of Hocking County.

  • Loganboundary: a polygon feature class of the boundary of the city of Logan.

  • RoadSegmenthocking: a line feature class of the county's roads.

  • HockingValleyHospital, HockingCoFairgrounds, LoganDam, and Platform: point feature classes with a digitized point representing the location of each place.

• Activate both the **Spatial Analyst** and the **3D Analyst** extensions. The **3D Analyst** is an ArcGIS extension used for viewing and analyzing data that has z-values (as digital elevation models do) in a 3D environment.

• Add the **3D Analyst toolbar**, the **Effects toolbar**, and the **Editor toolbar** to ArcMap.

• Add the NLCDhocking layer to the Table of Contents. Change the symbology of the various land-cover types to better represent each one, as you did in Chapter 12 (i.e., shades of green for the forested land-cover types, blue for the water, and so on).

• Next, use the following Geoprocessing Settings for this chapter (from the **Geoprocessing** pull-down menu, select **Environments...**).

  • Under the **Workplace** options, for **Current Workspace**, navigate to **D:\ GIS\Chapter15**.

  • Under the **Workspace** options, for **Scratch Workspace**, navigate to **D:\ GIS\Chapter15**.

  • Under the **Output Coordinates** options, for **Output Coordinate System**, select **Same as Layer "NLCDHocking"**.

3D Analyst An ArcGIS extension used for 3D visualization and working with data that has z-values.

- Under the **Processing Extent** options, for **Extent**, select **Same as Layer NLCDhocking**.

- Under the **Raster Analysis** options, for **Cell Size**, select the option for **Same as layer NLCDhocking**. This should automatically adjust the cell size to 30 (as in 30 meters).

- Under the **Raster Analysis** options, for **Mask**, navigate to your D:\GIS\ folder, then into the Hockingcounty geodatabase, and select **Maskhocking**.

- Click **OK** when you're done. All of your Environment Settings should now be in place for use in the rest of the chapter.

- The Coordinate System being used in this chapter is **Albers Conical Equal Area (NAD 83)**, using Map Units of **Meters**. Check the Data Frame to verify that this coordinate system is being used.

STEP 15.2 Working with Histograms

- The first thing to do as part of the planning process is to familiarize yourself with the landscape features of the county, both land cover and terrain-based. First, we'll focus on the land cover. NLCDhocking is a subset of the NLCD2006 data for the county (see Chapter 12 for more about NLCD, including which land-cover types are represented by each grid cell value). For further information on the distribution of land-cover types across the county, we'll examine a graph of the distribution of land-cover types. To begin, from the **View** pull-down menu select **Graphs**, then select **Create Graph....**

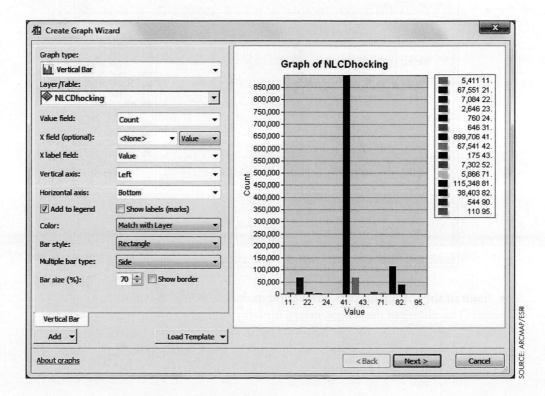

- For the Graph type: choose **Vertical Bar**.
- For Layer/Table: choose **NLCDhocking**.

histogram A graphical representation of a raster, showing the values of the raster on the *x*-axis and the count of those values on the *y*-axis.

- For the Value field: choose **Count**. This is the attribute that we will be charting on the *y*-axis.

- For the X label field: choose **Value**. This is to label the attribute present on the *x*-axis.

- Make sure the **Add to legend** box has a checkmark in it.

- Leave the rest of the options as the default and click **Next**.

- On the second screen of the wizard are more options for the presentation of the chart. Just accept the defaults and click **Finish**.

- A chart of the NLCDhocking raster will appear. (***Important Note:*** Expand the edges of the histogram window to see all of the detail and be sure that you can see all 15 categories being graphed.) The chart we've created is displaying a histogram of the data. A **histogram** is a graphical tool that displays the different categories or zones being displayed on the screen (drawn from the VALUE field of the VAT) on the *x*-axis, and the summation of each of these categories or zones (from the COUNT field of the VAT) on the *y*-axis. A color bar for each category or zone extends up from the *x*-axis to the summation level.

  - ***Important Note:*** ArcGIS has an interactive histogram tool for examining rasters on the Spatial Analyst toolbar. However, the tool gives different results depending on both the extent of the raster being shown and the screen resolution. To keep the results of your analysis consistent, you'll be using the Graph features instead of that particular tool.

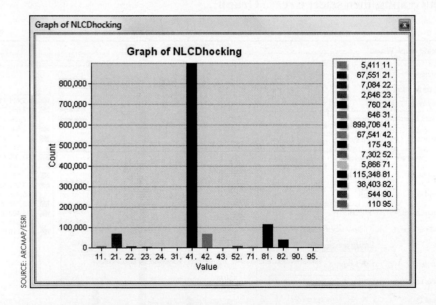

SOURCE: ARCMAP/ESRI

- Each of the histogram bars has a corresponding value shown at the right of the graph. The colors of the values match up with the colors of the NLCDhocking layer as shown in the TOC, and the values from the NLCD are also listed to the right of the count total in the chart's legend. Examine each of the land-cover categories and answer Questions 15.1 and 15.2.

QUESTION 15.1 Which land-cover type (i.e., open water, pasture/hay, etc.) is the most prevalent in Hocking County?

QUESTION 15.2 Which land-cover type (i.e., open water, pasture/hay, etc.) is the least prevalent in Hocking County?

STEP 15.3 Working with a National Elevation Dataset Layer

• Turn off the NLCDhocking layer for now and add the NEDhocking layer to the TOC. The NEDhocking raster is a digital elevation model of the county, in which each cell has a value for elevation. This raster was drawn from the National Elevation Dataset (NED). See **Smartbox 75** for more about NED.

Smartbox 75

What is the National Elevation Dataset?

The **National Elevation Dataset (NED)** is a digital elevation model (of equally spaced elevation values) compiled and distributed by the USGS. NED covers the United States and is compiled from a variety of different sources, including USGS DEMs and lidar products (see Chapter 17) with the goal of providing the best possible resolution elevation data for the United States. NED is available for free download as a layer of The National Map (see Chapter 5). For ease of delivery, The National Map has pre-packaged NED data as a series of tiles. A user can select an area of interest and then download all NED tiles that correspond with this area in the ".img" format (which ArcGIS will read as a raster layer). NED uses the NAVD 88 vertical datum and is available at multiple resolutions (Figure 15.2):

National Elevation Dataset (NED) A digital elevation model of the entire United States, provided by the USGS.

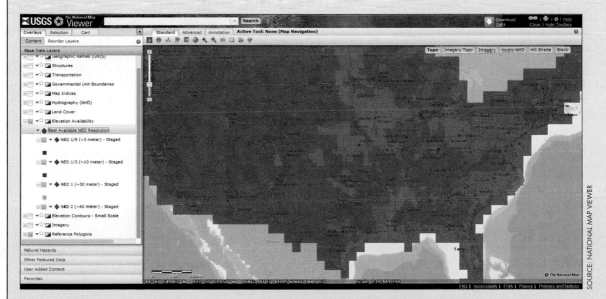

SOURCE: NATIONAL MAP VIEWER

FIGURE 15.2 The NED data available from The National Map.

• 1 arc-second (about 30 meters): This resolution is available for the entire United States (with the exception of Alaska, which is available at 2 arc-second resolution).

• 1/3 arc-second (about 10 meters): This resolution is also available for the entire United States (with the exception of Alaska).

• 1/9 arc-second (about 3 meters): This resolution is only available for certain areas within the United States.

- To help familiarize yourself with the terrain of Hocking County, you'll next create a hillshade of the NED. For more about hillshades, see **Smartbox 76**.

Smartbox 76

What is a hillshade?

hillshade A shaded relief map of the terrain created by modeling the position of an illumination source (e.g., the Sun) in relation to the terrain.

A **hillshade** is a representation of a surface depicting how the elevation features would look with a light source (such as the Sun) applied to them. Hillshades are used to better visualize features on a landscape, as well as examine what areas are illuminated (from the Sun) and what areas are in shadow under different Sun conditions or at different times of the day. Several hillshades can be created by altering the relative position of the Sun to the surface to model how the area would look or shadows would fall throughout the day.

Creating a hillshade relies on setting two parameters for the illumination source: azimuth and altitude (Figure 15.3). *Azimuth* is set up like a circle, with values from 0 to 360 degrees, and reflects the location of the sun in relation to the terrain. Assuming 0 degrees to be due north, 90 degrees would be due east, 180 degrees due south, and 270 degrees due west. *Altitude* refers to the elevation of the illumination source above the terrain, measured from 0 degrees (flat on the ground) up to 90 degrees (directly over top of the terrain).

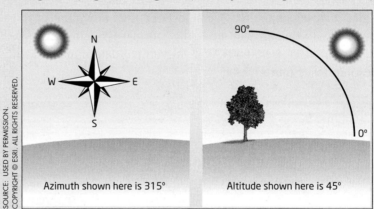

Azimuth shown here is 315°   Altitude shown here is 45°

SOURCE: USED BY PERMISSION. COPYRIGHT © ESRI. ALL RIGHTS RESERVED.

FIGURE 15.3 Settings for azimuth and altitude in a hillshade.

- To create a hillshade, from the **Spatial Analyst Tools** toolbox select the **Surface** toolset, then select the **Hillshade** tool.

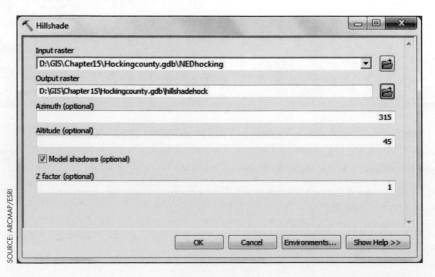

SOURCE: ARCMAP/ESRI

- For the Input raster, use the **NEDhocking** grid.

- Call the Output raster **hillshadehock** and place it in your Hockingcounty geodatabase in your **D:\GIS\Chapter15** folder.

- Use **315** degrees for the Azimuth (this is the ArcGIS default).

- Use **45** degrees for the Altitude (this is the ArcGIS default).

- Place a checkmark in the **Model shadows** box.

- Click **OK** to create the hillshade.

- Next, we want to examine the hillshade in relation to the NED. Turn off all of the other layers, except for hillshade-hock and NEDhocking. Use the **Swipe** tool on the **Effects toolbar** (see Chapter 14 for how to use this) to compare the two layers and answer Question 15.3.

> **QUESTION 15.3** How does the hillshade help in visually examining the landscape (in terms of seeing the hills, valleys, and flat areas in the county)?

STEP 15.4 Working with Slope Layers

- Elevation doesn't tell the whole story about what the terrain is really like; all it's showing you is the height above a vertical datum at each location. In answering Question 15.3, you likely saw areas of changes between lower and higher elevations as well as hills and flatter sections of lands. For information about how steep or flat areas are, you'll have to use the NED to create new grids, such as slope and slope aspect (see **Smartbox 77** for more about these two concepts).

Smartbox 77

What are slope and aspect measuring?

The slope of the land can be described using terms like "steep" or "flat" when applied to hills, mountains, or cliff faces. **Slope** represents the rate of change of elevations at a particular location and is calculated by measuring the vertical distance (the rise) over the horizontal distance (the run). The slope is computed as an angular degree (Figure 15.4) with values between 0 (completely flat with no slope) to 90 (a straight vertical drop).

slope A measurement of the rate of elevation change at a location found by dividing the vertical height (the rise) by the horizontal length (the run).

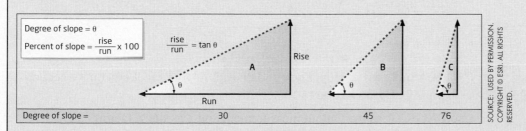

FIGURE 15.4 Calculating slope.

In a raster elevation model, each cell receives a value for slope, calculated as the maximum rate of change for that location in relation to its eight

aspect A determination of the direction the steepest slope is facing at a location.

neighboring cells, and a new slope raster will be computed. The direction of the steepest slope is referred to as the slope **aspect** and can also be computed as a separate raster layer. In the aspect grid, each cell receives a value that indicates in which of the eight cardinal directions the steepest slope faces for each cell (or it can receive a value indicating it is classified as "flat" and has no slope). See Figure 15.5 for an example of an aspect grid and how to interpret it.

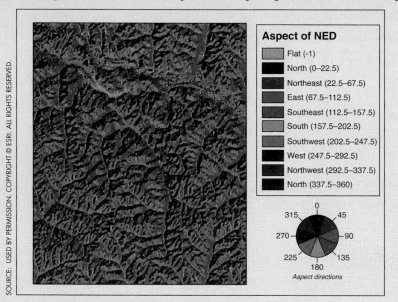

FIGURE 15.5 A raster of slope aspect, along with the corresponding aspect directions.

- To create a slope grid, first open ArcToolbox. Next, from the **Spatial Analyst Tools** toolbox, select the **Surface** toolset, then select the **Slope** tool.

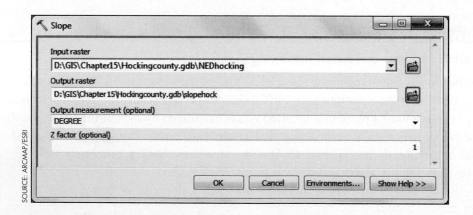

- For the input raster, use the **NEDhocking** grid.

- Call the Output raster **slopehock** and place it in your Hockingcounty geodatabase in your **D:\GIS\Chapter15** folder.

- Use **DEGREE** as the Output measurement and **1** as the Z factor.

- Click **OK** when all the options are set.

- Examine the slopehock grid—each cell contains the slope angle (measured in degrees) at that location. Answer Question 15.4, then turn off the slopehock grid.

QUESTION 15.4 Where, spatially, are the least-steep slopes in the county located?

• Turn off all of the other layers, except for hillshadehock and slopehock. Use the **Swipe** tool on the **Effects toolbar** to compare the two layers and answer Question 15.5.

QUESTION 15.5 Examine the hillshade in relation to the slope raster. What kinds of areas are those areas with the least-steep slopes in the county (i.e., your answer to Question 15.4)?

• Next, you'll create a grid of slope aspect. From the **Spatial Analyst Tools** toolbox select the **Surface** toolset, then select the **Aspect** tool.

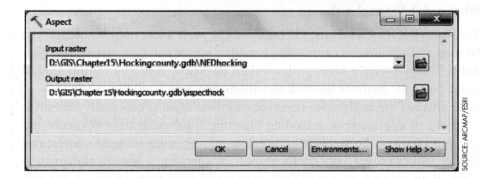

• For the Input raster, use the **NEDhocking** grid.

• Call the Output raster **aspecthock** and place it in your Hockingcounty geodatabase in your **D:\GIS\Chapter15** folder.

• Examine the aspecthock grid. It shows the direction of the steepest slope at that particular location. For instance, all grid cells with a value of "south" have their steepest slope as a south-facing slope at that cell.

• Next, we'll use these slope and aspect grids to get information about the potential building site for the observation platform. First, add the Loganboundary and RoadSegmenthocking layers to the TOC. These layers will allow you to see other developed areas in Hocking county, as well as the city of Logan. Change the Loganboundary's symbology from a solid color to hollow (so you can see the layers underneath it) and zoom in to its boundaries.

• Also add the LoganDam feature class to ArcMap, change its symbology so that it stands out from the other layers, and position it at the top of the TOC.

• By using the Identify tool, you can get information back about multiple layers at a single point (as you did in Chapter 14). Zoom in closely to the LoganDam location and use the Identify tool to answer Questions 15.6 and 15.7.

QUESTION 15.6 What direction is the steepest slope facing at the Logan Dam location?

QUESTION 15.7 What is the degree of slope at the Logan Dam location?

- Line of Sight allows you to draw a line between two points and get a result that shows what you can see (in green) and not see (in red) along that line. After you select the Line of Sight tool, the cursor will change to a crosshairs. Click once on the point location for Logan Dam and click again on the point location for Hocking County Fairgrounds. This line represents you standing at the Dam and looking straight in the direction of the Fairgrounds.

- The line will change colors. The areas in green represent locations that you can see along the line between the Dam and the Fairgrounds, while the red areas on the line represent areas that you cannot see. Most of the line between the two spots is red, with only a little portion in green, indicating that you can see part of the way from the Dam to the Fairgrounds, but that something on the terrain is blocking most of the view.

- To get a sense of what kind of terrain is blocking your view, you can create a profile graph for the LOS results that will show the terrain relief along the length of the line, as well as which areas of the terrain are seen and which are not seen. Select the **Profile Graph** button the on the **3D Analyst toolbar** (you may have to choose it from the available pull-down menu options next to its button on the toolbar).

- The Profile Graph will appear, showing distances on the *x*-axis, elevations on the *y*-axis, and a dashed line showing the line of sight between the Dam and the Fairgrounds. The profile of the landscape between the two points will be shown, along with the areas that can be seen (in green) and the areas that cannot be seen (in red). The black dot at the beginning of the line shows your current elevation (Figure 15.6).

FIGURE 15.6 The profile graph created from the line of sight.

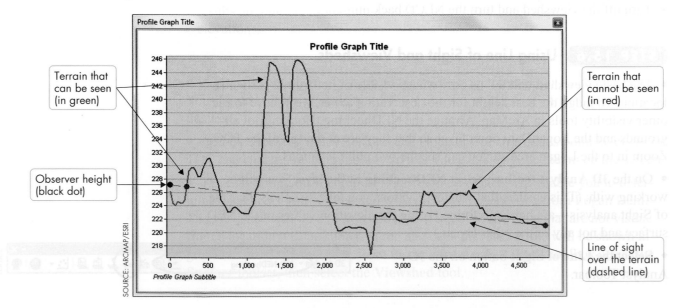

- Drag a corner of the Profile Graph out to expand it for better visualization, then answer Question 15.10.

QUESTION 15.10 Examine the information on the Profile Graph. At what elevation is the Dam? What is the elevation of the terrain feature blocking your immediate view of the Fairgrounds?

- Although LOS represents what types of terrain features are blocking a view, you can also see the type of land cover that is in the way on the terrain. Turn on the NLCDhocking layer. You'll see where the LOS line stops being green and turns to red in relation to the land-cover types. Answer Question 15.11.

QUESTION 15.11 What land-cover type on the terrain surface is blocking your immediate view from the dam?

- Use the cursor to select the line of sight itself on the screen, then right-click and select **Delete** to get rid of it.

- To see the Fairgrounds from the Dam, you're going to have to apply an "observer offset." This offset simulates you standing on a tall object (like the observation platform you're designing) in an effort to see over top of the terrain features in your way.

- Select the **Line of Sight tool** again. A pop-up window will open, allowing you to input a value for observer offset in the Z units of the NED (i.e., meters).

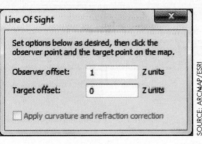

- You want to determine the minimum value necessary to use as an observer offset to be able to see the Fairgrounds from the Dam. Because we're designing an observation platform at the Dam site, and you want it to be the minimum height necessary to look down toward Logan to see the Fairgrounds. Keep in mind that the observer offset is the distance you have to go above the base of the terrain on which you're standing.

- Try an Observer offset of 5 (meters), and draw a new line of sight between the Dam and the Fairgrounds. Also create a Profile graph for this new line of sight. You'll see both the LOS line in ArcMap change, as well as the information on the Profile Graph generated from the LOS results. Specifically, the Profile Graph will show a dashed line representing the line of sight from the new 5 meter height with parts of the profile colored green or red to indicate whether you can see them or not see them from the new observer offset height. Answer Question 15.12.

QUESTION 15.12 You still cannot see the Fairgrounds from a 5-meter-high observation platform at the Dam, but more of the terrain is now visible. Based on the information on the LOS and its Profile Graph, what can you now see that you couldn't see before?

- What you're trying to determine is the minimum height of the platform (i.e., the lowest value for observer offset) needed to see the Fairgrounds from the platform location. Look carefully at the Profile Graph to try and gauge what a reasonable minimum height would be. Create a LOS and a corresponding Profile Graph to see if the height you've selected is the value to use. Try a few different observer offset values and draw LOS lines and create Profile Graphs that correspond to them. If the results of the line of sight are not what you need to see the Fairgrounds at the minimum observer offset (either the offset is too high or too short), delete them until you find the proper value.

• Once you're able to clearly see the Fairgrounds from the dam at the minimum height necessary, you'll know how tall the observer platform has to be. Answer Question 15.13.

QUESTION 15.13 What is the minimum observation platform height required to see the Fairgrounds from the Logan Dam? (*Important Note:* This should be the height of the platform itself—e.g., a 15-meter-tall platform—not the height the platform will be above the vertical datum.)

• Now that we know the height of the platform necessary to see the Fairgrounds from the potential area at the Dam, we'll also want to determine what other areas can and cannot be seen from the platform. Add the platform feature class to ArcMap. This feature class consists of a single point of a possible location for the platform. Change its symbology to stand out from the other layers.

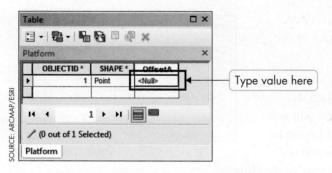

• We'll now compute a viewshed to determine all the areas that can be seen from the platform location, by simulating its height. The viewshed tool relies on specific attributes in the feature class it's computed for in order to control functions such as how high the observer offset of the point should be for the viewshed. For instance, if the source feature class's attribute table contains an attribute called "OffsetA," then the viewshed tool will interpret that as the observer offset, while an attribute called "OffsetB" will be interpreted as a target offset.

• Open the platform layer's attribute table. You'll see that it consists of a single record and that the new field of OffsetA has already been added for you (using the Add Field methods as described in Chapter 2). There is a currently a <Null> value in the table. What you'll have to do is add the value for OffsetA (i.e., the observer offset, or the minimum height of the platform) to the attribute table.

• You can edit the attribute table as you did in Chapter 2. From the Editor toolbar's **Editor** pull-down menu, select **Start Editing**.

• In the platform layer's attribute table, type the value for the minimum height of the platform (which you found in your answer to Question 15.13).

• After you've typed the value, from the Editor toolbar's **Editor** pull-down menu, select **Save Edits**, then from the **Editor** pull-down menu, select **Stop Editing**. Close the attribute table when you're done.

• To create the platform's viewshed, from the **Spatial Analyst Tools** toolbox select the **Surface** toolset, then select the **Viewshed** tool.

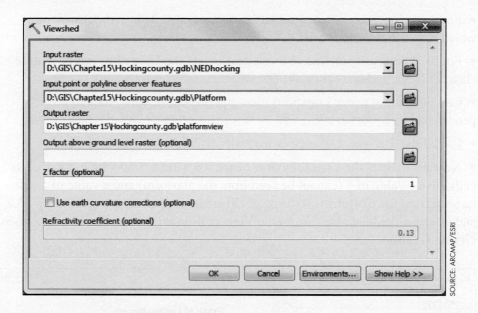

- Use the **NEDhocking** grid as the Input raster.

- Use the **Platform** as the Input point or polyline observer features.

- Call the Output raster **Platformview** and place it in your Hockingcounty geo-database in your **D:\GIS\Chapter15** folder.

- Click **OK**.

- Your new platformview layer will show you what areas can be seen and not seen (due to intervening terrain) from the platform location at the proposed height. The last analysis we want to perform is to determine how much of the entire city can be seen (or not seen) from the platform at that height. However, because your platformview viewshed covers the entire county, we want to examine only the parts of the viewshed within the boundaries of Logan. To do so, we'll have to create a subset of the viewshed raster (see Chapter 12) that conforms to the Logan boundaries. To begin, turn on the Loganboundary feature class and if it is still a solid color, change its symbology to hollow (so you can see the layers underneath it) and zoom in to its boundaries.

- To create a subset of the platformview viewshed raster to match this boundary, from the **Spatial Analyst Tools** toolbox, choose the **Extraction** toolset, then the **Extract By Mask** tool.

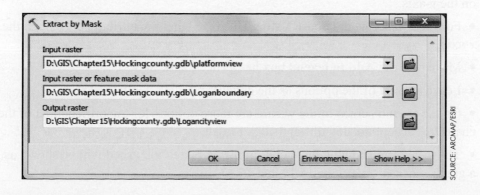

- For the Input raster, select the **Platformview** raster. This is the raster from which we want to extract cells.

- For the Input raster or feature mask data, select the **Loganboundary** feature class. This is the shape (or mask) that will be used to extract cells.

- Call the Output raster **Logancityview** and place it in your Hockingcounty geodatabase.

- Click **OK**. The new Logancityview raster will be added to the TOC.

- Next, to analyze the Logancityview raster, we want to know how many of its cells have a value of 0 (cannot be seen from the platform) and a value of 1 (can be seen from the platform). To do so, we'll create a graph as we did in Step 15.2 and examine the histogram. From the **View** pull-down menu, select **Graphs** then **Create Graph**.

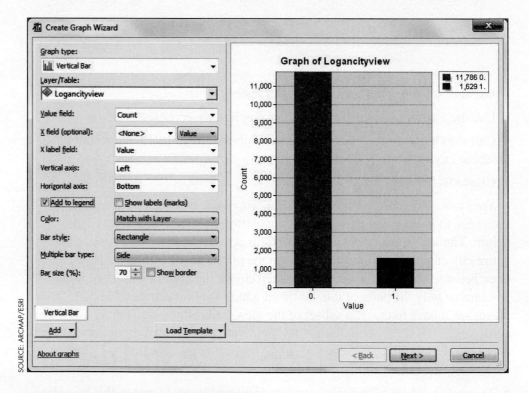

- For the Graph type: choose **Vertical Bar**.

- For Layer/Table: choose **Logancityview**.

- For the Value field: choose **Count**. This is the attribute that we will be charting on the *y*-axis.

- For the X label field: choose **Value**. This is to label the attribute present on the *x*-axis.

- Make sure the **Add to legend** box has a checkmark in it.

- Leave the rest of the options as the default and click **Next**.

- On the second screen of the wizard are more options for the presentation of the chart. Just accept the defaults and click **Finish**.

- The histogram showing the values of the extracted viewshed will be shown as a graph. Answer Questions 15.14 and 15.15.

QUESTION 15.14 What is the percentage of the area within the boundaries of Logan that can be seen from the height of the platform? (*Hint:* Use the histogram to find the total number of cells that have a 0 or a 1, then determine the percentage of the whole that can be seen.)

QUESTION 15.15 Why can the whole of Logan not be seen from the height of the platform? (*Hint:* You may want to refer to both the viewshed results and your final Profile Graph to answer this question.)

perspective view Viewing GIS data at an oblique angle that causes it to take on a "three-dimensional" appearance.

ArcScene The ArcGIS component of the 3D Analyst that allows for the visualization and analysis of data in a 3D environment.

STEP 15.7 Viewing Layers in a 3D Environment

- To really visualize what the terrain of Hocking County looks like, we can use a **perspective view** (such as from an angle) to see how the county's terrain looks in 3D. Viewing datasets in a 3D view can be done through the functions of the 3D Analyst extension and **ArcScene**, the component of ArcGIS used for pseudo-3D visualization. For more about 3D and Pseudo 3D visualization, see **Smartbox 79**.

- On the **3D Analyst toolbar**, press the **Launch ArcScene** button to activate ArcScene.

SOURCE: ARCMAP/ESRI

Smartbox 79

What is pseudo-3D visualization in ArcGIS 10.2?

Although data (such as the digital elevation model of Hocking County) can be viewed and examined in a perspective view and heights added to make it look 3D, the resulting model is not necessarily three-dimensional. Surfaces such as DEMs are considered to be **pseudo-3D** or **two-and-a-half-dimensional (2.5D)** models. In a two-dimensional (2D) surface, only x and y coordinates are measured, but no z-value is added. In a 2.5D surface, there can be only one z-value measured for each x and y coordinate. In the NED data, there is only a single value representing the elevation for that location above the vertical datum, so it is considered 2.5D. In a full **three-dimensional (3D)** surface, there can be multiple z-values assigned to each x and y coordinate (Figure 15.7). For instance, if a digital terrain model can simultaneously measure the elevation of the ground

pseudo-3D The term used to describe the perspective view of a digital terrain model because it is often a 2.5D model, not a 3D model.

two-and-a-half dimensional (2.5D) A model of the terrain that allows for a single z-value to be assigned to each x/y coordinate.

three-dimensional (3D) A model of the terrain that allows for multiple z-values to be assigned to each x/y coordinate.

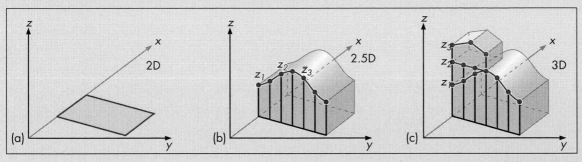

FIGURE 15.7 A comparison of a (a) 2D surface, (b) 2.5D surface, and (c) 3D surface.

functional surface A continuous representation of values that can have a single value assigned to each location.

and the elevation height of the building on the ground at that same location, it would be considered a 3D model. Because most digital terrain models are measuring a single *z*-value for each location, they are considered 2.5D. A digital elevation model is considered to be a **functional surface**, in that it is continuous so that you can measure a single elevation value at any location.

- ArcScene has the same interface as ArcMap (similar pull-down menus, a Table of Contents, and the Standard and Tools Toolbars). After ArcScene opens, select the option for a Blank Scene (this is similar to choosing a Blank Map in ArcMap).

- Next, add the NEDhocking grid to its TOC.

- When the NEDhocking grid appears, it will be shown in perspective view. Using ArcScene you can interactively move around the NED layer. Try navigating around the ArcScene environment. (For more information on the various controls used in maneuvering through ArcScene, see **Smartbox 80**.)

Smartbox 80

How can I navigate in ArcScene 10.2?

ArcScene has its own Tools toolbar for navigating around the pseudo-3D representation of Hocking County (Figure 15.8). *Important Note:* If this toolbar is not available, in ArcScene select the **Customize** pull-down menu, then select **Toolbars**, then select **Tools**.

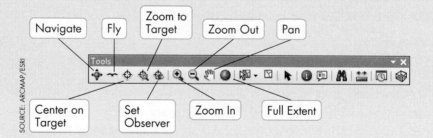

FIGURE 15.8 The tools available on the ArcScene Tools toolbar.

- The first icon (of the earth with the four points) is the **Navigate** icon, which allows you to turn, spin, and rotate the pseudo-3D layers.

- The bird is the **Fly** icon, which turns ArcScene into a first-person flight simulator. Fly allows you to move across the terrain (as if you were flying) or to also reposition your viewpoint for better observation than just zooming in or out. Flying can be a bit tricky at first, so keep the following in mind:

  - Clicking once on Fly changes your cursor to an image of a standing bird (wearing shoes).

  - With the bird icon on the screen, click the left button of the mouse once. This will change the icon to that of a flying bird.

  - Move the mouse slowly back and forth, and you'll see your perspective on the ground change, as if you were "hovering" over it.

  - Also notice that in the bottom left-hand corner of ArcScene, your flying speed will be clocked. You'll start at 0, and each time you left-click, your speed will increase.

- Right-clicking will slow your speed and eventually return you to a speed of zero (and the bird will return to its standing/shoe-wearing appearance).

  - *Important Note:* You will not be able to access any other tools in ArcScene until the bird is back to standing still. Keep careful track of your flying speed in the lower left-hand corner. The higher the value, the faster you will be flying forward (a positive value) or backward (a negative value).

  - If you ever fly off the screen into infinity, return to a fly speed of 0 and click the globe icon to return to the full extent of Hocking County.

- The crosshairs symbol is the **Center on Target** tool. When this tool is selected, you can click on a particular area in ArcScene and the camera viewpoint will change to center on that spot.

- The crosshairs with the magnifying glass is the **Zoom to Target** tool. With this tool, you can click on a location in ArcScene and the view will zoom down to that point.

- The crosshairs with the eye is the **Set Observer** tool. By selecting this tool and clicking on a particular point, you can set the camera angle as if you were standing at that point and looking outward.

- The magnifying glasses are the **Zoom In/Out** tools. You can also zoom in and out with the scroll wheel on your mouse.

- The hand is the **Pan** tool that allows you to move the scene.

- The globe icon is the **Full Extent** option. Like its ArcMap counterpart, it will reset the camera to the extent of all the displayed layers.

STEP 15.8　Examining Analysis Results in 3D

- One thing you probably noticed when you were testing out the various ArcScene navigation tools was that the NEDhocking layer was still two-dimensional and completely flat even though you were looking at it in perspective view. To be shown as a pseudo-3D layer in ArcScene, the layer needs a source for elevation values. These base heights can be drawn from a surface or from features. For more about base heights and how they're used in ArcGIS, see **Smartbox 81** .

Smartbox 81

What are base heights and how are they used for 3D visualization in ArcGIS 10.2?

When features are represented in perspective view (as in ArcScene), their height will be represented using the z-values of the terrain. These **base heights** represent the z-value drawn from a digital terrain model surface. When a digital terrain model is represented in ArcScene, the z-values of each x/y location can thus be shown at their proper elevations because ArcScene will extend these values to their proper heights.

　　You can show other types of data (rasters, feature classes, and imagery) in pseudo-3D by applying the base heights of an elevation surface to the corresponding locations of the other data. For instance, if the DTM covers the same

base heights The z-value obtained for a location from a digital terrain model surface.

geographic area as the NLCD raster, the *z*-values of the DTM may be matched with the *x/y* coordinates of the NLCD layer. Then, rather than extending the DTM locations up to their *z*-value heights, the NLCD layer can have its locations extended instead (Figure 15.9).

(a)   **(b)**

FIGURE 15.9 The Hocking County NED surface (a) and the NLCD layer draped over it (b).

> **draping** A process in which GIS layers are given *z*-values to match the corresponding heights in a digital terrain model.

This process is called **draping**, which can be thought of like taking a lumpy unmade bed and stretching a sheet over it. Wherever the lumps in the bed are, the sheet will also have the same lumps. Through draping, all types of GIS data can be shown in perspective view, as long as their geographic areas are within the boundaries of the DTM supplying the base heights. Draped satellite images or aerial photographs will show an area's hills and valleys, while draped roads will show how they traverse up and down steeply sloped areas.

- In this case, the NED provides the base heights for itself—each grid cell on the NED has an elevation value, so these values will be used as the source for the base heights. To set this up, right-click on **NEDhocking** in the TOC and select **Properties**. In the Layer Properties dialog, select the **Base Heights** tab.

- In the Elevation from Surfaces option, select the radio button next to **Floating on a custom surface**. For this option, select the **NEDhocking** raster in your Hocking-county geodatabase.

- Also click on the **Raster Resolution** button.

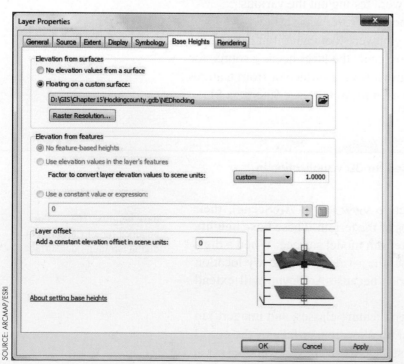

- For the options for Base Surface, type **30** for Cellsize X and also type **30** for Cellsize Y. (By doing so, you specify that the original resolution of the NED of 30 meters is used when displaying the layer.) Click **OK** to close the Raster Surface Resolution dialog.

- In the Layer Properties dialog, leave the other options alone and click **Apply**, then click **OK**. Your NED layer will now be shown in pseudo-3D. Use the ArcScene controls to move around the NED surface and get a better look at it in pseudo-3D. When you're done, click on the **Full Extent** icon to return to the full view of Hocking County. This ability to view layers in pseudo-3D isn't just limited to viewing digital terrain models like the NED. As **Smartbox 81** explained, ArcScene gives you the capability to use the base heights of a digital terrain surface and apply them to other layers as well, so we'll use this capability to examine the viewshed and LOS results in 3D, simulating a camera view of what can be seen from the proposed observation tower.

- We'll start with the Logancityview layer, so add this raster to ArcScene's TOC. You'll see that the Logancityview layer is also completely flat and "floating" underneath the NED layer. To drape the Logancityview, you'll assign the NED layer's base heights to their corresponding locations in the Logancityview raster. Right-click on the Logancityview layer and select **Properties**, then select the **Base Heights** tab.

- Just as you did when you assigned the base heights of the NED layer to itself, select the radio button next to **Floating on a custom surface**. To draw base heights from the NED layer, select the **NEDhocking** file in your Hockingcounty geodatabase.

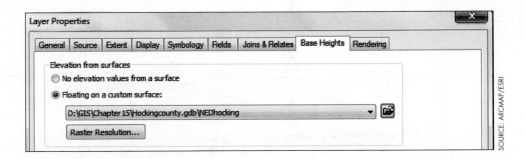

- Press the **Raster Resolution** button to set the Cellsize X and Cellsize Y to **30** (for the 30 meter resolution of the Logancityview layer) and click **OK** in the Raster Resolution dialog box. Leave the other options in the Layer Properties box alone and click **Apply**, then click **OK**. Your Logancityview layer is now draped over the NED.

- For now, turn the NEDhocking layer off so that the Logancityview layer can be clearly seen. Zoom in closer to the Logancityview layer so that you're focused on that area instead of the entire county.

- You can drape vector layers on the NED as well. (***Important Note:*** Even though the NED has been turned off and is not visible, it can still be used as a source of base heights.) Add the roadsegmenthocking layer to ArcScene and repeat the draping process with it (i.e., assign the NED's base heights to their corresponding locations for the roads layer). The roads layer should now also be draped over the NED. Change the symbology of the roadsegmenthocking layer to a thicker line of a solid color so it will stand out.

• Last, add the Loganboundary layer to ArcScene and drape it over the NED as well. Arrange the layers in the TOC so that you can see the roads and Loganboundary lying on top of the Logancityview layer. Also change the symbology of the Loganboundary layer to "no color" with a thicker line for its outline, so you can see the layers underneath it while the boundary can also clearly be seen.

• You should now have the Logancityview layer, the Loganboundary layer, and the RoadSegmenthocking layer shown in pseudo-3D and can begin to look around the 3D view.

• Next, add the LoganDam and HockingCoFairgrounds layers to the TOC. Repeat the draping process with both of these point layers by assigning the base heights of the NED layer to them.

• We'll want to change the two point layers to a 3D symbology to better show up in ArcScene. We'll start with the LoganDam layer. Right-click on it, select **Properties**, and then choose the **Symbology** tab.

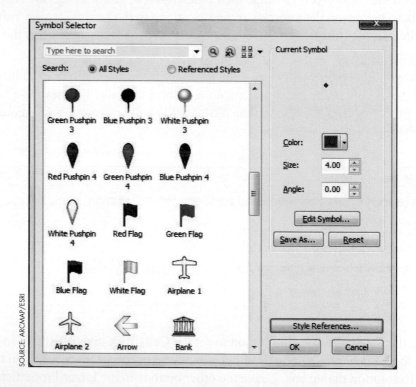

• Click on the button that shows the default symbol itself (likely a dot) and the Symbol Selector dialog will open, but it will now have basic symbols to be used in ArcScene. Choose an appropriate symbol for the dam (perhaps one of the pushpins or flags) and change the size so that it can be seen in the ArcScene view when zoomed in.

• Repeat this process to assign a 3D symbol to the HockingCoFairgrounds point layer as well.

• Finally, to visualize your line-of-sight result, you can see it in 3D in ArcScene. Return to ArcMap and use the cursor to select your final line-of-sight graphic. From the **Edit** pull-down menu, select **Copy**.

- Return to ArcScene. In ArcScene's **3D Analyst toolbar**, select **NEDhocking** as the Layer with which to work.

- From the **Edit** pull-down menu, select **Paste**. The line-of-sight graphic will be added to ArcScene and properly draped on the terrain. Click the cursor somewhere else to remove the "white" look of the graphic so you can see the red and green line of sight. If the line-of-sight graphic does not appear draped on the terrain, see **Troublebox 10** for more information on how to get a properly draped line.

Troublebox 10

Why did the line of sight not drape properly in ArcScene?

For the line of sight you created to drape properly in ArcScene, it needs to be created properly in ArcMap. When creating a line of sight in ArcMap, be sure that the NEDhocking layer (i.e., the layer with the elevation values) is selected in ArcMap's 3D Analyst pull-down menu. If another raster layer is chosen in this pull-down menu when you're creating the line of sight, then the LOS graphic will not have the proper elevation information (and will likely appear underneath the 3D version of the NED in ArcScene). Before creating the line (and thus, prior to copying and pasting it into ArcScene) be sure that the NEDhocking layer is properly selected (Figure 15.10).

FIGURE 15.10 Choosing the NEDhocking layer before creating the line of sight.

- Make sure you have only the Logancityview, Loganboundary, HockingCoFairgrounds, LoganDam, and RoadSegmenthocking displayed along with your final LOS. Also notice that the black dot representing the observer offset height used in the final LOS will be floating above the terrain. Use the Pan, Navigate, and Fly tools to best see this, as if you were hovering beside the observer offset point (at the Dam) and looking at an angle down toward the Fairgrounds (Figure 15.11).

FIGURE 15.11 A 3D view of Logan and Hocking County at the proposed platform height.

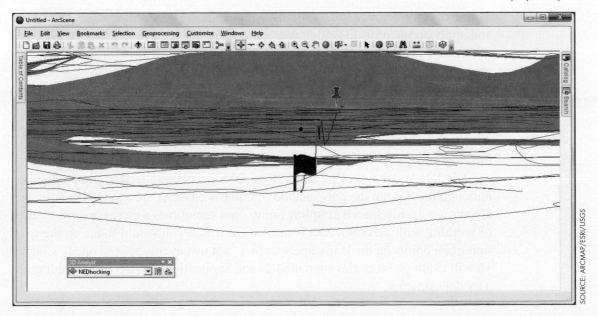

scene document (.sxd) The ArcScene equivalent of an ArcMap map document—a file that contains information about where all of the data layers used in an ArcScene session are located, as well as their appearance and settings.

STEP 15.9   Exporting, Saving, and Printing Your Results

- From the **File** pull-down menu in ArcScene, select **Save** to save your work in ArcScene as a **scene document (.sxd)** file. (This is the ArcScene equivalent of a map document and works the same way—it is a file with a .sxd extension.) Also, from the **File** pull-down menu, select **Scene Document Properties** to add the usual information concerning title, author, and description.

- To share your results with others (i.e., the 3D view from the observer point looking toward the fairgrounds), you'll have to export the scene to a graphic format. Unfortunately, ArcScene does not have Layout capabilities. Once you have the View looking how you want it in ArcScene, you can create a graphical image of the View and save it.

- From the **File** pull-down menu in ArcScene, select **Export Scene**, then select **2D**.

- In the Export Map dialog, choose the graphic type you want to save the View as (such as JPEG or TIFF), give the file a name, and save it into your Chapter15 folder.

- In ArcMap, compose a layout (see Chapter 3) of the platform location in relation to Logan showing (1) the platform layer, (2) your final line of sight on the terrain, (3) the Loganboundary polygon, and (4) the Logancityview layer. You would also want to add your final Profile Graph showing the minimum platform height and the simulated view from it to the layout as well. Right-click on the Profile Graph in ArcMap; a new menu of options will appear. Select **Add to Layout** to create a graphic of the final Profile Graph that can be placed and manipulated like other map elements. Last, you can add a graphic to the layout showing the view from the platform you just captured and saved. In the Layout View, from the **Insert** pull-down menu choose **Picture**.

- Save your work in ArcMap as a map document and add the appropriate information in the Map Document Properties dialog.

Closing Time

This chapter got you started working with elevation data and several derivations and applications of it. Elevation data are commonly used in many different types of GIS analysis, and we'll continue to use data from the NED in forthcoming chapters. The ability to visualize and interact with data in a 3D-style environment is a big advantage in working in ArcGIS, and several upcoming chapters will continue to utilize ArcScene for elevation-based data or analysis. Elevation data like NED is also useful for creating other points, lines, or polygons with z-values beyond lines of sight for use in ArcScene. See the *Related Concepts for Chapter 15* for more information.

Digital elevation models are very useful for modeling terrain and landscapes, although they're not the only method of doing so. DEMs are limited because they always use evenly spaced sampling points, and sometimes a more versatile method of working with elevation data is needed (in which you would focus on the most important points on the landscape's surface, not just evenly spaced ones). Chapter 16 will examine other elevation models and applications in ArcGIS, including the TIN data structure.

Related Concepts for Chapter 15 | Creating 3D Graphics with the 3D Analyst

> **graphics** Items (such as shapes) that can be drawn on the screen in ArcGIS.

Feature classes and rasters are not the only types of data that you can examine. ArcGIS also gives you the ability to create graphics to accompany your data. **Graphics** are basic shapes or items that you can draw directly onto the View. For instance, you may want to draw a circle around a result, place a box around a portion of a basemap, or add an arrow to point the viewer at a particular feature. Note that ArcGIS graphics are not feature classes—you're not digitizing a circle polygon or an arrow-shaped line into a geodatabase's feature class. Graphics don't have attribute tables or fields that can be joined to them. They're simply additional items that you can draw on the screen. (We'll do more with graphics in Chapter 20.)

The 3D Analyst tools give you the ability to draw 3D graphics for use in ArcScene. As long as you're working with an elevation surface (such as a digital elevation model) you can create graphics of points, lines, or polygons that will retain the z-values associated with where they're drawn. When these graphics are copied and pasted into ArcScene, they'll be drawn in pseudo-3D as if they were draped on the surface. The LOS lines you drew in this chapter are examples of this—you drew the line on the NED in ArcMap, but when it was copied and pasted into ArcScene, it properly draped over the pseudo-3D terrain.

On the ArcMap 3D Analyst toolbar are the three tools for drawing a point, line, or polygon graphic and having ArcGIS interpolate what the elevation values for these items would be (Figure 15.12). As you did in this chapter, you first have to select the elevation surface to draw on (which will be the source of the interpolated z-values). In addition to DEMs or NED, TINs (see Chapter 16) or terrain datasets (see Chapter 17) can be used as sources of z-values. If you're interpolating a line, you can also create a corresponding profile graph as well.

FIGURE 15.12 The tools of the ArcMap 3D Analyst toolbar used for creating 3D graphics.

See Figure 15.13 for an example. A polygon graphic is drawn in ArcMap using the interpolate polygon tool, using a NED layer as the source of the z-values. This graphic is then copied and pasted into ArcScene, where it retains the same x and y dimensions as the ArcMap polygon. The z-values of the NED have been interpolated along the boundaries of the polygon, giving it a pseudo-3D appearance.

FIGURE 15.13 A polygon created as an interpolated graphic in ArcMap (a) and its corresponding pseudo-3D appearance in ArcScene (b).

(a)

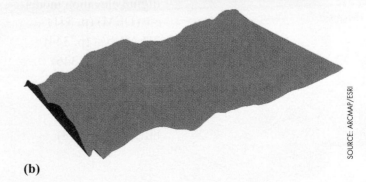

(b)

For More Information

For further in-depth information about the topics presented in this chapter, use the ArcGIS Help feature to search for the following items:

- Analyze Visibility
- Creating a line of sight
- Creating a profile graph from line-of-sight results
- Digitizing 3D graphics over a surface
- Exploring digital elevation models
- Fundamentals of Surfaces
- How Aspect works
- How Hillshade works
- How Slope works
- Line Of Sight (3D Analyst)
- Managing elevation data: Part 1: About elevation data
- Managing elevation data: Part 2: Design and data management plan
- Setting a raster layer's base heights using a surface in ArcScene
- Understanding visibility analysis
- Using Viewshed and Observer Points for visibility analysis
- Viewshed (3D Analyst)
- What are the World Elevation services?

For more about the National Elevation Dataset, see its FAQ online here: http://ned.usgs.gov/faq.asp.

Key Terms

digital terrain model (DTM) (p. 341)

z-value (p. 341)

vertical datum (p. 341)

NAVD 88 (p. 341)

digital elevation model (DEM) (p. 341)

3D Analyst (p. 344)

histogram (p. 346)

National Elevation Dataset (NED) (p. 347)

hillshade (p. 348)

slope (p. 349)

aspect (p. 350)

visibility analysis (p. 352)

line-of-sight (LOS) (p. 352)

viewshed (p. 352)

perspective view (p. 359)

ArcScene (p. 359)

pseudo-3D (p. 359)

two-and-a-half dimensional (2.5D) (p. 359)

three-dimensional (3D) (p. 359)

functional surface (p. 360)

base heights (p. 361)

draping (p. 362)

scene document (.sxd) (p. 366)

graphics (p. 367)

How to Work with Contours, TINs, and 3D Imagery in ArcGIS 10.2

Introduction

There are other ways of representing a surface as a digital terrain model beyond using the evenly spaced sample points of a digital elevation model or the NED. The data sources being used to produce an elevation surface may not be evenly distributed, or there may be certain areas (or smaller sections of the terrain) where the landscape undergoes rapid elevation changes and many more elevation points are necessary to accurately capture these features in GIS. In such cases, you would need a lot of points in the most important areas and fewer points for the less variable terrain areas. A data structure that accommodates irregularly spaced points would help to better represent the terrain.

A **TIN** (triangulated irregular network) represents this kind of a data structure in ArcGIS. TINs do not need to have evenly spaced sample points; they can be built from selected points (or those deemed the most important for representing the landscape). Unlike digital elevation models, TINs use points, lines, and a series of non-overlapping triangular polygons to represent the surface (Figure 16.1). TINs can often provide a better display of the terrain than DEMs can. They are also more versatile in representing landscape features, but they are more complicated to work with than evenly spaced sample points. Like the surfaces described in Chapter 15, a TIN acts as a functional surface, and it is also considered a 2.5D model of terrain. Also, as a TIN is a digital terrain model, all of the terrain analysis functions used in Chapter 15 (hillshades, slope, aspect, viewsheds, LOS, and draping) can also be performed using a TIN.

TIN Triangulated irregular network, a terrain model formed from non-overlapping triangles that allows for unevenly spaced elevation points.

SOURCE: NATIONAL MAP/USGS/ARCMAP/ESRI

FIGURE 16.1 A TIN representation of a surface.

TINs can be constructed from a variety of different data sources, including points drawn from DEMs (see Chapter 15), lidar (see Chapter 17), or contour lines from a map. Contours have long been a means of representing terrain features such as hills, peaks, or valleys on maps, and they can also be used within ArcGIS. Digitized versions of contours can be used as a basis for constructing digital terrain models. Using ArcScene, you can create detailed 3D representations of surfaces for analysis.

Chapter Scenario and Applications

This chapter puts you in the role of a member of the local chamber of commerce in Youngstown, Ohio. As part of a promotional effort to showcase the city to potential business investors and property developers, you will be creating a virtual 3D tour of the city and its nearby areas. To this end, you'll be showing high-resolution imagery draped over a 3D TIN of the city's landscape and creating an interactive version of the results that you can show on the Web.

The following are additional examples of other real-world applications of this chapter's skills.

• An archaeologist is beginning to reconstruct the original appearance and layout of a Mayan city in a 3D format. She could begin by building a detailed TIN of the surface and draping imagery of the environment over the TIN.

SOURCE: JIM RANKIN/TORONTO STAR VIA GETTY IMAGES

• An environmental scientist is performing analysis of the aftermath of a large-scale forest fire. She could use contours and a TIN to model the landscape and use draped imagery from before and after the fire to examine the extent and geographic spread of the burn.

• An urban planner wants to examine the potential changes in an area and present the results to a group of stakeholders. He can use elevation models and imagery to visualize the city layout and present the results using a 3D Web Scene.

ArcGIS Skills

In this chapter, you will learn:

• How to examine contour lines in ArcMap.

• How to construct a TIN from a feature class.

• How to work with the TIN data structure (including generating slope and slope aspect from a TIN).

• How to create a pseudo-3D visualization of a TIN landscape in ArcScene.

• How to drape and render high-resolution imagery on a digital terrain model.

• How to apply vertical exaggeration to a scene.

• How to create a 3D Web Scene.

• How to use a 3D Web Scene in ArcGIS Online.

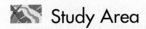 Study Area

For this chapter, you will be working with data from Youngstown, Ohio.

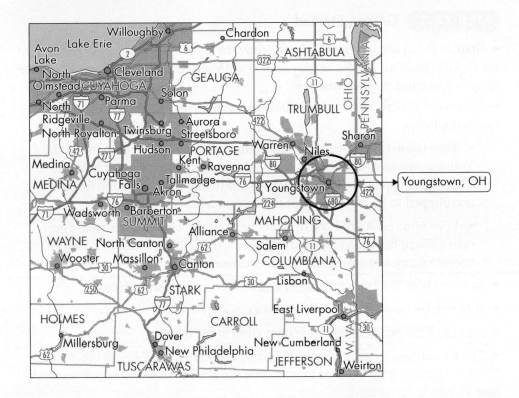

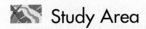 Data Sources and Localizing This Chapter

This chapter's data focus on features and locations within Youngstown, Ohio. However, you can easily modify this chapter to use data from your own city or local area instead. The contours and NAIP imagery used in this chapter were downloaded for free from The National Map and similar datasets are available for the entire country. For instance, if you were doing this same analysis for Blacksburg, Virginia, contours and NAIP imagery for that city are available via The National Map.

Important Note: When obtaining data for your own sites, keep in mind that The National Map contours and NAIP imagery are projected into UTM using meters as their linear units. The contours, however, have their elevations (*z*-values) measured in feet. By default, ArcScene will use the units of measurement of the horizontal projection for the vertical measurements as well. Thus, even though the contour elevations are in feet, ArcScene will treat these same *z*-values as meters (because the contour layer was in the UTM projection using meters). When working with such data, however, ArcScene can control these types of conversions for each layer without having to project them. It does this through the layer's properties under the **Base Heights** tab. In the **Elevation from features** section, you can select how to convert the units used for elevation to match those used for horizontal features (such as feet-to-meters or meters-to-feet conversions).

For ease of use in this chapter, both the contours and the NAIP imagery have been projected for you into a projection using feet as linear units (State Plane) to keep both the *x/y* values and the *z*-values consistent.

STEP 16.1   Getting Started

- Start ArcMap and use the Catalog to copy the folder called **Chapter16** from the C:\GISBookdata\ folder to your own D:\GIS\ drive. Be sure to copy the entire directory, not just the files within it.

- Chapter16 contains a file geodatabase called **Ytowncontours** that contains the following items:

  - **Ytownconsstateplane**: This line feature class contains contours for the Youngstown area. The feature class has been projected into the NAD 83 State Plane Ohio North coordinate system (using US Feet as the linear units) and clipped to the boundaries of the orthoimage.

  - **Ytownimage**: A 2011 NAIP orthoimage of the Youngstown area. The raster image has been projected into the NAD 83 State Plane Ohio North coordinate system (using US Feet as the linear units).

- Activate both the **Spatial Analyst** and the **3D Analyst** extensions.

- Add the **3D Analyst toolbar** to ArcMap.

- Add the Ytownconsstateplane feature class to the map.

- The Coordinate System being used in this chapter is **State Plane Ohio North (NAD 83)**, using Map Units of **Feet**. Check the Data Frame to verify this coordinate system is being used.

STEP 16.2   Examining Contour Lines

- Change the symbology of the Ytownconsstateplane feature class to a graduated color (see Chapter 3). Use the ContourElevation attribute for display and use the Natural Breaks method with five classes. Choose an appropriate and distinctive color ramp for the contours. Also, turn on the labels for the layer, using the ContourElevation field (see Chapter 2). For more about contour lines and contour intervals, see Smartbox 82.

Smartbox 82

How can contour lines be used in ArcGIS 10.2?

A **contour** is a line drawn on a map that joins points of equal elevation. For instance, if a contour line is labeled 800, then all points on that line are assumed to have an elevation value of 800 units above the vertical datum being used (such as 800 feet above the NAVD88 datum). If the contour line adjacent to it is labeled 810, then the points along those elevations are assumed to be 10 units higher than the points on the 800 contour (and there is 10 feet of elevation change between the two lines). Contour lines are imaginary in that they don't exist on real landscapes; they're just a device used to represent elevations on a map.

In theory, a map could have an infinite number of contours, but this would quickly make the map unreadable because it would be filled to the maximum

contour An imaginary line drawn on a map that connects points of common elevation.

with lines detailing every tiny change in the elevation. Thus, contours are drawn only at a certain distance apart, such as every 10 feet or every 50 feet of elevation change. This vertical distance between contours on a map is called the **contour interval**. For instance, if a contour was drawn at 800 feet and the next was drawn at 850 feet, the contour interval would be 50 feet. While any contour interval could be selected, in general maps of more mountainous terrain tend to use a wider contour interval, while maps of flatter terrain tend to use narrower contour intervals. Likewise, smaller-scale maps tend to use a wider contour interval, while larger-scale maps tend to use narrower contour intervals.

When displaying contours on a map or in ArcGIS, different types of contours are available. The thicker lines (which are often labeled on topographic maps) are called **index contours**, while the thinner (often unlabeled) lines drawn at the contour interval between the index contours are called **intermediate contours** (Figure 16.2). In ArcGIS, a digital terrain model (such as the NED) can be used as a source from which to derive contours, using tools available from the Surface toolset in the Spatial Analyst toolbox, as well as the Create Contours tool on the 3D Analyst toolbar. The Youngstown contours used in this chapter were derived from NED data.

contour interval The vertical distance between contour lines.

index contour The main contour lines on a map.

intermediate contour The contour lines drawn in between index contours at the contour interval.

FIGURE 16.2 Contour lines at a 10 foot interval—index contours are in black, while intermediate contours are in red.

SOURCE: NATIONAL MAP/ARCGIS/ESRI

- Pan and zoom around the Youngstown area to familiarize yourself with the elevation and landscape of the area. Answer Questions 16.1 and 16.2.

QUESTION 16.1 What is the contour interval of the Ytownconsstateplane layer?

QUESTION 16.2 Where are the highest (1060 feet) and lowest (820 feet) elevations located in the Youngstown area?

STEP 16.3 Creating a TIN

- To construct a 3D version of the area, you'll next construct a TIN from the contours.

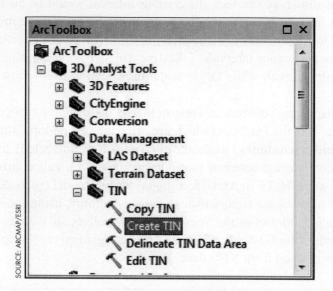

SOURCE: ARCMAP/ESRI

- Open **ArcToolbox**. From the **3D Analyst Tools** toolbox, open the **Data Management** toolset, then select the **TIN** toolset, then select the **Create TIN** tool.

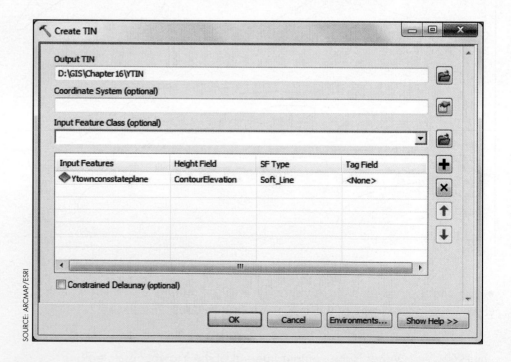

SOURCE: ARCMAP/ESRI

- For Output TIN, use the navigate button to go to your **D:\GIS\Chapter16** folder. Name the Output TIN **YTIN** and save it as a **TIN dataset**.

- For the Input Feature Class (optional), use the drop-down menu to select the **Ytownconsstateplane** layer.

- When Ytownconsstateplane is added to the Input Features menu, click in the field under **Height Field** next to the Shape.Z entry. A pull-down menu will appear—choose the **ContourElevation** option. By doing so, you tell ArcGIS which field in the Ytownconsstateplane layer to use as the source of the elevation values when it creates the TIN. The field called ContourElevation contains the elevation values (above the vertical datum) assigned to each contour (hence, why you labeled the contours using this field in Step 16.2).

- Click in the field under **SF Type**, and from the pull-down menu select **Soft_Line**. This option allows you to specify the type of features (mass points, soft lines, or hard lines) to construct the TIN from.

- Before pressing OK, you have to give your TIN spatial reference. Press the **browse button** next to the **Coordinate System (optional)** field.

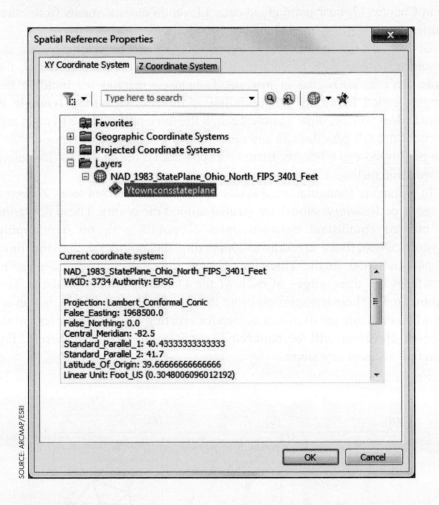

SOURCE: ARCMAP/ESRI

- You want your TIN to have the same coordinate system as the Ytownconsstateplane layer. Select the **XY Coordinate System** tab. From the **Layers** options, expand **NAD_1983_StatePlane_Ohio_North_FIPS_3401_Feet**. Next, select the **Ytownconsstateplane** layer.

- You'll see that the Current coordinate system will update to contain the same State Plane Ohio North information that the Ytownconsstateplane layer has. Click **OK**. You'll see that the Create TIN dialog has filled in the coordinate system information for you.

- When the spatial reference is properly set, click **OK**. For more information about how ArcGIS is constructing the TIN from the contours and these settings, see **Smartbox 83**.

Smartbox 83

How is a TIN created in ArcGIS 10.2?

mass points The selection of points used in creating a TIN.

Delaunay triangulation The process of connecting sets of points to build the triangles and edges of a TIN.

Delaunay triangle A triangle that can have a circle drawn through its three points and not pass through any additional points.

edges The lines that connect the points of the triangles of a TIN.

When you use the Create TIN tool in ArcGIS, a number of processes are happening behind the scenes to build the TIN, and there are several options that affect the end result. First, the TIN needs a set of points with z-values from which to be built (these points are often called **mass points**). A variety of sources can be used for the mass points: field point measurements or, as we'll see in Chapter 17, lidar point cloud data. Elevation measurements from other features, such as DEMs or contours, can be used as well.

Because the TIN is constructed from a series of non-overlapping triangles, the selected points must be connected to form these triangles. However, the points can't be connected in just any fashion—a method for building the triangles called **Delaunay triangulation** is used. A **Delaunay triangle** is created when you can draw a circle through the three points of the triangle and the circle doesn't pass through any other points. Creating these types of "fat" triangles (as close to 60-degree triangles as possible) is the goal of the Delaunay triangulation method.

In Delaunay triangulation, a series of polygons (referred to as *Thiessen polygons* or *Voronoi regions*) are created around the points. These polygons are created equidistant between pairs of points with no overlapping polygons. Next, lines are drawn connecting the points so that the lines are perpendicular to the Thiessen polygon boundaries. These connected lines form the three **edges** of each of the Delaunay triangles of the TIN (Figure 16.3). These triangles make up the faces of the TIN. Note, however, that when contours are used as a source for creating TINs, sometimes points of equal elevation will be connected, resulting in the faces being "flat triangles" without any slope.

FIGURE 16.3 (a) A sample set of mass points, (b) Thiessen polygons formed equidistant around the points, (c) Delaunay triangles constructed by drawing edges perpendicular to the Thiessen polygons.

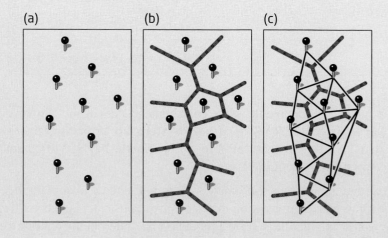

To add more realism to a TIN, breaklines can also be added. **Hard breaklines** represent areas on the landscape where the edges of the TIN should not cross. For instance, the faces of the triangles shouldn't cross areas such as stream boundaries, building footprints, or flat, paved roads. Incorporating hard breaklines to ensure that features like these do not have changing elevation values (when they should be flat, such as the surface of a road or stream) will make the TIN surface more realistic (Figure 16.4). **Soft breaklines** represent areas that should also be kept as edges in the TIN, but they do not represent the types of surface discontinuities that hard breaklines do. The last feature that is included in a TIN is a **hull**, or a polygon that defines the area of the TIN.

hard breaklines Lines used in a TIN to enforce surface discontinuities.

soft breaklines Enforced lines in a TIN that do not represent discontinuities.

hull The polygon boundary around the area of the TIN.

(a)

(b)

FIGURE 16.4 A comparison of two TINs—(a) one formed from only mass points, and (b) one formed from both mass points and breaklines.

After a TIN is created, more points or breaklines can be added to it in order to create more realistic representations of the terrain. The TIN Editing toolbar provides a variety of tools for making changes to a TIN in ArcGIS.

- Give ArcGIS some time to build the YTIN (a message will scroll across part of the bottom of the screen indicating the operation is working and a pop-up window will let you know when it's done). The YTIN will then be added to your Table of Contents.

- Turn off the contours (we won't be using them anymore).

- Also add the Ytownimage raster layer to the map. Be sure to add the composite image so that it appears in color, not just a single band (see Chapter 13).

- Note that Questions 16.3 through 16.7 will have you making reference to the TIN in relation to the city area. To do this, you could place the orthoimage on top of the TIN (you will have to select the **List By Drawing Order** button in the TOC to do this—see Chapter 1) in the Table of Contents and turn the orthoimage on and off, or use the Swipe tool (see Chapters 14 and 15) to look at both the TIN and the image. However, it might be easier to place the orthoimage on top of the

TIN in the Table of Contents and then make it semi-transparent (similar to how you did with a polygon layer in Chapter 7). To do so, do the following:

- Right-click on the **Ytownimage** layer and select **Properties**.
- In the Layer Properties dialog, click on the **Display** tab.

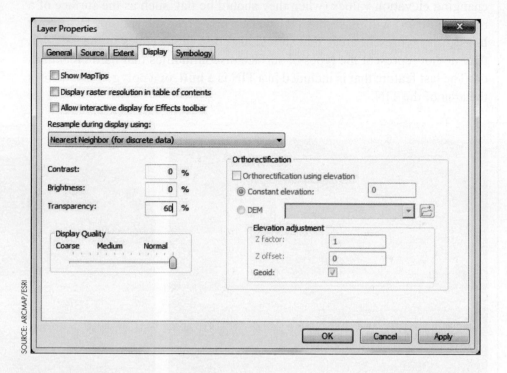

- Under the Transparency options, type another value higher than 0 (0% means the orthoimage will be completely solid and 100% means the ortho will be completely transparent or invisible). Try other values—like 50%, 60%, or 75%—until you can get a good "ghost" of the orthoimage positioned on top of the TIN so that you can see both the features of the TIN and the high-resolution orthoimage at the same time. Click **OK** to accept the new transparency level and adjust the values until you get something useful to you.

- Answer Question 16.3.

QUESTION 16.3 Locate the highest and lowest elevation areas again (your answer to Question 16.2) using the TIN. From your examination of these same areas on the orthoimage, what is actually at these areas in the Youngstown region?

STEP 16.4 Performing Surface Analysis of a TIN

- From here, you will be examining the slope and slope aspect of the TIN you've created. Slope and slope aspect function the same as in Chapter 15, but this time they are being computed for each triangular face of the TIN, rather than each grid cell of the NED.

- To compute slope, in ArcToolbox choose the **3D Analyst Tools** toolbox, then select the **Triangulated Surface** toolset, then select the **Surface Slope** tool.

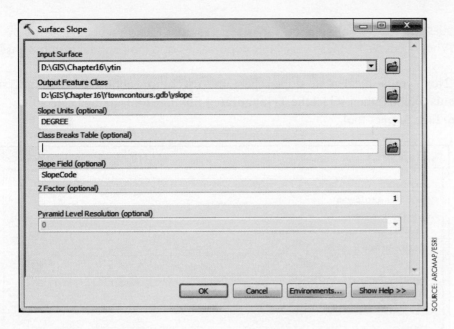

- Use your **ytin** as the input TIN.

- For Output Feature Class, navigate to your Ytowncontours geodatabase and call it **Yslope**.

- Choose **DEGREE** for your Slope Units.

- Leave the rest of the options at their default value.

- Click **OK**. Your slope map will be created and added to the TOC.

- Change the symbology of the Yslope layer so that you're looking at a set of seven classes of Quantiles and Graduated Colors instead of a single symbol and display by the field called Slopecode. Each of the slopecode values represents a different degree value for slope. Change the labels of the slopecode field as follows (so that, for instance, instead of a label of 3.0–4.0, it would read 2.66–5.71):

Slopecode	Label
1	0.57
2	1.43
3	2.66
4	5.71
5	12.13
6	24.89
9	90.0

- Examine your slope map carefully (in relation to your semi-transparent ortho-image of the city—you may have to turn off the TIN and rearrange the layers to examine the Yslope layer under the ortho). Keep in mind what the slopecode values represent (in terms of the degree of slope or steepness of that area). Answer Questions 16.4 and 16.5.

QUESTION 16.4 How is the slope of the terrain distributed around the area? (That is, what areas have very flat slopes, and what areas have steeper slopes?)

QUESTION 16.5 Where are the very steepest slopes found on your slope map? What accounts for these very high slope values and their spatial distributions?

• Next, you will calculate slope aspect. In ArcToolbox, choose the **3D Analyst Tools** toolbox, then select the **Triangulated Surface** toolset, then select the **Surface Aspect** tool.

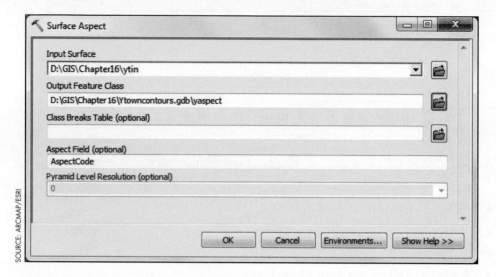

• Use your **ytin** as the input surface.

• For Output Feature Class, navigate to your Ytowncontours geodatabase and call it **Yaspect**.

• Leave the rest of the options at their default value.

• Click **OK**. Your aspect map will be created and added to the TOC.

• Change the symbology of the Yaspect layer so that you're looking at a set of Unique Values instead of a single symbol and display by the field called Aspect-code. You'll have to **Add All Values** of aspectcode to the symbology as well. Change the colors of the unique values that appear for aspectcode to better colors that you can use for interpretation. Also, change the labels for the classes in the symbology layer as the values of aspect are interpreted as follows:

Aspectcode	Label
−1	Flat
1 and 9	North
2	Northeast
3	East
4	Southeast
5	South
6	Southwest
7	West
8	Northwest

• Examine your aspect map carefully (in relation to your semi-transparent ortho-image of the city—you may have to turn off the TIN and rearrange the layers to examine the Yaspect layer under the ortho). Keep in mind what the aspect values represent (and that you have two values for north). Answer Questions 16.6 and 16.7.

> **QUESTION 16.6** How are your values for aspect distributed around the city? (That is, in what directions are the majority of the steepest slopes facing in the campus areas versus some surrounding areas, etc.?)

> **QUESTION 16.7** What is physically present (from your examination of the orthoimage) at the areas around the city that are considered flat with respect to aspect?

STEP 16.5 Viewing the TIN and Imagery in 3D

- For the remainder of the chapter, you'll be examining the TIN in 3D, draping the orthoimagery on it, and using this 3D visualization of the area to complete your virtual tour. On the 3D Analyst Toolbar, click the button that activates ArcScene (see Chapter 15) and select a new Blank Scene when ArcScene prompts you. Once ArcScene opens, you can close ArcMap.

- First, add the Ytownimage layer, and then second, add the YTIN layer to ArcScene's TOC.

- There are several different ways to display a TIN in ArcScene. Right-click on the YTIN layer and select its **Properties**, then select the **Symbology** tab.

- Under the **Show:** box, deselect the **Edge Types** and **Elevation** options.

- Click the **Add** button to choose some new TIN properties to display.

- Click on each option to see how the TIN's appearance changes each time, but for this chapter, select the option for **Face elevation with graduated color ramp**.

- Click **Add**.

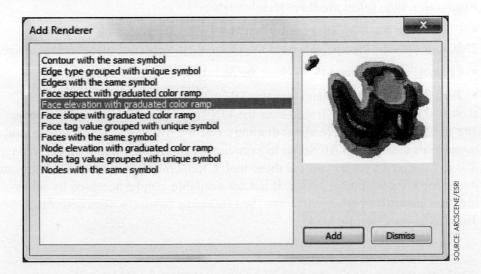

- Click **Dismiss** to close the dialog.

- Back in Layer Properties, make sure that the second **Elevation** option (the one you just added) is the only layer selected under Show. This will make sure that only the TIN elevations are being displayed (using the option you just selected).

- Click **Apply**, then click **OK**. (This will display the TIN faces by elevation as a graduated color.)

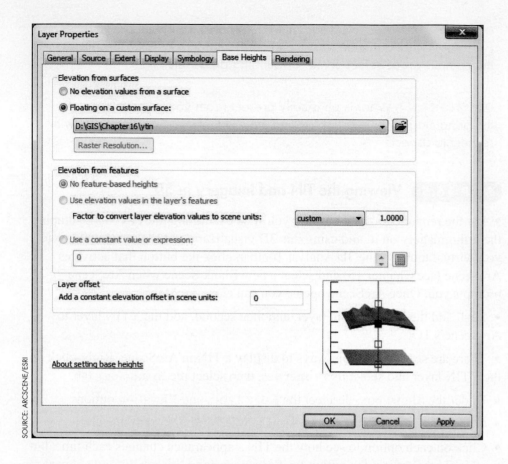

SOURCE: ARCSCENE/ESRI

- The next step is to drape the image onto the TIN to give some terrain relief to the flat orthoimage. Right-click on the Ytownimage layer and select its **Properties**, then select the **Base Heights** tab.

- Click on the **Floating on a custom surface** option and select the **YTIN** layer. This will apply the base heights from the TIN to their respective places on the image.

- Click **Apply** and then **OK** to close the dialog.

- Turn the orthoimage on and turn the TIN off. Your image should now be shown in pseudo-3D and draped over the TIN surface (see **Smartbox 81** on page 361 in Chapter 15 for more about draping). Use the fly, zoom, pan, navigate, and target tools available in ArcScene to examine the image (see **Smartbox 80** on page 360 in Chapter 15 for how to use these tools). Remember, they are available from ArcScene's Tools Toolbar which, if it's not available, can be accessed by selecting the **Customize** pull-down menu, then choosing **Toolbars**, then selecting **Tools**. Answer Question 16.8.

QUESTION 16.8 You'll see several bridges crossing the river on the southern side of the city, as well as several freeway overpasses in the surrounding areas. Focus your attention on each of these. How did the orthoimage drape over the bridges and overpasses, and why did this happen?

• Despite the changes in elevations, the Youngstown area is relatively flat; however, if you zoom in you'll be able to see some terrain relief. For visualization purposes, we'll add some vertical exaggeration to the scene (see **Smartbox 84** for more about vertical exaggeration).

Smartbox 84

What is vertical exaggeration?

Vertical exaggeration is a technique used in 3D visualization. With vertical exaggeration, the horizontal (x/y) scale remains the same, but the vertical (z) scale is altered by some amount. This method makes the differences in heights of features really stand out, such as making mountains look higher or valleys look deeper. Vertical exaggeration multiplies the z-values by a number. For instance, with a vertical exaggeration of 3, all elevation heights are shown at three times what they should be (see Figure 16.5 for two examples of vertically exaggerated terrain). Because vertical exaggeration alters the scale of the data, it's used only for visualization and should not be used for measurements.

vertical exaggeration A 3D visualization technique that alters the vertical scale but keeps the horizontal scale the same.

FIGURE 16.5 Draped imagery on a TIN shown at vertical exaggerations of (a) 1.5 and (b) 7.

(a)

(b)

SOURCE: NATIONAL MAP/ARCSCENE/ESRI

SOURCE: ARCSCENE/ESRI

- Right-click on the Scene's data frame and select **Scene Properties**. Select the **General** tab and use **3** as a value for Vertical Exaggeration. Click **OK** when you're done. The vertical exaggeration should have been added. Answer Question 16.9.

- For visual purposes of this chapter, return the Vertical Exaggeration to **3** after you've answered Question 16.9.

> **QUESTION 16.9** Test out several choices for vertical exaggeration—1.5, 2, 3, 7, and 10. Although the area is actually very flat, which of these exaggeration levels is best for visualizing the landscape features of Youngstown for this project (from what you've seen in examining the contours, slope, and slope aspect)?

STEP 16.6 **Rendering an Image for Crisper Viewing**

- Because one of the goals of 3D is to create a more pleasing visualization (and you want to create the best-quality visualization possible), this step of the chapter will allow you to create a better rendering of the image draped on the TIN. Zoom in on some portions of the image and you'll see that the image gets more heavily pixilated the closer you zoom in. ArcScene is "rendering" the image at a lower resolution. To make the imagery crisper and less pixilated, you'll need to render it at the maximum possible resolution. Right-click on the Ytownimage layer and select **Properties**, then select the **Rendering** tab.

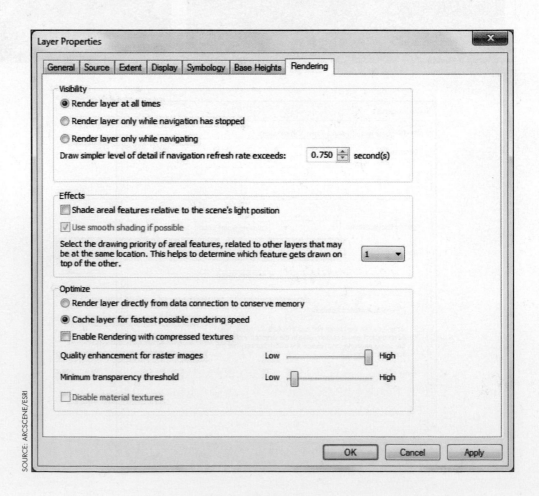

SOURCE: ARCSCENE/ESRI

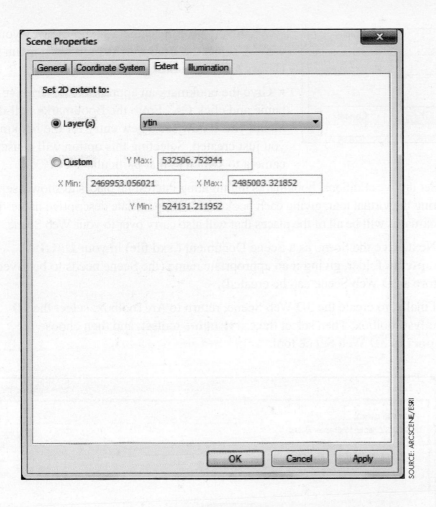

- To begin, we'll set the Extent of the area that will be shown in the Web Scene. Open the **Scene Properties** and select the **Extent** tab.

- Select the radio button for **Layer(s)** and choose **YTIN** from the pull-down menu. This will set the extent of the scene to the boundaries of the TIN.

- Click **Apply** and **OK** to close the dialog box.

- With the Extent set, press the **Full Extent** button on the **Tools** toolbar to return to the full view of the area.

- The next thing to do is to set up the places you want to highlight during your virtual tour. Think of the ArcScene view as a camera—whatever you focus the view on will be one "snapshot" of the camera. By taking several of these "snapshots," you can set up the places you want to automatically visit on the virtual tour without having to manually navigate to each one. You do this by setting up a series of bookmarks within ArcScene, and these bookmarks will act as the "snapshots" of the camera. To begin, use the fly and navigation tools to change the camera view to a part of the city you want to highlight on your virtual tour. Some potential suggestions for the Youngstown area include the YSU football stadium, the spot where the river and the lake come together, areas along the riverbank, the hill leading from campus to downtown, or a view from the top of the hill looking down into the city and river valley.

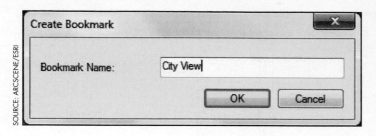

• When you have the camera set up the way you want it, select the **Bookmarks** pull-down menu and choose **Create Bookmark**.

• Give the bookmark an appropriate descriptive name and click **OK**. From the Bookmarks pull-down menu, you'll now see a new entry for the bookmark you just created. Selecting this option will cause the camera to switch to that particular view.

• Set up five additional bookmarks highlighting interesting places to showcase during the virtual tour, giving each bookmark an appropriate descriptive name. These bookmarks will be all of the places that will also carry over to your Web Scene.

• Next, save the Scene as a Scene Document (.sxd file) in your **D:\GIS\Chapter16** folder, giving it an appropriate name (the Scene needs to be saved before a 3D Web Scene can be created).

• Finally, to create the 3D Web Scene, return to ArcToolbox, select the **3D Analyst** toolbox. Then select the **CityEngine** toolset, and then choose the **Export to 3D Web Scene** tool.

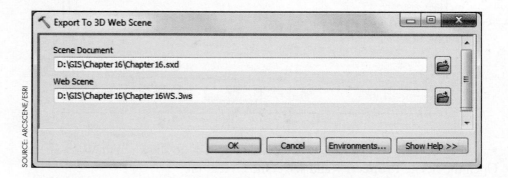

• For Scene Document, use the browse button to select the saved scene document (.sxd) file—in the graphic above, it is named **Chapter16.sxd**.

• For Web Scene, use the browse button to navigate to your **D:\GIS\Chapter16** folder and give the Web Scene a name. In the above graphic, it is named **Chapter16WS**—ArcScene will automatically add the .3ws extension.

• Click **OK** to begin the process. When the usual tool completion message appears in the lower-right-hand corner, the Web Scene has been created.

STEP 16.8 Viewing and Sharing Your Results

• Now that you have a Web Scene created, you'll want to access it (and share it) via ArcGIS Online. To begin, start a WebGL-compliant Web browser (see **Smartbox 85** on page 385 for a partial list, but this chapter was designed using the latest version of Google Chrome). Go to the Website for ArcGIS Online (http://www.arcgis.com) and log in with your Esri Global Account (see Chapter 4 for more about logging into ArcGIS Online). Note that you can work with a Web Scene using either the free version of ArcGIS Online or the subscription-level version.

• After logging in, select the option for **My Content**.

• In the My Content options, select **Add Item**. This will allow you to upload the Web Scene to ArcGIS Online.

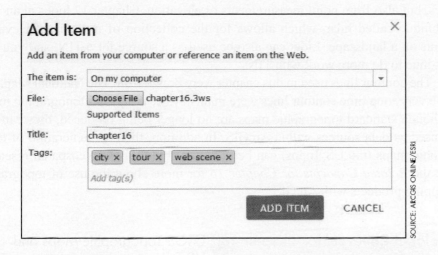

• In the Add Item dialog box that appears, for **The item is:** option, select **On my computer**.

• Click the **Choose File** button and navigate to your **D:\GIS\Chapter16** folder and select the Web Scene you created (it will be the .3ws file).

• Add some appropriate tags for your Web Scene.

• Finally, click **ADD ITEM**.

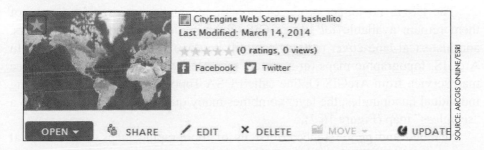

• On the next screen, click the **SHARE** button to share your Web Scene with Everyone (or the members of your organization if you're using the subscription-level version of ArcGIS Online).

• Finally, click **OPEN** and then select **View Application**. Your Web Scene will open using the CityEngine Web Viewer (see **Smartbox 85** on page 385 for more information about navigating the Web Scene and using the bookmarks you set up for your virtual tour). Use the bookmarks and Play options to view your virtual tour.

Closing Time

TINs are very useful and versatile data structures to use in modeling and in 3D landscape visualization. Being able to pick and choose the most important points on the landscape helps create a more detailed landscape where necessary, while breaklines can help in fashioning more realistic landscapes. TINs can be created from a variety of different sources, including digital elevation models or contour lines, but also from point measurements of elevation. Chapter 17 looks at another technique called lidar, which allows for the collection of millions of elevation points on a landscape. Lidar can also be used as a source for a TIN, and you will continue to do more work with TINs.

The contour lines used in this chapter were drawn from The National Map, and for a very long time contour lines were mainstays of the USGS topographic maps. Although standard topographic maps are no longer being produced, they can still be used as data sources within ArcGIS. In addition, the next generation of topographic maps (the US Topos) can be very valuable tools for geospatial research. See the *Related Concepts for Chapter 16* for more about the use of topographic mapping products with ArcGIS.

Related Concepts for Chapter 16   USGS Topographic Maps and US Topos

The traditional source for examining contour information was the USGS **topographic map** series. The function of topographic maps was to examine the features on Earth's surface with information about land cover and especially topography, as represented by contours on the map. Topographic maps were produced as large paper maps in a variety of formats. The 1:24000 scale topographic maps (sometimes referred to as "quadrangles" or "topoquads") covered a geographic area of 7.5 minutes of latitude by 7.5 minutes of longitude. Topographic maps were also produced as smaller-scale maps, such as 1:100000 and 1:250000.

The USGS no longer produces topographic maps, but digital versions of them remain available for use in GIS. Examination of landforms, contours, and historical land-cover patterns can still be done with topographic maps. In ArcGIS, topographic maps (at a variety of scales) are available through a free map service from ArcGIS Online called USA Topo Maps. Instead of showing individual quadrangles, the layer combines many quads of the same scale into a "seamless" map (Figure 16.7).

Scanned and georeferenced versions of topographic maps, called **digital raster graphics (DRGs)** are available and can be used as separate layers in ArcGIS (see Chapter 5 for sources for obtaining DRGs). Because they've been georeferenced, DRGs will align with other layers in ArcGIS. DRGs may be useful for draping over a terrain model in ArcScene to see how landforms match up with the features or contours of a topographic map. DRGs are considered "legacy" data as topographic maps are no longer being produced, thus neither are corresponding DRGs.

topographic map A map created by the USGS to show landscape and terrain as well as the location of features on the land.

digital raster graphic (DRG) A scanned version of a USGS topographic map.

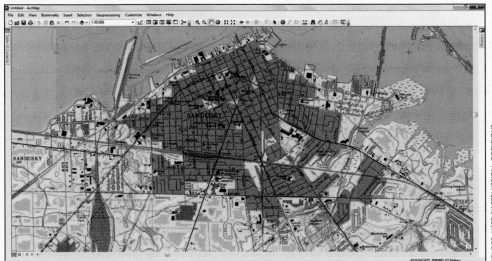

FIGURE 16.7 The USA Topo Maps map service from ArcGIS Online shown as a layer in ArcMap.

The current generation of topographic maps is the **US Topo** map series being produced by the USGS. US Topos are stand-alone products available for free digital download via the National Map and are being produced for areas on a three-year cycle. US Topos contain multiple layers, such as contours, transportation labels, and orthoimagery, all of which can be turned on and off like layers in ArcGIS's TOC. US Topos are available in the **GeoPDF** format, in which a PDF document has interactive capabilities and can return coordinate information (in GCS or UTM) when a user selects a location on the map (Figure 16.8) as well as allow a user to make measurements of lines or areas directly on the US Topos. The National Map also makes historic topographic maps available as GeoPDFs. Keep in mind, however, that a GeoPDF is a stand-alone document and cannot be used as a layer in ArcGIS. You can access the necessary GeoPDF plug-in and toolbar to work with US Topos from the TerraGo Website at http://www .terragotech.com/usgs-toolbar-download.

US Topo A digital topographic map series created by the USGS to allow multiple layers of data to be used on a map in GeoPDF file format.

GeoPDF A format for maps to be used as PDFs, but that can contain geographic information and multiple layers.

FIGURE 16.8 Finding coordinates on a US Topo map of Youngstown, Ohio.

For More Information

For further in-depth information about the topics presented in this chapter, use the ArcGIS Help feature to search for the following items:

- Breaklines in surface modeling
- Contours and isolines
- Delineate TIN Data Area (3D Analyst)
- Export to 3D Web Scene (3D Analyst)
- Fundamentals of creating TIN surfaces
- Fundamentals of Surfaces
- Fundamentals of TIN triangulation in ArcGIS
- Surface Slope (3D Analyst)
- TIN-based surface concepts
- Understanding the shape of a surface
- What is a TIN surface?
- What is sharing in 3D?

For further information about interpreting different contour features, see http://mapserver.mytopo.com/mapserver/topographic_symbols/Contours.html.

For more information about TINs, see the NCGIA Core Curriculum, Unit 56, available at http://www.ncgia.ucsb.edu/giscc/units/u056/u056.html.

For further information about US Topos and historic USGS topographic maps, see http://nationalmap.gov/ustopo/index.html.

For more detailed information about creating 3D Web Scenes (including information on limitations and suggestions for improving performance) see the Esri White Paper at http://support.esri.com/en/knowledgebase/whitepapers/view/productid/54/metaid/2018.

Key Terms

TIN (p. 369)
contour (p. 372)
contour interval (p. 373)
index contour (p. 373)
intermediate contour (p. 373)
mass points (p. 376)
Delaunay triangulation (p. 376)
Delaunay triangle (p. 376)
edges (p. 376)
hard breaklines (p. 377)

soft breaklines (p. 377)
hull (p. 377)
vertical exaggeration (p. 383)
Web Scene (p. 385)
WebGL (p. 385)
topographic map (p. 390)
digital raster graphic (DRG) (p. 390)
US Topo (p. 391)
GeoPDF (p. 391)

How to Work with Lidar Data in ArcGIS 10.2

Introduction

Obtaining accurate information about elevation and heights is critical for numerous GIS applications. Landscape elevations, terrain profiles, and height measurements of trees or buildings are necessary for forest managers, urban planners, archeologists, and environmental scientists (among others). While digital elevation models or contours can provide some of this information, a geospatial technique called lidar is used often to determine the elevation and height of the ground or objects on the surface. With **lidar**, an aircraft uses a laser beam to determine the heights of the terrain and objects over which it flies. The term "lidar" refers to a combination of the words "light" and "radar," though it has also been used as an acronym for "Light Detection and Ranging."

Lidar is a form of **active remote sensing** (see Chapter 13) in which the sensor generates its own energy source, fires it at a target, and measures the reflection of that particular type of energy. In the case of lidar, a laser beam on an aircraft is fired at the ground below, the laser bounces off a target, and the reflection is received back at the aircraft (Figure 17.1 on page 394). The times the laser is sent and received are precisely measured. With information regarding the time that it took for the beam to travel from the plane to a target and back again as well as the speed of the beam (i.e., the speed of light), the distance from the plane to the target can be calculated. The plane's GPS and inertial navigation system (INS) are used for accurately determining the plane's position and altitude. With the known altitude of the plane and the distance from the plane to the target (and information about the angle of the beam), the elevation of the target can be measured.

The strength of the laser pulse returned after striking an object is called the **intensity** of that pulse and millions of lidar measurements can be quickly taken over a small area. These lidar data are then processed into a series of points, where each point contains x and y measurements as well as a z-value for the elevation at that point. Because they have so many processed points, the lidar data comprise a **point cloud**, and this large dataset can be used in GIS. The point cloud represents the measured elevations not just of the ground surface, but also of all of the objects from which the laser beams were reflecting, such as buildings, bridges, or trees (Figure 17.2). Note that a different wavelength of laser beam is used for studies involving water or measuring the depths of bodies of water in a method called *bathymetric lidar.*

During the processing step, these points are classified according to what they represent (such as the ground or a building). Lidar point cloud data are accessed in the **LAS file** format, a standard that is commonly used with GIS. The LAS file

lidar A method for measuring elevation values using laser pulses sent to the ground from an aircraft.

active remote sensing When a sensor generates its own energy source and measures the reflection or backscatter of that energy.

intensity The strength of the reflected return of the laser pulse from an object.

point cloud The processed elevation points of a lidar dataset.

LAS file A standard file format for holding lidar data.

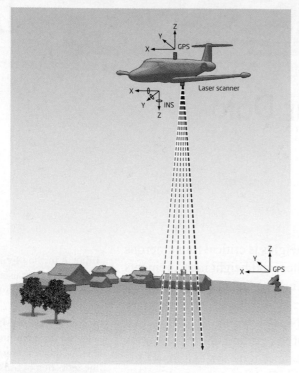

FIGURE 17.1 The operation of airborne lidar.

FIGURE 17.2 The different types of measurements made by lidar.

stores all of the information related to the point cloud data (including x, y, and z coordinates as well as the classification data).

Chapter Scenario and Applications

This chapter again puts you in the role of a member of the Youngstown, Ohio, chamber of commerce. As part of a marketing effort for the city, a 3D model of the urban landscape is being developed, and as part of this process, accurate heights of the city's buildings are needed. Your job is to investigate the use of a set of lidar data in measuring building heights as well as create a simplified 3D representation of the city to use as a basis for the model.

The following are additional examples of other real-world applications of this chapter's skills:

- An urban planner needs to know the height of the city's bridges as part of a redevelopment effort. He can use lidar data to quickly obtain the bridge height information.

- A biologist needs to know the heights of multiple tree stands as part of a larger bird migration study. She can use lidar data to quickly and efficiently measure tree-stand heights.

- A city council for a coastal community is assessing the visual impact of a policy that would allow much taller resort hotels to be built on the beachfront. The planners and developers can use lidar data to quickly determine the heights of existing structures as part of a viewshed analysis.

ArcGIS Skills

In this chapter, you will learn:

- How to create a LAS dataset and view lidar data (i.e., LAS files) in ArcGIS.

- How to examine elevation and class features of lidar data.

- How to edit lidar Class Codes.

- How to obtain information and measurements about lidar points.

- How to create a Profile View and a 3D View of lidar points.

- How to work with lidar data in ArcScene, examining derived contours as well as TIN representations of elevation and slope.

Study Area

For this chapter, you will be working with data from a section of the city of Youngstown, Ohio.

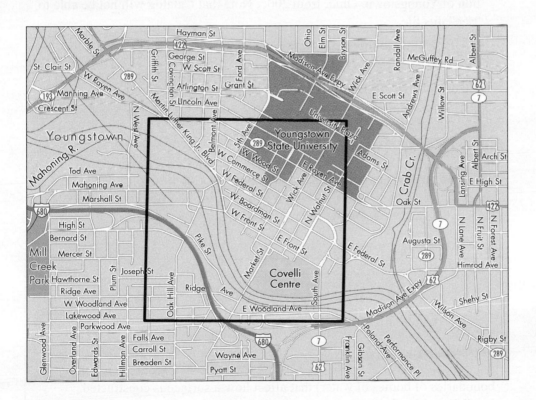

Data Sources and Localizing This Chapter

This chapter's data focus on a section of the city of Youngstown, Ohio. However, you can easily modify this lab to use data from your own local area instead if lidar data are available for free download for areas near you. The LAS dataset used in this lab was downloaded from EarthExplorer (http://earthexplorer. usgs.gov)—see Chapter 5. Free lidar datasets are also available through the USGS CLICK (Center for Lidar Information Coordination and Knowledge) Website (http://lidar.cr.usgs.gov).

Beyond what is available through EarthExplorer or CLICK, there may also be free lidar data available for specific states or counties. For example, if you were working with a section of a city in Iowa (say, Cedar Rapids), Iowa's Department of Natural Resources has LAS files available for download at http://www.iowadnr.gov/Environment/GeologyMapping/MappingGIS/LiDAR.aspx. You may also want to check out NOAA's Elevation Inventory Website for further details on where lidar data may be available: http://www.csc.noaa.gov/inventory/#.

The NAIP imagery used in this chapter is freely available for all U.S. states. (See Chapter 13 for more information about the NAIP program.)

STEP 17.1 Getting Started

- Start ArcMap and use the Catalog to copy the folder called **Chapter17** from the C:\GISBookdata\ folder to your own D:\GIS\ drive. Be sure to copy the entire directory, not just the files within it.

- Chapter17 contains the following items:

  - **OH_North_2006_015328.las**: This is a LAS file of lidar points of a section of Youngstown, Ohio, from 2006. Note that Catalog will not be able to "see" this file.

  - **OH_North_2006_015328.xml**: This is a metadata document for the LAS file.

- Activate the **3D Analyst** extension.

- Add the **3D Analyst toolbar** and the **LAS Dataset toolbar** to ArcMap.

STEP 17.2 Creating a LAS Dataset

- To begin analyzing your lidar data in ArcGIS, you'll need to create a LAS dataset and place your LAS files within in. For more information about just what a LAS dataset is, see **Smartbox 86**.

Smartbox 86

What is a LAS dataset and how is it used in ArcGIS 10.2?

LAS dataset An ArcGIS file structure used for storing, viewing, and analyzing one or more LAS files.

A **LAS dataset** is a specific file structure that allows for the storage of one or more LAS files for use in ArcGIS 10.2. The LAS dataset also allows users to easily view, analyze, query, edit, and visualize the point cloud data within the LAS files. The LAS dataset can also store any types of features representing surface constraints (such as hard breaklines or polygons representing the boundaries of bodies of water) that affect how a surface is constructed.

When a LAS dataset is created, it's empty (just as a new geodatabase is empty when it's created) and LAS files can be added to it. Typically, when LAS files are obtained, they are tiled (such as the LAS file of Youngstown you're using in this chapter). Tiles of data are small rectangular sections of point cloud data adjacent to one another. Thus, to look at the point cloud data for the immediate north of the area you're working with in this chapter, you would need to obtain the LAS file representing the next tile to the north. A LAS dataset can contain multiple LAS files, allowing you to work with all of their data together instead of individual data tiles (Figure 17.3).

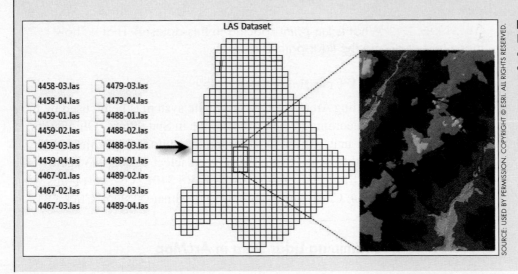

LAS Dataset

4458-03.las
4458-04.las
4459-01.las
4459-02.las
4459-03.las
4459-04.las
4467-01.las
4467-02.las
4467-03.las

4479-03.las
4479-04.las
4488-01.las
4488-02.las
4488-03.las
4489-01.las
4489-02.las
4489-03.las
4489-04.las

FIGURE 17.3 The relationship between LAS files, a LAS dataset, and the geospatial data they contain.

- In Catalog, right-click on the **Chapter17** folder and select **New**, then select **LAS Dataset**. You'll see a new file appear in the Chapter17 folder of Catalog called New LasDataset.lasd. Right-click on this and rename it **YLasDataset**.

- Right-click on **YLasDataset** and select **Properties**.

- In the LAS Dataset Properties dialog, select the **LAS Files** tab.

- Right now, the YLasDataset is empty. To add the LAS file to it, click on the **Add Files…** button. Navigate to your **D:\GIS\Chapter17** folder, select the **OH_North_2006_015328.las** file, and click **Open**. You'll see your OH_North_2006_015328.las file added as the first line. If you had multiple LAS files, you could add those here as well to combine several LAS files into a single LAS dataset.

- Answer Questions 17.1 and 17.2.

QUESTION 17.1 How many points does this LAS dataset (which is composed of a single LAS file) contain?

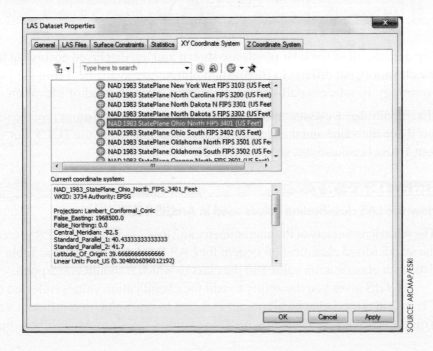

QUESTION 17.2 What is the point spacing in this dataset? That is, how many feet apart are the lidar points?

• Next, click on the **XY Coordinate System** tab.

• Here, you will be telling ArcGIS what coordinate system is being used by the LAS file you added. According to the metadata, it's in State Plane, Ohio North, using US Feet as linear units. To set this, expand the options for **Projected Coordinate Systems**, then select **State Plane**, then select **NAD 1983 (US Feet)**, then select **NAD 1983 StatePlane Ohio North FIPS 3401 (US Feet)**—this will have the WKID of 3734. Click **Apply** to set the coordinate system. Click **OK** to close the dialog box.

STEP 17.3   Examining Lidar Data in ArcMap

• Add YLasDataset to ArcMap's TOC. The point cloud of the area will appear displayed by default as points with an elevation value (note that the elevation values are in feet).

• In order to give your lidar points some context, add the Ohio 2013 1m NAIP orthoimagery from the USDA ArcGIS Server (at **http://gis.apfo.usda.gov/ arcgis/services**)—this is the same NAIP imagery you used in Chapter 13 (see Step 13.6 on page 314 in that chapter for how to properly add the imagery). Once the orthoimagery is added to the TOC, arrange the layers so that the YLasDataset is displayed on top of the imagery.

• There are hundreds of thousands of points in this one point cloud dataset, but the LAS Dataset toolbar gives you several options for which points you want to examine and how to do so. From the LAS Dataset toolbar's pull-down menu, select the **YLasDataset.lasd** (it should be the only option, but if you had multiple LAS datasets, you could use this option to choose which one to work with).

SOURCE: ARCMAP/ESRI

• By default, all of the lidar points are shown and are displayed according to their elevation, but different visualization options are available. From the **Point Symbology Renderers** pull-down menu, select **Class** instead of Elevation.

• Each point has a classification assigned to it for what that point represents (you'll see the value and the classification now displayed in the TOC). For more about LAS classifications, see **Smartbox 87**.

Smartbox 87

How are LAS classification values used in ArcGIS 10.2?

The American Society of Photogrammetry and Remote Sensing (ASPRS) created the standardized classification system for LAS 1.1 and later versions. Table 17.1 lists each classification value and the class to which it should correspond.

    ArcGIS gives you the ability to edit the classification values (referred to as Class Codes) and change all points assigned to one value to a different value. You do this with the Change LAS Class Codes tool from ArcToolbox, but as

Class Value	Class Description
0	Never Classified
1	Unassigned
2	Ground
3	Low Vegetation
4	Medium Vegetation
5	High Vegetation
6	Building
7	Noise
8	Model Key
9	Water
10–11	Reserved for ASPRS Definitions
12	Overlap
13–31	Reserved for ASPRS Definitions

Table 17.1 The ASPRS LAS classification system

you'll see in this chapter, a total attribute switch may introduce errors into the data. A different strategy would be to use the Set LAS Class Codes Using Features tool, in which feature classes are used when classifying points. For instance, you could digitize the footprints of buildings as a polygon feature class and any lidar points that fall within these boundaries could be reclassified as the "Buildings" Class Code.

You could also use the Profile View or 3D View (see Step 17.4) to examine a set of points, use the selection tools to select a set of points, and then manually edit the classification codes of the selected points (Figure 17.4).

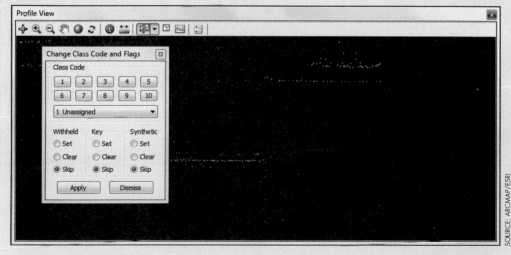

FIGURE 17.4 Manually editing the Class Codes of a selected set of lidar points in the Profile View.

SOURCE: ARCMAP/ESRI

- Zoom in on some of the points and carefully examine them according to their classification in relation to the NAIP imagery (keeping in mind the difference between the 2006 lidar data collection and the 2013 orthoimage). Answer Question 17.3.

QUESTION 17.3 The points classified with a value of "1" (Unassigned) should likely have been classified according to which of the LAS classifications? How can you tell?

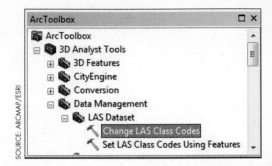

- ArcGIS allows you to edit the classification values assigned to a LAS file to update data or make necessary corrections. In this case, you'll want to change the value of the points classified as "1."

- To do so, open ArcToolbox. From the **3D Analyst Tools** toolbox, select the **Data Management** toolset, then the **LAS Dataset** toolset, then the **Change LAS Class Codes** tool.

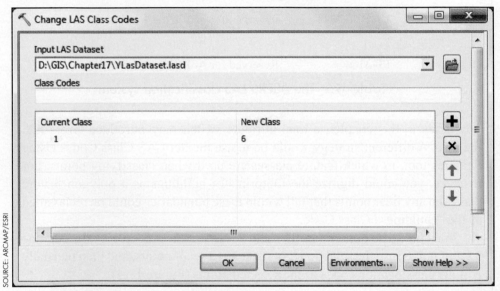

- Use **YLasDataset.lasd** as the Input LAS Dataset.

- Under Class Codes, type the value you want to change (in this case, type **1**).

- Next, press the + button. This will add 1 to the Current Class field.

- In the New Class column next to the value of 1, type the value to which you want the current class to be changed (in this case, type **6**).

- Click **OK** when you're ready.

- Re-examine your points and their new classifications. Answer Question 17.4.

QUESTION 17.4 Switching classifications from 1 to 6 did not solve all the misclassifications. What features are now incorrectly classified as "6"? (*Hint:* Zoom in on some areas and compare the lidar points with the NAIP imagery.)

- ArcGIS also gives you the option to examine whether a point is measuring a ground or non-ground feature. From the **Filters** pull-down menu on the LAS Dataset toolbar, select **Ground**. You'll see that the only points being displayed are those identified as measuring elevations on the ground. Next, from the **Filters** pull-down menu on the LAS Dataset toolbar, select **Non Ground**. You'll see that now the only points being displayed are those measuring points not on the "ground."

• Answer Question 17.5. For more information about what lidar measurements of ground and non-ground features represent, see **Smartbox 88**.

QUESTION 17.5 Aside from buildings, what other types of specific features on the landscape are considered "non-ground" in terms of lidar measurements?

Smartbox 88

What do lidar measurements represent?

When the laser pulses from the lidar system strike areas on the ground, they do so in "laser footprints" that are circular areas 30 centimeters in diameter. The **return** (or reflection) of this pulse is what is measured back at the airborne sensor. However, a pulse may have several returns, depending on the object that is struck with the pulse. For instance, see Figure 17.5, which shows a large tree illuminated by the laser pulse—different places (at different heights) on the tree will generate multiple returns. The **first return** is in many ways the most important, as it represents the top of the object (or the maximum elevation) and is used in determining the height of the tree. The intermediate returns can be used for other measurements or the profile of the tree, while the **last return** is often the one from the ground itself. All of these returns are used in constructing a **Digital Surface Model (DSM)** that will show the heights of the terrain and the objects on top of it (including buildings, tree canopies, and vegetation).

return The reflection of energy from a lidar laser pulse.

first return The initial reflection of the lidar laser pulse, which usually indicates the height of an object.

last return The final reflection of the lidar laser pulse, which usually indicates the elevation height of the ground.

Digital Surface Model (DSM) A representation of the surface and features generated from a set of lidar points.

FIGURE 17.5 The returns associated with lidar pulses.

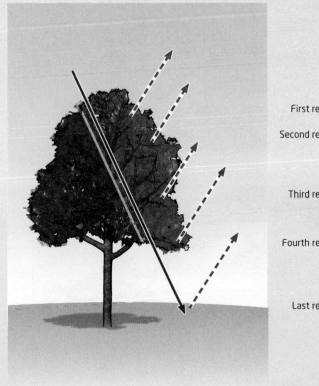

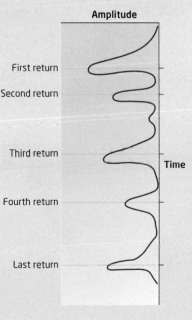

Amplitude

First return

Second return

Third return

Time

Fourth return

Last return

bare-Earth A model of the surface using only the elevation values of the ground (rather than features, structures, or vegetation on top of it).

During processing, a LAS file may have the points classified as whether they are first return or not (unfortunately, the data used in this chapter do not contain this information, though they do contain classifications for points to be separated as ground and non-ground), and one of the Filters options on the LAS Dataset toolbar allows you to examine only first-return points. Last-return data are utilized to create a **bare-Earth** digital elevation model, which shows only the ground elevations. Algorithms are used to remove return data (including last-return data) that contain vegetation, trees, or structures from the dataset to use only those points that represent the bare surface of the ground. Some of the data within the NED (see Chapter 15) consist of lidar-derived bare-Earth digital elevation models.

- The elevations value of an individual lidar point can be accessed with the Identify tool. For instance, zoom in closely on the area shown in Figure 17.6—this is the Lincoln Building on the Youngstown State University campus.

- Once you've zoomed in closely, change the **Filters** option back to **All**.

- Lincoln is a flat building with an air-conditioning unit on the roof (Figure 17.7).

- Use the Identify tool on one of the points on the flat part of the building's roof and answer Question 17.6.

FIGURE 17.6 Lidar points taken of the YSU campus.

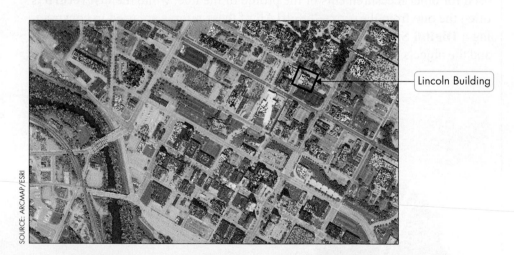

Lincoln Building

SOURCE: ARCMAP/ESRI

FIGURE 17.7 The lidar points obtained from the Lincoln Building and its surrounding areas.

SOURCE: ARCMAP/ESRI

QUESTION 17.6 What is the elevation height of the flat roof of the Lincoln Building (in feet)?

• Keep in mind that the elevation value for this point is the height above the vertical datum, not the height of the building itself. To obtain a relative value for the building's height, use the Identify tool on one of the lidar points on the flat sidewalk in front of the building. Compare the elevation value of the sidewalk and the elevation value of the flat roof of the building and answer Question 17.7.

QUESTION 17.7 Using these two measurements, what is the approximate height of the Lincoln Building?

STEP 17.4 Creating a LAS Dataset Profile View and 3D View in ArcMap

• It's one thing to look at points on the screen and get a sense of their elevations, but ArcGIS provides other tools for better visualization of lidar data. As in Chapter 15 when you created a profile graph of the Hocking County terrain, ArcGIS allows you to use the lidar points to create a detailed profile of the elevations of a section of the ground. In this example, however, you'll be creating a profile of non-ground elevations—in particular, this set of downtown buildings.

• To begin, select the **LAS Dataset Profile View** button from the LAS Dataset toolbar:

• Your cursor will change into crosshairs. Position the crosshairs where you want to begin the profile (in this case, click on a point somewhat past the middle of the southwestern side of the Lincoln Building) and drag a line to the width of the profile (in this case, somewhat past the middle of the northeastern side of the Lincoln Building—see Figure 17.8).

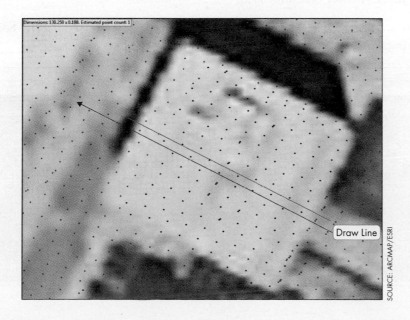

FIGURE 17.8 The first line of the LAS Dataset Profile View drawn through the Lincoln Building.

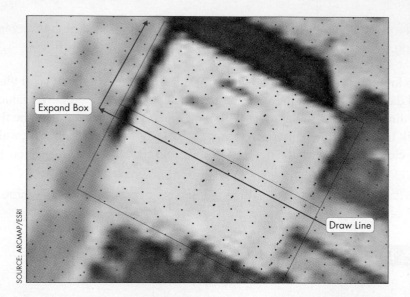

- Next, use the mouse to expand the line into a box, When you move the mouse to one side of the line you drew, a box will be expanded out in both directions from the line. All points within this box will be used in the profile. Expand the box so that the dimensions will cover the entire Lincoln Building and some of the sidewalk and surrounding areas (Figure 17.9).

- Once you have the box extended properly, click the left mouse button and a new window—the actual profile view—will open. You'll see the points used in profiling the Lincoln Building displayed in a profile of their height. Note that if the lidar dataset contained more points obtained from this area, the profile would be much more detailed. However, even with this profile, you should be able to visualize the flat roof of the Lincoln Building, the air-conditioning unit on the roof, the ground/sidewalk in front of the building, and some of the other buildings shown to the left of Lincoln.

- To make measurements from the Profile View, select the **Measure tool** from its toolbar. To use the Measure tool, click once on the starting location and again on the ending location. For instance, to measure the height of the Lincoln Building, click the Measure tool on the flat roof, then extend a vertical line down to the ground level. Answer Question 17.8.

QUESTION 17.8 From measurements made in the Profile View, what is the height of the Lincoln Building? (Compare with your answer from Question 17.6.)

- Close the Profile View, then zoom back out to the extent of the YLasDataset.

- Next, we'll examine a larger lidar profile of downtown. To begin, display your Point Symbology as **Elevation** and show only **Non Ground** points in the Filters.

- Zoom in to the section of downtown Youngstown shown in Figure 17.10. This section of downtown contains the Home Savings and Loan Building (the tallest structure on that block, it's the building with the pointed spire on the roof). See Figure 17.11 for an aerial image of what this section of downtown really looks like.

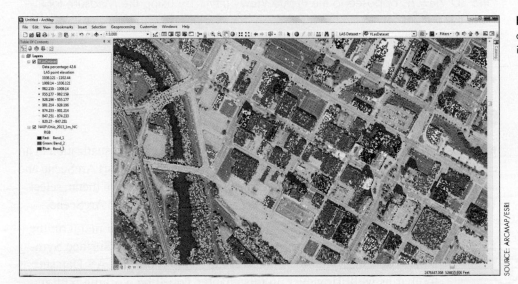

FIGURE 17.10 A section of downtown Youngstown as shown in NAIP imagery and lidar points.

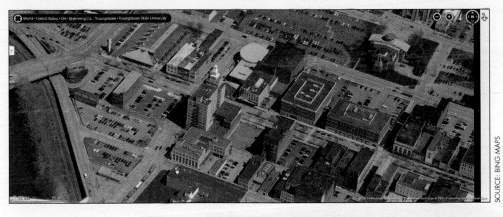

FIGURE 17.11 The same section of downtown Youngstown as seen in aerial imagery.

- Use the Profile View to examine the lidar profile of that block and the Measure tool to determine the height of the Home Savings and Loan Building (from the highest measured point of the spire to the ground below). Answer Question 17.9.

QUESTION 17.9 What is the height of the Home Savings and Loan Building?

- The Profile View lets you examine a 2D profile, but ArcGIS also gives you the capability to see a 3D profile of the lidar points. Stay zoomed in on the Home Savings and Loan Building area and press the **LAS Dataset 3D View** button on the LAS Dataset toolbar.

• A new window (called 3D View) will open, showing the lidar points placed at their elevated heights in a 3D environment. Use the pan, zoom, and navigate controls on the 3D View's toolbar to maneuver around the scene (these controls work like their ArcScene counterparts). Focus on the Home Savings and Loan Building again and navigate around it in 3D. Answer Question 17.10.

QUESTION 17.10 What are the advantages and disadvantages of working with the 3D View as opposed to the Profile View? (*Hint*: Try using the Measure tool and the options for Vertical Exaggeration from the pull-down menu to explore some 3D functions when answering this question.)

• Close the 3D View window when you're done.

STEP 17.5 Examining Lidar Data in ArcScene

• Lidar data can also be used in the ArcScene environment for 3D visualization in an environment with greater functionality than the 3D View window. Start ArcScene and add the YLasDataset to its TOC. Also, from the **Customize** pull-down menu, select **Toolbars**, then select **LAS Dataset** to add the LAS Dataset toolbar to ArcScene.

• From the **Surface Symbology Renderers** pull-down menu on the **LAS Dataset toolbar**, select **Elevation**. Note that the Surface Symbology is also available in the ArcMap version of the LAS Dataset toolbar as well. However, in this chapter you'll be working with its visualizations in ArcScene.

• The Surface Symbology options will convert the points over to a TIN representation of the surface (see Chapter 16) to create a Digital Surface Model (DSM)–see **Smartbox 88** on page 401. Note that by default the TIN uses all of the points in the LAS dataset, so you're able to see the various buildings and structures fashioned using the triangles and breaklines of a TIN.

- Use the various ArcScene tools (see Chapters 15 and 16) to examine both the Home Savings and Loan Building and the Lincoln Building in the TIN representation (and also look at them using the Elevation option under the Point Symbology Renderers to see the same representation as the 3D View window in ArcMap). Answer Question 17.11.

> **QUESTION 17.11** How are the two buildings (Lincoln and Home Savings) represented using the TIN data structure (compared to the point elevation representation)?

- Other settings are available from the Surface Symbology Renderer menu for representing the surface functions described in Chapters 15 and 16 (namely—slope, aspect, and contours). Select **Slope** from the **Surface Symbology Renderer** menu and examine the new representation of the area. Answer Question 17.12, then switch the Surface Symbology Renderer back to **Elevation**.

> **QUESTION 17.12** Why do all of the buildings have their sides as a solid red color? What is being measured and shown in this display?

- Next, we'll examine only the ground returns (instead of looking at points measuring non-ground features, such as buildings or trees). From the **Filters** pull-down menu, select **Ground**. The TIN representation will now be created from only the points taken from the ground itself.

- To view this as a contour representation, from the **Surface Symbology Renderers** pull-down menu select **Contour**. Use the ArcScene tools to examine the landscape, keeping in mind that this representation is being created from only the ground-based points. By default, ArcScene will show you the Index Contours in red and other contours in black with a five-foot contour interval. Answer Question 17.13.

> **QUESTION 17.13** Not counting the non-ground returns, what is the highest and lowest elevation being measured by the lidar dataset? (*Hint:* You can use the Identify tool on contour lines to get information about their elevation.)

STEP 17.6 Exporting and Saving Your Results

- To share your results with others (i.e., the view of the contours and the 3D TIN view of the structures), you'll have to export the scene to a graphic format. Set up the view to best show off the landscape as a set of contours and export the scene to a JPEG or TIFF graphic format (see Chapter 15 for how to export scenes).

- Do the same with the TIN-based 3D view of the buildings of Youngstown (focus on the Home Savings and Loan Building).

- Compose a layout that features these two graphics, as well as a title and some short descriptive text.

- Save your work in ArcScene as a scene document and add the appropriate information to the Scene Document Properties.

- Save your work in ArcMap as a map document and add the appropriate information to the Map Document Properties.

• To share your LAS dataset via ArcGIS Online, you would have to create a mosaic dataset (see Chapter 12), add the LAS dataset to it, and then publish the mosaic dataset as an image service (see Chapter 13) to ArcGIS Online. Unfortunately, at present, LAS datasets cannot be incorporated into 3D Web Scenes (see Chapter 16).

Closing Time

Lidar data provide a very rich and versatile dataset for use in measuring biomass, measuring building heights, or creating digital elevation models of an area. The LAS dataset allows you to easily work with several LAS files of point cloud data to work with height measurements and digital surface models. As lidar data are becoming more accessible, it's important to have new tools to take advantage of this type of geospatial data. For instance, ArcGIS 10.2 has a separate type of dataset for use in visualizing and working with large amounts of lidar data called the terrain dataset (see the *Related Concepts for Chapter 17*).

Although you were able to construct 3D profiles and TIN visualizations of the point cloud data, the buildings didn't retain all of their shape or detail, nor did the visualization resemble the exterior features of the structures. In Chapter 18, we'll examine further techniques for creating other types of three-dimensional visualizations of objects using another set of tools available through the 3D Analyst.

terrain dataset An ArcGIS file structure that can hold very large amounts of elevation data and render them quickly at a variety of scales and resolutions.

multipoint A geodatabase feature class that can hold very large amounts of point data.

Related Concepts for Chapter 17   Terrain Datasets in ArcGIS 10.2

Although lidar point cloud data can be used to create a TIN or DEM, a special file structure exists in ArcGIS specifically for containing, visualizing, and analyzing large quantities of elevation data. The **terrain dataset** is a TIN-based file structure that can "consume" very large amounts of elevation values to produce an elevation model. An advantage of the terrain dataset is that it can have pyramids generated for it, enabling visualization of the large amount of data quickly at different scales.

Terrain datasets are created from a feature dataset within a geodatabase (see Chapter 6). This feature dataset must contain the feature classes that will be used to create the terrain dataset. One caveat is that, while several LAS files will often be used as a source of elevation data to construct a terrain dataset, LAS files cannot be stored as a feature class. The LAS files must first be converted into a **multipoint** feature class in order to be used. Multipoint is a special type of feature class designed to hold arrays of millions of points (such as lidar point cloud data). Ordinarily, a point feature class will have a separate record for each individual point, something that would be unfeasible with millions (or billions) of points, but the multipoint feature class gathers these points into rows of data, thus allowing their use in a geodatabase.

A terrain dataset can then be constructed from lidar points stored as a multipoint feature class along with any other feature classes that would define surface considerations (such as polygon feature classes to show the boundaries of water bodies or line feature classes to show hard breaklines). The resultant terrain dataset will be able to quickly visualize the elevation values and surface features because it does not need to create and use a TIN, but rather references the feature class data itself and creates a TIN representation "on the fly" for the user. With the terrain dataset's pyramids, different scales of massive datasets can quickly be rendered (see Figure 17.12).

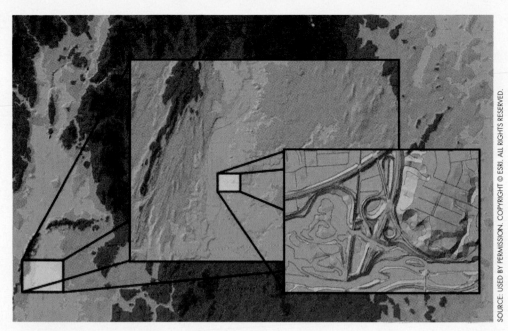

FIGURE 17.12 The use of a terrain dataset for quickly displaying a very large GIS dataset at various scales.

For More Information

For further in-depth information about the topics presented in this chapter, use the ArcGIS Help feature to search for the following items:

- An overview of displaying LAS datasets in ArcGIS (RC)
- Building a terrain dataset using the New Terrain wizard
- Change LAS Class Codes (3D Analyst)
- Common lidar data configurations
- Creating intensity images from lidar in ArcGIS
- Creating raster DEMs and DSMs from large lidar point collections
- Defining feature class properties
- Displaying terrain datasets in ArcGIS
- Fundamentals of Surfaces
- LAS dataset 3D View
- LAS dataset Profile View
- Sharing lidar datasets
- Storing lidar data
- Terrain dataset concepts
- The interactive LAS Dataset toolbar
- What is a terrain dataset?
- What is lidar data?

For more information about the functions of lidar and uses within GIS, see Jensen, J., and Jensen, R. 2012. *Introductory Geographic Information Systems.* Boston: Pearson, Chapter 3.

Key Terms

lidar (p. 393)

active remote sensing
   (p. 393)

intensity (p. 393)

point cloud (p. 393)

LAS file (p. 393)

LAS dataset (p. 396)

return (p. 401)

first return (p. 401)

last return (p. 401)

Digital Surface Model
   (DSM) (p. 401)

bare-Earth (p. 402)

terrain dataset (p. 408)

multipoint (p. 408)

How to Represent Geospatial Data in 3D with ArcGIS 10.2

Introduction

The last three chapters have focused on three-dimensional representations of Earth-based concepts. Land cover, terrain, and imagery and lidar gave us means of assessing heights of non-landscape objects and a rough visualization of them. Now it's time to focus on 3D visualizations of objects and items on the surface of Earth—buildings, structures, trees, cars, and pretty much anything you can envision (beyond what we saw with lidar in Chapter 17). It's one thing to represent a neighborhood as a series of two-dimensional polygons of houses' footprints, but it's another thing entirely to view the same neighborhood as three-dimensional polygons that show the height and dimensions of each house.

3D visualization communicates some geospatial concepts quickly and intuitively to observers and can have a strong impact on presenting ideas effectively. For instance, back in Chapter 6, you created a map of the Youngstown State University campus by digitizing a representation of its buildings. A newcomer to the campus using this map would have an idea of the relative size of each structure, but no feel for what the buildings themselves actually look like, nor how tall each one is. However, if you could take each of those digitized polygons and create three-dimensional blocks that show the height of the building, and use different textures and colors on the sides of those blocks to represent the physical appearance of the building, you would have a GIS dataset that would serve as a more useful map. If you then added some 3D objects like trees, cars, statues, or building signs to the map, you would have an even better visual representation of the campus.

This chapter serves as an introduction to taking two-dimensional GIS data and creating a three-dimensional representation from it. It's becoming more common to view and analyze geospatial data in a 3D format, and Esri has produced numerous tools for visualizing and working with 3D data. For instance, the **Esri CityEngine** program is designed to generate realistic-looking 3D visualizations of entire cities quickly and efficiently (Figure 18.1).

In this chapter, we won't be using anything as complex as CityEngine. Rather, we'll be working with the **ArcGlobe** component of ArcGIS 10.2 for creating

> **Esri CityEngine** A powerful 3D design software that allows for quickly creating large-scale 3D models (like cities).
>
> **ArcGlobe** The 3D virtual globe interface of ArcGIS.

FIGURE 18.1 An example of 3D design in GIS using Esri CityEngine.

multipatch The ArcGIS data format that allows for fully 3D objects and structures to be used.

a 3D representation of GIS data. Unlike ArcScene, ArcGlobe is a virtual globe interface (similar to Google Earth) that comes pre-loaded with imagery and terrain for the entire Earth and can be used for many different types of 3D visualizations. Note, however, that all of the techniques you'll be using in this chapter (extrusions, offsets, and flythroughs and animations) can also be performed using ArcScene.

As a note before we begin, keep in mind the previous chapters' distinction between 2.5D and 3D data: A layer or object can be considered 3D only if it can hold volume or contain multiple *z*-values for each *x/y* coordinate. For ease of use, this chapter will refer to any objects that contain a third dimension as "3D," but keep in mind that many of them will actually be "pseudo 3D" or 2.5D. To work with true 3D objects and structures in ArcGIS 10.2, you would use a different storage format referred to as a **multipatch** (see the *Related Concepts for Chapter 18* for more about working with multipatch data in ArcGIS).

Chapter Scenario and Applications

This chapter puts you in the role of a campus marketing manager at Youngstown State University. As part of a marketing program, your task is to create a new 3D version of the campus, as well as short videos to show off the 3D campus. Your goal is to start with a two-dimensional representation of the campus buildings and create a three-dimensional visualization of the data. You can then use this introductory 3D campus model as a starting point for creating other campus maps (or creating more detailed 3D visualizations).

The following are additional examples of other real-world applications of this chapter's skills:

SOURCE: JOSE LUIS PELAEZ/ ICONICA/GETTY IMAGES

- A city council of a coastal tourist town has to make decisions regarding changes to the size and height of new resort developments on the beach. Rather than looking at a two-dimensional map or architectural diagrams, they can use GIS to create a three-dimensional view of the current building arrangement, as well as the appearance of the proposed changes. They can then assess the visual impact on the area.

- Environmental planners are examining potential locations for the construction of wind turbines for a region's alternative energy sources. They want to construct a 3D visualization of the proposed sites that show a representation of the windmills (at their proper size and height) in relation to the structures and land use of the surrounding areas.

- An archaeologist wants to create a 3D visualization of a Mayan village that is the subject of a dig site. Her goal is to make a 3D version of the site being excavated to show what the area looked like centuries ago. She will then use this visualization to present the findings at the site.

ArcGIS Skills

In this chapter, you will learn:

- How to use the ArcGlobe environment for handling 3D data.

- How to extrude and offset polygon layers.

- How to apply base heights to vector layers.

- How to create and visualize representations of 3D objects.
- How to create 3D animations and flythroughs using ArcGlobe.
- How to use KMZ layers in ArcGlobe.

Study Area

For this chapter, you will be examining a portion of the Youngstown State University (YSU) campus in Youngstown, Ohio.

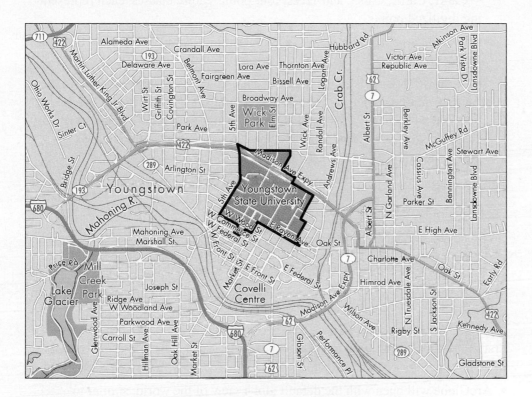

Data Sources and Localizing This Chapter

This chapter's data focus on features and locations within the YSU campus. However, you can easily modify this chapter to use data from your own campus or local area instead. Using the Imagery basemap, digitize a set of building footprints of your own campus (see Chapter 6), as well as points representing cars and trees. In the building footprints layer, create a new field called Height (using the Add Field techniques from Chapter 2) and add a value representing the heights of each building. Note that, for the purposes of this chapter, the heights of some buildings and features have been exaggerated or changed from their actual size to aid in visualization. To use exact values for building heights, you should consult blueprints or lidar height data, or make field measurements of the buildings.

STEP 18.1 Getting Started

- Start ArcMap and use Catalog to copy the folder called **Chapter18** from the C:\GISBookdata\ folder to your D:\GIS\ drive.

- Chapter18 contains a KMZ file called Courtyard_ApartmentsGE4.kmz. This is a 3D model of the YSU Courtyard Apartments created using SketchUp and converted to a KMZ format. It was created by members of the YSU 3D Campus Model team (see the Preface for more information and the names of the students involved). Note that the Catalog in ArcMap will not be able to "see" this file.

- Chapter18 also contains a file geodatabase called **Campus3D** that contains the following feature classes:

  - Bridge: a feature class containing a simplified digitized polygon of the bridge crossing Wick Avenue from the M1 parking deck.

  - Cars1, Cars2, Cars3, and Trees: four point feature classes, each representing digitized points.

  - Church: a polygon feature class representing a simplified outline of St. John's Church.

  - Campusbldgutm: a polygon feature class with digitized polygons of YSU campus buildings that has been projected into UTM Zone 17 (for use in this chapter). The original version of this campus data was created by YSU students Rob Carter, Paul Gromen, Sam Mancino, and Jaime Webber.

- Activate the **3D Analyst** extension.

- Also add the **3D Analyst toolbar** to ArcMap.

- Press the **ArcGlobe** button on the 3D Analyst toolbar.

- Once ArcGlobe opens, select the option for a new **Blank Globe** for working with ArcGlobe. After ArcGlobe opens, you can close ArcMap (note that you can also open ArcGlobe as a separate program without having to go through ArcMap first). A "Globe" is like a "Map" in ArcMap or a "Scene" in ArcScene—it's space to work with your geospatial data.

- ArcGlobe will open with the default globe view of the world, similar to programs like Google Earth. Maximize ArcGlobe to give yourself some room in which to work.

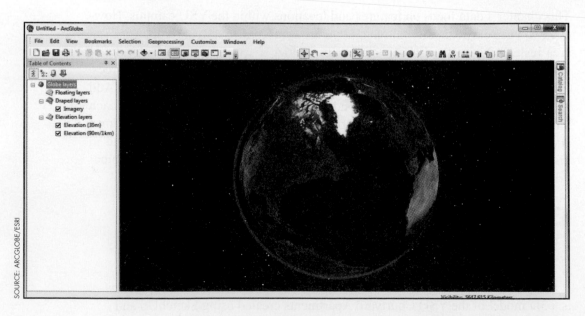

STEP 18.2 **3D Visualization of GIS Data in ArcGlobe**

- Add the campusbldgutm feature class to ArcGlobe.

- In ArcGlobe, not all layers will be displayed at all scales (for instance, if you were looking at the entire globe, you shouldn't be able to make out individual buildings in Youngstown). When adding data to ArcGlobe, you will be prompted with questions about the scale. For the campusbldgutm layer, just use the defaults (for the scales in the green range).

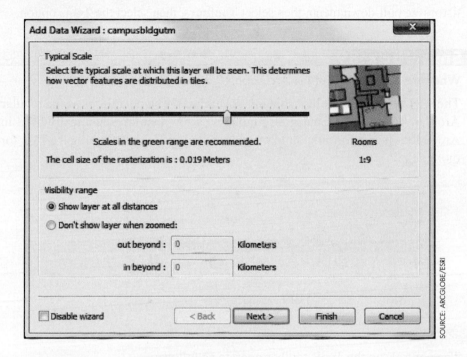

- Click **Next**.

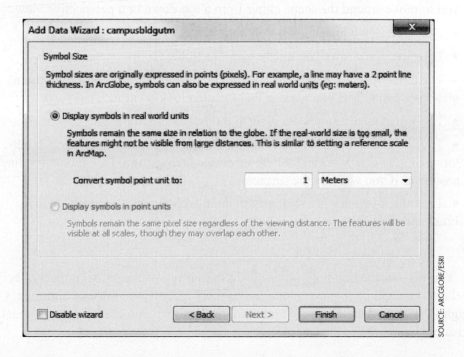

- ArcGlobe will prompt you again about symbol size. For now, accept the defaults and click **Finish**.

- Right-click on **campusbldgutm** and choose **Zoom to Layer**.

- The view of the globe will change to focus only on the Youngstown State University campus. The polygons of the campus buildings will appear as if you're looking straight down on them.

- By default, ArcGlobe's Tools toolbar should be available to you (see **Smartbox 89** for more information). ***Important Note:*** If the toolbar is not present, select the **Customize** pull-down menu, then select **Toolbars**, then select the **Tools** option.

Smartbox 89

What are the functions of the ArcGlobe 10.2 Tools toolbar?

The ArcGlobe Tools toolbar contains several of the same utilities as the regular ArcMap Tools toolbar, but it also contains a few that are specifically used in ArcGlobe (for performing similar functions in ArcScene). See Figure 18.2 for more details.

FIGURE 18.2 The ArcGlobe Tools toolbar.

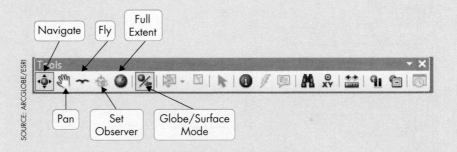

SOURCE: ARCGLOBE/ESRI

The controls on the Tools toolbar operate as follows:

- The first button with the four arrows is the **Navigate** button. This allows you to move around the scene either from a top-down or a perspective view.

- The hand icon allows you to **Pan** around the scene.

- The bird icon allows you to **Fly** over the globe.

- The target icon allows you to zoom in directly to a feature and set it as an **observer** point (as if you were at that location looking around).

- The **world** icon restores the view to the full extent of the globe.

- The sixth button allows you to **toggle** between **globe** mode (when the button is highlighted, you can look down on the area) and **surface** mode (moving around as if you were on the surface).

- The other buttons work similarly to their ArcMap counterparts (Identify, Find, and so on).

- You will see the building footprints on the surface of the global imagery. By default, you will be in globe mode (when the Toggle Globe/Surface button is highlighted). Use the Navigate button to move around and use the mouse wheel to zoom in and out of the campus area.

- Press the **Toggle Globe/Surface** button so that it's no longer highlighted to enter surface mode. Now you can use the Navigate button to tilt the scene so you're looking at a perspective view of the campus. Use the Navigate and Zoom controls to look around the campus area. Answer Question 18.1.

QUESTION 18.1 Why are the building footprints still flat in this 3D ArcGlobe environment?

STEP 18.3 Extruding Polygon Layers

- Right-click on **campusbldgutm** and select **Properties**. From there, select the **Globe Extrusion** tab.

- Put a checkmark in the **Extrude features in layer** box. This indicates to ArcGlobe that you want to extrude your building footprints into blocks. For more information about extrusion, see **Smartbox 90**.

Smartbox 90

What is extrusion and how does it work in ArcGIS 10.2?

Extrusion allows you to give a third dimension to objects by taking a two-dimensional item and extending it to a height. For example, a polygon shows the 2D view of a house, but an extruded polygon shows the polygon at 12 feet high. By taking a number of polygons, assigning an attribute value for the height to each, then extruding them to that height, the polygons can then be visualized in 3D (Figure 18.3).

extrusion Used to give an object height.

FIGURE 18.3 Flat 2D building footprint polygons and their pseudo-3D extruded versions.

2-dimensional building polygons

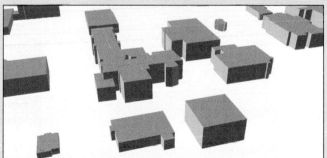
Extruded building polygons

SOURCE: ARCSCENE/ESRI

All three vector objects can be extruded. The extrusion process will turn points into poles, lines into walls, and polygons into blocks. ArcGIS 10.2 offers several different types of extrusion methods:

- Adding to each feature's base height: This method will add the extrusion value to the base height of the underlying terrain. Note that, as in Chapter 16, the height determination used here is based on the horizontal coordinate system (i.e., if the layer's coordinate system is measured in feet, then an extrusion height in feet will be used; if the layer's coordinate system is measured in meters, then meters will be used for the extrusion height).

- Adding to each feature's minimum height: This method will add the extrusion value to the smallest available height value and extrude up from there.

- Adding to each feature's maximum height: This method will add the extrusion value to the highest available height value and extrude up from there.

- Using the extrusion as a value to which features are extruded: This method will use the extrusion value as the height to extrude the objects to but not incorporate base heights into the equation.

- Next to the Extrusion value or expression box, press the **Calculation** button.

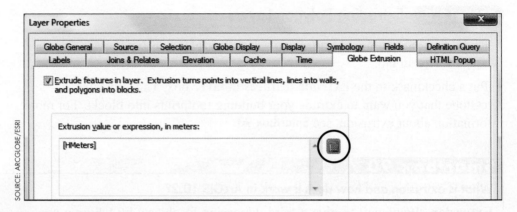

- In the Expression Builder that opens, make your expression equal to **[HMeters]**. This expression means that your buildings will be extruded to the height of the field called "HMeters" in the campusbldgutm attribute table.

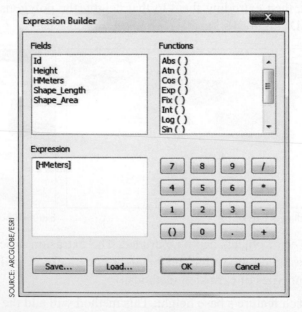

- Click **OK** to close the dialog. Answer Questions 18.2 and 18.3.

QUESTION 18.2 What would be the result of setting the expression builder to [HMeters] * 2?

QUESTION 18.3 What would be the result of setting the expression builder to 30?

- Back in the Layer Properties dialog, under the **Globe Extrusion** tab, select the option for Apply Extrusion by: **adding it to each feature's base heights**.

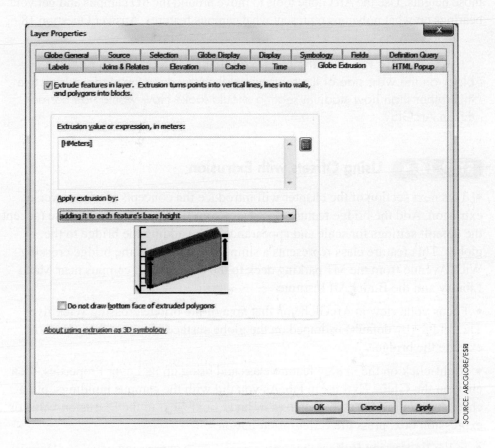

- Click **Apply**, then click **OK**. Your buildings are now extruded.

- Use the Navigate and Zoom tools (in globe mode and surface mode) to look around campus. Examine how the extruded buildings look. Answer Question 18.4.

QUESTION 18.4 Why do many of the campus buildings look so strange, like they're crooked or distorted?

- Return to the Layer Properties of the campusbldgutm layer and select the **Globe Extrusion** tab. This time, choose to Apply extrusion by: **adding it to each feature's minimum height**.

- Carefully examine the campus buildings with this new extrusion setting and answer Question 18.5.

QUESTION 18.5 How does this extrusion setting differ from the previous one? What is strange about the appearance of certain buildings in the center of the 3D campus using this setting?

- Return to the Layer Properties of the campusbldgutm layer one more time and select the **Globe Extrusion** tab. This time, choose to Apply extrusion by: **adding it to each feature's maximum height**.

• This 3D model is a basic, simplified representation of YSU's campus shown in ArcGlobe. The heights of buildings are approximated and the polygons extruded to those heights. Use the ArcGlobe tools to move around the 3D campus and get your bearings on what polygons represent what campus features. Answer Question 18.6.

QUESTION 18.6 Focus on the bleacher seating for Stambaugh Stadium (the block on the west side of the football field). Why does the stadium look like this rather than how stadium seating should look? How would you correct this in ArcGIS?

STEP 18.4 **Using Offsets with Extrusion**

• This next section of the chapter will introduce the concepts of offsets versus extrusion. Add the bridge feature class from your Campus3D geodatabase (accept the default settings for scale and appearance when adding the bridge to the globe). This feature class represents a simplified version of the bridge crossing Wick Avenue from the M1 parking deck to the main part of campus near Maag Library and the Butler Art Institute.

• Focus your view in ArcGlobe on that area of the bridge crossing Wick Ave. The bridge (by default) is draped on the globe surface. The first thing to do is extrude the bridge.

• Right-click on the **bridge** feature class and bring up its Layer Properties, then click on the **Globe Extrusion** tab. As you did with the campus buildings, put a checkmark in the **Extrude features in layer** box. Next to the Extrusion value or expression box, press the **Calculation** button.

• In the Expression Builder that opens, make your expression equal to **[Height]**.

• Also, choose to Apply extrusion by: **adding it to each feature's maximum height**.

• Click **Apply** and **OK**. The bridge will appear in ArcGlobe, now set at the proper elevation relative to the terrain. Examine how the bridge looks and answer Question 18.7.

QUESTION 18.7 Why does the bridge look like a wall stretching from the M1 deck through Wick Avenue?

• The last thing to do is assign the bridge an "offset" value to make it "float" above the terrain. For more information about offsets, see **Smartbox 91**.

Smartbox 91

How does an offset function in ArcGIS 10.2?

When visualizing geospatial data in 3D, there will often be features that you don't want to extrude from the ground level up. Instead, you'll want those features to "float" above the ground's surface. For instance, an elevated train track would first need to be raised off the ground surface before more modeling could be done with it; the same is true with a glass skywalk. You can make an object "float" in ArcGIS by assigning an **offset** value to the feature—the feature will then be placed in a floating position above the ground level at a height (in meters) equal to the value of the offset. If a monorail track should be 10 meters above the ground, assign the object representing the track an offset value of 10. ArcGIS will position the object above the ground surface; you can then extrude it or otherwise change the object's appearance.

offset Used to raise an object above the ground surface.

- Return to the bridge's Layer Properties and select the **Elevation** tab.

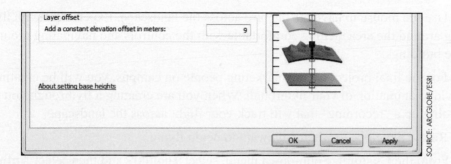

SOURCE: ARCGLOBE/ESRI

- Under Elevation from surfaces, be sure the **Draped on the globe surface** radio button is selected (this is the default option when adding things to the globe).

- Under Layer offset, type **9** in the **Add a constant elevation offset in meters** box. This will offset the bridge by a height of 9 meters.

- Click **Apply** then click **OK**. The bridge is now set to the offset value and floating above the terrain surface. Answer Question 18.8.

QUESTION 18.8 One element missing from this representation of YSU's campus is an elevated pedestrian bridge that connects Moser Hall and Cushwa Hall. What steps would be involved in properly adding this object to the model?

STEP 18.5   Creating an Animation in ArcGlobe

- It's time to go flying around the campus in preparation for the virtual tour you'll be creating. You used the Fly tool in Chapters 15 and 16 for positioning and movement. Here, you'll be able to utilize the full "flight simulator" capabilities of ArcGlobe. To recap the Fly tool from Chapter 15:

  - Clicking once on Fly changes your cursor to an image of a standing bird (wearing shoes).

  - With the bird icon on the screen, click the left button of the mouse once. This will change the icon to that of a flying bird.

  - Move the mouse slowly back and forth, and you'll see your perspective on the ground change, as if you were hovering over it.

  - Also notice that in the bottom left-hand corner of ArcGlobe, the program will clock your flying speed. You'll start at 0, but each time you left-click, your speed will increase.

- Right-clicking will slow your speed and eventually return you to a speed of zero (and the bird will return to its standing/shoe-wearing appearance).

- *Important Note:* You will not be able to access any other tools in Arc-Globe until the bird is back to standing still. Keep careful track of your flying speed in the lower left-hand corner. The higher the value, the faster you will be flying forward (a positive value) or backwards (a negative value).

- If you ever fly off the screen into infinity, return to a fly speed of 0 and right-click the **campusbldgutm** layer and select **Zoom to Layer** to return to the full extent of the campus.

• Use the mouse to navigate the bird across the landscape. Experiment with flying around the area, getting comfortable with the controls and navigating around the buildings.

• For the final project for the marketing people on campus, you will be creating a video animation of your flythrough. When you are creating a flythrough you are setting up a "recording" that will track your flight across the landscape.

• Rotate the globe to where you want to begin flying.

• From the **Customize** pull-down menu, select **Toolbars** and then select **Animation**. The Animation toolbar will appear.

SOURCE: ARCGLOBE/ESRI

- Press the **Open Animation Controls** button (to the right of the camera) to open the Animation Controls dialog.

SOURCE: ARCGLOBE/ESRI

• The Animation Controls dialog will act like a DVR with play, pause, stop, and record buttons. Pressing the **Options** button will expand the dialog for more options to control the recording and playback of animations. For more information about using animations in ArcGIS, see **Smartbox 92**.

Smartbox 92

How can animations be used in ArcGIS 10.2?

While 3D visualization is very cool and can effectively communicate concepts like elevations, heights, and depths, an exported view of a scene is still just a graphic. To capture the interactivity allowed by the 3D visualization tools of ArcGlobe (or ArcScene), you can create an animation of actions occurring in the scene. An **animation** captures movements, actions, or changes to objects or layers in a video format. One simple form of animation (that you'll be creating in this chapter) is a recording of what appears in the view as you move or fly over the layers. When creating an animation, ArcGlobe treats the View as a camera—wherever you direct the View, that is what will be recorded for the animation.

You can also create more complex animations by creating **keyframes,** or "snapshots" of single views, and then ArcGIS will fill in the necessary movements between these keyframes. For instance, if you wanted to create a smooth flyby of a mountain, you could capture a keyframe of the mountain from three different angles (one at the base of the mountain, one at the top, and one around the side). When the animation is created, ArcGIS will show the view of the first keyframe at the mountain base, then the view will move to the second keyframe to show the top of the mountain, then move around the mountain to the third keyframe.

Animations also allow you to show the changes to layers of data, permitting you to look at population changes over time, watch layers slowly becoming transparent to show other layers underneath them, or follow a pre-determined camera path (such as one simulating the movement of a car down a road). Many animation options are available through the Animation toolbar (and precise control of timings and the animations tracks is available through the Animation Manager option on the Animation toolbar). After animations are created, they can be played back in ArcGIS, or exported to Audio Video Interleave (AVI) format (.avi) or Apple's Quicktime format (.mov).

animation The capture of movements, actions, or changes to layers in a video format.

keyframe Snapshots of various frames used in an animation.

- Pushing the **Record** button (the red and black circle) will cause any actions taken in the window to be recorded. For this project, you are required to record a short flythrough (about 30 seconds is long enough) of a tour of the 3D version of campus.

  - Choose your flight path carefully. Because your goal is to show off the campus, you may want to use the target, pan, and navigate tools to find an appropriate starting point (perhaps flying into the campus over the football field) and touring various locations. With the campus buildings, try for something creative instead of just a generic fly-over of campus (try to fly under the bridge if you feel adventurous).

  - Whatever you show in the View will be captured by the animation, including hesitating, starting, and stopping.

  - To expand the area seen in the View for the recording, you may want to pin extra windows (such as the Table of Contents or the Catalog) to the sides of the View.

- After you have recorded a good tour, click the **Stop** button (the black square) to turn off the DVR and stop recording.

- To view your video, press the **Play** button (the triangle). Whatever you recorded will play in ArcGlobe's view. If you're unhappy with the results (or if it doesn't highlight the areas you want, or if the flying seems too erratic), re-record a new animation. By default, ArcGlobe will overwrite the previous animation with a new one.

- Right now, your tour exists as an animation within ArcGlobe. You could save it as an animation file and it could be re-opened and viewed using ArcGlobe only. However, since this tour is intended for use in presentations and promotional efforts, it needs to be converted to a format useable by everyone (and also a format that can be easily shared or posted online).

- To begin, from the **Animation** pull-down menu, select **Export Animation**.

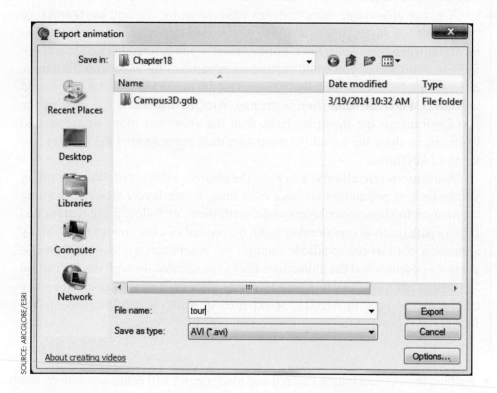

SOURCE: ARCGLOBE/ESRI

- Navigate to your **D:\GIS\Chapter18** folder as the location to which to save your video.

- Give your video a name (the graphic above calls it **tour**).

- Use **AVI** as the file type.

- Click **Export** when you're ready.

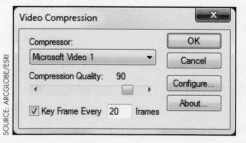

SOURCE: ARCGLOBE/ESRI

- Use the default options for Video Compression when ArcGlobe prompts you, so just click **OK**.

- ArcGlobe will start playing back your video in the View while it exports to AVI format. This may take a couple of minutes. Don't touch any controls or switch windows; just let the video play.

- When the playback is finished in ArcGlobe and the exporting is complete, open the AVI file using a video player on your computer (such as the Windows Media Player) and view your animation. If you're unhappy with it (for example, if there is too much hesitation or the flight is not smooth enough), return to ArcGlobe and re-record a new animation, then export it to a new AVI file. Answer Question 18.9.

> **QUESTION 18.9** Submit (or show) your AVI file to your instructor, who will check over your work for completeness and accuracy to make sure you get credit for this question.

STEP 18.6   Using 3D Objects in ArcGlobe

• A university campus is more than just a bunch of buildings. As a campus marketing person, you would want to showcase some unique items around the campus, as well as provide a better representation of the campus's greenery or parking areas. As such, the next visualization tools you'll use in this lab are the 3D object features. Rather than displaying vector objects like trees or cars as simple points, you'll use 3D objects to represent them in ArcGlobe.

• Add the church feature class to ArcGlobe, using the default settings for scale and appearance when it's added. Extrude it to the [Height] attribute, but this time use the **Adding it to each feature's base height** option. This is a (very) simplified version of St. John's church on Wick Avenue next to the M1 parking deck. Change the symbology so that the church polygon is a different color from the rest of campus.

• Add the trees feature class to ArcGlobe. A set of dialogs will appear for you to control the settings of the layer on the globe.

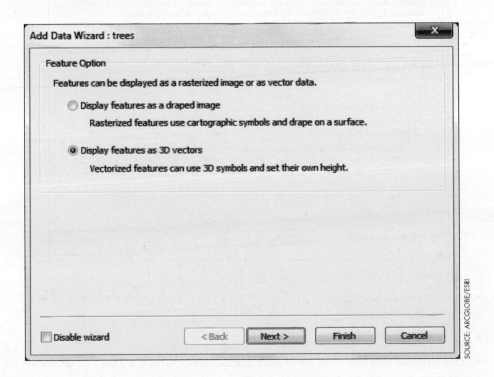

• In the Feature Option dialog, choose **Display features as 3D vectors**. Click **Next**.

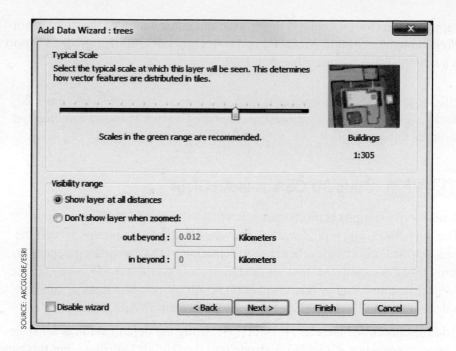

- In the Typical Scale section, slide the arrow to the option for **Buildings 1:305**.
- Make sure the **Show layer at all distances** radio button is selected. Click **Next**.

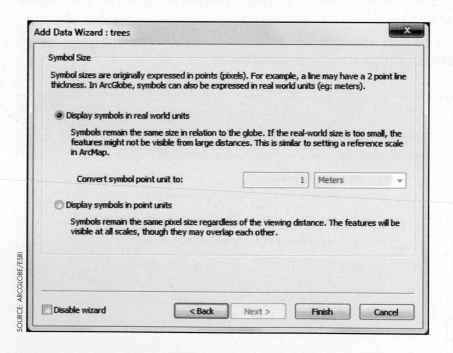

- In the Symbol Size dialog box, select the option for **Display symbols in real-world units**.
- Click **Finish**.
- You will see that the trees have been added to ArcGlobe's Table of Contents as a "Floating Layer" and drawn with a default red push-pin symbol. The first thing to do is switch the layer over to a draped layer (like your buildings) so they will sit properly on the globe.

- Right-click on the trees layer and bring up its **Layer Properties**. Select the **Elevation** tab.

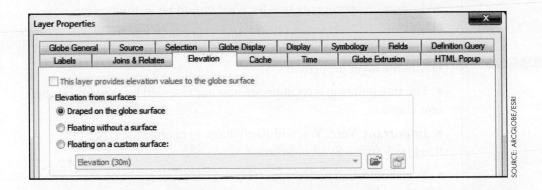

SOURCE: ARCGLOBE/ESRI

- In the Elevation from surfaces option, select the radio button for **Draped on the globe surface** (instead of a Floating option). This will fix the trees to the globe's surface. Click **Apply** then **OK**.

- Zoom in closer to the church parking lot. You should now have small spheres on the surface. The next step is to change their symbology to something that looks like trees.

STEP 18.7 Visualizing 3D Objects in ArcGlobe

- Unfortunately, creation of 3D objects such as trees isn't that simple—trying to create a tree in 3D with polygons (plus getting all of the heights correct) would be an incredibly time consuming and tedious task. To aid in this process, Esri has supplied numerous pre-made 3D objects. The trees have been represented with points, and those points can be changed into 3D objects.

- Right-click on the **trees** feature class and select **Properties**.

- Click on the **Symbology** tab. Normally this is where you would change the color or appearance of the points. Here, you will change the point into a 3D object.

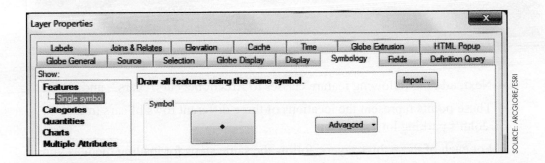

SOURCE: ARCGLOBE/ESRI

- Click on the point graphic itself.
- In the Symbol Selector box that opens, click on **Style References**.
- Select the option for **3D Trees**.

- This will expand the choices for what you want the trees to appear as. Click **OK**.

- Back in the Symbol Selector, scroll down through the list of options and select an appropriate tree for what would be in front of the church.

- Click **OK** when done to leave the Symbol Selector, and then click **OK** in the layer properties.

- The tree points in ArcGlobe are now represented by the tree symbol you chose.

- ***Important Note:*** You will likely have to change the size of the trees back in the Symbol Selector.

- You should also notice your trees "casting" shadows on the imagery—be sure to choose an appropriate style of tree to match the shadow in the image. Play around with the various tree options and the sizes until you get something that looks good and best represents the trees for that section of campus.

- Next, add the following feature classes to ArcGlobe: cars1, cars2, and cars3.

- These points represent the locations of three different rows of cars in the St. John's parking lot.

- For each of these three layers, follow the same steps for loading them as you previously did for the trees so that they are properly setting on the terrain. Use the Symbol Selector options for **3D Vehicles** to get some appropriate-looking cars.

- Choose an appropriate type of car for each of the three rows in the church parking lot.

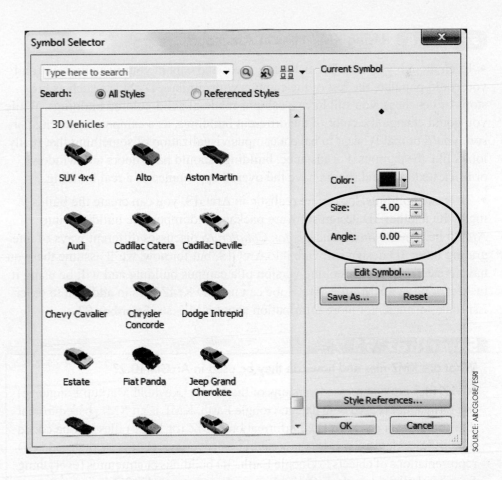

SOURCE: ARCGLOBE/ESRI

- If you need to adjust the angle the cars are parked at, return to the symbol selector for the car and adjust the values for **Angle** to change the angle of the cars in the parking spaces.

- Keep playing with the options until you get the cars parked correctly in their spaces.

- Also, adjust the **Size** of the cars so that the objects "cover" their corresponding cars on the image of the parking lot.

- Answer Question 18.10 (this will require some explanation on your part).

QUESTION 18.10 What steps would you take to create a more realistic 3D representation with respect to 1) the church, 2) cars, and 3) trees for this area?

- Note that you can also export the view of what you're seeing on the globe to an image just like you would a scene (from ArcScene—see Chapter 15). From the **File** pull-down menu, select **Export Globe** and follow the instructions to save your view as a graphic. Answer Question 18.11.

QUESTION 18.11 Submit the graphic of your final ArcGlobe 3D rendering of the church, trees, and cars to your instructor, who will check over your work for completeness and accuracy to make sure you get credit for this question.

STEP 18.8 Using KMZ Files in ArcGlobe

• Even though you have better-looking trees and cars in your visualization (and you could populate the rest of the campus with similar 3D objects such as bus stops or benches), you still have a campus made of solid-colored buildings. While you could change the color of the different buildings, as a campus marketing person, you'd probably want to have a campus visualization of something that really looks like the campus. For instance, buildings should have doors and windows, or brick textures, and should have the overall appearance of a real building.

• To make buildings look more realistic in ArcGIS, you can create the buildings with another 3D design software package and import the buildings into ArcGlobe. The *Related Concepts for Chapter 18* discusses different ways of integrating other 3D design software into ArcGIS, but for now, we'll assume that you have created a realistic-looking version of a campus building and will be using it in ArcGlobe as a KMZ file. ArcGlobe can use 3D KMZ files in addition to other Esri data formats. For more information about KMZ, see **Smartbox 93**.

Smartbox 93

What are KMZ files and how can they be used in ArcGIS 10.2?

KMZ A compressed version of a KML file.

KML Keyhole Markup Language. An XML-based file format that can hold georeferenced information and the file format used by Google Earth.

KMZ files are compressed versions of the **KML** (Keyhole Markup Language) file format that is used with data in Google Earth. KML is an XML-based format that can contain georeferenced information. KMZ (or KML) files can be added directly to Google Earth to show additional overlays or layers, or to add 3D representations of objects to Google Earth. 3D buildings or structures (everything from baseball fields to the Eiffel Tower) can be constructed in 3D design software, converted to a KML or KMZ, and then placed into Google Earth.

ArcGlobe can read KML and KMZ files that represent these 3D models (see Figure 18.4 for an example). In the design process, the geolocation of the building or structure being modeled is included so that, when the resultant KMZ is placed into Google Earth or ArcGlobe, the software knows where on Earth's surface it should end up and where it should be relative to the terrain. While there are several different 3D design programs available on the market, a common one

FIGURE 18.4 A KMZ version of the 3D Model of YSU's Stambaugh Stadium in ArcGlobe.

SOURCE: ARCGLOBE/ESRI

used for creating geospatial data is **SketchUp** (owned and distributed by Trimble). Realistic, detailed, and georeferenced models can be quickly designed using SketchUp, converted to KMZ format, and placed into ArcGlobe. The 3D KMZ of the Courtyard Apartments at YSU was created in this way as part of the YSU Campus 3D Model. By using SketchUp, you can create much more detailed and textured 3D representations of buildings and objects than you can through simple extrusion or symbology changes. For more about SketchUp (including downloading a free version of it) see the *Related Concepts for Chapter 18*.

> **SketchUp** A 3D design software owned and distributed by Trimble.

- From the **Customize** pull-down menu, select **Toolbars**, then select **KML**.

- The KML toolbar will appear:

SOURCE: ARCGLOBE/ESRI

- The first button on the KML toolbar is the **Add KML/KMZ** button. Press this button to add a new KMZ file. You'll be adding a 3D KMZ file that is a representation of the YSU Courtyard Apartments. This was created for the YSU 3D Campus Model, but the same file can be used in ArcGlobe. From the D:\GIS\Chapter18 folder, select the **Courtyard_ApartmentsGE4** file.

- The KMZ representation of the Courtyard Apartments will be added to the globe. Navigate and zoom over to it to see the apartments. Navigate, pan, and zoom around the apartment complex and check out the detail that is available for 3D representations created using SketchUp.

- Next, bring up the Animation tools (like you did in Step 18.5) and create a short (maybe 15–20 seconds) flythrough of the Courtyard Apartments and export it to an AVI file. Think of this as a short GIS promotional video of the apartments to show off their features in 3D. Answer Question 18.12.

> **QUESTION 18.12** Submit (or show) your AVI file of the Courtyard Apartments marketing video to your instructor, who will check over your work for completeness and accuracy to make sure you get credit for this question.

STEP 18.9 Saving and Sharing Your Results

- When you've finished putting together your animations, from the **File** pull-down menu, select **Save**. ArcGlobe will save your work as a **3D document** file (that has the extension .3dd). The 3D document is similar to a Map document in that it contains links to the various data layers used in the globe. 3D documents can be opened in ArcGlobe just like Map documents in ArcMap.

- You can also share your results with others via ArcGIS Online by saving your various layers as a **Layer Package**. In the Table of Contents, select all of the layers to combine in a Layer Package (hold down the **Ctrl** key on the keyboard while selecting to choose multiple layers). When you have selected all of the layers you want, right-click on one of them and select **Create Layer Package**. You can also include your AVI files as Additional Files when creating a Layer Package in ArcGlobe. Once the package is created, you can upload it to your ArcGIS Online account (or save it on the hard drive). To reopen a Layer Package in ArcGlobe, add it from ArcGIS Online, or if it's already downloaded, drag it from Catalog into ArcGlobe's Table of Contents.

> **3D document (.3dd)** The file that contains all the information about data layers in a globe.
>
> **Layer Package** A file containing a series of layers that can be shared via ArcGIS Online.

Closing Time

This chapter examined how geospatial data (such as trees, cars, bridges, and buildings) can be represented in 3D in ArcGIS 10.2 using the ArcGlobe environment. 3D GIS visualization is an excellent method for communicating and presenting geospatial concepts. Much more can be done with 3D visualization in ArcGIS. See the *Related Concepts for Chapter 18* for information about creating more realistic 3D data to use in ArcScene or ArcGlobe.

Starting in the next chapter, we'll be working with measuring distances as a surface, which requires understanding and using many concepts for modeling and problem solving, particularly with shortest paths over a landscape. You'll also be able to examine the results of your analysis in ArcScene (or ArcGlobe) because 3D visualization is also an ideal format for presenting the results of analysis. For example, an interactive 3D version of the wetlands in an area slated for construction of a new shopping center or a 3D visualization of a shortest path over mountainous terrain can carry more visual impact than a simple printed map.

Related Concepts for Chapter 18   Using More Detailed 3D Models in ArcGIS 10.2

As noted in **Smartbox 93**, there are several 3D design programs available for creating more detailed models to use in KMZ format in ArcGlobe. An easy-to-use design program is SketchUp (to download a free version, visit: http://www.sketchup.com). With SketchUp, you can start with georeferenced imagery, digitize building footprints, extrude and shape polygons, and add realistic textures and other features to create a rich 3D visualization of a building. SketchUp also gives you the option to start with a photo of a building and use the photo's dimensions to create the 3D model. SketchUp models can be as basic or complex as you want them to be and also allow you to create stand-alone 3D objects such as trees, cars, sculptures, statues, and the like.

Unfortunately, SketchUp models are not directly compatible with ArcGIS 10.2. However, you can easily integrate your SketchUp results into your ArcScene or ArcGlobe visualizations by exporting the SketchUp model to the **COLLADA** format. COLLADA (Collaborative Design Activity) acts as an "intermediate" file format to allow 3D models to move from one platform to another. For example, you could design your house as a 3D model using SketchUp, convert the model to COLLADA format, then use the COLLADA version in ArcScene or ArcGlobe.

ArcGIS uses COLLADA files in conjunction with the multipatch file format. As noted at the beginning of this chapter, multipatches are true 3D representations of objects. A multipatch object can hold volume and can contain multiple z-values for each x and y coordinate. As such, the complex 3D structures created using SketchUp can be represented in ArcGIS as multipatches. You would first convert your extruded buildings to multipatches. Then, using the 3D editing tools (on the 3D Editing toolbar) for ArcScene or ArcGlobe, you would select one of the multipatch buildings. Using the **Replace With Model** option, you can then exchange the multipatch blocks that comprise a building with the COLLADA version of that building created in SketchUp. ArcGIS will then have the fully visualized version of the building stored as a multipatch

COLLADA Collaborative Design Activity. An XML-based file type (with the extension .dae) that works as an interchange format from one kind of 3D modeling file type to another.

FIGURE 18.5 ArcGlobe renderings of extruded buildings as a multipatch versus imported COLLADA building as a multipatch—Bliss Hall at YSU Campus.

SOURCE: ARCGLOBE/ESRI

(Figure 18.5). By going through this process this for all of the campus buildings, you can create a rich, detailed, 3D version of the campus. This technique can be especially useful if you're working in 3D visualization in ArcScene because it does not directly support adding files to it in the KMZ format the way that ArcGlobe does.

For More Information

For further in-depth information about the topics presented in this chapter, use the ArcGIS Help feature to search for the following items:

- A quick tour of creating animations
- About using extrusion as 3D symbology
- About using extrusion as 3D symbology in ArcGlobe
- An overview of the To Collada toolset
- Creating a 3D layer package using ArcGlobe and ArcScene
- Creating keyframes of camera or map view properties
- Creating textured buildings from models
- Displaying 3D multipatch features
- Essential 3D Analyst vocabulary
- Essential animation vocabulary
- Extruding features by an attribute value
- Multipatch To Collada (Conversion)
- Obtaining elevation information for building footprints
- Publishing a globe service
- Using 3D styles to assign symbology
- What is an animation?
- What is KML?

For more information about 3D modeling and visualization, see the blog articles by Esri's Ken Field entitled "Making a large-scale 3D map" available online here:

- Part 1: http://blogs.esri.com/esri/arcgis/2012/01/05/making-a-large-scale-3d-map-part-1

• Part 2: http://blogs.esri.com/esri/arcgis/2012/01/10/making-a-large-scale-3d-map-part-2

• Part 3: http://blogs.esri.com/esri/arcgis/2012/01/12/making-a-large-scale-3d-map-part-3

For more information about multipatches and 3D objects, see the following article: Ford, A. 2007. "Visualizing Integrated Three-Dimensional Datasets." *ArcUser*. January–March. Available online at http://www.esri.com/news/arcuser/0207/files/petro.pdf.

Key Terms

Esri CityEngine (p. 411)

ArcGlobe (p. 411)

multipatch (p. 412)

extrusion (p. 417)

offset (p. 421)

animation (p. 423)

keyframe (p. 423)

KMZ (p. 430)

KML (p. 430)

SketchUp (p. 431)

3D document (.3dd) (p. 431)

Layer Package (p. 431)

COLLADA (p. 432)

How to Utilize Distance Calculations in ArcGIS 10.2

Introduction

So far, we've discussed many types of spatial analysis that relate to proximity (how far certain things are away from other things). We've used spatial joins, selection by location, network analysis, and buffers as measures of proximity, but we haven't yet discussed how to use ArcGIS to find the distance between locations not on a transportation network (aside from small uses of the Measure tool and one type of spatial join).

Let's start with an example. If you wanted to determine the distance from all locations in a metropark to the closest first-aid station, the tools we've used so far won't help you do that. What you need instead is some sort of a park map in which each location on the map is marked with the distance from that spot to the first-aid station(s) closest to it. However, a park has a nearly infinite number of locations—for instance, the park entrance, a spot one foot away from the park entrance, another spot two feet away from the entrance, and so on—and measurements need to be made from all of these to the closest aid station. Thus, we can think of a distance map as a continuous surface, where each location has a value denoting how far that location is from the nearest first-aid station. From previous chapters, we know we can represent this kind of continuous surface as a raster, and that's how ArcGIS represents distances—as raster surfaces, with each grid cell containing a value for the distance to the feature(s) we're trying to represent.

A distance raster is created by determining how far each cell is from a feature (such as the aid stations). Distance in a raster can be calculated as movement across a surface of grid cells from an origin cell (or several origin cells) to all other cells in the raster. Because this movement is measured in square blocks, these movements can either be lateral or diagonal.

Figure 19.1 shows movement across a raster from an origin point (the cell marked *A*). Movements are made from the center of the origin cell to the center of an adjacent cell. Moving in a lateral direction (up, down, left, or right) to an adjacent cell costs 1 unit and moving in a diagonal direction costs 1.414 units (because of the geometry measurements of the Pythagorean theorem). Moving to another cell adds an accumulative cost of either another 1 unit (for a lateral move) or another 1.414 units (for a diagonal move). In Figure 19.1, the distance from the origin to the grid cell marked *B* is 2 units away, while the distance from the origin to the grid cell marked *C* is 2.828 units.

FIGURE 19.1 The lateral and diagonal movement (and accumulation of distance) across a raster surface.

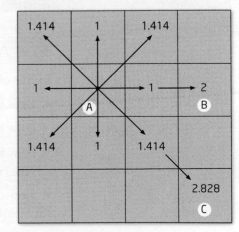

Each one of these grid cells represents geospatial data. As such, each grid cell represents a real-world size based on the raster resolution (see Chapter 12). For instance, if the raster in Figure 19.1 was 30 meter resolution, then the measured distance from the center of origin cell *A* to the center of destination cell *B* would be 60 meters (2 × 30 meters). Similarly, the measured distance from the center of origin cell *A* to the center of destination cell *C* would be 84.84 meters (2.828 × 30 meters). In this chapter, we'll expand this basic concept to use these kinds of distance measurements to determine the shortest path along a raster surface (such as finding the shortest distance from every grid cell location in a park to the grid cells representing the location of the first-aid stations).

Chapter Scenario and Applications

In this chapter, you will be taking the role of a ranger in the Hocking Hills region of Hocking County, Ohio, trying to find the shortest path from the ranger station to a group of stranded hikers and also to a separate campsite emergency. In the analysis, you'll need to take into account certain conditions in the park environment that will influence the overland path you'll have to take to rescue the hikers and get to the emergency site. These conditions include the park's terrain and land cover.

The following are additional examples of other real-world applications of this chapter's skills:

• An engineer is attempting to determine the best location for a new road through unpaved terrain. He needs to take into account factors such as the slope of the landscape when figuring out the "least cost" path of the new road.

• A park manager is laying out potential new trails for land acquired by a park to provide new routes for tourists and hikers. She can include factors such as the steepness of slopes and land cover in relation to determining the locations (and difficulty ratings) of these new paths.

• A game-control warden at a national park wants to identify likely areas for animal migrations and movements within the park's boundaries to aid in closing certain areas from tourists at certain times of the year. He can incorporate various landscape-based features of the park in determining the likely paths of animal movements.

ArcGIS Skills

In this chapter, you will learn:

• How to calculate a Euclidean distance surface.

• How to calculate an allocation grid.

• How to create a cost-distance surface (using different rasters for the various costs as different constraints on the problem).

• How to calculate a least-cost path based on the different costs for traversing each cell.

• How to reclassify the values of a raster to create a new raster with new values.

- How to use the Slice tool to create a new raster based on a classification scheme.

- How to examine the results of cost analysis in 3D.

 ## Study Area

For this chapter, you will be working with data from the Hocking Hills State Park region located in Hocking County, Ohio.

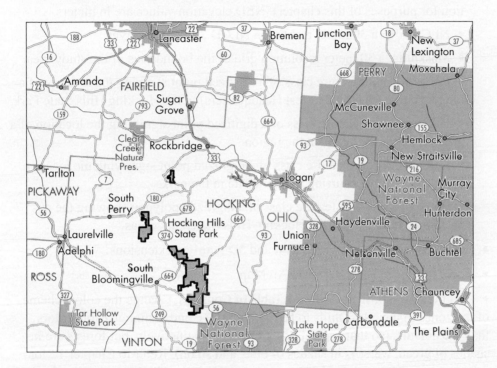

 ## Data Sources and Localizing This Chapter

This chapter's data focus on features and locations within the Hocking Hills State Park region. However, you can easily modify this chapter to use data from your local area instead (for instance, Petoskey State Park in Michigan).

The NED was downloaded from the USGS National Map. The NLCD 2006 dataset was downloaded (for free) from the Multi-Resolution Land Characteristics Consortium (MRLC), which is online at http://www.mrlc.gov. NLCD 2006 data for the entire United States can be downloaded from that Website and the data for your individual county (such as Emmet County, Michigan, where Petoskey State Park is located) can be clipped out using ArcGIS. Alternatively, NLCD 2006 data for a county can also be downloaded directly via The National Map. Both the NED and NLCD grids were clipped to fit the approximate boundaries of the study area.

The other layers of areas of the park were digitized using ArcGIS and the maps and information available from the Ohio Department of Natural Resources at http://parks.ohiodnr.gov/hockinghills. Other state parks will likely have similar maps and Websites available through your state's DNR. For instance, maps and information about Michigan's Petoskey State Park are available through the Michigan Department of Natural Resources at http://www.michigandnr.com/parksandtrails/details.aspx?id=483&type=SPRK.

STEP 19.1 Getting Started

• Start ArcMap and use the Catalog to copy the folder called **Chapter19** from the C:\GISBookdata\ folder to your own D:\GIS\ drive. Be sure to copy the entire directory, not just the files within it.

• Chapter19 contains a file geodatabase called **hockinghills** that contains the following items:

  • NEDhills: a 1 arc-second NED raster layer (resampled to 30 meter resolution for purposes of this chapter). NED elevation values are in meters.

  • NLCDhills: a 30 meter resolution NLCD 2006 raster layer.

  • Maskhills: a 30 meter resolution grid of the boundaries of the study area.

  • Ranger: a point feature class of a digitized point showing the location of a ranger station (the State Forest Headquarters) in the Hocking Hills State Park.

  • Hikers: a point feature class of a digitized point signifying the location of a group of stranded hikers in the region.

  • Campers: a point feature class of a digitized point signifying the location of an emergency occurring at a campsite in the park.

  • Bounds: a polygon feature class of a digitized polygon showing the approximate boundary of the Hocking Hills State Park areas.

• Activate both the **Spatial Analyst** and **3D Analyst** extensions.

• Add the **3D Analyst toolbar** to ArcMap.

• Add the **NLCDhills** layer to the Table of Contents. Change the color scheme of the zones in the NLCDhills to be more intuitive and distinguishable colors than the defaults (so that each class can be seen distinctively, but forests are in shades of green, water is in shades of blue, and so on). Refer to Chapter 12 for information as to which land-cover classes each of the NLCD grid-cell values represents.

• Add the **OpenStreetMap** basemap to put your study area into context.

• Next, use the following Geoprocessing Settings for this chapter (from the **Geoprocessing** pull-down menu, select **Environments...**):

  • Under the **Workplace** options, for **Current Workspace**, navigate to **D:\GIS\Chapter19**.

  • Under the **Workspace** options, for **Scratch Workspace**, navigate to **D:\GIS\Chapter19**.

  • Under the **Output Coordinates** options, for **Output Coordinate System**, select **Same as Layer "NLCDhills"**.

  • Under the **Processing Extent** options, for **Extent**, select **Same as Layer NLCDhills**.

  • Under the **Raster Analysis** options, for **Cell Size**, select the option for **Same as Layer NLCDhills**. This should automatically adjust the cell size to 30 (as in 30 meters).

  • Click **OK** when you're done. All of your Environment Settings should now be in place for use in the rest of the chapter.

- The Coordinate System being used in this chapter is **NAD 1983 (Albers)**, using Map Units of **Meters**. Check the Data Frame to verify that this coordinate system is being used.

STEP 19.2 Calculating a Euclidean Distance Surface

- This chapter has you working at a ranger station in the Hocking Hills region and receiving a call that hikers are stranded somewhere in the region. What you want to do is calculate the shortest overland path to reach them from the ranger station.

- To get started, add the NEDhills layer to the Table of Contents.

- Add the ranger point feature class and the hikers point feature class layers to ArcMap. Each layer consists of a single point, so those points may be difficult to see at a wide scale. Adjust the colors and symbology of each layer and zoom in to see where the two points (the ranger station and the stranded hikers) are located.

- The first calculation you'll make is a Euclidean distance calculation to determine the accumulated cost distance between the ranger station and the hikers. See **Smartbox 94** for more about using Euclidean distance in ArcGIS.

Smartbox 94

How is Euclidean distance calculated and used in ArcGIS 10.2?

A **Euclidean distance** measurement represents the straight-line distance between two points. Rather than accumulating the driving distance along many roads/edges in a network between two destinations (as in Chapter 11), Euclidean distance can be compared with taking a helicopter directly from one point to another. The old saying about "the shortest distance between two points is a straight line" applies to Euclidean distance measurements.

    In ArcGIS 10.2, Euclidean distance is applied as the measurement of a straight line between the centers of two grid cells. The Euclidean Distance tool will calculate this straight-line distance from each source cell to all other cells in the raster, with each cell assigned the shortest distance from a source. However, if the source from which you're trying to find Euclidean distance is a point, line, or polygon feature class, those features are first converted to a raster format (at whatever cell resolution is being used in the distance calculation). For instance, if you have a layer of all the lakes in the county and want to find how far each location in the county is from a lake, the Euclidean Distance tool will convert the lake polygons to raster format, then compute the straight-line distance from the cells that comprise the lakes to all other cells within the extent. The output of the Euclidean Distance tool will be a new surface in which every cell has a real-world value for the shortest straight-line distance from the source features/cells to each other cell in the raster.

    As mentioned in the Introduction, because grid cells are square and distances are calculated from the center of cells, the Pythagorean theorem is used to determine the Euclidean distance between cells. The Pythagorean theorem measures the sides and the hypotenuse (the diagonal) of a right-angle triangle as: $c^2 = a^2 + b^2$. See Figure 19.2—if the two sides of the triangle (a and b) are both equal to a value of 1, then the hypotenuse of the triangle (c) has a value of 1.414 ($c^2 = 1^2 + 1^2$, or $c = \sqrt{2}$, or 1.414). The application of the Pythagorean

Euclidean distance The straight-line distance between two points.

FIGURE 19.2 The basic setup of a triangle using the Pythagorean theorem.

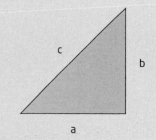

FIGURE 19.3 An example of Euclidean distance calculation of raster distance.

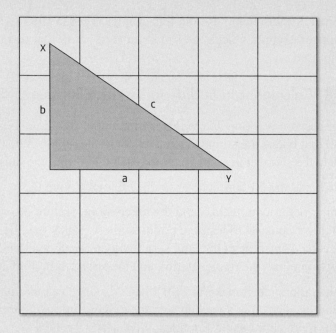

theorem explains why lateral movement on a raster has a value of 1, while a diagonal movement has a value of 1.414.

This same basic measurement is applied when calculating Euclidean distances between grid cells. For example, see Figure 19.3. What we want to do is determine the shortest distance between the source cell (X) and every other cell in the raster. To find the distance between cell X and cell Y, we use the Pythagorean theorem as follows: a (the bottom of the triangle) has a value of 3 (because it is three units between the centers of the grid cells on the bottom of the triangle), while b has a value of 2 (because it is two units between the centers of the grid cells on the side of the triangle). Thus, the value of c would be found as ($c^2 = a^2 + b^2$) or ($c^2 = 3^2 + 2^2$) or ($c^2 = 13$). Thus, c equals the square root of 13, or 3.61. The shortest distance between cell X and cell Y would thus be 3.61 units. As noted previously, the raster resolution determines the real-world size of the grid cells, so if the raster in Figure 19.3 was 30 meter resolution, the real-world Euclidean distance between cell X and cell Y would be 3.61 × 30 meters, or 108.3 meters.

For an example of calculating a Euclidean distance surface in ArcGIS 10.2, see Figure 19.4. An initial feature class representing streams in Columbiana County, Ohio, is used as the input to the Euclidean Distance tool. The resulting output will be a raster surface in which every grid cell has a value for the straight-line distance from any stream in the county. The raster surface's values reflect the real-world measurements of distance, based on the specified grid-cell size used by the tool.

FIGURE 19.4 A feature class of Columbiana County, Ohio, streams and a raster surface of Euclidean distance from each stream.

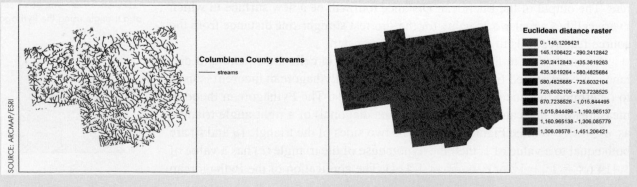

Columbiana County streams

— streams

Euclidean distance raster

- 0 - 145.1206421
- 145.1206422 - 290.2412842
- 290.2412843 - 435.3619263
- 435.3619264 - 580.4825684
- 580.4825685 - 725.6032104
- 725.6032105 - 870.7238525
- 870.7238526 - 1,015.844495
- 1,015.844496 - 1,160.965137
- 1,160.965138 - 1,306.085779
- 1,306.08578 - 1,451.206421

- Open ArcToolbox. From the **Spatial Analyst Tools** toolbox, select the **Distance** toolset, then select the **Euclidean Distance** tool.

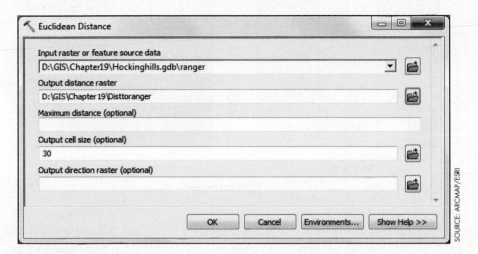

- Select **ranger** for the Input raster or feature source data.

- Call the Output distance raster **Disttoranger** and save it in your D:\GIS\Chapter19 folder.

- Leave the other options blank (but make sure that the cell size is **30**).

- Click **OK** to calculate the distance surface.

- In the new Disttoranger grid, each grid cell will have a value that represents the distance from the state forest headquarters (HQ) to every cell in the grid. Answer Question 19.1.

QUESTION 19.1 Zoom in closely on the grid cell on which the hikers' point lies. Use the Identify tool to obtain the grid-cell value of the Disttoranger grid at that location. How far (using the value from the Euclidean distance grid) are the hikers from the state forest HQ?

- Next, you will calculate a Euclidean allocation grid, which will show each cell's value that represents the closest source (the state forest HQ) to which it is allocated (see **Smartbox 95** for more about allocation).

Smartbox 95

What is allocation and how is it used in ArcGIS 10.2?

The Euclidean Distance tool allows you to find the shortest straight-line distance to each cell from a source. However, if you have multiple sources, that Euclidean distance surface will show the shortest distance from any of the sources. By looking at the value of an individual cell, you can't tell to which of the sources it is closest. It may be more useful knowing to which source a particular cell is nearest. For instance, if an accident occurs in a park, knowing which of several different first-aid stations is closest to the incident location is going to be critical.

An **allocation** process will create a new grid that shows to which of the sources a cell is closest. For example, see Figure 19.5. The Source_Ras grid has two different sources (these could be the first-aid stations in a park), with

allocation The determination of which source is closest in distance to each grid cell.

FIGURE 19.5 The Euclidean Allocation calculation in ArcGIS 10.2, used for assigning grid cells to their nearest source cell.

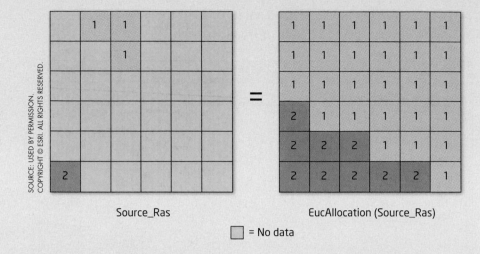

Source_Ras   EucAllocation (Source_Ras)

☐ = No data

each source consisting of one or more cells. The Euclidean Allocation tool will calculate the Euclidean distance to each cell (as described in **Smartbox 94**), but rather than assigning the distance value to the grid cell, it will assign a value indicating to which source that cell is closest (based on the Euclidean distance to that cell). The EucAllocation grid in Figure 19.5 shows the output of the allocation process. Each cell has a value of 1 if it is closest to source #1 and a value of 2 if it is closest to source #2.

- From the **Spatial Analyst Tools** toolbox, select the **Distance** toolset, then select the **Euclidean Allocation** tool.

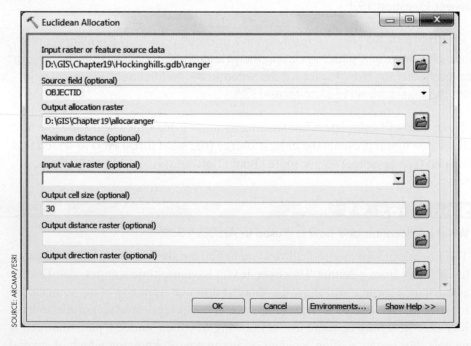

- Select **ranger** for the Input raster or feature source data.
- Call the Output allocation raster **Allocaranger** and save it in your D:\GIS\Chapter19 folder.
- Leave the other options blank (but again make sure that the cell size is **30**).

- Click **OK** to calculate the allocation surface.

- Examine this new "Allocation to Ranger" grid. Answer Question 19.2.

> **QUESTION 19.2** Why does every cell of the "Allocation to Ranger" grid have the same value?

STEP 19.3 Shortest Paths over Raster Surfaces

- Now you will set up the constraints on the problem in terms of the "cost" of moving across the surface (see **Smartbox 96** for more about cost distances). In this case, the accumulated distance to travel will be the "cost" of traveling from the state forest HQ to the stranded hikers.

Smartbox 96

How is cost distance used in ArcGIS 10.2?

In ArcGIS 10.2, the **cost** of moving from cell to cell is the number of units it takes to do so. For instance, a basic lateral movement will cost 1 unit, while a diagonal movement will cost 1.414 units. If you move three grid cells in a lateral direction, the total amount of movement (or the **accumulated cost**) will be 3 units. If you travel two grid cells laterally, then one grid cell diagonally, the accumulated cost will be 3.414 units. However, these basic costs assume that nothing is impeding the progress of movement from one cell to another and the cost of moving from one cell to another will always remain at constant values of 1 or 1.414. This assumption is not always true in reality.

For example, if you were hiking in a park, a flat section of land would take less exertion (and time) to cross than hiking up a steep hill. Similarly, walking across a paved section of road would take less exertion or time than walking through dense forest or over rocky, uneven terrain. If you were simulating movement through a park to a first-aid station, many things could slow your progress; navigating through a forest, climbing steep terrain, or fording streams would all add to the time it would take to reach the first-aid station. Thus, if you were trying to find the shortest way to the first-aid station, you'd want to take these factors into account and travel along the path that impedes your progress the least.

In ArcGIS 10.2, you can simulate these kinds of factors by assigning additional costs (or impedances) to traveling from cell to cell. For instance, rougher terrain or swampy land may have a higher cost value across which to move. Grid cells with steeper slopes can be assigned a higher cost value. Similarly, different types of land cover can be assigned higher cost values depending on the difficulty of traversing them. If you have multiple grids of different costs (for instance, one grid of higher costs based on slope conditions and another grid of higher costs based on land-cover conditions), they need to be combined into a single cost grid for use in ArcGIS 10.2.

The number of units (x) for laterally moving from one grid cell to an adjacent one is calculated as:

$$x = 1 \times \frac{(\text{cost of starting grid cell}) + (\text{cost of ending grid cell})}{2}$$

cost The number of units it takes to move from one grid cell to the next.

accumulated cost The total cost of traveling from a source to a grid cell.

The number of units (x) for diagonally moving from one grid cell to an adjacent one is calculated as:

$$x = 1.414 \times \frac{(\text{cost of starting grid cell}) + (\text{cost of ending grid cell})}{2}$$

For an example of how this all works in practice, see Figure 19.6, which shows a sample raster with additional costs assigned to each grid cell. The cost to move laterally from the cell in the upper-left corner (with a value of 5) to the adjacent cell on its right (with a value of 1) would be 1 × (5 + 1)/2, or 3. The cost of diagonally moving from the cell in the upper-left corner (with a value of 5) to its diagonally lower-right cell (with a value of 6) would be 1.414 × (5 + 6)/2, or 7.77. (Note that if values of 1 are used in the starting and ending cells, the calculated costs will be values of 1 for lateral movements and 1.414 for diagonal movements.)

FIGURE 19.6 A sample raster of various cost values and the corresponding computed lateral and diagonal movement costs.

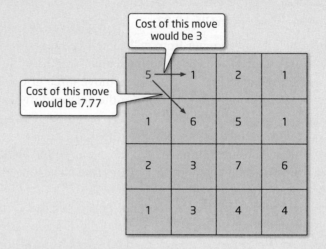

The Cost Distance tool will use this method to calculate a new grid of output values, with each one containing the lowest accumulative cost value of movement from one cell to another. The algorithm used by the Cost Distance tool requires an initial set of sources (and if these sources are feature classes, they will be internally converted to a raster format for use) and a raster of the cost values associated with each grid cell (Figure 19.7).

FIGURE 19.7 The two layers necessary for computing cost distance—a layer of source cells from which to find the distance and a second layer of the costs associated with each cell.

Source raster

1	3	4	4	3	2
4	6	2	3	7	6
5	8	7	5	6	6
1	4	5		5	1
4	7	5		2	6
1	2	2	1	3	4

Cost raster

☐ = No data

The first step of the cost-distance algorithm is calculating the cost of moving from each source cell to its adjacent cells. The cost formulas listed above are used (Figure 19.8). Thus, the cost of moving from the source cell labeled with a "2" to the adjacent cell on its right would be a lateral move with a cost of 1.5, or $(1 \times (1 + 2)/2)$. The cost of moving from that same source cell to the cell diagonally adjacent to it would be 5.7, or $(1.414 \times (1 + 7)/2)$. These new values are the total accumulative costs of moving from a source cell to adjacent cells. ArcGIS 10.2 will keep track of a list of all cells that now have an accumulated cost computed for them.

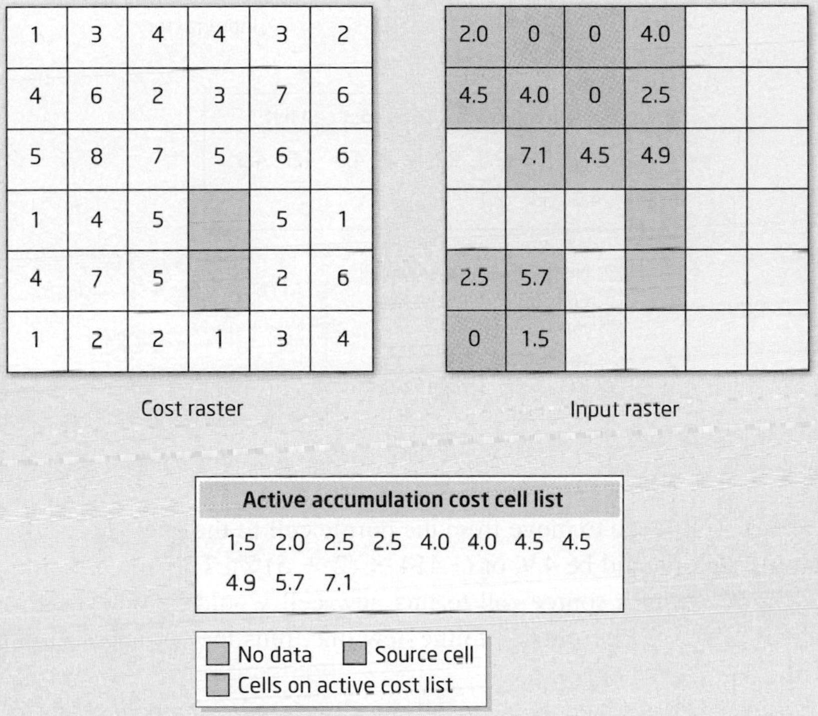

Cost raster

1	3	4	4	3	2
4	6	2	3	7	6
5	8	7	5	6	6
1	4	5		5	1
4	7	5		2	6
1	2	2	1	3	4

Input raster

2.0	0	0	4.0		
4.5	4.0	0	2.5		
	7.1	4.5	4.9		
2.5	5.7				
0	1.5				

FIGURE 19.8 The movement costs computed for cells adjacent to all source cells and the initial active accumulated cost list that is generated.

Active accumulation cost cell list

1.5 2.0 2.5 2.5 4.0 4.0 4.5 4.5
4.9 5.7 7.1

□ No data   ■ Source cell
□ Cells on active cost list

The source cells are now considered "exhausted," as there is no need to perform any further cost-distance calculation from them. What the algorithm needs to do now is select a new grid cell from which to begin calculating cost distance. It will select the cell that corresponds with the lowest value on the active accumulated cost list and use that as the new cell to use as a new starting point for calculations (Figure 19.9). In this case, the grid cell that corresponds with the lowest active accumulative cost value of 1.5 is chosen (shown in purple).

Cost distances from this cell to its adjacent cells are calculated using the values from the initial cost raster. For instance, the lateral movement from the purple cell (with a value of 1.5) would be 2, or $(1 \times (2 + 2)/2)$. The accumulative cost value of the purple cell that's being used for the calculations (1.5) is added to this value of 2 (because it has already cost 1.5 units to travel from a source cell to the purple cell). Thus, the final accumulative cost of moving laterally to the cell to the right of the purple cell would be 3.5 (movement cost of 2 plus the already accumulated cost of 1.5).

FIGURE 19.9 The computed accumulative cost values for the cells connected to the new searching cell.

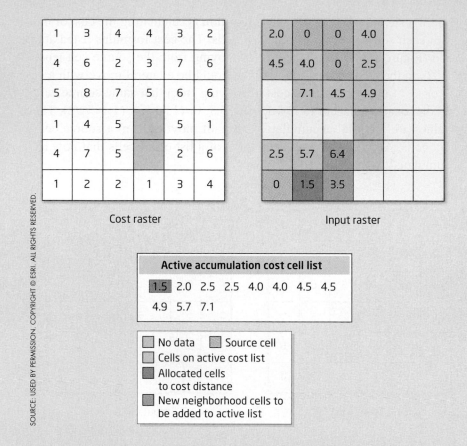

The cost distance to move from the purple cell to the one diagonally adjacent to its right would be 4.9, or $(1.414 \times (2 + 5)/2)$. The total accumulative cost to move from a source cell to this new cell would be 6.4 (4.9 to move diagonally from the purple cell to the new one, plus the already accumulated cost of 1.5).

If a cost-distance calculation results in a lower value than one that's already been calculated for a cell, this new lower value is placed into the cell instead. The higher value is removed from the active accumulative cost-cell list, and this new lower value is placed there instead. Because we're trying to find the least-cost value that it takes to reach each cell from a source, these newly calculated lower values represent movement options that are shorter (or "cost less") than previously computed options.

Because we've looked at all possible connections from the purple cell, it will be removed from the active accumulation cost-cell list and will now be considered "exhausted." The newly calculated accumulative cost values of 3.5 and 6.4 will be added to the accumulative cost-cell list in its place.

This process continues until all cells on the active list are searched from and calculations made from them. The end result of this process will be a new raster in which each cell contains a value representing the shortest calculated cost distance (or the "least cost" value) of traveling from a source location to that cell (Figure 19.10). The Cost Distance tool will also generate another raster, called a *backlink*, which is used in determining the shortest path from a source to each cell.

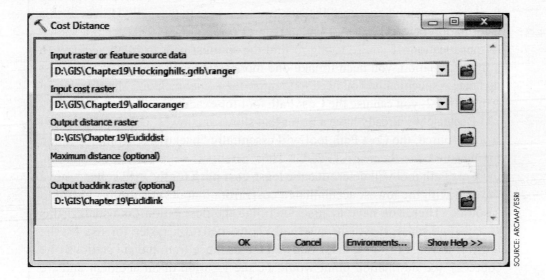

Cost raster

Cost distance output values

FIGURE 19.10 The initial cost raster and the final computed cost distance raster, showing the least cost of moving from a source cell to each other cell.

- From the **Spatial Analyst Tools** toolbox, select the **Distance** toolset, then select the **Cost Distance** tool.

- Select **ranger** for the Input raster or feature source data.

- Select **allocaranger** for the Input cost raster.

- Call the Output distance raster **Eucliddist** and save it in your D:\GIS\Chapter19 folder.

- Call the Output backlink raster **Euclidlink** and save it in your D:\GIS\Chapter19 folder.

- Leave the other option blank.

- Click **OK** to calculate the cost distance based on these parameters.

- In Eucliddist, each cell will have a value that represents the cost distance from the state forest HQ to every cell in the grid. Answer Questions 19.3 and 19.4.

QUESTION 19.3 The "allocaranger" grid has a value of 1 for each grid cell. Why was this value used for the "input cost raster" in this calculation? What is the cost of traversing each grid cell in this step?

QUESTION 19.4 What (specifically) were the constraints (in terms of cost-distance factors) placed on this shortest-path calculation?

- The last step is to calculate the shortest overland path from the state forest HQ to the hikers. In ArcGIS, this is done by calculating a least-cost path (see **Smartbox 97** for more information on how a least-cost path is calculated). From the **Spatial Analyst Tools** toolbox, select the **Distance** toolset, then select the **Cost Path** tool.

Smartbox 97

How is a cost path calculated in ArcGIS 10.2?

The raster created by the Cost Distance tool has grid-cell values that represent the lowest accumulative cost distance from a source to that cell. However, this information does not tell you which path of cells to take to get from a source to that location. If you're trying to find the shortest path from an accident to a first-aid station, you need to take one more step beyond generating the cost distance—specifically, you need to identify the actual path you should take. In ArcGIS 10.2, you can use the Cost Path tool for exactly this purpose.

Because we already have a raster that shows the lowest cost to reach a cell from a source, the Cost Path tool will essentially "backtrack" from a location on the surface to the nearest source. Thus, after you've selected a destination, the Cost Path tool will determine the **least-cost path** (or the grid cells to move through with the lowest accumulated cost) from that destination backward to a source. Thus, you have to first determine the destination (as usual, if this is a feature class, it will be internally converted to a raster for use by the tool). If the cost-distance raster showed the distance from the aid stations (the sources), the destination point would be the location of an accident and we will determine the path backward from the accident destination to the first-aid station source.

This process is enabled by the use of the backlink grid that is also part of the output from the Cost Distance tool. The **backlink** grid will examine the cost grid and compute which direction you would need to move from each cell to reach a neighboring cell with the least cost to continue toward the nearest source. Each cell of the backlink grid is then coded with a number from 1 through 8, with each value indicating what direction the movement should take. Figure 19.11 shows the assigned values and their backlink direction positions. For instance, if movement from the destination cell in a least-cost manner toward a source should be to the north, a value of 7 is saved for that cell in the backlink grid. Or, if the movement should go to the southeast, a value of 2 is saved for that cell in the backlink grid. A value of 0 indicates a source cell (and thus no movement is necessary).

least-cost path The set of grid cells that represents the lowest value for accumulated cost to move from a source to a destination.

backlink The grid that shows the direction to be traveled in for the least-cost amount of movement from cell to cell.

FIGURE 19.11 The specific values that correspond with cardinal directions in the backlink grid.

6 = NW	7 = N	8 = NE
5 = W	0	1 = E
4 = SW	3 = S	2 = SE

The Cost Path tool will then use the directions from the backlink grid to "trace" the least-cost path from the destination back to the source. For instance, if the destination is at a cell with a backlink value of 5, then movement to the west is indicated. That cell to the west becomes part of the least-cost path. The cell to the immediate west of that one has a value of 3, indicating movement to the south. Thus, the cell to the south becomes the next grid cell in the least-cost path. This process continues until a source is reached. All of the adjacent cells found in this way become the set of cells extracted to form the tool's output, which will represent the least-cost path from a destination to a source.

Note that the least-cost path generated in this way by ArcGIS will be a new raster surface, consisting of the cells making up the least-cost path (they will be assigned a value of 3) and another cell denoting the location of the destination (assigned a value of 1). All other cells on the surface will receive a value of "No Data" (because they are not cells that are part of the least-cost path).

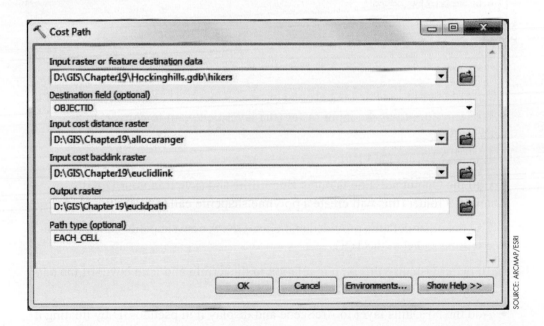

- Select **hikers** for the Input raster or feature destination data.

- Select **allocaranger** for the Input cost distance raster.

- Select **Euclidlink** for the Input cost backlink raster.

- Call the Output raster **Euclidpath** and save it in your D:\GIS\Chapter19 folder.

- Select **EACH_CELL** for the Path type.

- Click **OK** to calculate the cost distance based on these parameters.

- The shortest path between the state forest HQ and the hikers will be displayed by a new raster.

STEP 19.4 **Visualizing the Cost-Path Results**

• Rather than working with this new shortest path as a group of raster cells, we'll now convert it to a line. From the **Conversion** toolbox, select the **From Raster** toolset, then select the **Raster to Polyline** tool.

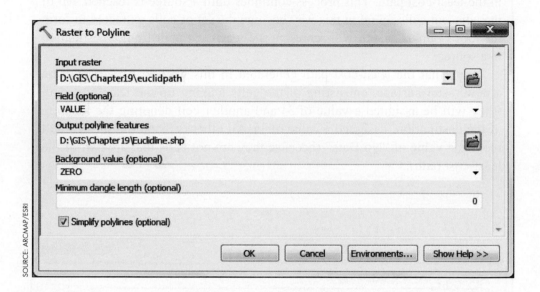

• Use **Euclidpath** as the input raster (the layer you want to convert to a vector line).

• Use **VALUE** for the Field.

• Call the Output polyline features **Euclidline** and save it in your D:\GIS\ Chapter19 folder (this will create a polyline shapefile called Euclidline).

• Use the other defaults provided and click **OK**. A new shapefile called Euclid-line will be added to the TOC.

• To better visualize this path in relation to the terrain and land cover of the park, we'll use ArcScene. Open **ArcScene**.

• Add the NEDhills layer to ArcScene and display it in pseudo-3D by floating it on a custom surface—apply the base heights of the NEDhills layer to itself (see Chapter 15 for how to do this).

• You may also want to adjust the Vertical Exaggeration of the scene itself to a factor of 2 for visualization purposes (see Chapter 16 for instructions on adjusting the vertical exaggeration).

• Next, add the Euclidline layer to ArcScene and drape it on the NEDhills layer (see Chapter 15 for instruction on how to drape a layer in ArcScene). You may want to change the symbology to a thicker line and different color that will really stand out. If necessary, you may want to offset the draped line by 1 to 3 units for visualization purposes (see Chapter 18 for instructions on how to offset a vector layer). Zoom in and examine the draped calculated shortest path, and answer Questions 19.5 and 19.6.

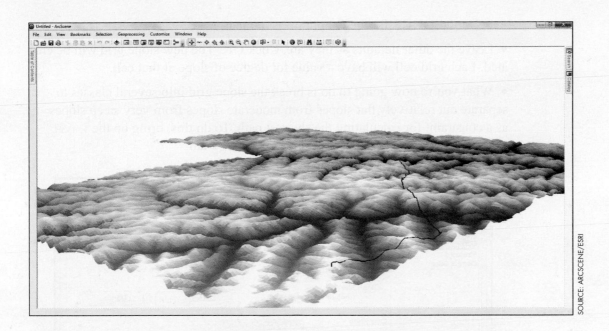

SOURCE: ARCSCENE/ESRI

QUESTION 19.5 What factors were unaccounted for in calculating this shortest path?

QUESTION 19.6 Is this a path that you would use to rescue the hikers? Why or why not?

STEP 19.5 **Further Cost-Distance Calculations**

• Return to ArcMap and turn off all layers except for the NEDhills layer. In this step, you'll add more complexity to the problem—specifically, the steepness of the terrain (see Chapter 15 for more about slope calculation from an elevation surface).

• From the **Spatial Analyst Tools** toolbox, select the **Surface** toolset, then select the **Slope** tool.

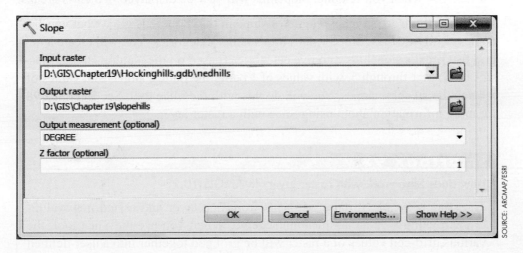

SOURCE: ARCMAP/ESRI

• Select **NEDhills** as the Input raster.

• Call the Output raster **slopehills** and save it in your D:\GIS\Chapter19 folder.

• Select **DEGREE** for the Output measurement.

• Leave the other defaults as they are. Click **OK**. The new slope grid will be created. Each grid cell will have a value for degree of slope at that cell.

• What you're now going to do is break the slope grid into several classes to separate out relatively flat slopes from moderate slopes from very steep slopes as a constraint for calculating the shortest path. To do this, bring up the **Layer Properties** of slopehills and select the **Symbology** tab.

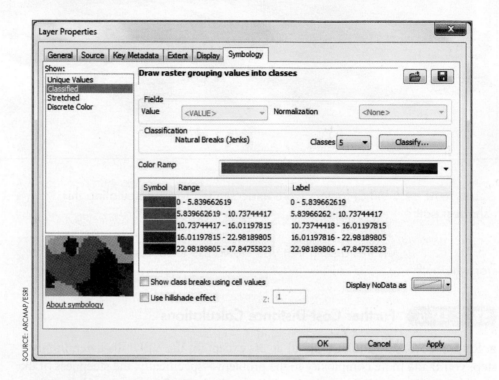

SOURCE: ARCMAP/ESRI

• Under the Classification heading, change to **5** Classes instead of 9 (but still using the **Natural Breaks** method).

• Click **OK** when you're done. Slopehills will now be displayed in 5 class breaks.

• Because there are so many values of slope in the Slopehills grid, what you'll do next is divide these slope values into smaller categories. You will create a new grid that breaks slopehills into five categories (based on the Natural Breaks classification) of 1 through 5, with values of 1 being the flattest slopes and values of 5 being the steepest slopes. You will do this with the Slice tool. See **Smartbox 98** for more information about using Slice with a raster layer.

Smartbox 98

How does Slice work with raster layers in ArcGIS 10.2?

slice Grouping several grid-cell values together according to a classification scheme.

The **slice** process will take grid cells from one layer and assign new values to them using a data classification scheme. When you use the Slice tool, the various grid-cell values of a raster will be grouped together into a user-defined number of classes. The slice process can be used to group all of the unique distance values (potentially thousands of different values) into a smaller number of classes, with each class assigned a new value. In Figure 19.12, the first raster

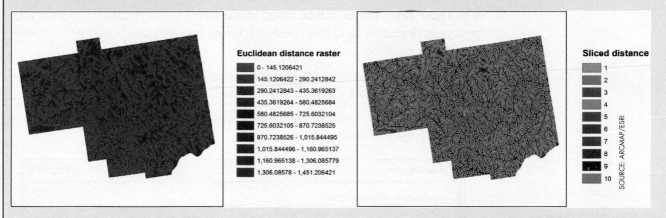

Euclidean distance raster

- 0 - 145.1206421
- 145.1206422 - 290.2412842
- 290.2412843 - 435.3619263
- 435.3619264 - 580.4825684
- 580.4825685 - 725.6032104
- 725.6032105 - 870.7238525
- 870.7238526 - 1,015.844495
- 1,015.844496 - 1,160.965137
- 1,160.965138 - 1,306.085779
- 1,306.08578 - 1,451.206421

Sliced distance

- 1
- 2
- 3
- 4
- 5
- 6
- 7
- 8
- 9
- 10

SOURCE: ARCMAP/ESRI

FIGURE 19.12 A Euclidean distance raster that has been sliced into 10 different categories.

depicts each grid cell containing a value for the Euclidean distance away from a set of streams. The second raster shows that same raster after a slice process—all grid cells now have a value of 1 to 10, with 1 as the lowest distance value and 10 as the highest distance value.

The classifications available through slice are Equal Interval and Natural Breaks (which work as described in Chapter 3, where we used them for classifying values on choropleth maps) and Equal Area (which is similar to the Quantiles method from Chapter 3, in that it seeks to have an equal number of cells—not data values, but a count of cells—in each category).

- From the **Spatial Analyst Tools** toolbox, select the **Reclass** toolset, then select the **Slice** tool.

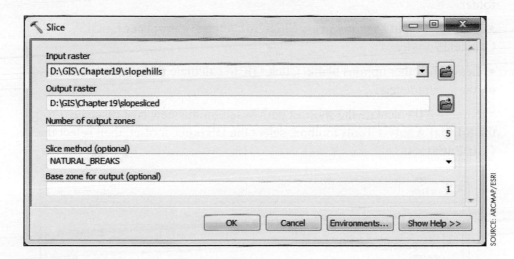

SOURCE: ARCMAP/ESRI

- Select **slopehills** as the input raster.

- Call the output raster **slopesliced** and save it in your D:\GIS\Chapter19 folder.

- Use **5** for the number of output zones.

- Use **NATURAL_BREAKS** for the Slice method.

- Click **OK**. The new grid will be added to the TOC and will show the slope values placed in one of five new classes, with the lowest slope values represented by cells with a value of 1 and the highest slopes represented by cells with a value of 5. These slope values will be based on the five Natural Breaks categories you have your slopehills layer displayed as (change between the display of the slopehills and slopesliced grids to see this).

- You'll now create a new shortest path, using this newly recategorized slope as a constraint on the cost-distance calculation.

- From the **Spatial Analyst Tools** toolbox, select the **Distance** toolset, then select the **Cost Distance** tool.

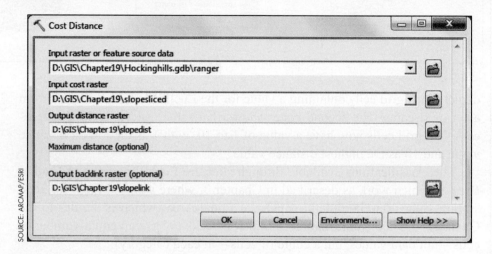

- Select **ranger** for the Input raster or feature source data.

- Select **slopesliced** for the Input cost raster.

- Call the Output distance raster **Slopedist** and save it in your D:\GIS\Chapter19 folder.

- Call the Output backlink raster **Slopelink** and save it in your D:\GIS\ Chapter19 folder.

- Leave the other options blank. Click **OK** to calculate the cost distance based on these parameters.

- Next, we'll calculate the shortest path based on this new cost distance. From the **Spatial Analyst Tools** toolbox, select the **Distance** toolset, then select the **Cost Path** tool.

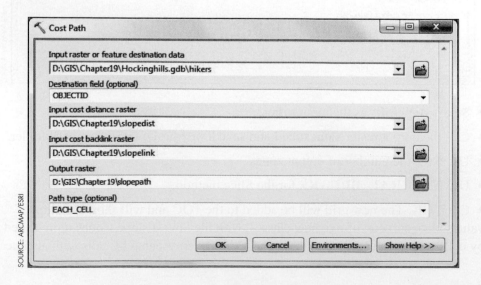

- Select **hikers** for the Input raster or feature destination data.

- Select **slopedist** for the Input cost distance raster.

- Select **slopelink** for the Input cost backlink raster.

- Call the Output raster **slopepath** and save it in your D:\GIS\Chapter19 folder.

- Select **EACH_CELL** for the Path type.

- Click **OK** to calculate the cost distance based on these parameters.

- The new shortest path between the state forest HQ and the hikers will be displayed by a new raster called slopepath.

- To better visualize the results of this cost-distance calculation, do the following (as you did in Step 19.4):

  - Convert the slopepath grid into a polyline called slopeline.

  - Switch back to ArcScene and add the new slopeline layer to the Scene (you may have to Refresh your Catalog in order to view the new slopeline layer).

  - Drape slopeline over the NEDhills layer to see it in 3D and change the color of slopeline to make it more distinctive than the previous Euclidline layer. You should now have two paths (slopeline and Euclidline) being displayed in 3D. Answer Questions 19.7 and 19.8.

QUESTION 19.7 How did this new cost distance change the shortest path?

QUESTION 19.8 Is this new path one that you would use to rescue the hikers? Why or why not?

STEP 19.6 Using More Than One Cost Layer for Cost Distance

- Return to ArcMap and turn off all layers except the NEDhills layer. You'll now be creating a third cost-distance surface, but this time you'll use two factors—the steepness of the terrain you already calculated and the type of land cover one would travel across. The assumption being made now is that certain types of land cover will "cost" more to travel through than others (i.e., it's easier/less costly to traverse developed terrain—like paved surfaces—than it is to traverse forested terrain or water).

- To create this new cost-distance surface, you'll be creating a new grid and reclassifying the land-cover types into cost values. The higher the cost value, the more difficult (i.e., the more costly) it will be to traverse the terrain. You can accomplish these goals with the Reclassify tool. (See **Smartbox 99** for more about reclassifying a raster layer.)

Smartbox 99

How are raster layers reclassified in ArcGIS 10.2?

The Slice tool will assign new values to grid cells by placing them in different categories according to a classification scheme. However, you will sometimes want to change the value of cells yourself, not group them into classes. For instance, if you have a grid with values of 1, 2, and 3, but you want to create a grid that

Reclassify The ArcGIS tool that allows you to create a new grid by assigning new grid-cell values to an existing raster.

assigns all of these cells a value of 1, you need a way to tell ArcGIS what new values need to be assigned to these cells. The Reclassify tool allows you do this.

Reclassify allows you to create a new raster by changing the values assigned to grid cells in another raster. With Reclassify, you specify what new values are assigned to old values. These new values must be numerical, or you can also type NoData for the assigned new value. Reclassify is commonly used to make changes to a grid or update its information. It's also used to change values from one classification to another. For instance, an NLCD raster has all of its developed categories classified as different values of 21, 22, 23, and 24. If you are simply examining all built areas, you may want to reclassify the NLCD grid, so all four of these values are assigned a new value of 1, while all other NLCD values are assigned a value of "NoData." The end result will be a new grid that contains only grid cells representing developed areas, with no other values in the grid.

- From the **Spatial Analyst Tools** toolbox, select the **Reclass** toolset, then select the **Reclassify** tool.

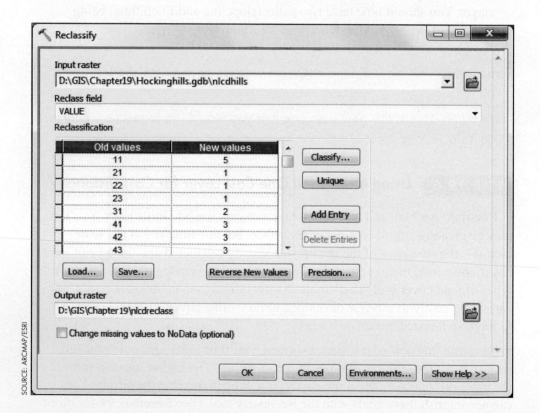

SOURCE: ARCMAP/ESRI

- Select **nlcdhills** as the Input raster.

- Select **VALUE** as the Reclass field.

- Press the **Unique** button to separate out each of the NLCD 2006 values into its own separate number to be reclassified.

- Reclassify the NLCD 2006 values as follows (***Important Note:*** Remember that the higher the new value, the more costly it is to traverse that particular land cover.):

Old Value	New Value	NLCD 2006 Classification
11	5	Open Water
21	1	Developed, Open Space
22	1	Developed, Low Intensity
23	1	Developed, Medium Intensity
31	2	Barren
41	3	Deciduous Forest
42	3	Evergreen Forest
43	3	Mixed Forest
52	3	Shrub Scrub
71	2	Grassland/Herbaceous
81	2	Pasture/Hay
82	2	Cultivated Crops
95	4	Emergent Wetlands

- Call the Output raster **nlcdreclass** and save it in your D:\GIS\Chapter19 folder.

- Click **OK** when you're done. A new grid called nlcdreclass will be created with the new cost values.

- Now you'll combine your two cost grids (the reclassified slope and the reclassified land cover) to form a new cost grid. This new grid will take both weights into account. To combine your grids, you'll have to overlay them. Raster overlay works differently than the polygon overlays of Chapter 9 and we'll do much more with raster overlay and get into how it works in Chapter 20. For now, however, follow the next few steps to combine the two layers by simply adding them together.

- From the **Spatial Analyst Tools** toolbox, select the **Map Algebra** toolset, then select the **Raster Calculator** tool.

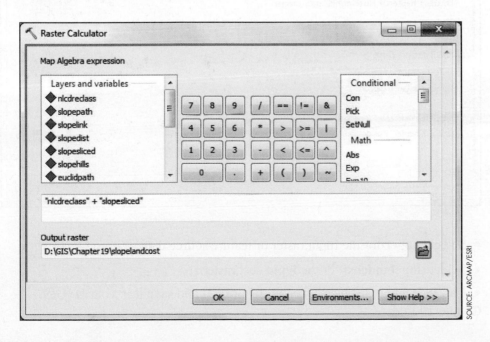

SOURCE: ARCMAP/ESRI

- Build a raster calculation to combine the two grids by adding them together. From the Layers and variables list, first double-click on **nlcdreclass**. Next, press the + button. Last, double-click on **slopesliced** from the Layers and variables list. Your raster calculation should read **"nlcdreclass" + "slopesliced"**.

- Call the Output raster **slopelandcost** and save it in your D:\GIS\Chapter19 folder.

- Click **OK** when you're done.

- Examine the new slopelandcost grid. You've now combined the two rasters by adding the cell values together for cells that would be on top of each other. For instance, a location on the slopesliced grid may have a value of 2. That same location on the nlcdreclass grid may have a value of 3. Thus, because 2 + 3 = 5, the same location on the slopelandcost grid would have a value of 5. Answer Questions 19.9 and 19.10.

> **QUESTION 19.9** What (specifically) do values of "2" and values of "8" represent in the new slopelandcost grid (beyond simply values added together)?

> **QUESTION 19.10** Why are there no values of "10" in the slopelandcost grid?

- Now you'll compute the cost-weighted distance and shortest path using this new cost-distance grid. From the **Spatial Analyst Tools** toolbox, select the **Distance** toolset, then select the **Cost Distance** tool.

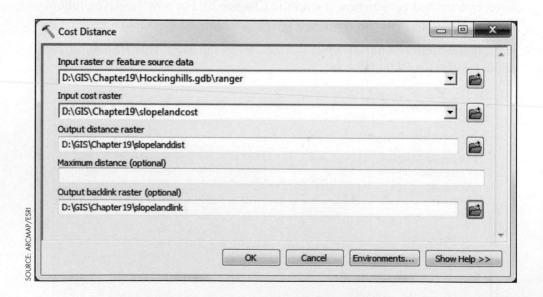

- Select **ranger** for the Input raster or feature source data.

- Select **slopelandcost** for the Input cost raster.

- Call the Output distance raster **slopelanddist** and save it in your D:\GIS\ Chapter19 folder.

- Call the Output backlink raster **slopelandlink** and save it in your D:\GIS\ Chapter19 folder.

- Leave the other options blank.

- Click **OK** to calculate the cost distance based on these parameters.

- Next, we'll calculate the shortest cost path based on this new cost distance. From the **Spatial Analyst Tools** toolbox, select the **Distance** toolset, then select the **Cost Path** tool.

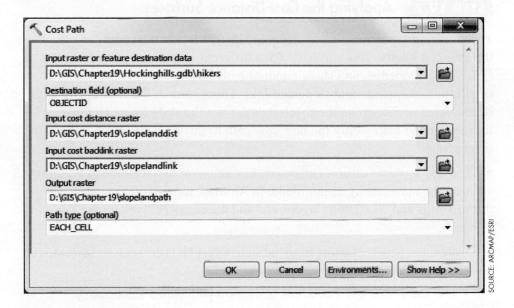

- Select **hikers** for the Input raster or feature destination data.

- Select **slopelanddist** for the Input cost distance raster.

- Select **slopelandlink** for the Input cost backlink raster.

- Call the Output raster **slopelandpath** and save it in your D:\GIS\Chapter19 folder.

- Select **EACH_CELL** for the Path type.

- Click **OK** to calculate the cost distance based on these parameters.

- The new shortest path between the state forest HQ and the hikers will be displayed by a new raster.

- To better visualize the results of this cost-distance calculation, do the following:

  - Convert the slopelandpath grid into a polyline called slopelandline.

  - Switch back to ArcScene and add the new slopelandline layer to the Scene.

  - Drape slopelandline over the NEDhills layer to see it in 3D and change the color of slopelandline to make it more distinctive than the previous slopeline and Euclidline layers. You should now have three paths (slopelandline along with slopeline and Euclidline) being displayed in 3D.

  - Also add the NLCDhills grid to ArcScene and drape it over the NEDhills layer (turn off NEDhills when you're done). Rearrange the layers so that your three paths are shown on top of the NLCDhills grid. Answer Questions 19.11 and 19.12.

> **QUESTION 19.11** How (specifically) did this third cost distance change the shortest path away from the one generated using the slope as a constraint on the problem? Why did the path change this way?

> **QUESTION 19.12** Which of these three cost paths would you select to rescue the hikers? Why?

STEP 19.7 **Applying the Cost-Distance Surfaces**

• At this point, you have generated three different "shortest paths" to reach the stranded hikers. You also have three different cost-distance surfaces (Eucliddist, slopedist, and slopelanddist) with corresponding backlink grids (Euclidlink, slopelink, and slopelandlink). Thus, none of these grids needs to be recreated to determine the shortest path from the state forest HQ to another location on the map—only the new cost path to this new point would have to be generated.

• This next scenario assumes an emergency situation has occurred at a campsite near the main family campground. In ArcMap, turn off all the layers except the NEDhills and NLCDhills layers, and add the campers layers to the Table of Contents.

• Create three new cost paths from the state forest HQ (the rangers layer) to the destination of the campers point as follows:

  • Use Eucliddist and Euclidlink to create a path called Euclidcamp.

  • Use slopedist and slopelink to create a path called slopecamp.

  • Use slopelanddist and slopelandlink to create a path called slopelandcamp.

• Convert these paths to polylines and drape them over the NEDhills layer in ArcScene to examine each of these paths, then answer Question 19.13.

> **QUESTION 19.13** How does each of these three paths to reach the campers differ from the other two? Be specific when referring to specific terrain features or land-cover features near the campsite that the paths avoid or do not avoid. Also describe which path you would take to reach the campers and why.

STEP 19.8 **Exporting, Saving, and Sharing Your Results**

• Save your work in ArcScene as a scene document and add the appropriate information in the Scene Document Properties dialog.

• To share your results with others (i.e., the 3D view of the three paths to the hikers as well as a separate 3D view of the three paths to the campers), you'll have to export the scene to a graphic format—see Chapter 15 for more about exporting a scene. Create two different graphics, one for each set of paths.

• Save your work in ArcMap as a map document and add the appropriate information in the Map Document Properties dialog.

- In ArcMap, show only the locations of the ranger, the hikers, and the campers, as well as the six polyline versions of the paths. Either print a layout (see Chapter 3) to show the point locations and the paths (including all of the usual map elements and design for a layout) or share only these layers as a map service through ArcGIS Online (see Chapter 4).

Closing Time

This chapter explored the concept of representing distance in a GIS as a continuous surface. By creating a surface of Euclidean distance measurements, you can easily determine how far away every location is from a source. Through the use of the cost-distance and least-cost path processes, you can then generate the shortest path across the surface from any location to a source. While the cost-distance tools are very versatile, ArcGIS also provides an additional set of tools for modeling other types of surface movement (see the *Related Concepts for Chapter 19* for more information).

One thing that you did in this chapter was to combine two different raster layers to create a single new raster. By combining the two cost layers (the sliced slope layer and the reclassified land-cover layer) you created a new cost raster that had elements of both. Essentially, what you did is overlay these two rasters, which is a very powerful tool to have at hand when working with raster data. In the next chapter, we'll do much more with raster overlay in the context of a new concept called Map Algebra.

Related Concepts for Chapter 19   Using Path Distance in ArcGIS 10.2

The cost-distance functions in ArcGIS 10.2 allow for a wide variety of different applications involving movements across a surface. However, ArcGIS 10.2 contains another set of functions that allow you to apply cost-distance concepts with additional complexity for use in dispersion modeling or more detailed least-cost path studies. The functions of **path distance** use the same types of accumulative costs, allocation, backlink, and least-cost path algorithms as their cost-distance counterparts but allow you to model many other factors that would influence movement across a surface. Rather than the Cost Distance tool, you would use the Path Distance tool when modeling these additional forces.

Path distance allows for the inclusion of the actual surface distance being traveled as well as other horizontal-based factors that would influence movement. For instance, a strong wind behind you would allow you to move faster, while a strong headwind would slow your movement. Vertical factors can be included as well. For example, if you were walking uphill on a steeper slope, your progress would be slowed (as we simulated in this chapter). However, if you were walking downhill on the same steep slope, your movement would speed up. Many types of frictions can be used, depending on the phenomena being modeled with path distance. For instance, smoke plumes, pollution concentrations, and movement of contaminants can all be modeled as continuous surfaces in ArcGIS.

path distance A cost-distance function that allows for a variety of other frictions (such as horizontal or vertical factors) to be included in the analysis.

For More Information

For further in-depth information about the topics presented in this chapter, use the ArcGIS Help feature to search for the following items:

- An overview of the Distance toolset
- Cost Back Link (Spatial Analyst)
- Cost Distance (Spatial Analyst)
- Cost Path (Spatial Analyst)
- Creating a cost surface raster
- Creating the least-cost path
- Euclidean Allocation (Spatial Analyst)
- Euclidean Distance (Spatial Analyst)
- How cost-distance tools work
- How the horizontal and vertical factors affect path distance
- How the path-distance tools work
- Reclassify (Spatial Analyst)
- Slice (Spatial Analyst)
- Understanding cost-distance analysis
- Understanding Euclidean distance analysis
- Understanding path-distance analysis

Key Terms

Euclidean distance (p. 439)

allocation (p. 441)

cost (p. 443)

accumulated cost (p. 443)

least-cost path (p. 448)

backlink (p. 448)

slice (p. 452)

Reclassify (p. 456)

path distance (p. 461)

How to Perform Map Algebra in ArcGIS 10.2

Introduction

In the last chapter, we combined two raster layers (representing slope and land-cover rankings) to form a new, third layer. By adding the two layers together, we performed a raster overlay. In ArcGIS, rasters can be overlaid through the use of **Map Algebra**, a means of using Spatial Analyst tools and operators to perform actions on rasters. As its name implies, Map Algebra also represents combining datasets through mathematical operations, such as adding or multiplying rasters to create a new raster. Map Algebra can take the values of cells from multiple grids that share the same location and combine those values mathematically.

Suppose, for instance, you want to rank locations on their difficulty in traversing the terrain based on the steepness of the slopes and the type of land cover, and you find those locations that are the most difficult (or easiest) to cross based on these two factors. See Figure 20.1 for an example of this scenario: Two grids are featured—a raster called sloperank (which ranks the difficulty of traversing the slope of the land on a scale from 1 to 5) and a second raster called landrank (which ranks the difficulty of traversing the land cover, also on a scale from 1 to 5). By combining the sloperank and landrank grids, Map Algebra forms the new output grid. The value for the cell in the upper-left corner in sloperank is 5, and the value for the cell in the upper-left corner in landrank is 3, so when these values are added, the cell in the upper-left corner of output is 8. The output grid represents the overlay of the sloperank and landrank grids using an addition operation to show the cumulative difficulty of crossing a part of the landscape.

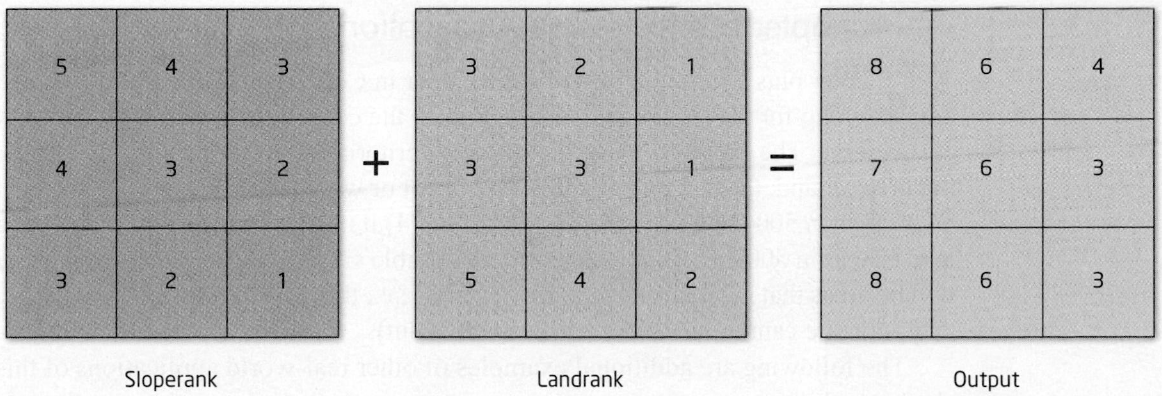

> **Map Algebra** Operation performed on rasters, such as overlaying two or more grids.

FIGURE 20.1 An example of using an addition operator in Map Algebra to combine two grids.

5	4	3
4	3	2
3	2	1

Sloperank

+

3	2	1
3	3	1
5	4	2

Landrank

=

8	6	4
7	6	3
8	6	3

Output

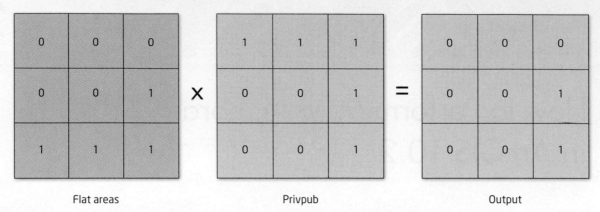

Flat areas   Privpub   Output

FIGURE 20.2 An example of using a multiplication operator in Map Algebra for combining two grids.

Other applications for combining grids are also available. For instance, if you're trying to find an initial set of locations for new housing developments, two criteria would be to build on relatively flat sections of land that are privately owned. See Figure 20.2 for an example of this scenario: There are two grids—a raster called flatareas (which features a value of 1 if the slope of the land is flat enough to build on and a value of 0 if it is not) and a second raster called privpub (which features a value of 1 if the section of land is privately owned and a value of 0 if it is publicly owned). By multiplying these two grids together, Map Algebra creates a new output grid. As in the previous addition example, the grid cells that correspond with each other (such as the value of 1 in the upper-left corner of the flatareas grid and the value of 0 in the upper-left corner of the privpub grid) are multiplied. The output grid shows that those areas that meet both criteria have a value of 1 (because only those cells that have a value of 1 in both the flatareas grid and the privpub grid could be multiplied together to get an output value of 1), and those areas that meet one or neither criterion have a value of 0. Thus, only those sections of the landscape with an output value of 1 are suitable as the initial areas to build on.

Beyond math operators, there are several other forms of Map Algebra available (such as relational or Boolean operations) that we'll discuss in this chapter. In addition, this chapter will introduce the ArcGIS ModelBuilder environment for laying out the various steps you'll use with Map Algebra. While all of the analysis in this chapter can be performed without using ModelBuilder, you'll likely find its visual interface very helpful in conceptualizing the **workflow**, or the sequence of actions you're performing with various layers and tools.

workflow The sequence of actions performed using GIS layers and tools.

Chapter Scenario and Applications

This chapter puts you in the role of a developer in Columbiana County, Ohio. You are searching for the most suitable locations in the county for building an ecological preserve. The site must meet the following criteria: (1) it must be on a relatively flat area of land, (2) it must contain either forest or wetlands land cover, (3) it must be more than 500 meters from a highway, and (4) it must be within 500 meters of a lake or within 500 meters of a stream. The suitable sites for the ecological preserve will be areas that meet all of the above criteria (i.e., these conditions are mandatory and your site cannot leave any of these items out).

The following are additional examples of other real-world applications of this chapter's skills.

- An emergency preparedness center for a coastal city is simulating the effects of the amount of flooding generated by different categories of hurricanes. The center's staff members can use Map Algebra to examine selected levels of elevation models according to the amount of expected flooding as a starting point for their analysis.

- An environmental planner is examining a multi-state distribution of forests and wetlands as part of a larger modeling initiative. She can use Map Algebra techniques with a large NLCD dataset to extract separate 30 meter layers of different forest and wetland types to use as model inputs.

- An urban planner is examining a statewide distribution of seasonal homes. He can use Map Algebra techniques in conjunction with layers such as NLCD to determine some of the most suitable places in the state for the locations of seasonal home developments.

ArcGIS Skills

In this chapter, you will learn:

- How to use the ArcGIS ModelBuilder environment for setting up and executing workflows.

- How to use the Raster Calculator to perform Map Algebra and raster overlay.

- How to combine raster layers with various Boolean and Map Algebra functions.

- How to use a relational operator with a raster surface.

- How to use the Draw toolbar functions to add graphics to a layout.

Study Area

For this chapter, you will be working with data from Columbiana County, Ohio.

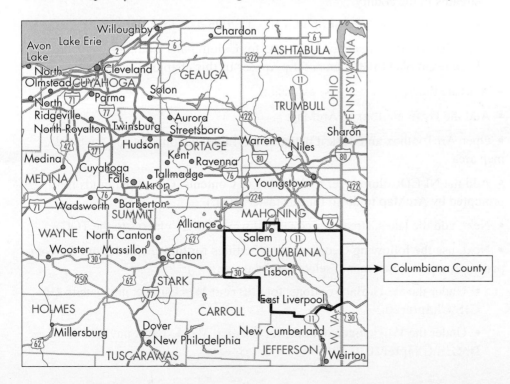

Data Sources and Localizing This Chapter

This chapter's data focus on features and locations within Columbiana County, Ohio. However, you can easily modify this chapter to use data from your own local county instead—for instance, if you wanted to perform this chapter's applications in Beaver County, Pennsylvania. The NED and NLCD2006 that were downloaded from The National Map would also be available for Beaver County; you could clip the layers to the county boundaries and process them to keep consistent 30 meter cell sizes for this exercise.

The lakes, streams, and UShighways layers were also derived from data downloaded from The National Map. If you were going to use these for Beaver County, Pennsylvania, you would download the Structures geodatabase for that county and use the Trans_Roadsegment layer as the source for UShighways. Only those road segments that had a non-null entry in the US_Route field were used. For lakes and streams, you would download the Hydrography geodatabase for that county and use the NHD_waterbody layer for lakes and the NHD_Flowline layer for streams.

STEP 20.1 Getting Started

• Start ArcMap and use the Catalog to copy the folder called **Chapter20** from the C:\GISBookdata\ folder to your own D:\GIS\ drive. Be sure to copy the entire directory, not just the files within it.

• Chapter20 contains a personal geodatabase called **Colummodeldata** that contains the following items:

  • **UShighways**: A line feature class representing the U.S. highways that run through the county.

  • **Lakes**: A polygon feature class representing the lakes, ponds, and reservoirs in the county.

  • **Streams**: A line feature class representing the various types of rivers and streams in the county.

  • **NLCDColum**: A raster of 30m NLCD 2006 data for Columbiana County.

  • **NEDColum**: A raster of NED data for Columbiana County (which has been resampled to 30 meters for purposes of this chapter).

• Activate the **Spatial Analyst** extension.

• Add the **Draw toolbar** to ArcMap.

• Open ArcToolbox and dock it between the Table of Contents and the blank map area.

• Add the NLCDColum layer to the Table of Contents. (Build pyramids if prompted by ArcMap to do so for this raster layer.)

• Next, add the lakes, streams, and UShighways feature classes to the TOC.

• Next, use the following Geoprocessing Settings for this chapter (from the **Geoprocessing** pull-down menu, select **Environments...**):

  • Under the **Workplace** options, for **Current Workspace**, navigate to **D:\ GIS\Chapter20**.

  • Under the **Workspace** options, for **Scratch Workspace**, navigate to **D:\GIS\Chapter20**.

- Under the **Output Coordinates** options, for **Output Coordinate System**, select **Same as Layer "NLCDcolum"**.

- Under the **Processing Extent** options, for **Extent**, select **Same as Layer NLCDColum**.

- Under the **Raster Analysis** options, for **Cell Size**, select the option for **Same as layer NLCDColum**. This should automatically adjust the cell size to 30 (as in 30 meters).

• Click **OK** when you're done. All of your Environment Settings should now be in place for use in the rest of the chapter.

• The Coordinate System being used in this chapter is **Albers Conical Equal Area (NAD 83)**, using Map Units of **Meters**. Check the Data Frame to verify that this coordinate system is being used.

STEP 20.2 Creating a New Model

• You'll be setting up your workflow as a model and ArcGIS will treat a new model as a tool because it's composed of one or more ArcToolbox tools. The first thing to do is to create a new Toolbox in which to place your model. In Catalog, right-click on the **Colummodeldata** geodatabase, select **New**, and select **Toolbox**. A new toolbox will be added to your geodatabase and given the name "Toolbox."

• Rename the "Toolbox" to be called **MapAlg** (for "Map Algebra").

• To create a new model to be stored in the MapAlg toolbox, right-click on **MapAlg**, select **New**, then select **Model**. A new item called Model will appear in the MapAlg toolbox.

• The blank ModelBuilder dialog will also appear. For more information about ModelBuilder, see **Smartbox 100**. Refer to **Troublebox 11** if you run into later difficulties with saving and re-opening a model in ArcGIS.

Smartbox 100

What is ModelBuilder and how is it used in ArcGIS 10.2?

ModelBuilder is an ArcGIS graphical environment that allows you to build workflows and then run them to create outputs. ModelBuilder works as a drag-and-drop environment in which you select data layers from the TOC or the Catalog (or tools from ArcToolbox) and place them into a visual layout for designing a workflow. In ModelBuilder, by default, the data layers used in the model will be displayed as a blue oval, the tools being used will be displayed as a yellow rectangle, and the outputs from the tools will be displayed as green ovals. Lines and arrows show the connections among these elements. In Figure 20.3, the operation being performed starts with a data layer called NLCDColum, uses the Raster Calculator tool on this layer, and ends up with a resulting layer called rastercalc.

When you place a layer or tool in the ModelBuilder environment, you can move the element to anywhere inside the window. Think of using ModelBuilder the way you would think about designing a flowchart that shows how various layers and tools fit together. ModelBuilder then allows you to run the processes

ModelBuilder The interface of ArcGIS used in laying out workflows and creating models.

FIGURE 20.3 A simple Model-Builder workflow showing a layer connected to a tool connected to an output.

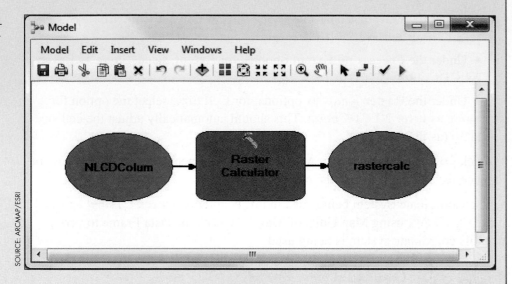

SOURCE: ARCMAP/ESRI

FIGURE 20.4 The tools of the ModelBuilder interface.

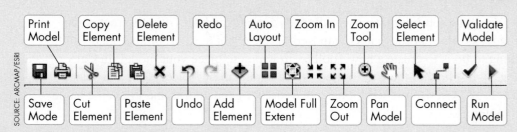

SOURCE: ARCMAP/ESRI

and create the various outputs (as you'll do in this chapter). See Figure 20.4 for an explanation of the various icons in the ModelBuilder interface.

This chapter asks you to design and run a basic workflow with ModelBuilder. In Chapter 21, you'll learn many more aspects and features of ModelBuilder.

Troublebox 11

Why can't I re-open my saved model?

Models can be saved and then re-opened later when you want to continue designing or run them. If you have saved and closed your model and want to re-open it to continue to work, navigate to the toolbox containing your model and expand it (in this chapter it will be called the MapAlg toolbox and be located in the D:\GIS\Chapter 20 folder). You'll see the icon for the model inside the toolbox. Right-click on the model and a new set of options will appear. If you select **Open**, however, the ModelBuilder dialog will appear with a statement saying "This tool has no parameters" and all of your design will not appear.

Don't panic—all of your model work is still there. Until you formally assign model parameters to the model (which we will do in Chapter 21, when we make use of them in building a model), the model can still be opened. When you right-click on the model icon, select **Edit**. Your model will re-open for you to continue to work with. If you save it and close it again, just use the Edit option to re-open the model.

- ModelBuilder works as a drag-and-drop environment, so position it to the right of ArcToolbox for ease of use.

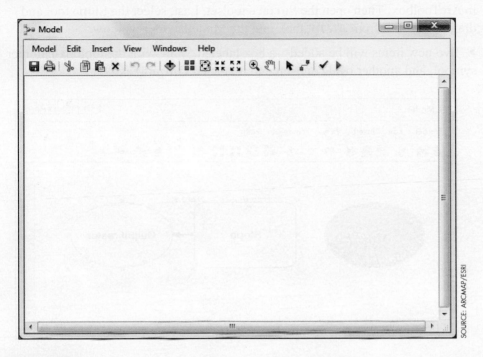

SOURCE: ARCMAP/ESRI

STEP 20.3   **First Criterion: Must Be on a Relatively Flat Piece of Land**

- Add the NEDColum grid to the Table of Contents. (Build pyramids if prompted by ArcMap to do so.) The NED layer contains elevation information, with each cell coded with an elevation value. To get information about relatively flat (versus steep) land, you would first need to calculate the slope of the land. Thus, the NED will be the first source grid to use in the model.

- In the Table of Contents, left-click once on the name of the **NEDColum** grid, hold down the mouse button, then drag the mouse over to the blank ModelBuilder window. Release the mouse button and the first source will be added to the Model—a blue oval called "NEDColum."

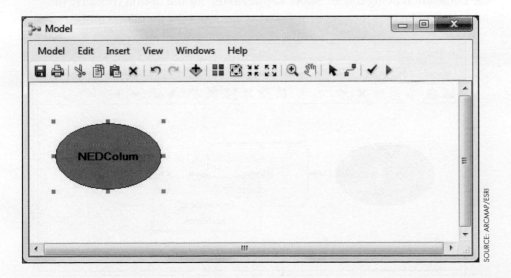

SOURCE: ARCMAP/ESRI

- Next, you'll want to calculate the slope of the DEM to find those places of relatively flat and relatively steep slope. Open the **Spatial Analyst Tools** toolbox in ArcToolbox. Then open the **Surface** toolset. Last, select the **Slope** tool and drag and drop it from ArcToolbox into the ModelBuilder window.

- Two new items will be added—a box labeled **Slope** with the toolbox hammer symbol, and another oval connected to it labeled **Output raster**.

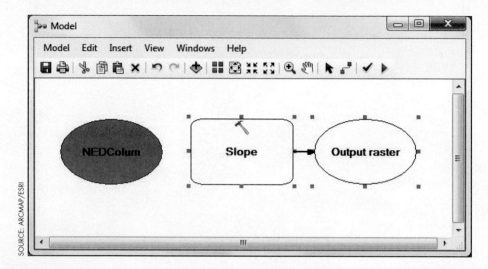

SOURCE: ARCMAP/ESRI

- The last step is to connect the NEDColum grid to the Slope tool. From the ModelBuilder window's toolbar, select the **Connect** tool.

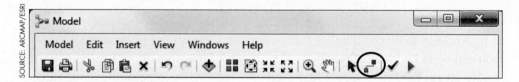

SOURCE: ARCMAP/ESRI

- The cursor will change to a magic wand icon. Use the new icon to draw a line between the NEDColum oval and the Slope tool. A blue line will appear between the two.

- A new menu will pop up asking for more information about the nature of the connection being made. Select **Input raster** for the option (because the NEDColum layer is the one being used as the input for the Slope tool).

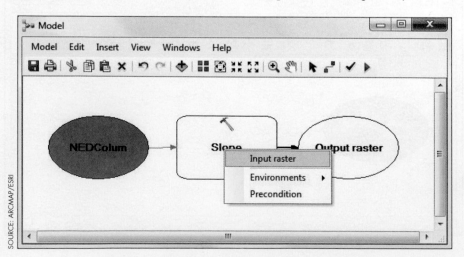

SOURCE: ARCMAP/ESRI

- ModelBuilder will establish the link between the items and the blue line will turn black.

- The slope toolbox will turn yellow, indicating that all its initial parameters are set. Similarly, the Output raster will turn green (and now have a default name added to it), indicating that an output is ready to be created.

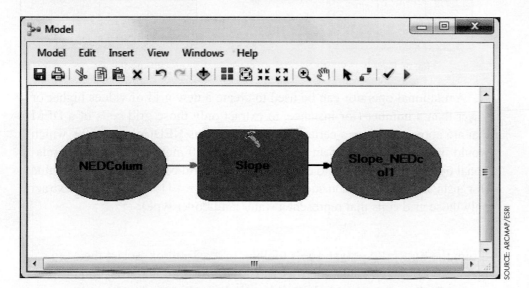

- Double-click on the yellow Slope box in the ModelBuilder dialog to see its settings. Answer Question 20.1.

QUESTION 20.1 The model (such as it is) could be run right now. What would happen and what would the model output be (at this stage)?

- What you now want to find are areas that have less than a 10-degree slope (to meet the first modeling condition). To do so, you'll have to query the Output raster from the Slope tool to find values of less than 10. A relational tool is needed to do this. See **Smartbox 101** for more about using relational operators in Map Algebra.

Smartbox 101

What is a relational operator and how is it used in Map Algebra?

A **relational operator** is one of six different connectors used in a Raster Calculator expression (similar to what we used back in Chapter 2 in a SQL statement): equal to (==), not equal to (!=), greater than (>), greater than or equal to (>=), less than (<), and less than or equal to (<=). In Map Algebra, each relational operator could be used to build an expression that will create a new grid containing a value of 1 for each grid cell for which that expression is true and a value of 0 for each grid cell for which that expression is false. For instance, in Figure 20.5 the expression being evaluated is "rasterlayer >= 5." Every grid cell in the raster layer contains a value and that value will be compared with 5. An output raster will be created with the same dimensions as the raster layer, but it will contain only values of 0 and 1. For each cell in the raster layer greater than or equal to a value of 5, the output raster will contain a value of 1. For each cell in the raster layer less than a value of 5, the output raster will contain a value of 0.

relational operator One of the six operators (equal, not equal, greater than, greater than or equal to, less than, less than or equal to) used when building a Raster Calculator expression.

FIGURE 20.5 An example of a relational operation in Map Algebra creating a new grid.

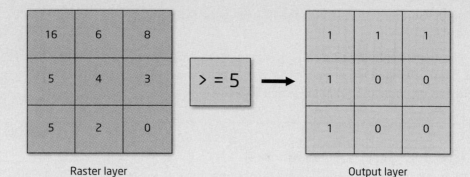

Raster layer   Output layer

A relational operator can be used to create a new grid of values higher or lower than a number (for instance, to extract only those grid cells of a DEM that are above or below a certain elevation, such as NEDColum > 300, which would find all grid cells of an elevation above 300 meters). Similarly, a relational operator can be used to extract only those grid cells with a certain value (for instance, an expression such as NLCDColum == 11, which would extract only those grid cells that represent a water land-cover type).

Raster Calculator ArcGIS tool used for Map Algebra operations.

• From the **Spatial Analyst Tools** toolbox in ArcToolbox, select the **Map Algebra** toolset, then select the **Raster Calculator** tool and drag and drop it from ArcToolbox into the ModelBuilder window. Choose the Select Elements tool and use that to position the Raster Calculator box to the right of the Output raster from the Slope tool (you may have to readjust the size of the ModelBuilder window). The **Raster Calculator** is the ArcGIS tool used for performing Map Algebra.

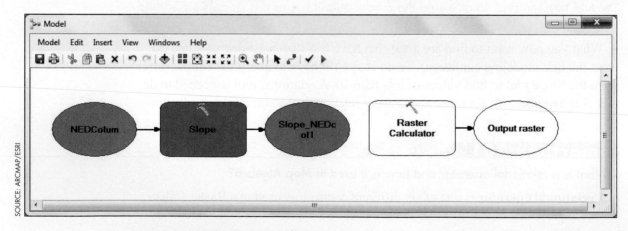

SOURCE: ARCMAP/ESRI

• Use the Connect tool to draw a line connecting the Output raster from the slope tool to the Raster Calculator box.

• Select the option for **Map Algebra expression** from the menu that appears. Answer Question 20.2.

QUESTION 20.2 Why did the Raster Calculator tool not turn yellow?

• You'll see the connecting line did not appear yet, but it will after you finalize the raster calculation. Switch back to the regular cursor instead of the wand by clicking on the **Cursor** button in the ModelBuilder window button bar. Then, in the ModelBuilder window itself, double-click on the **Raster Calculator tool** (the white box). The Raster Calculator tool dialog will open, so you can make changes to its parameters.

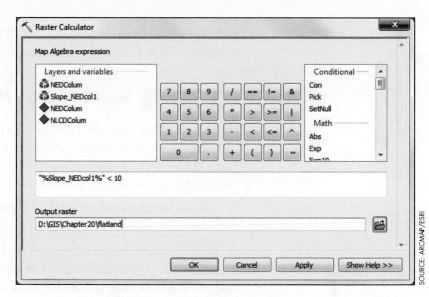

• Build an expression in the Raster Calculator of the Slope_NEDcol1 grid less than 10 (because you want slopes of less than 10 degrees). You build Raster Calculator expressions similar to the way you built a query in Chapter 2—first double-click the layer you want to work with (in this case, **Slope_NEDcol1**), then select the operator you want to use (in this case, the < operator for a less than operation), and finally use the number buttons of the Raster Calculator to type the value to use (in this case **10**).

• Name the Output raster **flatland** and save it in your D:\GIS\Chapter20 folder.

• Click **OK** when done.

• Back in ModelBuilder, you'll see the Raster Calculator tool properly connected and ready to go.

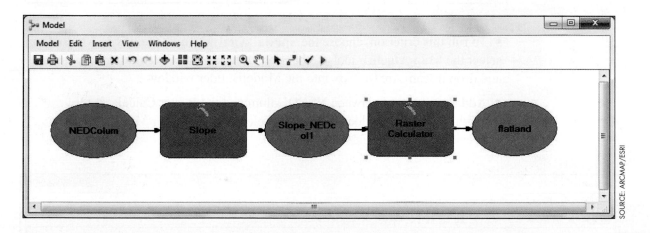

• Even though there are more criteria to be modeled, for now press the **Run** button in ModelBuilder.

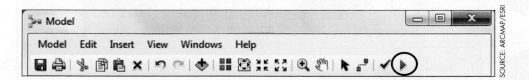

• The two processes (the yellow boxes) will be executed and the final output raster (called **flatland**) will be created.

• In ArcMap, add the flatland grid to the Table of Contents. Answer Question 20.3.

QUESTION 20.3 What do the values of 1 and 0 in the flatland grid represent?

STEP 20.4 **Second Criterion: Must Contain Forests or Wetlands**

- The NLCDColum grid contains NLCD classification regarding land-cover information. For the model, you are interested in those grid cells representing a forested land-cover class (class 41, class 42, and class 43) and those grid cells representing a wetlands land-cover class (class 90 and class 95).

  - *Important Note:* Refer to **Smartbox 61** on pages 282–283 in Chapter 12 for what, specifically, each of these land-cover classifications represents.

- Drag and drop the NLCDColum grid from ArcMap's Table of Contents into the ModelBuilder window. You may want to rearrange the window or expand it to give yourself more room to work with (after all, there are still more criteria to be modeled).

  - You can also use the zoom tools as well as Full Extent and Autolayout tools on the ModelBuilder toolbar to readjust items in the window.

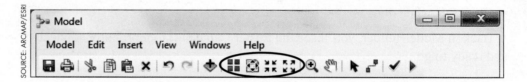

- To run this criterion, choose the **Spatial Analyst Tools** toolbox in ArcToolbox, select the **Map Algebra** toolset, then select the **Raster Calculator** tool and drag and drop it from ArcToolbox into the ModelBuilder window.

- Add a connection between NLCDColum and the Raster Calculator tool (though the black line doesn't appear). Select the option for **Map Algebra expression** when making the connection.

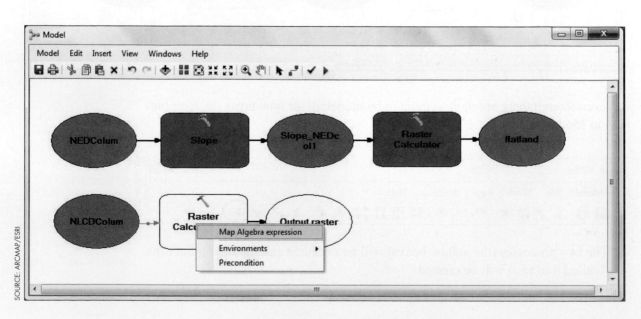

- The Raster Calculator tool is still white, so double-click on it so you can set its parameters. Note that the connection line will not be built until after you enter a Map Algebra expression. (While it's not strictly necessary to draw a connection line when using the Raster Calculator tool, it's strongly suggested you do so for best practices and good modeling habits.) For more about building more complex expressions in Map Algebra, see **Smartbox 102**.

Smartbox 102

How are more complex expressions built in Map Algebra?

Rasters can be overlaid using the basic mathematical operations, but many types of overlays or analysis require you to use more complex Map Algebra expressions. For instance, if you want to create a new grid that contains all of the input layer's values of 2 and all of the input layer's values of 3, just using a single relational operator or mathematical expression will not work. In essence, you would need ArcGIS to evaluate the input grid with two expressions at the same time (input grid == 2 or input grid == 3) to create an output grid that has a value of 1 where the input grid either had a value of either 2 or 3. Several Map Algebra expressions can be evaluated simultaneously using a Boolean operator.

When a **Boolean operator** is used, one of four connectors is used with the expressions or rasters (AND, OR, XOR, or NOT). These operators work as follows:

- **AND**: This Boolean operator is used to find the intersection of two expressions. Where both expressions are true, the output grid will have a value of 1, and where only one (or neither) of the expressions is true, the output grid will have a value of 0. For instance, Figure 20.6 contains two rasters—elevation and landcover. We want to find all grid cells with an elevation value of less than 300 and also all landcover values equal to 11. The Map Algebra expression (elevation < 300) AND (landcover == 11) will create a new grid with values of 1 for cells that meet both criteria and values of 0 for cells that meet one or neither of the criteria. Note that, in the Raster Calculator, the AND operation is represented with the **&** symbol.

Boolean operator (Map Algebra) One of the four operators (AND, OR, XOR, NOT) used when building a Raster Calculator expression.

AND (Map Algebra) The Boolean operator used with an intersection of two grids.

OR (Map Algebra) The Boolean operator used with a union of two grids.

FIGURE 20.6 An example of using the AND expression in Map Algebra

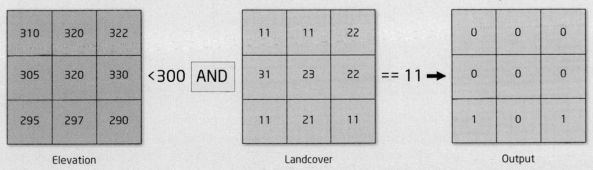

Elevation   Landcover   Output

- **OR**: This Boolean operator is used to find the union of two expressions. Where one or both of the expressions are true, the output grid will have a value of 1, and where neither of the expressions is true, the output grid will have a value of 0. For instance, Figure 20.7 contains two rasters—elevation and landcover. We want to find all grid cells with an elevation value of less

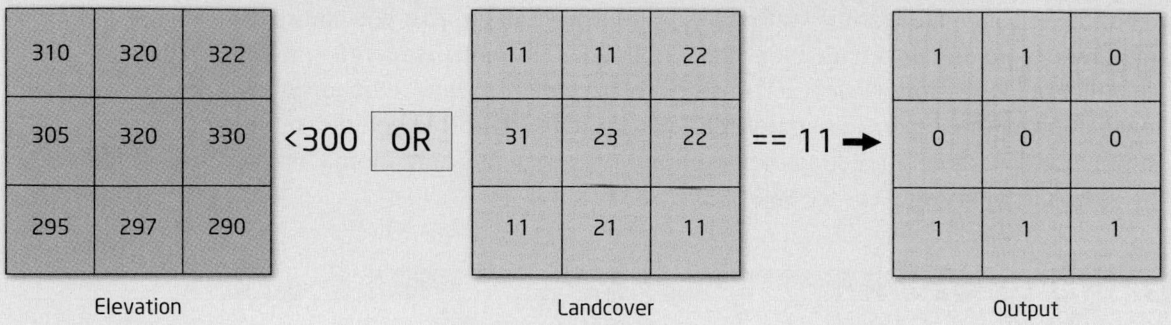

310	320	322
305	320	330
295	297	290

Elevation

<300 **OR**

11	11	22
31	23	22
11	21	11

Landcover

== 11 ➡

1	1	0
0	0	0
1	1	1

Output

FIGURE 20.7 An example of using the OR expression in Map Algebra.

than 300 as well as all landcover values equal to 11. The Map Algebra expression (elevation < 300) OR (landcover == 11) will create a new grid with values of 1 for cells that meet one or both criteria and values of 0 for cells that meet neither of the criteria. Note that, in the Raster Calculator, the OR operation is represented with the | symbol.

• **XOR**: This Boolean operator is used to find the "exclusive or" overlay of two expressions. "Exclusive or" refers to finding everything except what the two expressions have in common with each other. When one of the expressions is true, the output grid will have a value of 1, and where both (or neither) of the expressions are true, the output grid will have a value of 0. For instance, Figure 20.8 contains two rasters—elevation and landcover. We want to find all grid cells with an elevation value of less than 300 or all landcover values equal to 11, but not both conditions. The Map Algebra expression (elevation < 300) XOR (landcover == 11) will create a new grid with values of 1 for cells that meet one criteria and values of 0 for cells that meet neither (or both) of the criteria. Note that, in the Raster Calculator, the XOR operation is represented with the ^ symbol.

FIGURE 20.8 An example of using the XOR expression in Map Algebra.

310	320	322
305	320	330
295	297	290

Elevation

<300 **XOR**

11	11	22
31	23	22
11	21	11

Landcover

== 11 ➡

1	1	0
0	0	0
0	1	0

Output

XOR (Map Algebra) The Boolean operator used with an "exclusive or" of two grids.

NOT (Map Algebra) The Boolean operator used with a negation of a grid or expression.

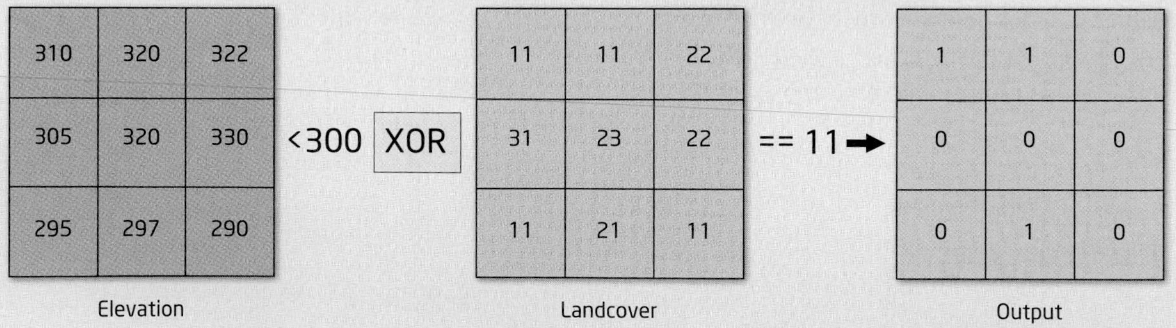

• **NOT**: This Boolean operator is used to find the negation of an expression. Where the expression is false, the output grid will have a value of 1, and where the expression is true, the output grid will have a value of 0. For instance, Figure 20.9 contains one raster—landcover. We want to find all grid cells that have a landcover value not equal to 11. The Map Algebra expression NOT (landcover == 11) will create a new grid with values of 1 for cells that are any value other than 11 and 0 for cells that have a value of 11. Note that, in the Raster Calculator, the NOT operation is represented with the ~ symbol.

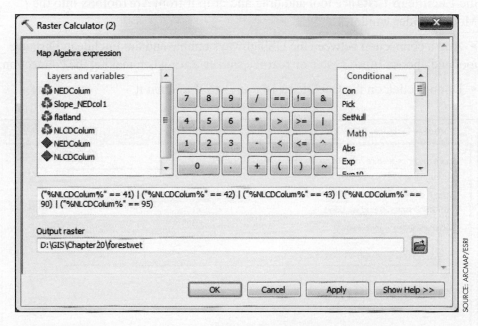

FIGURE 20.9 An example of using the NOT expression in Map Algebra.

- Build a query to find values from the NLCDColum layer that are equal to 41, 42, 43, 90, or 95.

- Use the == button for an equals sign.

- Use the | button for OR.

- Call the resulting Output raster **forestwet** and save it in your D:\GIS\ Chapter20 folder.

  - *Important Note:* Use the model version of NLCDColum (the one with the three arrows in a triangle, not the layer version, which is the one with the yellow diamond next to it).

  - *Important Note:* You must use parentheses around each expression in the Raster Calculator.

- Answer Question 20.4.

QUESTION 20.4 What would the output grid result be if you used "and" instead of "or" in each case of this Map Algebra expression? Why?

- Click **OK** to complete the setup of the tool. Back in the ModelBuilder, the new criterion will be completed. You should see the arrow stretching from the NLCDColum layer to the Raster Calculator tool.

STEP 20.5 Third Criterion: Must Be More Than 500 Meters from a Highway

- Drag and drop the **UShighways** feature class from ArcMap's Table of Contents into the ModelBuilder window. You may want to rearrange the window, zoom, or expand it to give yourself more room to work (since there are still more criteria to be modeled).

- First, you have to compute distance from highways. Choose the **Spatial Analyst Tools** toolbox in ArcToolbox, select the **Distance** toolset, then select the **Euclidean Distance** tool and drag and drop it from ArcToolbox into the ModelBuilder window.

- Add a connection between the UShighways bubble and the Euclidean Distance tool and choose **Input raster or feature source data** when making the connection.

- Double-click on the **Euclidean Distance tool** to open it.

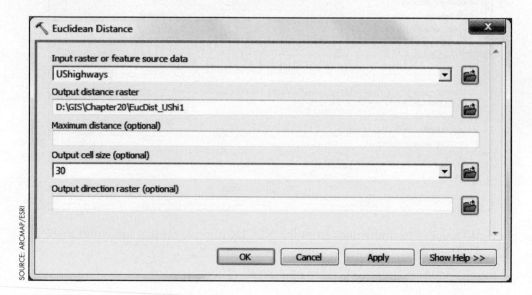

- Make sure that all parameters (especially the cell size) are set correctly:
  - **UShighways** should be set as the Input raster or feature source data.
  - The default name for the distance calculation should be **EucDist_UShi1**.
  - The output cell size should be **30** (this should be set because of your Geoprocessing Settings from Step 20.1).
- Click **OK**.
- The tool will now be properly set up in ModelBuilder.

- Now that you've calculated distance, you have to find those areas greater than 500 meters from a highway, and you need another relational tool to do this. From the **Spatial Analyst Tools** toolbox in ArcToolbox, select the **Map Algebra** toolset, then select the **Raster Calculator** tool and drag and drop it from ArcToolbox into the ModelBuilder window.

- Add a connection between the EucDist_UShi1 raster and the Raster Calculator tool. Select **Map Algebra expression** when prompted, then open the tool itself.

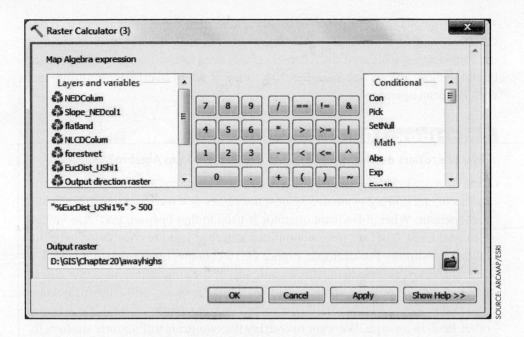

- Build an expression in the Raster Calculator of the EucDist_UShi1 grid greater than 500 (because you want to find places more than 500 meters from a highway). First, select the layer with which you want to work (in this case, **EucDist_UShi1**), then select the operator you want to use (in this case, the > operator for a greater than operation), and last, select the value to use (in this case, **500**).

- Name the Output raster **awayhighs** and save it in your D:\GIS\Chapter20 folder.

- Click **OK** when done.

STEP 20.6 Fourth Criterion: Must Be Within 500 Meters of a Lake or of a River

- Repeat the same process as the third Criterion to compute distance from the Columbiana County streams. This time, use the less than (<) operator to select those areas within 500 meters of streams. Name your final grid **nearstreams**.

  - Make sure that all grid cell sizes are 30 meters.

  - Repeat the same process by computing distance from the Columbiana County lakes. This time, use the less than (<) operator to select out those areas within 500 meters of lakes. Name your final grid **nearlakes**.

  - Make sure that all grid cell sizes are 30 meters.

STEP 20.7 Completing and Running the Map Algebra Model

- By this point, you should have the results of five sets of tools:
  - **Flatland**: Places on the landscape with a relatively flat slope
  - **Forestwet**: Land uses that are forests or wetlands
  - **Awayhighs**: Areas that are more than 500 meters from a highway
  - **Nearstreams**: Areas that are within 500 meters of a river
  - **Nearlakes**: Areas that are within 500 meters of a lake

• You'll now have to combine (or overlay) all five of these grids in order to find the most suitable sites for the ecological preserve, using the criteria laid out in the Introduction section of this chapter. You'll need to use Boolean operators to combine these layers. See **Smartbox 103** for more about overlaying multiple grids with Boolean operators.

Smartbox 103

How are rasters overlaid with Boolean operators in Map Algebra?

Two or more grids can be overlaid using Boolean operators, in which case the values of one grid are evaluated against the values of a second grid using a Boolean operator. When a Boolean operator is used in this fashion, non-zero values are considered to be the "true" condition and zero values are considered to be the "false" condition. For example, Figure 20.10 contains two grids—lowland and water. The lowland grid contains a value of 1 if the cell represents a low elevation and a value of 0 for higher elevation, while the water grid contains a value of 1 if the grid cell is classified as a water land-cover type and a 0 if it is some other land-cover type. We want to overlay the two grids to find only those cells that are both low elevation and are classified as water. The expression (lowland AND water) will produce an output grid with a value of 1 for those cells that have a non-zero value in the lowland grid and a non-zero value in the water grid.

FIGURE 20.10 Using an AND expression to overlay two grids.

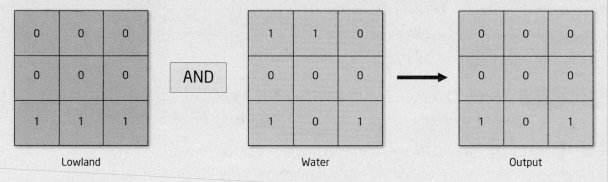

Lowland   Water   Output

Multiple Boolean operators may be used in a single Map Algebra expression to overlay more than two grids. Figure 20.11 contains three grids—lowland, water, and nearroads. The lowland grid contains a value of 1 if the cell represents a low elevation and a value of 0 for higher elevation. The water grid contains a value of 1 if the grid cell is classified as a water land-cover type and a 0 if it is some other land-cover type. The nearroads grid contains a value of 1 if the grid

FIGURE 20.11 Using multiple Boolean operators to overlay three grids.

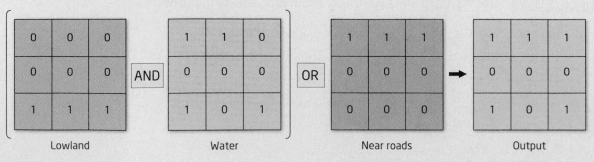

Lowland   Water   Near roads   Output

cell is close to a road and a value of 0 if the grid cell is far from a road. We want to overlay the three grids to find only those cells that are both low elevation and classified as water, in addition to those cells that are near a road. The expression ((lowland AND water) OR nearroads) will produce an output grid with a value of 1 for those cells that have a non-zero value in the lowland grid and a non-zero value in the water grid, but also have either a zero or non-zero value in nearroads.

- *Important Note:* Think very carefully about how you combine layers using the Boolean operators. Also keep in mind how parentheses are used in an expression as well as how the order of operations of mathematics is evaluated. Before doing any more in ModelBuilder, answer Question 20.5.

> **QUESTION 20.5** Write down the operation needed to find the suitable sites using the five final output grids and appropriate logical operators. For example, you would write something like *flatland XOR (awayhighs OR nearrivers)* — or whatever set of grids and operators will find the suitable sites.

- From the **Spatial Analyst Tools** toolbox in ArcToolbox, select the **Map Algebra** toolset, then select the **Raster Calculator** tool and drag and drop it from ArcToolbox into the ModelBuilder window. Position this tool to the side of all of the five criteria outputs.

- Make a connection from each of the five criteria outputs to the Raster Calculator tool, using the Map Algebra expression for each one.

- Next, open the Raster Calculator tool and construct the Map Algebra expression to properly combine these five grids using the Boolean operators (AND, OR, NOT, XOR) and parentheses.

- Name the Output raster **finalsites** and save it in your D:\GIS\Chapter20 folder.

- Click **OK** to close the Raster Calculator tool—you should see that all five of your criteria outputs now have black arrows connecting them to the tool.

- Now, from the ModelBuilder environment's **Model** pull-down menu, select **Run Entire Model**. ModelBuilder will run each tool and create the output from each. You'll see each tool turn red as it is run. When the Model finishes running, you can minimize it.

STEP 20.8 Examining Map Algebra Results

- Add the finalsites grid to the Table of Contents. Change this grid to have values of 1 (the suitable sites) in a distinctive color (red, blue, black, etc.) and transparent (or "no color") for values of 0 (the unsuitable sites). Put finalsites at the top of the Table of Contents and turn off all other layers.

- You'll see that there are many, many grid cells that meet your criteria for where to place the ecological preserve. Answer Question 20.6.

> **QUESTION 20.6** How many grid cells are considered "suitable sites" for the ecological preserve? In real-world values, how many square meters are considered "suitable sites"?

- The suitable sites are spread across the map. You will now assess the locations of the ecological preserve in relation to the other land-cover features.

- Turn on the NLCDColum layer underneath the finalsites layer. If necessary, use the layer properties to change the colors of the suitable sites and land-cover classifications to make analysis easier on you (i.e., to make the suitable sites really stand out from the land uses).

> **graphics** Elements (such as shapes or text) that can be added to a view or layout but are not data layers.

- In answering Questions 20.7 and 20.8, you'll need to refer to some specific locations in the results. To reference specific areas (for example, to draw a circle or polygon around them or add some informative text) you can create graphics using the Draw toolbar. **Graphics** are elements that can be added to a view or layout but are not data layers. For example, you could draw a polygon shape as a graphic, but it would not be saved as a polygon feature class; it would simply be an additional picture on the screen, not geospatial data. Several useful functions of the Draw toolbar are highlighted in Figure 20.12.

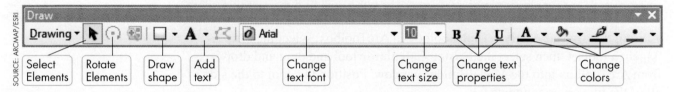

FIGURE 20.12 Some of the properties of the Draw toolbar in ArcGIS 10.2.

- Using the **Draw shape** tool you can draw various graphics (from several options in the Draw shape pull-down menu). The **Add text** tool will create a text box into which you can place text, and the other tools will allow you to change the color or properties of your items. Each graphic is treated like a map element in a layout (see Chapter 3); you can choose the **Select Elements** cursor from the Draw toolbar, click on an element (for instance, a circle you drew), and adjust the properties of that element (such as the fill color or line thickness of the circle).

- Answer Questions 20.7 and 20.8. You should reference specific locations of your results when answering these questions.

> **QUESTION 20.7** What sites would you (as the developer in this scenario) consider particularly non-viable in relation to other land uses? When answering this question, refer to specific sites on your final results (and highlight these on the screen with graphics). For example, are there any sites (specifically) that you feel are too close to urban areas, too close to agricultural areas, or too small for an ecological preserve? Explain (using specific highlighted or circled sections of your results) why these areas are particularly non-viable.

> **QUESTION 20.8** What sites would you (as the developer in this scenario) consider particularly viable in relation to other land uses? When answering this question, refer to specific sites on your final results (and highlight these on the screen with graphics). Explain (using specific highlighted or circled sections of your results) why these areas would be particularly viable (in contrast with the non-viable sites you chose in Question 20.7). Make sure the graphics you use here can be visually distinguished from those you used in Question 20.7.

STEP 20.9 **Saving and Printing Your Results**

- Save the model. Also save your work as a map document. Include the usual information in the Map Document Properties.

- Compose and print a layout (see Chapter 3) of your final version of your sites (including all of the usual map elements and design for a layout) so that potential sites can be clearly seen in relation to the land cover. Your graphics and text should be readable and the sites that correspond to your answers to Questions 20.7 and 20.8 should be easily distinguished from one another. Also make sure that the final layout has land-cover classifications properly labeled (i.e., make sure your legend says things like "commercial" or "residential" instead of things like "21" or "22").

- You will also want to print out your final geoprocessing workflow and attach it to your layout. Under the ModelBuilder's Model pull-down menu are options for Print Setup, Print Preview, and Print. Only print out one final version of the flowchart, and have the entire geoprocessing workflow on a single page.

Closing Time

This chapter examined how to use mathematical, relational, and Boolean operators on rasters via Map Algebra for a variety of different applications. Map Algebra allows you to work with grids in several ways, including querying rasters and performing raster overlay. However, many more options are available when you are overlaying rasters. For instance, this chapter assumed that all five of the criteria were of equal importance when combining them to find potential sites for the ecological preserve. If one of the criteria had been more important than the others (if, for example, the presence of forests or wetlands was twice as important as the other factors), you would have to give more weight to that layer than to the others. See the *Related Concepts for Chapter 20* for more information about performing weighted overlays with rasters in ArcGIS.

In this chapter, you worked with ModelBuilder to visualize the workflow of taking the various layers, performing actions on them, and overlaying them using the Raster Calculator. ModelBuilder is an excellent tool for setting up a workflow to track how layers and tools connect with one another, and in essence you created a simple "model" to identify possible ecological preserve sites. However, much more can be done with ModelBuilder; as the name implies, it's also used for creating GIS models, something that we'll do in the next chapter.

Related Concepts for Chapter 20   Using Weighted Raster Overlay

When combining layers, there are times you'll find that some layers are more important, or should carry more weight, than other layers. For instance, a historical preservationist is attempting to determine which sections of the landscape are under the greatest development pressures, based on a variety of conditions (such as proximity to existing urban areas or the slope of the land). Based on key informant interviews and other research, she determines that the proximity to water bodies is 80 percent of the overall importance, while the slope of the land is merely 20 percent of the overall importance in selecting a site. Thus, she can't simply add

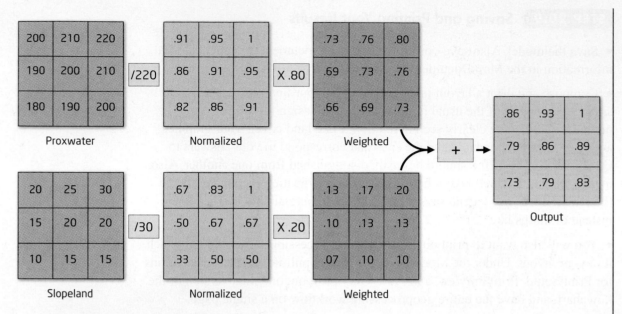

FIGURE 20.13 The weighted overlay process, showing how two grids are normalized, weighted, and then combined together using Map Algebra techniques.

the two grids together. Rather, she needs to give a greater weight to the proximity-to-water-bodies grid and then combine the two grids. This process of assigning relative weights to layers during overlay is referred to as **weighted overlay**.

Figure 20.13 shows an example of weighted overlay. In this example, two sample grids (proxwater and slopeland) simulate two factors. The first has values that represent the distance in meters to a body of water and the second shows the degree of slope of the land. However, you'll see that the values for the proxwater and the slopeland grids are completely different—a value of 15 in the slopeland grid indicates a 15-degree slope, while a value of 15 in the proxwater grid indicates 15 meters from a body of water. Because a numerical value of 15 means two completely different things, we can't simply add these two values together.

The same holds true for a value of 220. In proxwater, this number indicates a location 220 meters from a body of water, while in slopeland, this number means a 220-degree slope (which is impossible). Before we can combine the rasters, we must first bring these raster values to a level at which they mean the same thing, such as expressing them as a percentage. Altering all values so they are on the same level results in values that have been **normalized** (see Chapter 3 for more about normalizing numerical values). A simple way to normalize these raster values is to divide each grid cell value by the highest overall value in the raster (using a Map Algebra division operation). After normalizing, the values in both grids mean the same thing.

Next, from research, we know that the proxwater grid should make up 80 percent of the analytical weight, and the slopeland grid should consist of 20 percent of the weight. The process of weighting each grid involves multiplying the (normalized) values by the assigned weight. In this case, all of the normalized proxwater grid values will be multiplied by 0.80, while the normalized slopeland grid values will be multiplied by 0.20 (using a Map Algebra multiplication operation). Proper selection of weights is often an involved process, sometimes informed by expert opinions on the subject at hand or crafted through a complicated mathematical or computerized process.

Once each grid has been weighted, the grids can be added together (with a Map Algebra addition operation) to create a ranking of the site locations. The final output can be interpreted as "The higher the grid value, the more appealing that location is for urban development, based on the concept that sites farthest from water are much more desirable than sites with a higher degree of slope of the land." In ArcGIS, the **Weighted Overlay** tool will help with these tasks.

For More Information

For further in-depth information about the topics presented in this chapter, use the ArcGIS Help feature to search for the following items:

- A quick tour of using Map Algebra
- Boolean AND (Spatial Analyst)
- Boolean NOT (Spatial Analyst)
- Boolean OR (Spatial Analyst)
- Boolean XOR (Spatial Analyst)
- Building expressions in Raster Calculator
- How Weighted Overlay works
- How Raster Calculator works
- Raster Calculator (Spatial Analyst)
- What is Map Algebra?
- What is ModelBuilder?

Key Terms

Map Algebra (p. 463)
workflow (p. 464)
ModelBuilder (p. 467)
relational operator (p. 471)
Raster Calculator (p. 472)

Boolean operator (Map Algebra) (p. 475)
AND (Map Algebra) (p. 475)
OR (Map Algebra) (p. 475)
XOR (Map Algebra) (p. 476)

NOT (Map Algebra) (p. 476)
graphics (p. 482)
weighted overlay (p. 484)
normalize (p. 484)

How to Build a Model in ArcGIS 10.2

Introduction

When you're working with a GIS project, chances are it'll involve the use of multiple types of datasets, tools, and methods. For example, a GIS analysis for determining which sections of coastal and inland areas are most likely to be converted to an urban land use through development would need to examine a wide variety of factors. Proximity to the shoreline, proximity to existing urban areas, and the designation of private versus public lands are just a few of the factors that would be required in this kind of analysis. With GIS, you would use a variety of tools and different spatial datasets to design and then combine all of these factors. You would then analyze the results to find which areas of land are likeliest to be developed.

A **GIS model** examines all of these factors in an attempt to describe or explain a process or to predict results. Numerous types of GIS models can be found in the literature for a variety of different applications. You can use the ModelBuilder utility in ArcGIS 10.2 to design and share your own models. In ArcGIS 10.2, a **model** is a workflow using different layers and tools that can also be used as its own separate tool. In Chapter 20, you used ModelBuilder to design a simple model to implement a workflow of various Map Algebra concepts. You can use that same graphical design interface to create a standalone tool that can be executed to combine multiple tools and datasets, which you can then share with others.

A model consists of a set of **variables**, or different types of factors, that are combined as part of the overall analysis. For instance, in the example in the opening paragraph, "proximity to the shoreline" would be a variable; so would "proximity to existing urban areas." In ModelBuilder, these variables can be created through the use of input layers and tools. To determine "proximity to the shoreline" in ModelBuilder, you would take a layer representing the shoreline and use the Euclidean Distance tool to create a new raster in which each grid cell has a value of that cell's proximity to the features of the shoreline layer (similar to what you did in Chapter 19).

Keep in mind that this is just one simple operation and models will often contain many types of processes used together. See Figure 21.1 for an example: An initial layer called "all urban areas" first must be projected to the coordinate system used by the other layers. Next, the distance to all the features in the urban areas layer is computed, and that distance layer is then sliced into 10 subsets. All of these tools and layers can be chained together in ModelBuilder. As each tool is used, a new layer is generated. This concept of taking a layer, applying a process to it, and generating a new layer (which can then have another process applied to it, if necessary, to generate a third layer) is called **cartographic modeling**. (Note that the modeling you're doing in this chapter involves raster layers, but models can be designed using both vector and raster layers.)

GIS model A representation of the factors used for explaining the processes that underlie an event or for predicting results.

model An ArcGIS tool that combines data layers, geoprocessing tools, and outputs in a workflow.

variables Different factors that are combined into an overall analysis.

cartographic modeling The concept of taking a layer, applying a process or action to it, and creating a new layer as an output.

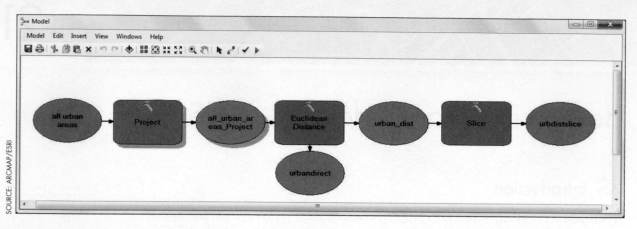

SOURCE: ARCMAP/ESRI

FIGURE 21.1 One part of a sample model created using ModelBuilder.

Chapter Scenario and Applications

In this chapter, you will be taking the role of a planner in Columbiana County, Ohio. You are searching for the most suitable site in the county for relocating a specific animal species. The relocation area has to meet the following criteria: (1) It must be in a forested area of land, and (2) it must be at a low elevation. In addition, the relocation area ideally should be (3) as close as possible to a water area, and (4) as far away as possible from an urban or developed area. Last, the relocation area must be on a section of land that is at least 25 acres. The relocation area has to meet the first two criteria, and be the best option while taking the third, fourth, and fifth criteria into account.

The following are additional examples of other real-world applications of this chapter's skills.

SOURCE: ALINA G/SHUTTERSTOCK

- A historical preservationist is trying to determine which sections of unprotected Civil War battlefield land are under the greatest development pressures in order to best direct preservation efforts or funds. She can build and run a GIS model to help locate the sites under the greatest pressures.

- An environmental planner is trying to find the most suitable sites in a state for designation as a protected animal habitat based on a variety of different factors. He can build a GIS model to combine these factors to aid in determining which sites are most suited for the habitat.

- An urban planner is attempting to determine how many residents of a city live on the 100-year and 500-year floodplains. He can use multiple layers in a GIS model to establish the floodplains, then compute the residential population in these zones.

ArcGIS Skills

In this chapter, you will learn:

- How to construct, validate, and run a model using ModelBuilder.

- How to add model parameters to a model.

- How to use multiple tools in a model.

- How to use NoData in Reclassification.

- How to use the Slice tool to create rankings.

- How to display rasters based on attributes other than VALUE.

- How to create and evaluate a suitability index.

- How to create a geoprocessing package from your model for sharing online.

Study Area

For this chapter, you will be working with data from Columbiana County, Ohio.

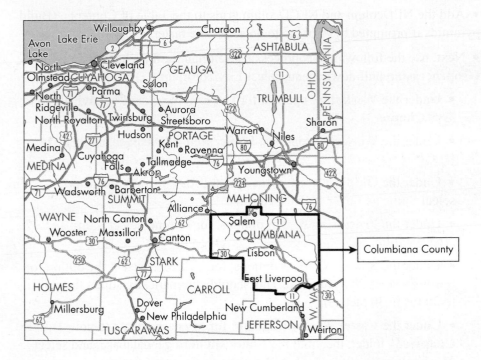

Data Sources and Localizing This Chapter

This chapter's data focuses on features and locations within Columbiana County, Ohio. However, you can easily modify this chapter to use data from your own local county instead—for instance, if you wanted to perform this chapter's applications in Marlboro County, South Carolina. The NED and NLCD2006 that were downloaded from The National Map are also available for Marlboro County; you could clip the layers to the county boundaries and process them to keep consistent 30 meter cell sizes for this exercise.

The Maskcolum grid was created by taking the NLCD grid and performing Map Algebra to extract all grid cells with a value of greater than 1, then reclassifying all values of 0 as No Data. The same could be done to create a mask grid for Marlboro County using its NLCD layer.

STEP 21.1 Getting Started

- Start ArcMap and use the Catalog to copy the folder called **Chapter21** from the C:\GISBookdata\ folder to your own D:\GIS\ drive. Be sure to copy the entire directory, not just the files within it.

- Chapter21 contains a personal geodatabase called **Columsitedata** that contains the following items:

  - **NLCDcolum**: A raster of 30 meter NLCD 2006 data for Columbiana County.

  - **NEDcolum**: A raster of NED data for Columbiana County (which has been resampled to 30 meters for purposes of this chapter).

  - **Maskcolum**: A 30 meter raster that conforms to the boundaries of Columbiana County.

- Activate the **Spatial Analyst** extension.

- Add the NEDcolum and NLCDcolum grids to the Table of Contents. (Build pyramids if prompted by ArcMap to do so for the raster layers.)

- Next, use the following Geoprocessing Settings for this chapter (from the **Geoprocessing** pull-down menu, select **Environments…**):

  - Under the **Workplace** options, for **Current Workspace**, navigate to **D:\ GIS\Chapter21**.

  - Under the **Workspace** options, for **Scratch Workspace**, navigate to **D:\GIS\Chapter21**.

  - Under the **Output Coordinates** options, for **Output Coordinate System**, select **Same as Layer "NLCDcolum"**.

  - Under the **Processing Extent** options, for **Extent**, select **Same as Layer NLCDcolum**.

  - Under the **Raster Analysis** options, for **Cell Size**, select the option for **Same as layer NLCDcolum**. This should automatically adjust the cell size to 30 (as in 30 meters).

  - Under the **Raster Analysis** options, for **Mask**, navigate to your **D:\GIS\ Chapter21** folder, then into the **Columsitedata** geodatabase, and select **Maskcolum**.

    - Click **OK** when you're done. All of your Environment Settings should now be in place for use in the rest of the chapter.

- The coordinate system being used in this chapter is **Albers Conical Equal Area (NAD 83)**, using Map Units of **Meters**. Check the Data Frame to verify that this coordinate system is being used.

- Open ArcToolbox and dock it between the Table of Contents and the blank map area.

- You will also be using the ModelBuilder environment (see Chapter 20) to build the suitability model in this Chapter. As in Chapter 20, the first thing to do is create a new Toolbox in which to place your model. In Catalog, right-click on the **Columsitedata** geodatabase, select **New**, and select **Toolbox**. A new toolbox will be added to your geodatabase and given the name "Toolbox."

- Rename the "Toolbox" as **Sitemodeltools** (for containing your site suitability model).

- To create a new model to be stored in the Sitemodeltools toolbox, right-click on **Sitemodeltools**, select **New**, then select **Model**. A new item called Model will appear in the Sitemodeltools toolbox and a black ModelBuilder dialog will appear. For now, close the ModelBuilder dialog named "Model."

• In Catalog, rename the new item called Model as **SiteSuitModel**.

• Right-click on the **SiteSuitModel** and select **Edit** to begin working with the model—the blank ModelBuilder dialog will reopen, but this time named "Site-SuitModel." As you saw in Chapter 20, ModelBuilder works as a drag-and-drop environment, so position it to the right of ArcToolbox for ease of use. Remember, each item you'll be using in ModelBuilder will be a model element. For further information about model elements, see **Smartbox 104**.

Smartbox 104

What are model elements and how are they represented in ModelBuilder?

In ModelBuilder, each of the items you'll work with is considered a **model element** (Figure 21.2). There are three main items in ModelBuilder that are model elements:

• **Variables**: This is the ModelBuilder terminology for either data layers or values. For instance, an NLCD layer could be added to ModelBuilder as a data variable, while an input for a query threshold could be used as a value variable. By default, data variables appear as dark blue ovals and value variables appear as lighter blue ovals. The output from a tool can also be considered a variable, and is referred to as *derived data* (and by default will appear as a green oval).

• **Tools**: These are the various items from ArcToolbox that are used to perform an action on a variable. For instance, the Project tool or the Slice tool from ArcToolbox can be added as Tool model elements in Model-Builder. Python scripts and completed models with parameters can also be added as Tool model elements. By default, tools appear as yellow rectangles. A special type of tool called an *Iterator* (by default shown as a six-sided orange box) allows you to repeat actions several times or perform multiple iterations of an action.

• **Connectors**: These are the lines that link variables and tools. For instance, to project a layer called "roads" a connector is drawn between the data

model element Each of the items (Variables, Tools, and Connectors) that can be placed into the ModelBuilder interface.

Variable (ModelBuilder) Specific term used in ModelBuilder for a data layer or value input (by default they appear as blue ovals in ModelBuilder).

Tool (ModelBuilder) Specific term used in ModelBuilder for one of the geoprocessing tools, models, or scripts (by default they appear as yellow rectangles).

Connector (ModelBuilder) Specific term used in ModelBuilder for one of the lines that link Variables and Tools (by default they appear as one of four types of lines).

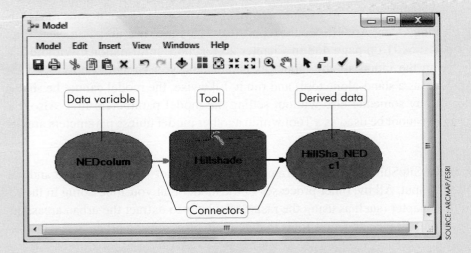

FIGURE 21.2 Various model elements inside the ModelBuilder environment.

SOURCE: ARCMAP/ESRI

variable "roads" and the "Project" tool. By default, a solid black line joins data to a tool, a dashed black line indicates a connector based on a Geo-processing Environment Setting, a dotted black line indicates some sort of precondition that must be met before the tool will run, and dashed blue lines are feedback connectors that loop the output of a tool back into the tool again.

• Drag and drop the **NLCDcolum** grid from ArcMap's Table of Contents into the ModelBuilder window. Also drag and drop the **NEDcolum** grid from ArcMap's Table of Contents into the ModelBuilder window. These will be your two model elements from which everything else in the model is built.

STEP 21.2 **Setting Model Parameters**

• Before we begin building the model, we'll set up the model's parameters. (See **Smartbox 105** for more about model parameters.)

Smartbox 105

What are model parameters and how are they used in ArcGIS 10.2?

model parameters The user-defined inputs or variables of a model.

One advantage of creating a model in ArcGIS is the flexibility to use (or re-use) it with different inputs. For instance, the model you're creating in this chapter is reliant on two different inputs—the NLCD layer and the NED layer. Everything else you'll be working with is created or derived from these two layers. In theory, after completing this chapter, you could take an NLCD and a NED layer for another county and plug them into the model and examine the output for animal-relocation areas for that county. You wouldn't have to remake the model—all of those processes would remain the same; all that would have changed would be these two inputs. The ability to set up differing inputs (whether data variables or value variables) makes these two layers the **model parameters**—they are the only items that the user has to specify when running the model.

Using model parameters also allows you to re-run a model by using differ-ent inputs. For instance, if a model input was a value for population, you could set this as a model parameter and allow the users of the model to type in what-ever population value they want. In the ModelBuilder dialog, the variables that are used as model parameters will have a letter "P" next to them.

For a model to open and run as an independent item, the parameters must be set. For instance, in Chapter 20 when you tried to re-open your model by selecting Open, ArcGIS would report that "This tool has no parameters" (see **Troublebox 11** on page 468 in Chapter 20 for more information). Without set-ting up the inputs of the model as parameters, you cannot simply open the model (as a stand-alone tool) and run it. Likewise, the model cannot be shared or run by someone else without setting the model parameters first. Also, the model cannot be used as a Tool within another model unless parameters are set.

• In the SiteSuitModel, two parameters need to be set—the NED input and the NLCD input. All the other processes of the model that you'll be doing in the rest of this chapter (such as using the raster calculator to extract the urban areas, or using the distance and slice tools to determine proximity to urban areas) are all derived from these two initial parameters. The first thing we'll do is to rename these two items to be more useful and intuitive for model inputs than NEDcolum or NLCDcolum. In the Site-SuitModel itself, right-click on the **NLCDcolum model element** and select **Rename**.

- In the Rename dialog box, type **NLCD input layer** and click **OK**. You'll see that the NLCDcolum element now is labeled as NLCD input layer.

- Use the same process to rename the NEDcolum model element as **NED input layer**.

- Save the SiteSuitModel (see Chapter 20 for how to save a model).

- From the **SiteSuitModel's Model** pull-down menu, select **Model Properties**. In the SiteSuitModel Properties dialog box, select the **Parameters** tab. Here is where you will add the layers that are model parameters. Press the **Plus** button and, from the new list of items that appears in the Add Model Parameter dialog box, select both **NED input layer** and **NLCD input layer** and click **OK** in the Add Model Parameter dialog. You'll see that those two layers are now added as model parameters.

- Accept the default settings (because both of these layers are required to run the model, "Required" is already the default state for them as model parameters) and click **Apply**.

- Next, select the **Environments** tab. These settings allow the model to override any of the Geoprocessing Environment Settings that are already in place for when the model is run. To set up which environments will be overridden, place a checkmark in the box for the particular option and then press the **Values** button. This will bring up another dialog like the Environment Settings you put in place in Step 21.1. Make sure that the following settings are in place (and if they are not, change them):

- Under the **Output Coordinates** options, for **Output Coordinate System**, select **Same as Layer "NLCDcolum"**.

- Under the **Processing Extent** options, for **Extent**, select **Same as Layer NLCDcolum**.

- Under the **Raster Analysis** options, for **Cell Size**, select the option for **Same as layer NLCDcolum**. This should automatically adjust the cell size to 30 (as in 30 meters).

- Under the **Raster Analysis** options, for **Mask**, navigate to your **D:\GIS\ Chapter21** folder, then into the **Columsitedata** geodatabase, and select **Maskcolum**.

- When all settings are in place, click **Apply**, then click **OK** to close the dialog.

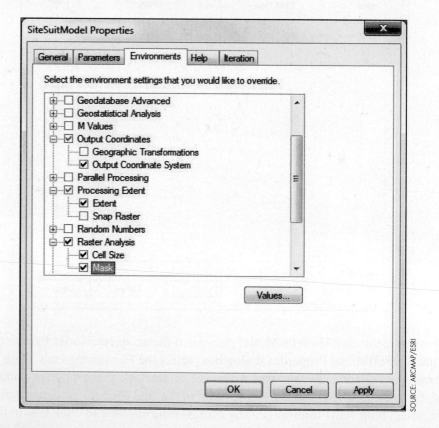

SOURCE: ARCMAP/ESRI

STEP 21.3 First Criterion: Must Be in a Forested Area

- The first criterion for the animal-relocation area is that it must be in a forested area. To extract these areas from the NLCD grid, you will have to build an expression to do so using the Raster Calculator. From the **Spatial Analyst Tools** toolbox in ArcToolbox, select the **Map Algebra** toolset, then select the **Raster Calculator** tool and drag and drop it from ArcToolbox into the ModelBuilder window.

- Add a connection between NLCD input layer and the Raster Calculator tool (though the black line doesn't appear). Select the option for **Map Algebra expression** when making the connection.

- When the Raster Calculator opens, use it to create an expression that will allow you to select all the grid cells representing forested land-cover types (those values of 41, 42, and 43).

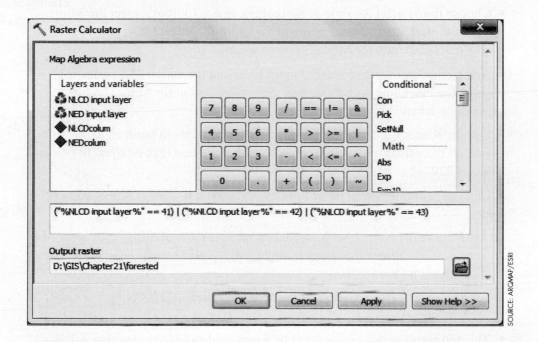

- Build a query to find values from the NLCD input layer variable that are equal to 41, 42, or 43 (see Chapter 20 for how to build a Map Algebra expression using the Raster Calculator).

- Use the == button for an equals sign.

- Use the | button for OR.

- Call the resulting output raster **forested** and save it in your D:\GIS\Chapter21 folder.

  - *Important Note:* Use the model variable version of NLCD input layer (the one with the three arrows in a triangle, not the layer version called NLCDcolum with the yellow diamond next to it).

  - *Important Note:* Use parentheses around each expression in the Raster Calculator.

- The end result of this process will be a new grid called forested that will have all grid cells that met the criteria from the Raster Calculator expression (i.e., all NLCD grid cells with values of 41, 42, or 43) assigned a value of 1 and all cells that did not meet the criteria (i.e., all NLCD grid cells with a value other than 41, 42, or 43) assigned a value of 0.

- Click **OK** to close the Raster Calculator. Back in ModelBuilder, you'll see the NLCD input layer map element, Raster Calculator tool, and forested derived data output grid all properly connected.

STEP 21.4 **Second Criterion: Must Be at a Low Elevation**

• The second criterion for the animal-relocation area is that it must be at a low elevation. From examining the values of the NEDcolum grid, you'll see that the county's elevations run from 200 meters to 440 meters. To determine lower elevations, you will now build a second Raster Calculator expression to find all of the areas in the county that are below 320 meters.

• Choose the **Spatial Analyst Tools** toolbox in ArcToolbox, select the **Map Algebra** toolset, then select the **Raster Calculator** tool and drag and drop it from ArcToolbox into the ModelBuilder window.

• Add a connection between NED input layer and the Raster Calculator tool (though the black line doesn't appear). Select the option for **Map Algebra expression** when making the connection.

• Use the Raster Calculator's buttons and menu options to build an expression to find all the values of the NED input layer that are less than or equal to (< =) a value of 320.

• Call the resulting output raster **lowelev** and save it in your D:\GIS\Chapter21 folder.

  • *Important Note:* Use the model variable version of NED input layer (the one with the three arrows in a triangle, not the NEDcolum layer with the yellow diamond next to it).

  • *Important Note:* Use parentheses around each expression in the Raster Calculator.

• The end result of this process will be a new grid called lowelev that will have all grid cells that met the criteria from the Raster Calculator expression (i.e., all NED grid cells that were either less than or equal to a value of 320) assigned a value of 1 and all cells that did not meet the criteria (i.e., all NED grid cells with a value greater than 320) assigned a value of 0.

• Click **OK** to close the Raster Calculator. Back in ModelBuilder, you'll see the NED input layer model element, Raster Calculator tool, and lowelev derived data output grid all properly connected.

STEP 21.5 **Must Be at Both a Forested Area and at a Low Elevation**

• Next, you'll want to find the areas that are both forested and at a low elevation. Because both of these grids only contain values of 0 (does not meet the criteria) or 1 (meets the criteria), we can simply multiply the two grids together in a Map Algebra expression (see Chapter 20 for further information on Map Algebra). The output from this process will be a grid with values of 1 that are both forested and low elevation, and values of 0 everywhere else.

• Choose the **Spatial Analyst Tools** toolbox in ArcToolbox, select the **Map Algebra** toolset, then select the **Raster Calculator** tool and drag and drop it from ArcToolbox into the ModelBuilder window. Position this Raster Calculator tool so that both the forested and lowelev grids can clearly connect to it. Open the Raster Calculator tool.

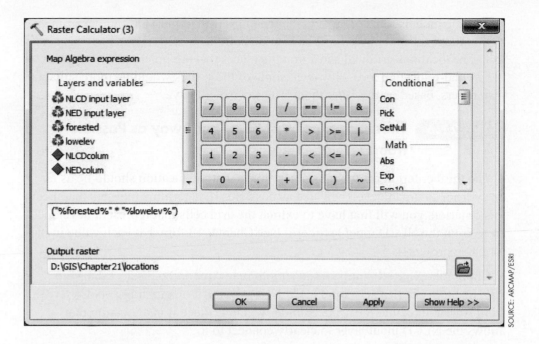

SOURCE: ARCMAP/ESRI

- Build a Map Algebra expression to multiply the forested grid and the lowelev grid.

- Call the Output raster **locations** and save it in your D:\GIS\Chapter21 folder.

- Click **OK**. Answer Question 21.1.

QUESTION 21.1 Why did you use multiplication to overlay the two grids instead of addition? What would be the end result of this expression if you had used addition instead of multiplication?

- Back in ModelBuilder, you'll see that your locations grid is set up as the product of the forested and lowelev grids.

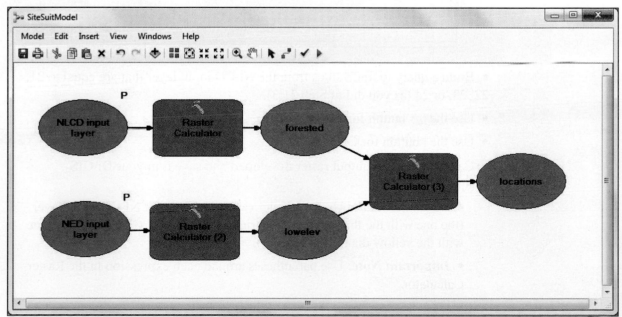

SOURCE: ARCMAP/ESRI

- Run the model at this point, add the locations grid to the TOC, and examine it. Your animal-relocation area will have to be built in an area with grid cells equal to 1. The locations grid will show you the places that are potentially viable for the animal-relocation area. However, some of these sites would be more suitable than others, based on the other three criteria you still have.

STEP 21.6 Third Criterion: Must Be as Far Away as Possible from Developed Areas

- The third criterion for the relocation area is that the location should be as far away from an urban/developed area as possible. To determine the sites that meet this criterion, you will first have to extract the grid cells that represent developed areas from the NLCD input layer grid (see Chapter 12 for what each value in the NLCD represents).

- Choose the **Spatial Analyst Tools** toolbox in ArcToolbox, select the **Map Algebra** toolset, then select the **Raster Calculator** tool and drag and drop it from ArcToolbox into the ModelBuilder window. Place it in a location that allows the NLCD input layer to clearly connect to it.

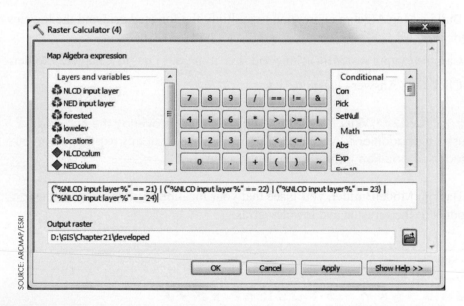

SOURCE: ARCMAP/ESRI

- Build a query to find values from the NLCD input layer that are equal to 21, 22, 23, or 24 (as you did in Step 21.3).

- Use the == button for an equals sign.

- Use the | button for OR.

- Call the resulting output raster **developed** and save it in your D:\GIS\ Chapter21 folder.

  - *Important Note:* Use the model variable version of NLCD input layer (the one with the three arrows in a triangle, not the NLCDcolum layer one with the yellow diamond next to it).

  - *Important Note:* Use parentheses around each expression in the Raster Calculator.

- Click **OK**.

- **Run** the model at this point to generate the developed grid. The developed grid will consist of values of 1 for grid cells assigned a developed land-cover type, and values of 0 for non-developed areas.

- The next step is to find the distance away from each developed grid cell. As in Chapters 19 and 20, you can do a Euclidean distance calculation to find this information. However, you only want to find the distance from those cells with a value of 1, not all cells (1's and 0's) in the grid. To do this, you'll first have to reclassify the grid (see Chapter 19) so that all cells with a value of 0 get assigned a value of No Data.

- From the **Spatial Analyst** toolbox in ArcToolbox, select the **Reclass** toolset, then select the **Reclassify** tool and drag and drop it from ArcToolbox into the ModelBuilder window. Place it in a location that allows the developed grid to clearly connect to it.

- Make a connection between developed and the Reclassify tool, selecting **Input raster** when making the connection.

- Open the **Reclassify** tool.

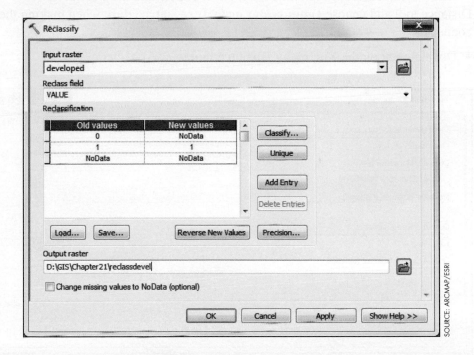

- For the Input raster, choose the **developed** variable grid.

- For the Reclass field, choose **VALUE**.

- Press the **Unique** button (so that you can assign a new value to each old value in the grid).

- For Old values of 0, make them New values of **NoData**.

- For Old values of 1, leave them as New values of **1**.

- For Old values of NoData, leave them as New values of **NoData**.

- Call the Output raster **reclassdevel** and save it in your D:\GIS\Chapter21 folder.

- Click **OK** when all settings are correct.

- Run the model again to create a new grid called reclassdevel. Add it to the TOC. It should consist only of values of 1 (and NoData) that show the locations of developed areas in the county.

- Now you can calculate distance from each of these developed grid cells. Answer Question 21.2.

> **QUESTION 21.2** Why was it necessary to reclassify the grid so that values of 1 remained the same but values of 0 were assigned to NoData? (*Hint:* Consider that the next step is calculating distance and think about how the distance calculation will be performed from this source raster.)

- Choose the **Spatial Analyst Tools** toolbox in ArcToolbox, select the **Distance** toolset, then select the **Euclidean Distance** tool and drag and drop it from ArcToolbox into the ModelBuilder window.

- Add a connection between the **reclassdevel** bubble and the Euclidean Distance tool and choose **Input raster or feature source data** when making the connection.

- Double-click on the **Euclidean Distance tool** to open it.

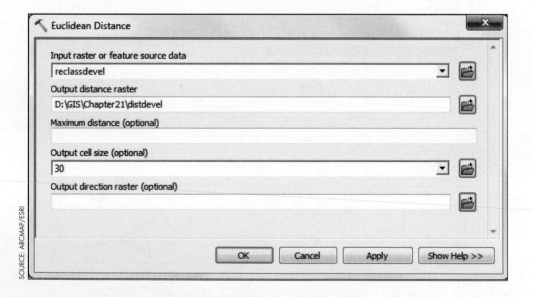

- For the Input raster, select the **reclassdevel** grid.

- Call the Output distance raster **distdevel**.

- Make sure the Output cell size is **30**.

- Click **OK** to calculate the new distance raster.

- The new distdevel raster will have a value showing the distance each cell is from a developed cell. However, this distdevel grid will have thousands of different values. What we want to find is which areas are closest to an urban area and which areas are farthest away. The next step is to calculate this by breaking up these thousands of distance values into more manageable groups on a 1-to-10 scale, where the most desirable places (those farthest away) get a value of 10 and the least desirable places (those closest to developed areas) get a value of 1. In ArcGIS, you can do this using the Slice tool (see Chapter 19).

• From the **Spatial Analyst** toolbox in ArcToolbox, select the **Reclass** toolset, then select the **Slice** tool and drag and drop it from ArcToolbox into the ModelBuilder window. Place it in a location that allows the distdevel grid to clearly connect to it.

• Make a connection between distdevel and the Slice tool, selecting **Input raster** when making the connection.

• Open the **Slice** tool.

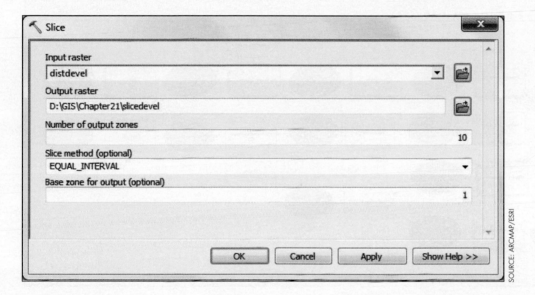

SOURCE: ARCMAP/ESRI

• For the Input raster, select **distdevel**.

• Call the Output raster **slicedevel** and save it in your D:\GIS\Chapter21 folder.

• Use **10** for the Number of output zones.

• Select **EQUAL_INTERVAL** for the Slice method.

• Use **1** for the Base zone for output.

• Click **OK** when all settings are correct.

• Run the model and add the **slicedevel** grid to the TOC. Each cell will have a value from 1 through 10, with 10 in those cells farthest from a developed area and 1 in those cells closest to a developed area.

STEP 21.7 Fourth Criterion: Must Be as Close as Possible to a Body of Water

• The fourth criterion for the relocation area involves being as close as possible to a body of water. Repeat the above steps to find these areas as follows:

• Use the Raster Calculator to extract all cell values of 11 (indicating an "Open Water" land cover) from the NLCD input layer variable. Call this new grid **water**.

- Reclassify the water grid so it only has values of 1 and NoData instead of 1, 0, and NoData. Call this new grid **reclasswater**.

- Calculate Euclidean distance from the reclasswater grid. Call this new grid **distwater**.

- Last, Slice the distwater grid into 10 equal-interval output zones. Call this new grid **slicewater**.

- Your model is getting more complicated, so you'll probably have to adjust the position of the model elements or resize the model accordingly, as we've still got more to do.

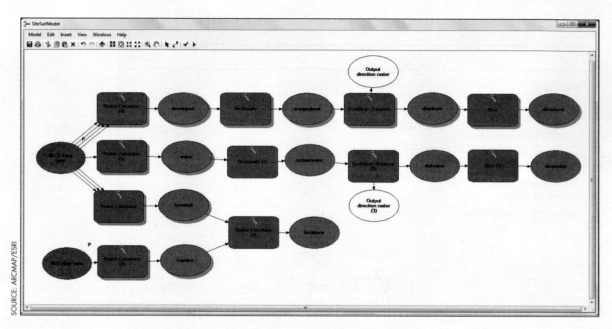

SOURCE: ARCMAP/ESRI

- **Run** the model to generate the slicewater grid and add it to the TOC.

- You'll notice that the slicewater grid has its highest values (of 10) as those places farthest from a body of water and its lowest values (of 1) as those places closest to a body of water, which is the opposite of what we want. Thus, you will have to reclassify the slicewater grid to reorder the values.

- From the **Spatial Analyst** toolbox in ArcToolbox, select the **Reclass** toolset, then select the **Reclassify** tool and drag and drop it from ArcToolbox into the Model-Builder window. Place it in a location that allows the slicewater grid to clearly connect to it.

- Make a connection between slicewater and the Reclassify tool, selecting **Input raster** when making the connection.

- Open the **Reclassify** tool.

- Select **slicewater** for the Input raster.

SOURCE: ARCMAP/ESRI

- Use **VALUE** as the Reclass field.

- Press the **Unique** button to assign individual new values to each of the old values.

- Call the Output raster **reslicewater** and save it in your D:\GIS\Chapter21 folder.

- To make those areas closest to the water bodies higher values and those areas farthest away into lower values, Reclassify your new values as follows:

Old Values	New Values
1	10
2	9
3	8
4	7
5	6
6	5
7	4
8	3
9	2
10	1

- Click **OK** to close the tool. Run the model again and add the reslicewater grid to the TOC. Examine both slicewater and reslicewater and answer Question 21.3.

QUESTION 21.3 How do the slicewater and the reslicewater grids compare with each other? How do they relate to the water grid?

STEP 21.8 Creating a Suitability Index

- Now you have one grid (slicedevel) that shows the distance rankings from 1 to 10 for proximity to developed areas, and a second grid (reslicewater) that shows the rankings from 1 to 10 for proximity to water bodies. We are assuming that values of 10 in each grid are the most desirable areas (they are the values farthest from developed lands in the slicedevel grid and the closest to water in the reslicewater grid). The next step will be to overlay these two grids.

- To overlay these two grids, we use the Raster Calculator. Choose the **Spatial Analyst Tools** toolbox in ArcToolbox, select the **Map Algebra** toolset, then select the **Raster Calculator** tool and drag and drop it from ArcToolbox into the ModelBuilder window. Position this Raster Calculator tool so that both the slicedevel and reslicewater grids can easily connect to it. Open the Raster Calculator tool.

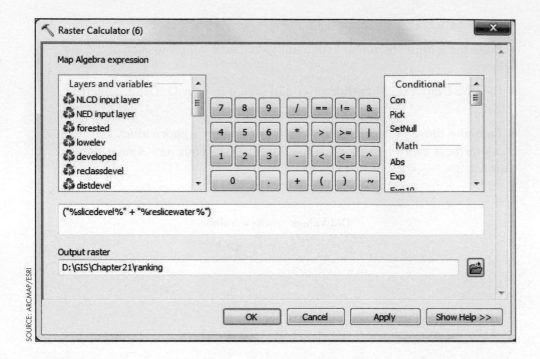

SOURCE: ARCMAP/ESRI

- Use the Raster Calculator to build an expression to add the slicedevel grid and the reslicewater grid together. Call the output raster **ranking**. After calculating, run the model and add this new ranking grid to the TOC. What you have created here is a suitability index for assessing your suitable sites—for more information about suitability indexes, see **Smartbox 106**. Answer Questions 21.4, 21.5, and 21.6.

QUESTION 21.4 What do values of 2, 3, and 4 represent in the ranking grid? What do values of 17, 18, and 19 represent?

QUESTION 21.5 How many grid cells have a value of 2? How many grid cells have a value of 19?

QUESTION 21.6 As a result of adding two grids of values of 1 to 10 together, the ranking grid should have values from 2 to 20. However, there are no values of 20 in the ranking grid. Why?

Smartbox 106

What is a suitability index?

In Chapter 20, you identified the best locations in the county for an ecological preserve, a type of site suitability analysis. The results of that analysis were locations that were either suitable (cells with a value of 1) or unsuitable (cells with a value of 0). Similarly, the "locations" grid you created in this chapter shows which locations in the county will be suitable for animal relocation (those that are forested and at a low elevation, and that have a value of 1) and all other locations in the county (that are not forested or wetlands, or at a higher elevation, and have a value of 0 and are thus unsuitable). The disadvantage of this type of site-suitability analysis is it tells you only which sites are desirable and which are not. Sometimes, this may be all of the information that you need, but in some studies you may want additional information on just how suitable a particular site is compared with other suitable sites.

To obtain a ranking of site suitability, you would want to build a **suitability index**. This type of result will help you develop a wider range of options for suitable sites beyond the basic "suitable or not suitable" binary output. For example, Figure 21.3 shows three sample grids, each one a variable for determining a suitable site for a vacation home: distance to forests, distance to rivers, and distance to roads. In each raster, the cells have a value of 1 through 5, with this value representing a ranking of the suitability of that site (1 being the least desirable and 5 being the most desirable). By adding all of these ordinal values (see Chapter 2) together, you can obtain a ranking of the relative suitability of each site. For instance, in the suitability index shown in Figure 21.3, those grid cells with values of 12 or 13 would be the most suitable sites for a vacation home, those cells with values of 7 or 10 would be the next step down in terms of suitability, and those cells with values of 3 or 4 would be the least suitable choices.

> **suitability index** A ranking of locations according to a set of criteria.

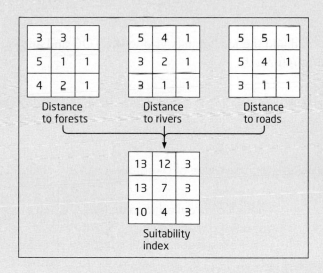

FIGURE 21.3 An example of overlaying three rasters to create a basic suitability index.

- You now have the information you need to determine the third and fourth criteria for finding the animal-relocation areas. It's time to combine this ranking of all grid cells in the county with the chosen locations identified from the first two criteria by overlaying the ranking and locations grids.

- The next step is to overlay these two grids using the Raster Calculator. Choose the **Spatial Analyst Tools** toolbox in ArcToolbox, select the **Map Algebra** toolset, then select the **Raster Calculator** tool and drag and drop it from ArcToolbox into the ModelBuilder window. Position this Raster Calculator tool so that both the ranking and locations grids can easily connect to it.

- Open the Raster Calculator tool in ModelBuilder and build an expression multiplying **location** and **ranking**. Call the results from this calculation **locateranks**.

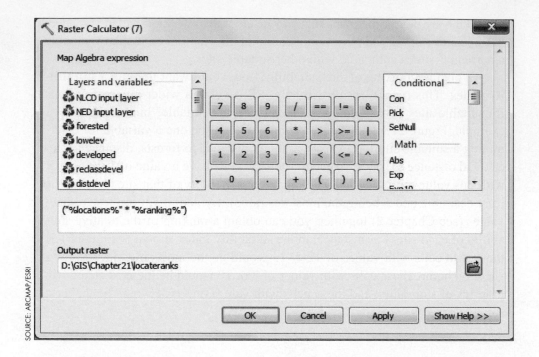

SOURCE: ARCMAP/ESRI

- Run the model and add the locateranks grid to the TOC and answer Questions 21.7 through 21.10.

> **QUESTION 21.7** Why are there now no values of 19 in the locateranks grid? (That is, why is the value of 18 the highest value in the locateranks grid?)

> **QUESTION 21.8** What does the value of 0 represent in the locateranks grid?

> **QUESTION 21.9** Create a new grid called locateranks2, calculated by adding the locations and ranking grid together with the Raster Calculator in ModelBuilder. Why is the output of this grid not useful in determining the final sites for the animal-relocation areas? Delete the locateranks2 grid (and any tools and connects for it in the final model) when you're done.

> **QUESTION 21.10** Create a new grid called locateranks3, calculated by overlaying the locations and ranking grid together with the AND operator in the Raster Calculator in ModelBuilder. Why is the output of this grid only 1's and 0's compared with the results you obtained from multiplying the two grids? Delete the locateranks3 grid (and any tools and connects for it in the final model) when you're done.

STEP 21.9 **Final Criterion: Find Large Sections of Best Possible Land Area**

- The last step is to find the best sections of land on the largest possible areas to relocate the animals to (because we need a minimum of 25 acres of land). To do this, you'll have to regiongroup the ranked sites together (see Chapter 12) to form contiguous sections of land to examine.

• From the **Spatial Analyst Tools** toolbox in ArcToolbox, select the **Generalization** toolset, then select the **Region Group** tool and drag and drop it from ArcToolbox into the ModelBuilder window. Position this Region Group tool so that the locateranks grid can easily connect to it. Open the Region Group tool.

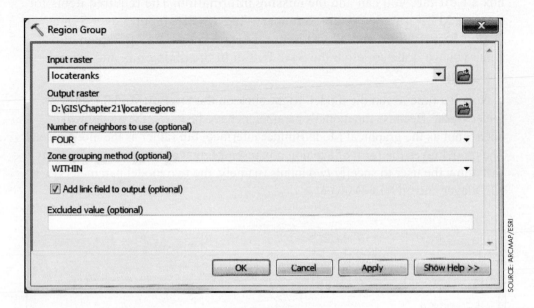

• **locateranks** should be the Input raster.

• Call the Output raster **locateregions** and save it in your D:\GIS\Chapter21 folder.

• Use **EIGHT** for the Number of neighbors to use.

• Use **WITHIN** for the Zone grouping method.

• Be sure to put a checkmark in the **Add link field to output** box.

• Click **OK** when you're ready.

• Run the model and add the locateregions grid to the TOC. It will show all the regions created by grid cells of similar ranking values that have been merged together (i.e., all cells of value 8 that are contiguous will be joined to form a region, all cells of value 9 that are contiguous will be joined to form another, different region, and so on).

STEP 21.10 Validating and Running the Model

• In this step, you will validate and run the complete model. When a model is validated, ArcGIS will check to be sure that all of the model's variables and parameters are correct. When a model is validated all of the variables will be reset as if they haven't been run while ArcGIS checks all of the model's elements to make sure everything is okay. To validate the model, in the **Model** pull-down menu of the **SiteSuitModel**, select **Validate Entire Model**.

- Before proceeding, save the SiteSuitModel.

- Next, close the SiteSuitModel box because we're now going to run the entire model. In Catalog, navigate to the D:\GIS\Chapter21 folder and expand the Sitemodeltools toolbox. Before running the model, we want to provide documentation for it. In the Sitemodeltools toolbox, right-click on **SiteSuitModel** and select **Item Description**. A new dialog box will appear with information about the model. By selecting the Item Description dialog box's **Edit** tab, you can add the missing information. The required items for you to add information to are: Summary, Tags, NED_Input_Layer Syntax Dialog Explanation, and NLCD_Input_Layer Syntax Dialog Explanation. Once you've done so, click the **Save** button in the dialog box and close the dialog.

- Now it's time to run the model. Right-click on the **SiteSuitModel icon** and select **Open**. Because the model's parameters have been set, the model will open—not as the graphical ModelBuilder interface, but rather as the model itself. The model takes the form of a dialog box resembling a geoprocessing tool that will allow the user to specify two inputs (namely, the two model parameters)—a NED layer and an NLCD layer.

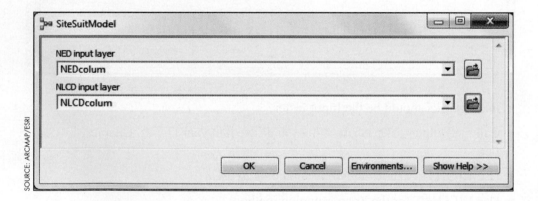

- Click **OK**. A new dialog box will appear showing each process of the SiteSuit-Model as it runs. When it completes, close the dialog box.

- At this point, the entire model has been rerun and any previously created layers have been overwritten. The areas in the locateregions grid have met the first four of the model criteria as well as being brought together into large contiguous areas. In the next step, we'll evaluate the results of the model and find which of these large areas best meets our criteria as well as the minimum size requirement.

STEP 21.11 Analyzing Site Suitability Model Results

- Display the locateregions grid using **Unique Values** and the **LINK** attribute as the **Value Field**. As noted in Chapter 12, Link will show you the original values assigned to the grid cells (i.e., their original ranking) prior to the Region Group process. Change the Color Scheme to a ramp of colors that will show the lower-ranked grid cells in a more distinctive color than the higher-ranked grid cells.

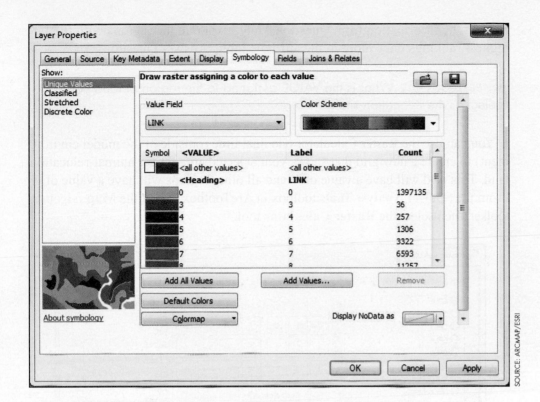

• Open the attribute table for the locateregions grid and sort the LINK field so that the higher values are sorted at the top of the attribute table. You should have over 5000 different regions, but only a few of them would be the most desirable places to use as a relocation area for the animals.

• Carefully examine the regions with the highest rankings to determine which one you will select as the relocation area. The values in the attribute table can be read as follows:

  • The number under VALUE indicates the grid-cell identifier for a particular region (i.e., the first region created will have a VALUE of 1, the second region created will have a VALUE of 2, etc.).

  • The number under LINK indicates the ranking of that region (this is the original value from the locateranks grid).

  • The number under COUNT indicates the number of grid cells that comprise that particular region.

• Keep in mind that you're looking at the number of grid cells in each region, and you need a plot of land with at least 25 acres. Before deciding on the best region to which to relocate the animals, you should calculate how many grid cells are equivalent to 25 acres (and thus, you can disregard all regions of less than this number). Answer Question 21.11.

QUESTION 21.11 The size of each grid cell is 900 square meters (since they are 30 meter resolution grid cells). One square meter (m^2) is equal to 0.000247105 acres. How many grid cells represent 25 acres? (Show your work.)

• With the information from Question 21.11, examine the regions with the highest rankings once more in the attribute table. Answer Question 21.12.

> **QUESTION 21.12** What is the VALUE assigned to the region of land you will select as the relocation site?

• You can use the Raster Calculator one final time (outside of the model environment) to create a new grid that shows your selected area for the animal-relocation land. This grid will have a value of 1, and all other regions will have a value of 0. From the **Spatial Analyst Tools** toolbox in ArcToolbox, select the **Map Algebra** toolset, then select the **Raster Calculator** tool.

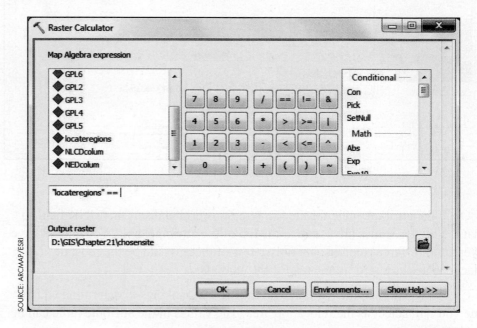

• Build a query to find values from the locateregions grid (note that now you're using the layer itself, so use the one with the yellow diamond, rather than the model version with the three triangles) that are equal to the VALUE of the chosen region.

• Use the == button for an equals sign.

• Call the resulting output raster **chosensite** and save it in your D:\GIS\ Chapter21 folder.

• Click **OK**. The chosensite grid will consist of values of 1 for the region you've selected and values of 0 for all other areas.

• Zoom in closely on this area and answer Question 21.13.

> **QUESTION 21.13** Why was this one site chosen over other regions that were ranked higher?

STEP 21.12 Printing or Sharing Your Results

• Save your work as a map document. Include the usual information in the Map Document Properties.

- Next, turn off all layers except the chosensite layer and change the symbology so that the values of 1 are clearly distinguished from the values of 0.

- Compose and print a layout (see Chapter 3) of your final version of your sites (including all of the usual map elements and design for a layout) so that the dimensions of the final chosen site can be clearly seen. Also, you want to add a graphic of the model to the layout as well. In the Sitemodeltools toolbox, right-click on **SiteSuitModel** and select **Edit**. When the SiteSuitModel ModelBuilder dialog re-opens, select the **Model** pull-down menu, then select **Export**, and then select **Graphic**. Save a graphic of the model as a .jpg file. Next, in the Layout, use the Insert Picture option (see Chapter 15) to place a large, readable graphic of the model on the layout as well.

- An advantage of creating a model with parameters is that others can utilize it as well. If you want to share your model, you can accomplish this with a **geoprocessing package** (.gpk) file. A geoprocessing package can take any tool (or a completed and run model with parameters) and save the model or tool (along with its data) as a single file that can be shared (either to disk or to ArcGIS Online). To save your SiteSuitModel as a geoprocessing package and share it, do the following:

  - From the **Geoprocessing** pull-down menu, select **Results**.

  - A new, pinnable window called Results will open. In it, expand the option for **Current Session** and you'll see your model.

  - Right-click on the **SiteSuitModel icon** and, from the menu of options, select **Share As** and then select **Geoprocessing Package**. A new dialog box will open with several tabs of required information to fill out. When the information has been entered, click **Analyze**. A new box called "Prepare" will appear at the bottom of ArcGIS and show you any errors that the geoprocessing package may have. If there are no errors in the Prepare box, then click the **Share** button in the Geo-processing Package dialog box and follow the prompts to log into ArcGIS Online to share your geoprocessing package.

geoprocessing package A file that contains a model, its inputs, and results that can be shared.

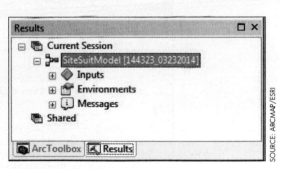

SOURCE: ARCMAP/ESRI

Closing Time

This chapter examined how to create a functioning stand-alone model using ModelBuilder that can be used as a separate tool or shared via a geoprocessing package. Models are very powerful and versatile items and can be as simple or as complex as needed for problem solving. By using model parameters, you can tweak and fine-tune your models for a variety of different inputs and variables. In the next chapter, we'll continue using ModelBuilder to design a specific type of model that will use several new tools for hydrologic applications.

All of the tasks you performed in this chapter could have been performed using individual tools, such as Slice, Reclassify, and the Raster Calculator, but ModelBuilder provides a more intuitive visual interface for conceptualizing and running workflows. The Python scripting language can be used for modeling as well—see the *Related Concepts for Chapter 21* for more about using Python with models.

Related Concepts for Chapter 21 Models and Python

As noted in Chapter 9, geoprocessing can be done in ArcGIS through the Python scripting language. For example, all of the modeling done in this chapter could have also been accomplished through programming using Python. In fact, your model can be converted over to a Python script—you can do this for your own benefit so that you can see how certain functions can be coded using Python, but you can also do it for distribution of your model. To convert a model to a Python script, in the box of the Model itself, select the **Model** pull-down menu, then choose **Export**, then select **To Python Script**. From there, you can give it a name and your model will be converted to a text file with the .py extension, creating a Python script from it.

You can open this script using a utility like PythonWin (see Chapter 9). The script can also be opened in the ArcGIS Python window itself. To examine your python script in the Python window, first open the Python window. Next, in the left-hand pane of the window, right-click, and from the menu that appears, select **Load**. In the Open dialog box that appears, navigate to where you saved your exported script, select it, and open it. Your exported model script will now be added to the Python window (Figure 21.4). Scroll through it to see how your model variables and processes (as well as the use of the Spatial Analyst extension) are coded in Python.

FIGURE 21.4 An ArcGIS model exported to a Python script.

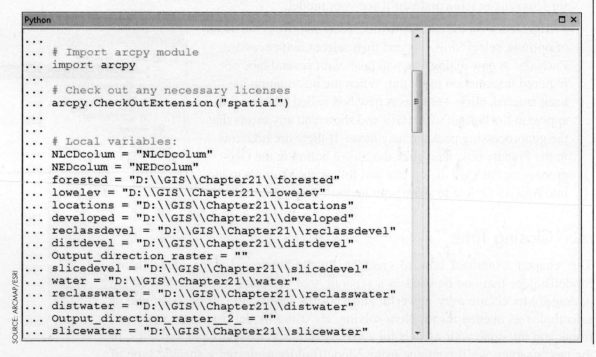

SOURCE: ARCMAP/ESRI

For More Information

For further in-depth information about the topics presented in this chapter, use the ArcGIS Help feature to search for the following items:

- A quick tour of geoprocessing packages
- Essential geoprocessing vocabulary

- Essential ModelBuilder vocabulary
- Exporting a model to a Python script
- Model elements
- Renaming model parameters
- Tutorial: Creating tools with ModelBuilder
- Validating a model
- What is a geoprocessing package?

Key Terms

GIS model (p. 487)

model (p. 487)

variables (p. 487)

cartographic modeling (p. 487)

model element (p. 491)

Variable (ModelBuilder) (p. 491)

Tool (ModelBuilder) (p. 491)

Connector (ModelBuilder) (p. 491)

model paramaters (p. 492)

suitability index (p. 505)

geoprocessing package (p. 511)

How to Use Hydrologic Modeling Tools in ArcGIS 10.2

Introduction

Numerous types of models can be constructed using the various ArcGIS tools. One set of tools with numerous applications related to modeling water-related issues is in the Hydrology toolset available in the Spatial Analyst toolbox. **Hydrologic modeling** refers to examining factors related to the movement of water across a landscape. As water falls on the terrain, it moves across the land, and an elevation surface is used in GIS to model the effects of the direction and accumulation of the flow of water, as well as to delineate streams and catchment areas. Like the networks discussed in Chapter 11, a stream is defined by its length (referred to as a *link*) and where it meets another stream (referred to as a *junction*). In ArcGIS, the Hydrology toolset allows you to start with an elevation surface of an area, model the flow and accumulation of water, and extract the area's streams and watershed boundaries.

A **watershed** is an area in which all of the water moving through it flows to a common location. For example, the Chesapeake Bay Watershed covers a large multi-state area because the drainage location for the region's water will end up as the Chesapeake Bay. Smaller watersheds can be defined by a river system. For instance, the Mahoning River Watershed covers portions of eight northeast Ohio counties and is formed by several streams flowing into other streams and eventually into the river. The boundaries of the watershed (referred to as the *drainage divide*) define the area for which all water will drain to a common location (such as the river, the bay, or a culvert). In this chapter, we will use the links of the various streams in a county and delineate a watershed area (also called a *sub-basin*) for each stream link (the portion of the stream between junctions, Figure 22.1).

hydrologic modeling Examining the movement of water over a surface.

watershed The area for which all locations within it flow to a common drainage area.

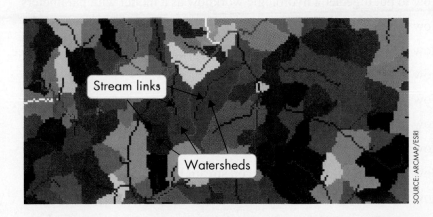

FIGURE 22.1 A portion of a stream network and the watershed area computed for each link of a stream (each watershed and stream link is a different color).

SOURCE: ARCMAP/ESRI

The source for the various hydrologic processes we'll be working with in this chapter is a digital elevation model. As such, these hydrologic modeling processes assume that all locations receive an equal amount of rainfall and that the soil and land-cover characteristics don't affect the overland flow of water. Our work in deriving factors like the watersheds or the stream network will be based on the DEM and created as raster surfaces. We'll thus end up with raster surfaces in which we can assess the direction of the flow of water or the accumulation of water at each grid cell location.

Chapter Scenario and Applications

This chapter puts you in the role of an environmental scientist examining potentially polluted areas in relation to the stream network of Columbiana County, Ohio. In particular, you want to determine what areas of the landscape will flow or drain into specific streams in the county. To this end, you'll start with a digital elevation model of the county and use it to compute watersheds for each section of a stream in the county and then analyze your results.

The following are additional examples of other real-world applications of this chapter's skills:

- A county engineer wants to model the amount of water run-off for a study area. She can use the hydrologic modeling tools to determine the direction and accumulation of water flow as layers for her analysis.

- An urban planner is modeling the flood potential of an area where a new subdivision is being planned. She can use the hydrologic modeling tools to examine the watershed area and the amount of flow accumulation for the proposed site.

- An archeologist is working onsite and wants to determine the drainage basin area that would collect rainfall. He can use the watershed tools to generate this area as another layer to add to the GIS data for the site.

ArcGIS Skills

In this chapter, you will learn:

- How to put together a hydrologic workflow as a model with parameters.
- How to calculate flow direction.
- How to locate and fill sinks to create a "depressionless" elevation surface.
- How to calculate flow accumulation.
- How to extract stream channels.
- How to set a value variable in a model.
- How to set up and evaluate stream ordering.
- How to extract watersheds based on stream links.
- How to work with intermediate data in a model.

Study Area

For this chapter, you will be working with data from Columbiana County, Ohio.

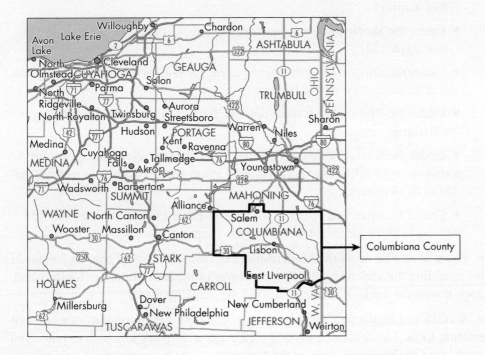

Data Sources and Localizing This Chapter

This chapter's data focus on features and locations within Columbiana County, Ohio. However, you can easily modify this chapter to use data from your own county instead, such as Harper County, Kansas. The NED was downloaded from The National Map and processed to align the layer to the county boundaries and keep consistent 30 meter cell sizes for this chapter; the same NED data are available for Harper County. The streams data were downloaded from The National Map hydrography data—use the NHD_Flowline layer for Harper County for streams.

STEP 22.1 **Getting Started**

• Start ArcMap and use the Catalog to copy the folder called **Chapter22** from the C:\GISBookdata\ folder to your own D:\GIS\ drive. Be sure to copy the entire directory, not just the files within it.

• Chapter22 contains a personal geodatabase called **Columhydrodata** that contains the following items:

   • **NEDcolum**: A raster of NED data for Columbiana County (which has been resampled to 30 meters for purposes of this chapter).

   • **Streams**: A line feature class representing the various types of rivers and streams in the county.

• Activate the **Spatial Analyst** extension.

• Add the NEDcolum grid to the Table of Contents. (Build pyramids if prompted by ArcMap to do so.)

- Next, use the following Geoprocessing Settings for this chapter (from the **Geoprocessing** pull-down menu, select **Environments...**):

    - Under the **Workplace** options, for **Current Workspace**, navigate to **D:\ GIS\Chapter22**.

    - Under the **Workspace** options, for **Scratch Workspace**, navigate to **D:\ GIS\Chapter22**.

    - Under the **Output Coordinates** options, for **Output Coordinate System**, select **Same as Layer "NEDcolum"**.

    - Under the **Processing Extent** options, for **Extent**, select **Same as Layer NEDcolum**.

    - Under the **Raster Analysis** options, for **Cell Size**, select the option for **Same as layer NEDcolum**. This should automatically adjust the cell size to 30 (as in 30 meters).

    - Click **OK** when you're done. All of your Environment Settings should now be in place for use in the rest of the chapter.

- You will also be using the ModelBuilder environment (see Chapters 20 and 21) for modeling the various hydrologic processes of the area. Start ArcToolbox and dock it with the Table of Contents next to the blank map area.

- ArcGIS will treat a new model as a tool because it's composed of one or more existing tools. The first thing to do is create a new Toolbox in which to place your model. In Catalog, right-click on the **Columhydrodata** geodatabase, select **New**, and select **Toolbox**.

- A new toolbox will be added to your geodatabase and given the name "Toolbox." Rename the "Toolbox" as **Hydrotools** (for containing your hydrologic model).

- To create a new model to be stored in the Hydrotools toolbox, right-click on **Hydrotools**, select **New**, then select **Model**.

- A new item called Model will appear in the Hydrotools toolbox. Rename this item as **Hydromodel**. The blank ModelBuilder dialog will also appear—position it to the right of ArcToolbox and the TOC for ease of use.

STEP 22.2   Setting the Hydro Model Parameters and Environments

- Drag and drop the **NEDcolum** grid from ArcMap's TOC into the ModelBuilder window. This grid will be the model element from which everything else in the model is built.

- Before we begin building the model, we'll set up the model's parameters (see Chapter 21 for more about setting model parameters). There are two parameters for the hydrologic model that you'll be creating in this chapter—a data variable and a value variable.

- First, in the ModelBuilder window, rename NEDcolum as **NED Input Layer**.

• Next, bring up the Hydromodel's Model Properties and add the NED Input Layer as a required model parameter (see Chapter 21 for instructions on how to do this). The first parameter (the data variable—the NED Input Layer) is now set. You'll set the value variable later in this chapter.

• Next, in the Model Properties, set the following Environments (see Chapter 21 for more about setting the Environments within Model Properties):

  • Under the **Output Coordinates** options, for **Output Coordinate System**, select **Same as Layer "NEDcolum"**.

  • Under the **Processing Extent** options, for **Extent**, select **Same as Layer NEDcolum**.

  • Under the **Raster Analysis** options, for **Cell Size**, select the option for **Same as layer NEDcolum**. This should automatically adjust the cell size to 30 (as in 30 meters).

STEP 22.3 Examining Slopes in Relation to Streams

• The first thing you'll generate for analysis is a slope grid of the NED Input Layer. From ArcToolbox's **Spatial Analyst Tools** toolbox, select the **Surface** toolset, then select the **Slope** tool and drag and drop it into the Model dialog.

• Build a connection between the NED Input Layer and the Slope tool. Call the new Output raster **slopegrid** and save it in your D:\GIS\Chapter22 folder.

• Run the model to create the slopegrid layer, then add it to ArcMap's Table of Contents.

• This slopegrid layer will give you some indication of how water will flow along the terrain surface of Columbiana County. Examine the layer to get a feel for where the stream channels are likely to be found.

• Add the streams feature class to the Table of Contents and position it so that you can see the streams layer on top of the slopegrid layer.

• Answer Question 22.1, then turn off both slopegrid and the streams layer for now (but we'll use the streams in later steps).

QUESTION 22.1 Without doing any further hydrologic modeling, and just from visual inspection, how are the streams in the Columbiana County area distributed with respect to slope?

STEP 22.4 Calculating Flow Direction

• The flow direction can be determined by using the Flow Direction tool. From the **Spatial Analyst Tools** toolbox, select the **Hydrology** toolset, then select the **Flow Direction** tool and drag and drop it into the Model dialog.

• Build a connection between the NED Input Layer and the Flow Direction tool. Double-click on the **Flow Direction tool** in the Model dialog to see its parameters.

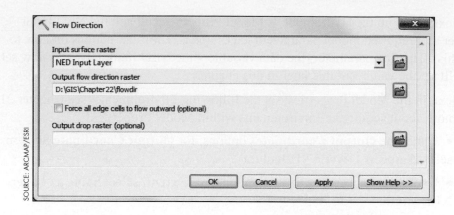

- Call the output flow direction raster **flowdir** and save it in your D:\GIS\ Chapter22 folder.

- Make sure the parameters are set correctly. For now, don't use the drop raster or force edge cells options. Click **OK** to close the dialog box.

- Run this portion of the model.

- This will create a new grid called **flowdir**, which shows the direction of water flow at each cell along the NED surface. For more information about flow direction, see **Smartbox 107**.

Smartbox 107

How does flow direction operate in ArcGIS 10.2?

Flow direction describes how water will move overland and in which direction it will move. The Flow Direction tool in ArcGIS 10.2 will determine in which of the eight cardinal directions water will move at each cell of the raster elevation surface. The computation is performed in a manner similar to computing slope aspect (see Chapter 15), as it is assumed that water will flow downhill via the steepest slope. The slope is computed from a grid cell to its adjacent grid cell in each of the eight directions. The grid cell with the steepest slope is chosen as the one to which water will flow.

ArcGIS uses the **D8** (deterministic 8) algorithm to calculate flow direction. This algorithm indicates that one (and only one) of the eight directions must be selected for water flow. With D8, water cannot move in multiple directions at once, nor is any randomness involved in which way water will flow. The flow direction is calculated by subtracting the elevation value of each of the eight adjacent cells from the elevation value of the grid cell for which flow direction is being determined. The four lateral directions (N, W, E, S) have this difference divided by a value of 1, and the four diagonal directions (NE, NW, SE, SW) have this difference divided by a value of 1.414 (because these differences are measured from the center of each cell—see Chapter 19 for further information). The grid cell with the overall greatest positive value is chosen as the cell to which water should flow. ArcGIS will then assign a numerical value to the grid cell to indicate which of the eight cells to which the water should flow, based on the direction from the originating cell. These newly assigned values are placed into a new "flow direction grid." The values assigned indicate which direction the water will flow from (Table 22.1):

flow direction ArcGIS tool that determines the direction in which overland water flow will move.

D8 An algorithm for determining flow direction that will output which of the eight cardinal directions will be the steepest slope for the flow of water.

Assigned Value in Flow Direction Grid	Direction of Flow
1	East
2	Southeast
4	South
8	Southwest
16	West
32	Northwest
64	North
128	Northeast

Table 22.1 The assigned values in the flow direction grid and their corresponding flow directions

Figure 22.2 shows an example of calculating the flow direction for one grid cell—in this case, the center cell in blue. The Elevation Raster contains elevation values at each grid cell (like the NED). The "Calculation" shows the values internally calculated by the flow direction process for each cell. These values are determined by subtracting the value of each cell from the center cell and dividing by the appropriate distance value (1 for lateral calculations, 1.414 for diagonal ones). For instance, the adjacent cell to the northeast is calculated to be 2.83 or ((10 – 6)/1.414). The steepest slope is found to be to the adjacent grid cell on the east (the value of 6 is the highest calculated value and this value will be direction of the steepest slope). Thus, at the cell's location in the flow direction grid a value of 1 is written, indicating that water at that cell will flow to the east. A flow direction value is then calculated for all cells in the elevation raster.

FIGURE 22.2 The initial elevation raster, the internal calculation of the steepest slope, and the corresponding output written to the flow direction grid.

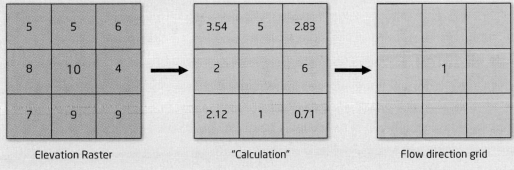

| Elevation Raster | "Calculation" | Flow direction grid |

- Display the flowdir grid as a graduated color (using value as your display field) and using Unique Values. Note the values computed for flow direction. Answer Question 22.2.

QUESTION 22.2 Why are there values other than 1, 2, 4, 8, 16, 32, 64, and 128 in your flow direction grid? (*Hint:* You may wish to refer to the ArcGIS Desktop Help entry for "Flow Direction (Spatial Analyst)" to answer this question.)

STEP 22.5 Locating Sinks

- You will now examine the surface for the sinks that exist on it. From the **Spatial Analyst Tools** toolbox, select the **Hydrology** toolset, then select the **Sink** tool and drag and drop it into the Model dialog.

- Build a connection between the flowdir layer and the Sink tool.

- Double-click on the **Sink tool** to open its dialog.

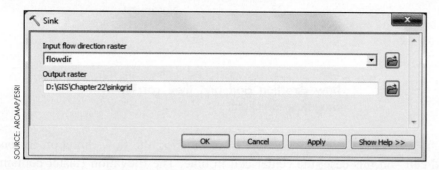

- Name your Output Raster **sinkgrid** and store it in your D:\GIS\Chapter22 folder. Click **OK** to close the Sink dialog.

- Run this portion of the model.

- Add the sinkgrid layer to the Table of Contents. This layer will show the location of the sinks in the Columbiana County NED layer. See **Smartbox 108** for more information about sinks.

- Answer Questions 22.3 and 22.4.

> **QUESTION 22.3** How many sinks are in the Columbiana County NED layer?

> **QUESTION 22.4** What is the total number of cells that make up the sinks?

Smartbox 108

What are sinks in hydrologic modeling?

sinks Low areas in a digital-elevation model surrounded on all sides by high areas.

An elevation surface will often have low-elevation values surrounded on all sides by high-elevation values (Figure 22.3). These low-elevation locations are called **sinks** and are often the result of errors in the elevation surface. Sinks are problematic in hydrologic modeling because they represent grid cells for which flow direction can't be properly computed. The direction of flow is determined by the direction of the steepest slope, but this condition isn't present at a sink (because all eight cells around the sink cell contain higher values, thus indicating that water would have to flow uphill at these spots). When the flow direction algorithm tries to act on a cell with a sink, it can't properly determine which way the water should go and the resultant flow direction grid ends up with improper values.

To accurately determine flow direction (which serves as the basis for many other hydrologic functions), these sinks need to be removed from the elevation surface. ArcGIS can "fill" these sink cells by artificially adding elevation

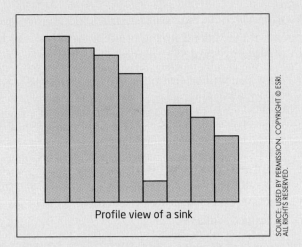

FIGURE 22.3 The appearance of a sink (low elevation cells among higher elevation cells).

values to them so that these cells are no longer surrounded by higher elevations (Figure 22.4). The end result of "filling" the sinks in a DEM is the creation of a "sinkless" DEM, or **depressionless DEM**. Without sinks, this new depressionless elevation surface is used in the remainder of the hydrologic modeling processes.

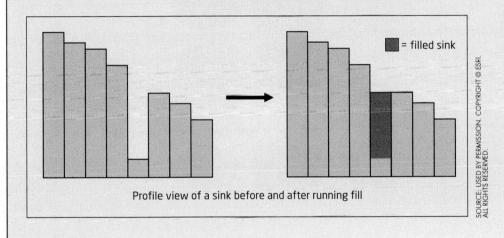

FIGURE 22.4 An example of "filling" a sink so that water flow across a surface can be determined.

depressionless DEM DEM that has had its sinks filled.

STEP 22.6 Filling Sinks

• To create a continuous flow network, you first have to fill in the sinks back in the original elevation layer (in this case, the NED Input Layer).

• From the **Spatial Analyst Tools** toolbox, select the **Hydrology** toolset, then select the **Fill** tool and drag and drop it into the Model dialog.

• Build a connection between the NED Input Layer and the Fill tool.

• Double-click on the **Fill tool** to open its dialog.

• Name your Output surface raster **NEDfilled** and store it in your D:\Chapter22 folder. Click **OK** to close the Fill dialog.

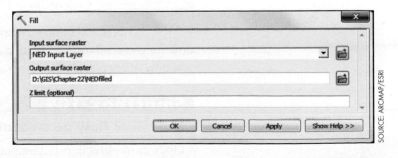

- Add the NEDfilled layer to the TOC. This is a "depressionless" or "sinkless" DEM (in which all of the sinks from the original layer have been filled—see **Smartbox 108** on pages 522 and 523 for more information).

- *Important Note:* You will be working with the "sinkless" DEM (**NEDfilled**) and derivations of it for the remainder of the chapter.

- Use the ModelBuilder tools to create a new flow direction grid (call it **newflowdir**) from the sinkless NEDfilled layer.

- Run this part of the model and add the newflowdir layer to the TOC. Note the values that are now computed for flow direction at each cell compared with the previous flow direction grid (which was created from the original NED Input Layer that still contained sinks).

- At this point, your Hydromodel dialog should look something like this:

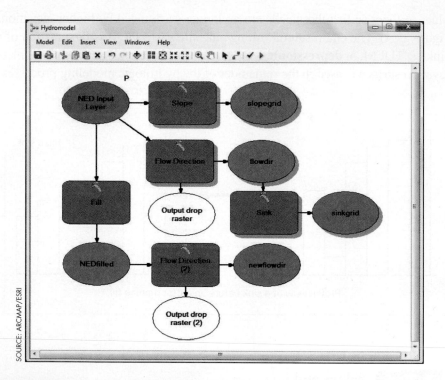

SOURCE: ARCMAP/ESRI

STEP 22.7 Calculating Flow Accumulation

- With the direction of flow computed, you can now determine the accumulation of flow for each cell on the elevation surface. The Flow Accumulation tool will show you how many pixels "flow" into a location (see **Smartbox 109** for more information).

Smartbox 109

How does flow accumulation work in ArcGIS 10.2?

flow accumulation ArcGIS tool that will determine how many cells will flow into a location.

Once flow direction is computed (and a coded value indicating in which of the eight directions that flow will occur is assigned to the grid), ArcGIS can determine **flow accumulation**, or a count of how many cells have their flow connected to a location cell. The more cells that are connected, the greater the amount of water that flows into those cells. As the amount of connected cells grows, higher values for flow accumulation will be calculated.

Figure 22.5 provides an example of calculating flow accumulation. The flow direction grid is coded with the specific values indicating the direction that water will move from each cell. The "flow network" diagram graphically shows what the flow direction grid represents. For instance, three cells connect to the middle grid cell. Thus, its value for flow accumulation is 3. The grid cell in the bottom-left corner has only one cell that directly connects to it, so its value for flow accumulation is 1. However, the flow accumulation values build up as more and more cells connect to one another and a total cumulative value is calculated. For instance, the grid cell in the bottom center has a flow accumulation value of 8 because there is a cumulative total of 8 cells that flow into it. Three cells flow into the middle cell and then the middle cell flows into the bottom center, for a total of four. Then each of the bottom corner cells have a total of two cells each that connect to the bottom center. Thus, the total flow accumulation value of the bottom-center cell is 8.

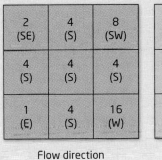
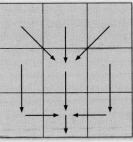
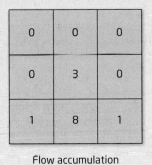

Flow direction   "Flow network"   Flow accumulation

FIGURE 22.5 An example of calculating flow accumulation from a flow direction grid.

- From the **Spatial Analyst Tools** toolbox, select the **Hydrology** toolset, then select the **Flow Accumulation** tool and drag and drop it into the Model dialog.

- Build a connection between the newflowdir layer and the Flow Accumulation tool. Double-click on the **Flow Accumulation tool** to open it.

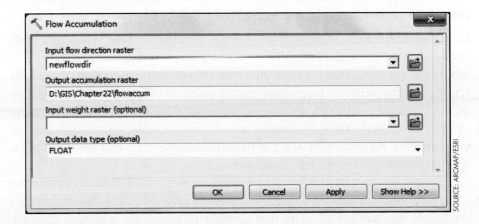

- Name your Output accumulation raster **flowaccum** and save it in your D:\GIS\Chapter22 folder. Click **OK** to close the dialog.

- Run this portion of the model and add the flowaccum grid to the TOC.

- The next step will be to display the flowaccum grid visualized with a different classification method. To start, from flowaccum's layer properties, choose the **Symbology** tab.

- In the Show: options, select **Classified** (if asked about computing a histogram, click **Yes**).

- Press the **Classify** button to change the classification away from the default (Natural Breaks) to a different classification. From the classification dialog that appears, choose **Quantiles** as your method and **10** as the number of Classes.

- Change the color ramp to something other than black and white.

- Click **OK** to close the dialog, then **Apply** and **OK** to close the Layer Properties.

- Flowaccum is now displayed using Quantiles, with 10 different categories. Turn on the streams layer and place it over top of the flowaccum grid. Answer Question 22.5 and turn off the streams layer when you're finished.

> **QUESTION 22.5** Keeping in mind that flowaccum is showing only the accumulation of overland water flow, how does it compare with the streams layer? What differences are there between the two datasets? What does this say about the channels shown in the streams layer? *Hint:* Select some of the areas on the flowaccum grid that have the highest values and zoom in closely to see those areas in relation to the streams layer.

STEP 22.8 Creating a Stream Network with a Threshold Value

- What we will now do is extract those cells from the flowaccum grid that represent the stream network—see **Smartbox 110** for more information about this process. To begin, from the **Spatial Analyst** toolbox, select the **Conditional** toolset, then select the **Con** tool and drag and drop it into the Model dialog.

Smartbox 110

How can a stream network be extracted in ArcGIS 10.2?

Each cell of the flow accumulation grid has a value for how many grid cells connect to that particular cell. As flow accumulation values increase, more and more cells are connected together. In terms of hydrologic modeling, these cells represent areas on the surface where greater amounts of water flow link together and accumulate, and cells with large values of flow accumulation represent streams and rivers. Thus, by examining only those cells with a high enough value of flow accumulation, we can determine those cells that are representative of streams and model the stream network.

The tricky part of this process is determining what flow accumulation values are high enough to constitute a stream network. This *threshold* value represents the number of cells that connect to constitute a stream channel. Knowledge of the study area and its various factors (such as the climate or the soil conditions) will help in determining the proper threshold value for accurate modeling of streams. For purposes of this chapter's activities, we're using a sample value of 100 grid cells as a threshold and comparing the results with a

National Hydrology Dataset (NHD) layer from The National Map. Figure 22.6 shows an example of identifying areas that have a greater amount of water accumulating at each location and extracting them as streams.

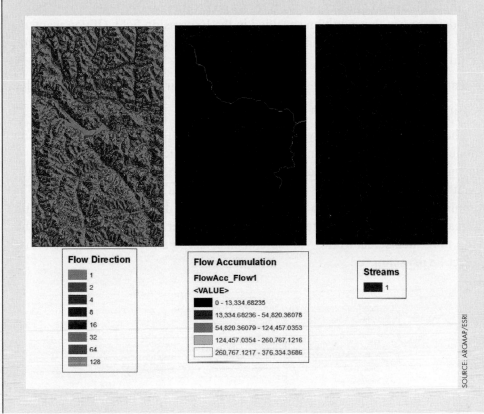

FIGURE 22.6 Flow direction and flow accumulation grids, and the resultant stream network of grid cells extracted from them using a threshold value.

SOURCE: ARCMAP/ESRI

- Build a connection between the flowaccum layer and the Con tool (and select **Input Conditional Raster** from the available choices when building the connection). Double-click on the **Con tool** to open it.

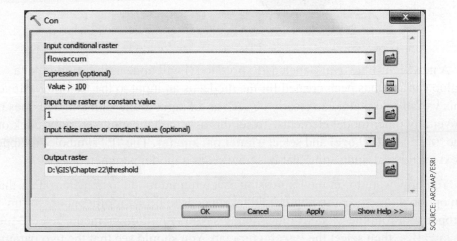

SOURCE: ARCMAP/ESRI

- In the Expression (optional) field, type **Value > 100**. This indicates that we want to take all grid cells in the flowaccum raster that have a value higher than 100 and assign them a new value and all cells that have a value of 100 or less and assign them a different value.

- For the Input true raster or constant value, type **1**. This value indicates that all cells that meet the condition specified in the Expression field will receive a value of 1.

- Leave the Input false raster or constant value (optional) field blank. By leaving this field blank, you instruct ArcGIS to assign a value of "No Data" to all cells that do not meet the condition specified in the Expression field. (If you wanted all cells that didn't meet the condition to receive a value other than "No Data," you could type that value here).

- Call the Output raster **threshold** and save it in your D:\GIS\Chapter22 folder.

- Click **OK** to close the Con dialog.

- The Expression that you typed ("Value > 100") will also become the second model parameter. As noted in **Smartbox 110**, the value used for the flow accumulation threshold can change, depending on the project. As such, we'll want the model's users to be able to define their own threshold value when running the model. To set up this part of the Con tool as its own variable (which we can then set as a parameter), in the ModelBuilder dialog, right-click on the **Con tool**, then select **Make Variable**, then select **From Parameter**, then finally select **Expression**.

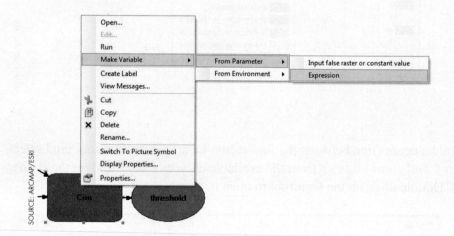

- A new light-blue oval (labeled "Expression") will appear, indicating that a value variable has been created for the model as an input to the Con tool. Move this variable so that it can clearly be seen as an input to the Con tool and does not cover any other model elements. To set this as a model parameter, right-click on the **"Expression" oval** and select **Model parameter**. The "P" symbol will appear next to the oval, indicating that it is now set as a model parameter.

- You now want to set the Expression as a required model parameter rather than an optional one (because a threshold level needs to be set for the model to run properly). From the ModelBuilder dialog's **Model** pull-down menu, select **Model properties**, then select the **Parameters** tab. You should see that the two parameters are set up—NED Input Layer as a Required parameter and Expression as an Optional parameter. Click next to **Optional** and from the pull-down menu that appears, change its state to **Required**. Click **OK** when both model parameters are set as Required.

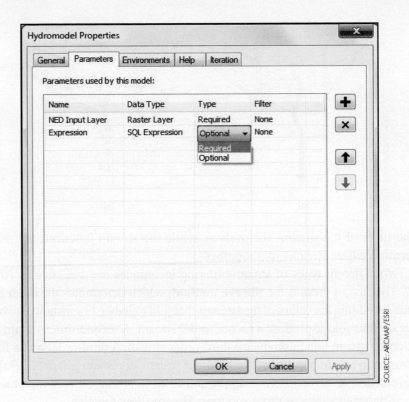

SOURCE: ARCMAP/ESRI

• Run this portion of the model and add the threshold grid to the TOC. Turn off the other layers, except for threshold and streams, and answer Question 22.6. Turn off both threshold and streams when you're done.

> **QUESTION 22.6** How does the threshold layer compare with the streams layer? What differences are there between the two datasets? *Hint:* Select some of the areas on the threshold grid and zoom in closely to see those areas in relation to the streams layer.

STEP 22.9   Stream Ordering

• Now that you have a set of cells that represent the layout of the stream network, we'll examine the stream ordering—see **Smartbox 111** for more about the two methods of stream ordering used in ArcGIS 10.2. From the **Spatial Analyst Tools** toolbox, select the **Hydrology** toolset, then select the **Stream Order** tool and drag and drop it into the Model dialog.

Smartbox 111

How does stream ordering work in ArcGIS 10.2?

Once you've identified the grid cells that represent the set of streams with which you'll be working, you can assign an ordering to these streams. **Stream order** represents a way of ranking streams—lower orders will be very minor streams with little in the way of cells that flow into them, while higher orders will be those with a larger number of smaller streams that flow together to produce larger streams. In other words, stream ordering is a means of classifying streams by how many other streams flow into them. This ordering is assigned

stream order A computed ranking of streams based on how they flow into one another.

FIGURE 22.7 A stream network ordered using (a) the Shreve method and (b) the Strahler method.

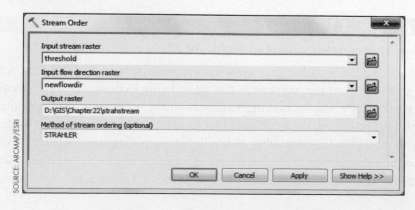

to the links of the streams themselves, while the stream junctions are used to determine where streams join together.

Two different types of stream-ordering techniques are available in ArcGIS 10.2. The first of these is the **Shreve method**, which determines the order of the stream by adding the values of the streams that join together. For instance, two first-order streams join to produce a second-order stream. A second-order stream and a first-order stream join to produce a third-order stream. A third-order and a fourth-order stream join to become a seventh-order stream, and so on (Figure 22.7a). The Shreve ordering method is very simple and tends to produce a very high range of stream-ordering values as more and more streams flow together.

The second stream-ordering technique is the **Strahler method**, which allows an increase in stream ranking only when two streams of the same order join. For instance, two first-order streams join to form a second-order stream. Two second-order streams join to form a third-order stream. However, when a first-order and a second-order stream join, the result is a second-order stream (because the two initial streams are of different rankings, the ranking of the new stream does not increase). The Strahler method tends to produce an overall lower range of values for the ranking of streams. Figure 22.7b shows an example of the Strahler method.

Shreve method A stream-ordering method that allows for the stream-order ranking to be computed by adding the ranks of the two streams flowing into each another.

Strahler method A stream-ordering method that allows for stream order to increase in rank only if two streams of the same order of magnitude flow into each another.

- Build a connection between the newflowdir variable and the Stream Order tool, and select the option for **input flow direction raster** when doing so.

- Also, build a connection between the threshold variable and the Stream Order tool, and select the option for **input stream raster** when doing so.

- Double-click on the **Stream Order tool** to open it.

- Make sure **threshold** is used for your Input stream raster.

- Make sure **newflowdir** is used for your Input flow direction raster.

- Call the Output raster **strahstream** and place it in your D:\GIS\Chapter22 folder.

- Use **STRAHLER** as your method of ordering.

- Click **OK** to close the Stream Order dialog.

- Use the ModelBuilder tools to create a second, separate stream-order grid called **shrevestream**, but this time using the **SHREVE** method of stream ordering. Again, use **threshold** as your input stream raster and **newflowdir** as your input flow direction raster.

- Run this section of the model, then add both the strahstream and shrevestream grids to the Table of Contents and examine them. Answer Question 22.7.

> **QUESTION 22.7** What orders of magnitude streams are represented by both the Strahler and Shreve stream-ordering methods?

STEP 22.10 Watershed Delineation

- ArcGIS can delineate a series of watersheds for each one of the streams calculated from the NED surface (based on the surface flow). First, however, it needs to compute the Stream Links (the junctions of each stream) as part of the Watershed delineation process. From the **Spatial Analyst Tools** toolbox, select the **Hydrology** toolset, then select the **Stream Link** tool and drag and drop it into the Model dialog.

- Build a connection between the newflowdir variable and the Stream Link tool, and select the option for **input flow direction raster** when doing so.

- Also build a connection between the threshold variable and the Stream Link tool, and select the option for **input stream raster** when doing so.

- Double-click on the **Stream Link tool** to open it.

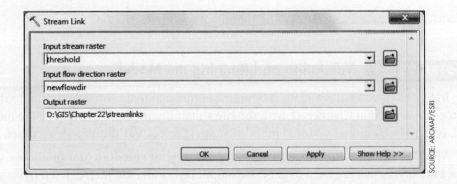

- Make sure **threshold** is used for your Input stream raster.

- Make sure **newflowdir** is used for your Input flow direction raster.

- Call the Output raster **streamlinks** and place it in your D:\GIS\Chapter22 folder.

- Click **OK** to close the Stream Links dialog.

- Next, from the **Spatial Analyst Tools** toolbox, select the **Hydrology** toolset, then select the **Watershed** tool and drag and drop it into the Model dialog.

- Build a connection between the streamlinks variable and the Watershed tool, and select the option for **input raster or feature pour point data** when doing so.

- Also, build a connection between the newflowdir variable and the Watershed tool, and select the option for **input flow direction raster** when doing so.

- Double-click on the **Watershed tool** to set its parameters.

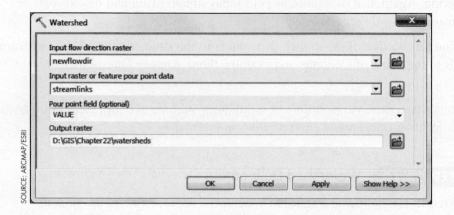

- Make sure **newflowdir** is used for your Input flow direction raster.

- Make sure **streamlinks** is used for your Input raster or feature pour point data.

- Use **VALUE** for Pour point field (optional).

- Call the Output raster **watersheds** and place it in your D:\GIS\Chapter22 folder.

- Click **OK** to close the Watershed dialog.

- Run this portion of the model and add the resultant grid (called watersheds) to the TOC. Remember, this watershed grid is based on the streamlinks extracted from the value of 100 threshold previously set. Answer Question 22.8.

QUESTION 22.8 How many watersheds are extracted?

STEP 22.11 Validating and Running the Model

- In this step, you will validate and run the completed hydro model (similar to the process we used in Chapter 21 with the site suitability model). To validate the model, in the **Hydromodel's Model** pull-down menu select **Validate Entire Model**.

- ModelBuilder also designates some variables as **intermediate data**—these data are layers that are used to generate output layers and are then deleted by ModelBuilder. When you run a model from the ModelBuilder interface, the intermediate layers are not deleted, but when you run the Model as a stand-alone tool (as we did in Chapter 21 and as we will do in the next several steps), these intermediate layers are automatically removed. Because we will want to preserve one of the intermediate layers (the streamlinks layer) for further analysis, we need

intermediate data Layers created using ModelBuilder that are used to generate outputs and are removed when the model is run as a stand-alone tool.

to make ModelBuilder understand that it should not treat it as an intermediate layer. To do so, right-click on the green **streamlinks** bubble in the ModelBuilder dialog.

- In the pop-up menu that appears, click on the option for **Intermediate** to remove the check mark next to that option. If a check mark is next to Intermediate, that layer will be treated as intermediate data, and if there is no check mark, it will not. Right-click on the green **streamlinks** bubble again to verify that there is no check mark next to Intermediate.

- Before proceeding, save the Hydromodel.

- Next, close the Hydromodel dialog box because we're now going to run the entire model. In Catalog, navigate to the D:\GIS\Chapter22 folder and expand the Hydrotools toolbox. Before running the model, we want to provide documentation for it. In the Hydrotools toolbox, right-click on **Hydromodel** and select **Item Description**. A new dialog box will appear with information about the model. By selecting the Item Description dialog box's **Edit** tab, you can add the missing information. The items to which you are required to add information are: Summary, Tags, NED_Input_Layer Syntax Dialog Explanation, and Expression Syntax Dialog Explanation. Once you've done so, click the **Save** button in the dialog box and close the dialog.

- Now it's time to run the model. Right-click on the **Hydromodel icon** and select **Open**. Since the model's parameters have been set, the model will open—not as the graphical ModelBuilder interface, but the model itself. The model takes the form of a dialog box resembling a geoprocessing tool that will allow the user to specify two inputs (namely, the two model parameters)—an NED layer and the Expression used to determine the flow accumulation threshold.

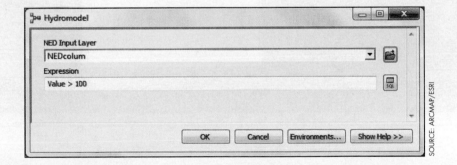

- You'll see that NEDcolum and Value > 100 are the defaults used when running the model. If you wanted to use a different NED layer or a new value for the threshold, you could specify it here. Click **OK** to run the model with these two default parameters. A new dialog box will appear showing each process of the Hydromodel as it runs. When it completes, close the dialog box.

- At this point, the entire model has been rerun and any previously created layers have been overwritten.

STEP 22.12 Analyzing the Results of Hydrologic Modeling

- We'll now take a look at the watershed results of the model and how they relate to the streams. As part of this chapter's scenario, you're examining potentially polluted areas around Columbiana County streams. What you want to do is find the area of the watershed around McCormick Run for use in future analysis related to pollution sources near that stream. For now, turn on the streams, streamlinks, and watersheds layers and arrange them so that streams is on top and watersheds is on the bottom.

• Change the symbology of the streams and streamlinks layers so that you can easily distinguish the two layers.

• Change the symbology of the watersheds layer to Unique Values using the Value attribute. This will assign each of the watersheds its own color to distinguish them from one another.

• Open the attribute table for the Streams layer and locate the McCormick Run stream (you may want to sort the GNIS_Name field and find "McCormick Run"). There will be two links—select the main one (it has the OBJECTID field equal to 3532) and zoom in to that stream. You'll see the stream line laid over its 30-meter-resolution raster counterpart in the streamlinks layer along with the computed watershed for that stream link. Answer Questions 22.9 and 22.10.

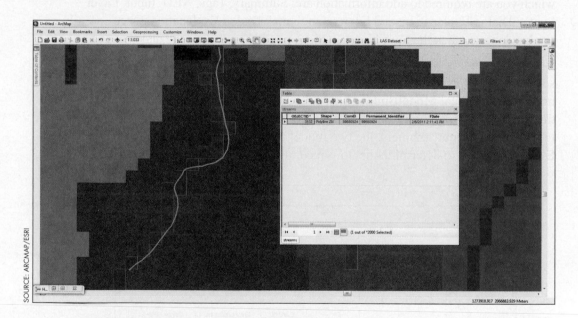

SOURCE: ARCMAP/ESRI

QUESTION 22.9 How does the vector version of McCormick Run (from the National Hydrology Dataset layer from The National Map) compare with its 30-meter-resolution counterpart derived from the NED?

QUESTION 22.10 What order of magnitude stream is McCormick Run according to both the Strahler method and the Shreve method of stream ordering? (*Hint*: Turn on your strahstream and shrevestream grids to answer this question.)

• By using the Identify tool and the watersheds layer's attribute table, answer Questions 22.11 and 22.12.

QUESTION 22.11 Which watershed number (in terms of its OBJECTID value) has been computed for McCormick Run?

QUESTION 22.12 How large (in terms of acreage) is the McCormick Run watershed? (*Hint:* The size of each grid cell is 900 square meters—because they are 30-meter-resolution grid cells—and one square meter (m²) is equal to 0.000247105 acres.) Show your work.

STEP 22.13 **Saving, Printing, or Sharing Your Results**

• Save your work as a map document. Include the usual information in the Map Document Properties.

• Compose and print a layout (see Chapter 3) of your final version of the McCormick Run stream, the streamlinks version of the same, and the computed watershed for that stream link (including all of the usual map elements and design for a layout). You'll want to add a graphic of the model to the layout. In the Hydrotools toolbox, right-click on **Hydromodel** and select **Edit**. When the Hydromodel ModelBuilder dialog re-opens, select the **Model** pull-down menu, then select **Export**, and then select **To Graphic**. Save a graphic of the model as a .jpg file. Next, in the Layout, use the Insert Picture option (see Chapter 15) to place a large, readable graphic of the model on the layout as well. Make sure that the workflow of your model is uncluttered and easy to follow.

• If you want to share your Hydromodel, you can do so as a geoprocessing package (see Chapter 21 for how to create and share a model as a geoprocessing package).

Closing Time

This chapter explored how to work with the various hydrology tools available in ArcGIS by starting with a digital elevation model and ending with a stream network and a set of watersheds delineated from the streams. Hydrologic modeling techniques allow for many types of analysis related to the overland flow of water and are commonly utilized by engineers, water resource managers, and environmental scientists. Beyond the Hydrology toolset, Esri has an extended set of hydrologic tools available through the Arc Hydro data model and tools (see online here for more information: http://resources.arcgis.com/en/communities/hydro/01vn0000000s000000.htm).

Watershed delineation can be performed for locations other than just stream links. For instance, if you have identified specific drainage points throughout the area and want to determine the watershed for these areas, you'll need to use a different technique than the one you use for computing stream links. See the *Related Concepts for Chapter 22* for more information.

Related Concepts for Chapter 22   Using Pour Points in ArcGIS 10.2

In hydrologic modeling, you're not limited to delineating watersheds based on stream links. For instance, if you have a set of culverts throughout a region, you may want to determine the catchment area for each one. In this case, you could use a point feature class of the culvert locations (rather than the stream links) as the input into the Watershed tool. ArcGIS 10.2 refers to a feature class that represents an outlet for a drainage system as a **pour point**. By using pour points, you can specify the sub-basin for each of these locations (Figure 22.8).

FIGURE 22.8 A sample set of watersheds (sub-basins) created using pour points.

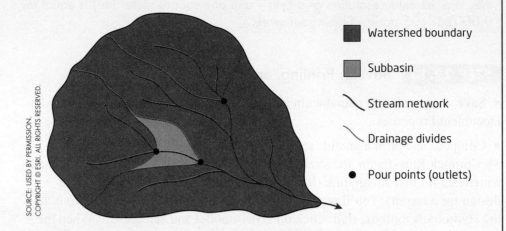

- ⬛ Watershed boundary
- ⬛ Subbasin
- ╲ Stream network
- ╲ Drainage divides
- ● Pour points (outlets)

Pour points will be those locations to which all of the water in the catchment area flows. In the hydrologic modeling process, using pour points allows you to go straight from creating the flow direction grid (from a depressionless DEM) into the watershed delineation process. The Watershed tool has two required inputs—a flow direction raster and a layer representing the sources for the water outlets. In this chapter, we derived the stream links for use as the outlet source layer, but pour points can be used instead. If you use pour points in this fashion, the Snap Pour Point tool (in the Hydrology toolset) may be useful. It will locate the highest-value cell of the flow accumulation grid that is near the point and then move the point to that location. In this way, the location of the point will be in the area of the largest nearby water accumulation.

pour point A feature class that represents the outlet location for the drainage of water.

For More Information

For further in-depth information about the topics presented in this chapter, use the ArcGIS Help feature to search for the following items:

- Arc Hydro Overview (RC)
- Con (Spatial Analyst)
- Conditional evaluation with Con
- Creating stand-alone variables
- Deriving runoff characteristics
- Identifying stream networks
- How Fill works
- How Flow Accumulation works
- How Flow Direction works
- How Sink works
- How Stream Order works
- How Watershed works

- Hydrologic analysis sample applications
- Snap Pour Point (Spatial Analyst)
- Stream Link (Spatial Analyst)
- Stream Order (Spatial Analyst)
- Watershed (Spatial Analyst)

For further information about hydrologic modeling in ArcGIS (and access to an additional set of Arc Hydro tools) see the ArcGIS Resource Center online at http://resources.arcgis.com/en/communities/hydro/01vn0000000s000000.htm.

For further information about the Mahoning River Watershed, see online at https://web.archive.org/web/20120204053716/http://www.ysu.edu/mahoning_river/basic_info.htm.

For more in-depth information on extracting stream channels from an elevation surface (which is also recommended by Esri), see the following article: Tarboton, D. G., R. L. Bras, and I. Rodriguez-Iturbe. 1991. "On the Extraction of Channel Networks from Digital Elevation Data." *Hydrological Processes* 5: 81–100.

Key Terms

hydrologic modeling (p. 515)

watershed (p. 515)

flow direction (p. 520)

D8 (p. 520)

sinks (p. 522)

depressionless DEM (p. 523)

flow accumulation (p. 524)

stream order (p. 529)

Shreve method (p. 530)

Strahler method (p. 530)

intermediate data (p. 532)

pour point (p. 536)

How to Customize Toolbars in ArcGIS 10.2

While ArcGIS provides you with numerous menus, tools, and toolbars, you also have the ability to customize the existing toolbars and build your own. For instance, if you're working on a project that requires tools from several different toolboxes, menu items, and buttons on existing toolbars, it may be convenient to create a custom toolbar. This process could involve making a new toolbar containing only the tools you use, or adding or removing tools and buttons from an existing toolbar.

For instance, Chapter 1 has you use several different tools: Project, Define Projection, and Identify. It also asks you to use buttons to open Catalog, ArcToolbox, and Search. This Appendix will take you through the steps required to build a new toolbar that holds only these six tools. You can use the same steps to design other toolbars for your projects (or for other chapters in this book).

To create a new toolbar (or to make adjustment to an existing one), from the **Customize** pull-down menu, select **Customize Mode**.

In the Customize dialog box, the **Toolbars** tab will show you all of the available toolbars. If you want to make alterations to one of the toolbars, you can select it from here. To create a brand-new toolbar, press the **New** button.

In the New Toolbar dialog box, you can give your toolbar a name. In this case, we've called it **Chapter1**. Click **OK**. Back in the Customize dialog, you'll see a toolbar called Chapter1 added into its alphabetical location in the list of toolbars. You'll also see that a new, empty, unlabeled toolbar has been added to ArcMap along the top, which is docked with any

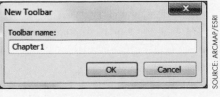

other toolbars you may have added. Locate this empty toolbar, undock it, and drag it out into the View to work with it.

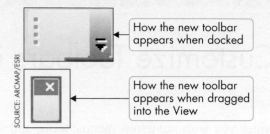

SOURCE: ARCMAP/ESRI

How the new toolbar appears when docked

How the new toolbar appears when dragged into the View

In the Customize dialog box, select the **Commands** tab.

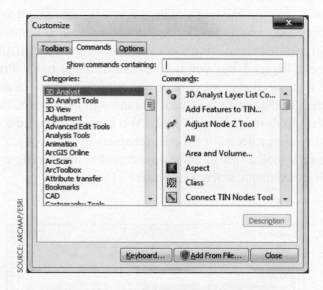

SOURCE: ARCMAP/ESRI

The Categories list will show all available sources for tools, buttons, or other items that can be added to a toolbar. Select one of these Categories (for instance, the 3D Analyst category), and the Commands list will show all tools associated with it.

You can drag and drop an item from the Commands list into the new, empty toolbar you've created to add it.

SOURCE: ARCMAP/ESRI

For example, to add the **Project** tool to the new toolbar, from **Categories** select **Data Management Tools**, and then from **Commands** select **Project**. Drag the **Project tool** from the Commands box onto the new toolbar.

Note that with so many different available commands, it can sometimes be hard to find what you're looking for. To filter the Commands list, you can type the name of the tool you're searching for into the **Show commands containing** box.

To complete the buttons used in the Chapter1 toolbar, add the **Define Projection** tool (available under the Data Management category), and the **Identify** tool (available under the Map Inquiry category) to the new toolbar. Reorder these three tools and (if necessary) stretch the toolbar so that they're all displayed in a horizontal row.

You'll see the three tools added to the toolbar, but you'll see that both Project and Define Projection have the default "hammer" symbol carried over from ArcToolbox. To change the tool's icon, right-click on the icon itself on the toolbar, select **Change Button Image**, and choose an appropriate icon from the default list. You can also select **Browse for Image** to load an image from your computer to use as the picture for the icon.

You can also add a pull-down menu to your toolbar to include further options. Scroll to the bottom of the Categories list in the Customize dialog and there will be two options—one for [Menus], which would allow you to select and add an existing ArcGIS menu to your toolbar, and a second option for [New Menu]. Choose the **[New Menu]** option from Categories and an item called New Menu will be added to the Commands list. Drag and drop the New Menu command like any other command onto the Chapter1 toolbar.

Right-click on the **New Menu** option on the **Chapter1 toolbar** and from the new available list select **Name**. Then, in place of "New Menu," type a new name (in this case, we've named it **Window Options**).

Now click on the new **Window Options** menu; you'll see it's empty. You can drag and drop commands directly into this new menu and organize them. For instance, place the **ArcToolbox** command (located in the ArcToolbox category), the **Catalog** command (available in the File category), and the **Search** command (also in the File category) into the Window Options menu.

Stretch or adjust the toolbar to one horizontal row. You'll now have a new toolbar called Chapter1 that has three buttons (Project, Define Projection, and Identify) and a menu (called Window Options) with three choices (ArcToolbox, Catalog, and Search). By touching each one with the mouse, you'll make their tooltips appear.

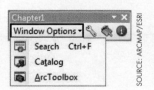

Your new Chapter1 toolbar will be saved and available like other ArcGIS toolbars until you delete it (which you can do from the Customize dialog by selecting the **Toolbars** tab, then choosing **Chapter1** from the list of toolbars, and clicking the **Delete** button).

Using Coordinate Systems in ArcGIS 10.2

In ArcGIS, coordinate systems come in two varieties: a geographic coordinate system and a projected coordinate system. Chapter 1 describes some of the basics behind each system and how they're used in ArcGIS to define the coordinates of geospatial data. This Appendix describes some of the technical details of using a geographic coordinate system and two common projected coordinate systems (UTM and SPCS) for measurements in ArcGIS.

Part 1: Geographic Coordinate Systems

A geographic coordinate system (GCS) is a global reference system for determining the exact position of a point on Earth. GCS consists of lines of latitude and longitude that cover the entire planet and are used to find a position. Lines of latitude (also referred to as *parallels*) run in an east-to-west direction around the globe. The Equator serves as the starting point of zero, with measurements north of the Equator being numbers read as north latitude, and measurements south of the Equator being numbers read as south latitude. Lines of longitude (also called *meridians*) run north to south from the North Pole to the South Pole. The Prime Meridian (the line of longitude that runs through Greenwich, England) serves as the starting point of zero. Measurements made east of the Prime Meridian are numbers read as east longitude, while measurements made west of the Prime Meridian are numbers read as west longitude (Figure B-1). The breaking point between east and west longitude is the 180th meridian.

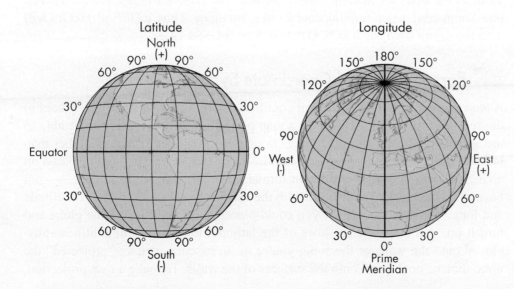

FIGURE B-1 Using latitude and longitude for locating coordinates in GCS.

With this system (where measurements are made north or south of the Equator, and east or west of the Prime Meridian), any point on the globe can be located. GCS measurements are not made in feet or meters or miles, but rather in degrees, minutes, and seconds (DMS). A degree is a unit of measurement that can be broken into 60 subsections (each one referred to as a minute of distance). A minute can be subsequently broken down into 60 subsections (each one referred to as a second of distance). A single minute of latitude is equivalent to roughly one nautical mile (or about 1.15 miles). However, the distance between degrees of longitude varies across the globe as lines of longitude are closer together at the poles and farther apart at the Equator.

When GCS coordinates are cited for a location, they are written in terms of their distance (in terms of degrees, minutes, and seconds) from the Equator and the Prime Meridian. Latitude values run from 0 to 90 degrees north and south of the Equator, while longitude values run from 0 to 180 degrees east and west of the Prime Meridian. For example, the GCS coordinates of a spot at the Mount Rushmore National Memorial are 43 degrees, 52 minutes, 53.38 seconds north of the Equator (43°52'53.38" N latitude) and 103 degrees, 27 minutes, 56.46 seconds west of the Prime Meridian (103°27'56.46" W longitude).

Negative values can also be used when expressing coordinates using latitude and longitude. When distance measurements are made west of the Prime Meridian, they are often listed with a negative sign to note the direction. For instance, the location at Mount Rushmore, being west longitude, could also be written as -103°27'56.46" longitude. The same use of negative values holds true when making measurements of south latitude (below the Equator).

GCS coordinates can also be expressed in their decimal equivalent, referred to as decimal degrees (DD). In the decimal degrees method, values for minutes and seconds are converted to their fractional number of degrees. For instance, 1 degree and 30 minutes of latitude (in DMS) would equate to 1.5 degrees in DD (because 30 minutes is equal to 0.5 degrees). Using the Mount Rushmore coordinates as an example, the latitude measurement in decimal degrees would be 43.881494 north latitude (since 52 minutes and 53.38 seconds is equal to 0.881494 of one degree) and 103.465683 west longitude.

GCS in ArcGIS also includes the datum, or the mathematical model of Earth used, as the basis for making measurements. As Chapter 1 notes, there's not just one datum used in determining coordinates, but many. Thus, a GCS in ArcGIS will note the datum used, such as NAD83 GCS or WGS84 GCS.

Part 2: Projected Coordinate Systems

A *map projection* is a translation of coordinates and locations from their places in the real world to a flat surface. On a map projection, it is necessary to be able to use geospatial data to measure things in units such as feet or meters, rather than latitude and longitude. A projected coordinate system in ArcGIS will be based on a GCS, then projected to a different format. Think of a projection like this—you have a see-through globe of Earth with the various landforms and lines of latitude and longitude painted on it. If you could place a light bulb inside the globe and turn it on, you'd see the shadows of the latitude, longitude, and continents displayed onto the walls of the room you're in. In essence, you've "projected" the three-dimensional world onto the surfaces of the walls. To make a map projection, you'd have to wrap a piece of paper around the globe and "project" the lines onto that paper. Then, when you unroll the paper, you'd have the globe translated onto the map.

The real process is a much more complicated mathematical translation of coordinates, but the basic concept is similar. A map projection is created through the use of a *developable surface*, or a flat area onto which the coordinates from Earth can be placed. In this way,

Cylindrical   Conical   Azimuthal

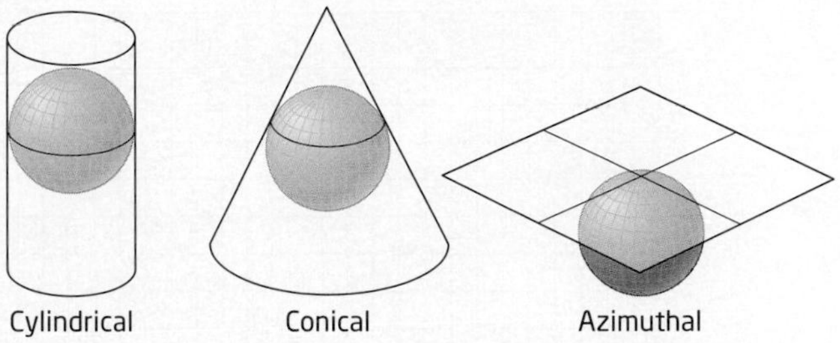

FIGURE B-2 The three types of developable surfaces in map projections.

the locations on Earth's surface get translated onto the flat surface. The three main developable surfaces (Figure B-2) are a cylinder (which creates a cylindrical projection), a cone (which creates a conical projection), and a flat plane (which creates an azimuthal surface). The place where Earth "touches" the surface is called the *point of tangency*, which is where the map's distortion is minimized.

There are many different types of map projections used with geospatial data, usually chosen to minimize the projection distortions for the area being studied. Chapter 1 describes some common projected coordinate systems, but there are many others. The Project commands in ArcGIS allow you to change geospatial data from the GCS to a projected coordinate system (or vice versa) or from one projected coordinate system to another. There are more options available than can be listed here. In ArcGIS Help, see the entry for "*What are projected coordinate systems?*" for a link to a PDF file that lists all of the myriad projected coordinate systems supported by ArcGIS.

Part 3: Universal Transverse Mercator (UTM)

The Universal Transverse Mercator (UTM) grid system is a projected coordinate system. It works by dividing the world into a series of zones, then determining the *x* and *y* coordinates for a location in that zone. The UTM grid is set up by translating real-world locations to their corresponding places on a two-dimensional surface (using the Transverse Mercator projection). However, the UTM only covers Earth between 84° N latitude and 80° S latitude (and thus couldn't be used for mapping the polar regions—a different system, the Universal Polar Stereographic Grid System, is used to find coordinates on the poles).

The first step in determining UTM coordinates is to find the UTM zone the location is in. The UTM divides the world into 60 zones; each zone is 6° of longitude wide (Figure B-3), and is set up as its own cylindrical projection. The zones are numbered 1 through 60, beginning at the 180th meridian and moving east. For instance, Mount Rushmore, outside of Keystone, South Dakota, is located in Zone 13; Beijing, China, is located in Zone 50, and Sydney, Australia is in Zone 56.

The next step is to determine the *x* and *y* grid coordinates of the location. Unlike GCS, which uses degrees, minutes, and seconds, UTM uses meters as its measurement unit. The UTM's *x* and *y* coordinates are referred to as an easting and a northing. The UTM northing is found by counting the number of meters the point is north of the Equator. For instance, a particular location at Mount Rushmore is located 4,859,580 meters to the north of the Equator.

Locations in the southern hemisphere handle northing a little differently because UTM does not contain negative values. For southern hemisphere locations, count the number of meters south of the Equator to the point (as a negative

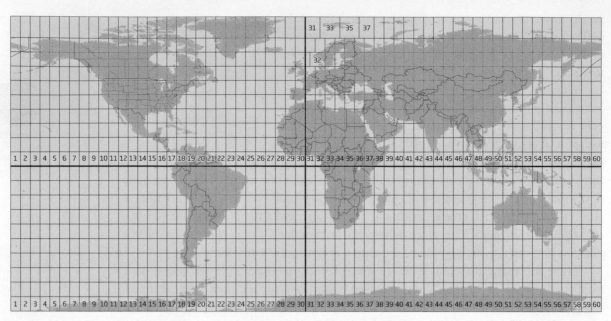

FIGURE B-3 The 60 zones that comprise the UTM system.

value), then add 10,000,000 meters to the number. For example, a location at the Opera House in Sydney, Australia (in the southern hemisphere), has a northing of 6,252,268 meters. This indicates it would be approximately 3,747,732 meters south of the Equator (−3,747,732 + 10,000,000). Another way of conceptualizing this (for southern hemisphere locations) is to assume that rather than measuring north of the Equator, you're measuring north of an imaginary line drawn 10,000,000 meters south of the Equator. This extra 10 million meters is referred to as a false northing because it's a value set up by the system's designers to avoid negative values.

The next step is to measure the point's easting. Unlike GCS, there is no one Prime Meridian from which to make measurements. Instead, each UTM zone has its own central meridian that runs through the zone's center. This central meridian is given a value of 500,000 meters, and measurements are made to the east (adding meters) or west (subtracting meters) from this value. For example, a specific spot at Mount Rushmore is located 123,733.22 meters to the east of Zone 13's central meridian—so the easting for this location at Mount Rushmore is 623,733.22 meters (500,000 + 123,733.22). A location in the town of Buffalo, Wyoming, is also in Zone 13 and has an easting value of 365,568 meters. This indicates this location is approximately 134,432 meters to the west of Zone 13's central meridian (500,000 − 134,432). No zone is more than 1,000,000 meters wide, so no negative values are assigned to eastings in a zone. This value for easting is referred to as a "false easting" because the easting calculation is based on a central meridian for each zone.

When UTM coordinates are written, the false easting, northing, and zone must all be specified. If a false northing is used (that is, the coordinates are found in a location south of the Equator), then a notation that the coordinates are from the southern hemisphere is included. For example, the UTM coordinates of a location at Mount Rushmore would be written: 623733.22 E, 4859580.26 N, Zone 13. (See Figure B-4 for a simplified graphical example of finding the location at Mount Rushmore's coordinates using UTM.)

In ArcGIS, several types of UTM coordinate systems are available, beyond NAD83 and NAD27 versions of UTM. These include UTM versions for various parts of Asia, Africa, Indonesia, and many other parts of the world. When using

UTM data in ArcGIS, it's important to know which version of UTM you're using (with United States data, it will likely be NAD83 UTM).

Part 4: State Plane Coordinate System (SPCS)

A second grid-based system for determining coordinates is the SPCS (State Plane Coordinate System), a measurement system designed back in the 1930s. Today, SPCS is a data coordinate system used in the United States, particularly for city or county data and measurements. SPCS is a projected coordinate system that uses a translation method similar to UTM to set up a grid coordinate system covering the entire United States. SPCS data was originally measured using the NAD27 datum, but there are now data sets using the NAD83 datum. Like UTM, SPCS requires you to specify a zone, an easting, and a northing in addition to the state in which the measurements are being made.

SPCS divides the United States into a series of zones. Unlike UTM zones, which follow lines of longitude, SPCS zones are formed by following state or county boundaries. A state is often divided into multiple zones (e.g., Ohio has two zones: Ohio-North and Ohio-South; Nevada has three zones: Nevada-West, Nevada-Central, and Nevada-East; Texas has five zones; California has six zones; while all of Montana is represented by a single zone (Figure B-5)). To determine

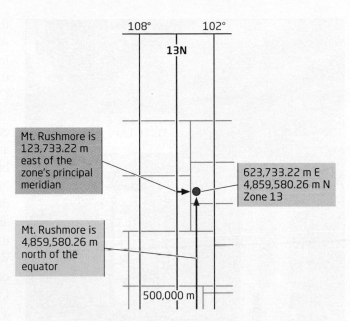

FIGURE B-4 A simplified representation of determining the UTM coordinates (easting, northing, and zone) of a location at Mount Rushmore.

FIGURE B-5 The SPCS zones for all states.

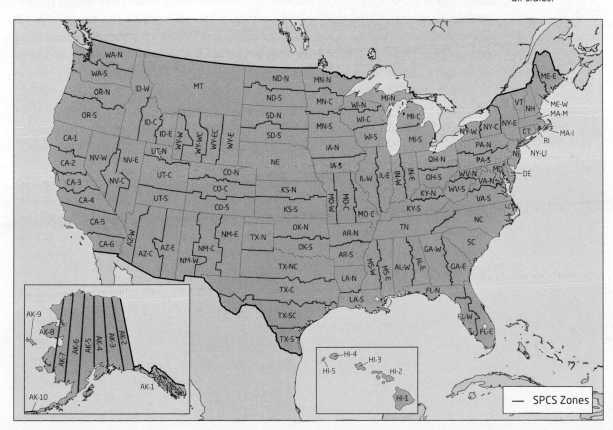

2478424.96 E,
531216.13 N,
Ohio, N

FIGURE B-6 A simplified representation of SPCS measurements and coordinates for a location on the YSU football field in Youngstown, Ohio.

the coordinates for a point in SPCS, you must first know which state the point is in, and then identify the zone in that state. For example, to map data from Miami, the state is Florida and the zone is Florida-East, while Cleveland is in the state of Ohio in the Ohio-North zone.

The next step is to determine the northing and easting. As in UTM, both of these are measured north and east from a predetermined baseline and meridian, but in SPCS, measurements are made in feet (using the NAD27 datum) or in feet or meters (using the NAD83 datum). Each of the SPCS zones has its own baseline and its own principal meridian. Two projected versions of SPCS are used—the Transverse Mercator version, which is usually used for states that have a north-south orientation (such as Indiana), and the Lambert Conformal Conic version, which is usually used for states that have a more east-west orientation (such as Ohio). Depending on the version used, the zones' baselines and meridians are positioned differently. Baselines are placed a distance under the zone's southern border and measured north from there (to ensure all positive values). In the NAD27 version, states using the Transverse Mercator format have their zones' origins 500,000 feet to the west of the principal meridian, while states using the Lambert Conformal Conic projection have their zones' origins 2,000,000 feet to the west of the meridian. Coordinates are determined by measuring distance using these false origin points, creating a false northing and a false easting. Thus, as each of the zones uses its own coordinate measurements, each zone has a different false northing and false easting from all other zones. Note that in the NAD83 version of SPCS, different values are often used for each zone's origin.

SPCS coordinates are referenced by listing the false easting, false northing, state, and zone. For example, the SPCS (NAD83 feet) coordinates of a location on the football field of Stambaugh Stadium in Youngstown, Ohio, are approximately: 2478424.96 E, 531216.13 N, Ohio, N (North Zone; see Figure B-6 for a simplified graphical representation of SPCS measurements).

In ArcGIS, each of these zones can be specified, and different datum are also available. For instance, if your SPCS data uses the NAD27 datum, that can be selected. There are many types of NAD83 datum available for the SPCS zones (including NAD83 US Feet, NAD83 International Feet, and NAD83 Meters, among others). So, when using SPCS data, it's important to know which datum (and measurement systems) are being used.

GLOSSARY

3D Analyst An ArcGIS extension used for 3D visualization and working with data that has z-values. (p. 344)

3D document (.3dd) The file that contains all the information about data layers in a globe. (p. 431)

accumulated cost The total cost of traveling from a source to a grid cell. (p. 443)

accuracy How closely a measurement matches up with its real-world counterpart. (p. 163)

active remote sensing When a sensor generates its own energy source and measures the reflection or backscatter of that energy. (p. 393)

Address Locator An ArcGIS utility that specifies the reference data, geocoding style, and necessary attributes for geocoding. (p. 234)

address locator style The format that will be used for geocoding addresses. (p. 234)

aerial photography The acquisition of imagery of the ground taken from an airborne platform. (p. 295)

Albers Equal Area Conic A projected coordinate system used for east-west data at mid-latitudes. (p. 19)

algorithm A set of steps used in a process. (p. 263)

allocation The determination of which source is closest in distance to each grid cell. (p. 442)

American FactFinder A Web resource run by the U.S. Census that allows for access to U.S. Census data in tabular form. (p.57)

AND (Map Algebra) The Boolean operator used with an intersection of two grids. (p. 475)

AND (Query) The Boolean operator wherein the chosen features are those that meet both criteria in the query. (p. 48)

animation The capture of movements, actions, or changes to layers in a video format. (p. 423)

annotation A method of adding text to a map that allows for individual pieces of text to be edited separately. (p. 87)

append A geoprocessing operation that combines one or more layers of data together with an already existing layer. (p. 211)

ArcGIS 10.2 The version of the ArcGIS software package released in 2013 and the basis for the activities in this book. (p. 2)

ArcGIS Advanced The version of ArcGIS for Desktop with the greatest number of functions. (p. 2)

ArcGIS Basic The version of ArcGIS for Desktop with the fewest functions. (p. 2)

ArcGIS for Desktop 10.2 The version of ArcGIS that runs on a personal computer. (p. 2)

ArcGIS for Server An Esri software product used for distributing maps, data, and services via a user's own servers via the Internet, as well as designing powerful Web applications. (p. 108)

ArcGIS Online Esri's cloud-based GIS platform where data and Web mapping software can be accessed via the Internet. (p. 89)

ArcGIS Standard The version of ArcGIS for Desktop with a moderate number of functions. (p. 2)

ArcGlobe The 3D virtual globe interface of ArcGIS. (p. 411)

ArcMap The main component of ArcGIS for Desktop. (p. 4)

ArcPad An Esri software product for mobile devices that allows for GPS integration and field data collection. (p. 161)

ArcPy The Python add-in that allows for ArcGIS tools and settings to be used in Python scripts. (p. 226)

ArcScene The ArcGIS component of the 3D Analyst that allows for the visualization and analysis of data in a 3D environment. (p. 359)

ArcToolbox A separate window in ArcGIS containing multiple tools and functions. (p. 21)

aspect A determination of the direction the steepest slope is facing at a location. (p. 350)

attribute accuracy How closely the non-spatial features of a dataset match their real-world counterparts. (p. 175)

attribute table A spreadsheet-style form where the rows consist of individual objects and the columns are the attributes associated with these objects. (p. 33)

attributes The non-spatial data that can be associated with a geospatial location. (p. 33)

backlink The grid that shows the direction to be traveled in for the least-cost amount of movement from cell to cell. (p. 448)

band A narrow range of wavelengths being measured by a remote sensing device. (p. 301)

bare-Earth A model of the surface using only the elevation values of the ground (rather than features, structures, or vegetation on top of it). (p. 402)

base heights The z-value obtained for a location from a digital terrain model surface. (p. 361)

basemap An image layer that serves as a backdrop for the other layers used in ArcGIS Online. (p. 102) Also, a map service available from ArcGIS Online (such as topographic maps or high resolution imagery) that can be used in ArcGIS for Desktop. (p. 126)

batch geocoding Matching a group of addresses together at once. (p. 229)

blue Electromagnetic energy with wavelengths between 0.4 and 0.5 micrometers. (p. 302)

Boolean operator (Map Algebra) One of the four operators (AND, OR, XOR, NOT) used when building a Raster Calculator expression. (p. 475)

Boolean operator (Query) One of the connectors used in building a compound query. (p. 48)

brightness value Another term used synonymously with digital number. (p. 302)

buffer A zone of spatial proximity around a feature or set of features. (p. 217)

cartographic modeling The concept of taking a layer, applying a process or action to it, and creating a new layer as an output. (p. 487)

cartography The art and science of making maps. (p. 59)

Catalog The component of ArcGIS used for copying, moving, and organizing GIS data. (p. 9)

channel The term used for the display of imagery in shades of red, green, or blue. (p. 302)

choropleth map A type of thematic map in which data are displayed according to one of several different classifications. (p. 59)

clip A geoprocessing operation that extracts objects from one layer based on the geometry of a second layer. (p. 215)

closest facility A process used to determine which locations on a network are nearest (via network distances) to a different set of locations. (p. 273)

cloud A computing structure where data, content, and resources are all stored at another location and served to the user via the Internet. (p. 89)

COLLADA Collaborative Design Activity. An XMLbased file type (with the extension .dae) that works as an interchange format from one kind of 3D modeling file type to another. (p. 432)

color composite An image formed by placing a band of imagery into each of the three channels (red, green, and blue) to view a color image instead of a grayscale one. (p. 303)

color ramp A range of colors that are applied to the thematic data of a map. (p. 70)

completeness A measure of the wholeness of a dataset. (p. 176)

compound query A query that contains more than one operator. (p. 48)

connectivity The linkages between edges and junctions of a network. (p. 257)

Connector (ModelBuilder) Specific term used in ModelBuilder for one of the lines that link Variables and Tools (by default they appear as one of four types of lines). (p. 491)

Content Standards for Digital Geospatial Metadata (CSDGM) A set format created by the FGDC for what items a metadata file should contain. (p. 133)

continuous field view The conceptualization of the world that all items vary across Earth's surface as constant fields, and values are available at all locations along the field. (p. 277)

contour An imaginary line drawn on a map that connects points of common elevation. (p. 372)

contour interval The vertical distance between contour lines. (p. 373)

control points Point locations where the coordinates are known—these points are used in aligning the unreferenced image to the source. (p. 316)

cost The number of units it takes to move from one grid cell to the next. (p. 443)

coverage A data layer represented by a group of files in a workspace consisting of a folder structure filled with files and also associated files in an INFO folder. (p. 143)

credits A system used by Esri to control the amount of content that can be served by an organization in the ArcGIS Online subscription model. (p. 95)

crowdsourcing Leveraging the knowledge and resources of multiple individual users for a larger project. (p. 128)

D8 An algorithm for determining flow direction that will output which of the eight cardinal directions will be the steepest slope for the flow of water. (p. 520)

data classification Various methods used for grouping together (and displaying) values on a choropleth map. (p. 66)

Data Frame Component of ArcMap that contains and displays all data layers. (p. 6)

Data Interoperability A concept that allows GIS data of many different file types to be imported into ArcGIS or ArcGIS file formats to be converted to another file type. (p. 132)

datum A reference surface of Earth. (p. 18)

Delaunay triangle A triangle that can have a circle drawn through its three points and not pass through any additional points. (p. 376)

Delaunay triangulation The process of connecting sets of points to build the triangles and edges of a TIN. (p. 376)

depressionless DEM DEM that has had its sinks filled. (p. 523)

deterministic An interpolation method that does not take spatial variation into account. (p. 337)

digital elevation model (DEM) A representation of a terrain surface created by measuring elevations at evenly spaced sampling points. (p. 341)

Digital Line Graphs (DLGs) The digitized features from USGS topographic maps. (p. 130)

digital number The energy measured at a single pixel according to a predetermined scale. (p. 302)

Digital Raster Graphics (DRGs) Scanned and georeferenced versions of USGS topographic maps. (p. 131, 390)

Digital Surface Model (DSM) A representation of the surface and features generated from a set of lidar points. (p. 401)

digital terrain model (DTM) A representation of a terrain surface calculated by measuring elevation values at a series of locations. (p. 341)

digitizing The process of creating digital GIS data from a secondary source. (p. 137)

Dijkstra's algorithm A mathematical process that will determine the shortest path along a network from an origin node to the other nodes on the network. (p. 263)

discrete object view The conceptualization of the world that all items can be represented by objects. (p. 12)

dissolve A geoprocessing operation that combines polygons with similar attributes together. (p. 213)

DOQQ A Digital Orthophoto Quarter Quad showing orthoimagery covering 3.75 degrees of latitude by 3.75 degrees of longitude. (p. 313)

draping A process in which GIS layers are given z-values to match the corresponding heights in a digital terrain model. (p. 362)

dynamic text Descriptive information about map properties that will change as those properties change or are updated. (p. 83)

edges The links of a network. (p. 253)

edges The lines that connect the points of the triangles of a TIN. (p. 376)

environment settings Geoprocessing options that control activities in a map document, such as the extent of outputs, the coordinate system of outputs, or the cell size of outputs. (p. 210)

Equal Interval A dataclassification method that selects class break levels by taking the total span of values (from highest to lowest) and dividing by the number of desired classes. (p. 67)

Erase A type of GIS overlay that retains all of the features from the first layer except for what they have in common with the second layer. (p. 221)

error of commission When extra items are added to a dataset that should not be there. (p. 176)

error of omission When items are left out of a dataset that should be there. (p. 176)

Esri Environmental Systems Research Institute—the market leader in GIS software and the developer of ArcGIS. (p. 2)

Esri CityEngine A powerful 3D design software that allows for quickly creating large-scale 3D models (like cities). (p. 411)

ETM+ The Enhanced Thematic Mapper instrument onboard Landsat 7. (p. 304)

Euclidian distance The straight-line distance between two points. (p. 439)

Event Layer A GIS data layer created from tabular data. (p. 145)

extension An add-on set of functions for ArcGIS. (p. 256)

extent The viewable dimensions of a layer. (p. 15)

extrusion Used to give an object height. (p. 417)

facility Locations from which service areas are created. (p. 271)

false color composite An image arranged by not placing the red band in the red channel, the green band in the green channel, and the blue band in the blue channel. (p. 303)

feature class A GIS layer stored inside a geodatabase—it can contain multiple items of the same data type. (p. 142)

feature dataset A grouping of related feature classes stored within a geodatabase. (p. 142)

feature service A hosted service that allows a user to share GIS data layers that can also be displayed, queried, or edited. (p. 95)

Federal Geographic Data Committee (FGDC) A committee that is responsible for setting GIS metadata content standards. (p. 133)

fields The columns of an attribute table. (p. 33)

file geodatabase A geodatabase that can hold an unlimited amount of data and has the file extension .gdb. (p. 142)

first return The initial reflection of the lidar laser pulse, which usually indicates the height of an object. (p. 401)

flow accumulation ArcGIS tool that will determine how many cells will flow into a location. (p. 524)

flow direction ArcGIS tool that determines the direction in which overland water flow will move. (p. 520)

functional surface A continuous representation of values that can have a single value assigned to each location. (p. 360)

geocoding The process of using the numbers and letters of an address to plot a point at that location. (p. 229)

geocoding service An online utility that allows for geocoding one or more addresses. (p. 250)

geodatabase An ArcGIS structure for a single item that contains multiple datasets, each as its own feature class. (p. 13, 142)

geodatabase topology A set of rules applied to the feature classes of a feature dataset in order to remove errors. (p. 180)

geographic coordinate system (GCS) A set of global latitude and longitude measurements used as a reference system for finding locations. (p. 18)

geographic information systems (GIS) The hardware and software that allow for computer-based analysis, manipulation, visualization, and retrieval of location-related data. (p. 1)

geographic scale The real-world size or extent of an area. (p. 76)

geometric network An ArcGIS structure of a series of connected junctions and edges, commonly used for utility-related problems. (p. 275)

GeoPDF A format for maps to be used as PDFs, but that can contain geographic information and multiple layers. (p. 391)

geoprocessing When an action is taken to a dataset that results in a new dataset being created. (p. 207)

geoprocessing package A file that contains a model, its inputs, and results that can be shared. (p. 511)

georeferencing A process whereby spatial referencing is given to data without it. (p. 315)

geospatial or spatial Referring to items that are tied to a specific real-world location. (p. 1)

geospatial technologies A term that encompasses many types of methods and techniques for the collection, analysis, modeling, and visualization of geospatial data. (p. 1)

geostatistical An interpolation method that accounts for spatial variation. (p. 337)

Geostatistical Analyst The ArcGIS extension that contains several functions related to examining and working with spatial statistics. (p. 338)

geotag A process whereby non-spatial media are linked to geospatial features. (p. 53)

GIS model A representation of the factors used for explaining the processes that underlie an event or for predicting results. (p. 487)

global interpolation A method that uses all known values in the dataset when approximating an unknown value. (p. 320)

Global Navigation Satellite System (GNSS) An overall term for the technologies that use signals from satellites for finding locations on Earth's surface. (p. 161)

Global Positioning System (GPS) A technology that uses signals broadcast from satellites for position determination on Earth. (p. 160)

GPX The standard format for data collected by a GPS receiver. (p. 160)

graduated colors The use of various hues in representing ranges of values on a map. (p. 71)

graduated symbols The use of different sized symbology to convey thematic information on a map. (p. 71)

graphics Items (such as shapes) that can be drawn on the screen in ArcGIS. (p. 367) Also, elements (such as shapes or text) that can be added to a view or layout but are not data layers. (p. 482)

green Electromagnetic energy with wavelengths between 0.5 and 0.6 micrometers. (p. 302)

grid cell A single square unit of a raster. (p. 277)

hard breaklines Lines used in a TIN to enforce surface discontinuities. (p. 377)

heads-up digitizing Using a map or imagery as a backdrop in the digitizing process. (p. 137)

hillshade A shaded relief map of the terrain created by modeling the position of an illumination source (e.g., the Sun) in relation to the terrain. (p. 348)

histogram A graphical representation of a raster, showing the values of the raster on the x-axis and the count of those values on the y-axis. (p. 346)

hosted services Esri's term for the different types of map services that can be utilized through ArcGIS Online. (p. 95)

hot spot analysis A technique that determines spatial clusters of high values and low values. (p. 205)

hull The polygon boundary around the area of the TIN. (p. 377)

hydrologic modeling Examining the movement of water over a surface. (p. 515)

hyperlink A function that allows the user to connect non-spatial media such as documents, files, photos, URLs, or scripts to a geospatial feature and interactively trigger them. (p. 53)

hyperspectral Sensing hundreds of bands of energy at once. (p. 302)

Identity A type of GIS overlay that retains all the features from the first layer along with the features it has in common from the second layer. (p. 221)

image service An ArcGIS Online format used for the distribution of raster data. (p. 308)

impedance A value that represents how many units (of time or distance) are used in moving across a network edge. Also referred to as transit cost. (p. 258)

incident The locations for which the closest facility function tries to determine the nearest facility. (p. 273)

index contour The main contour lines on a map. (p. 373)

intensity The strength of the reflected return of the laser pulse from an object. (p. 393)

intermediate contour The contour lines drawn in between index contours at the contour interval. (p. 373)

intermediate data Layers created using ModelBuilder that are used to generate outputs and are removed when the model is run as a stand-alone tool. (p. 532)

Intersect A type of GIS overlay that retains the features that are common to both layers. (p. 221)

interval data A type of numerical data in which the difference between numbers is significant but there is no fixed non-arbitrary zero point associated with the data. (p. 39)

inverse distance weighted (IDW) An interpolation process that assigns a higher weight to the values of known points closer to the location being interpolated and lower weights to the values of known points that are farther away. (p. 325)

ISO 19115 An international metadata standard being adopted in the United States as the North American Profile of ISO 19115. (p. 133)

join A method of linking two or more tables together. (p. 41) key The field that two tables have to have in common with each other in order for the tables to be joined. (p. 41)

junctions The nodes of a network (or the places where edges come together). (p. 253)

key The field that two tables have to have in common with each other in order for the tables to be joined. (p. 41)

keyframe Snapshots of various frames used in an animation. (p. 423)

KML Keyhole Markup Language. An XML-based file format that can hold georeferenced information and the file format used by Google Earth. (p. 430)

KMZ A compressed version of a KML file. (p. 430)

Kriging A local geostatistical interpolation method. (p. 337)

Lambert Conformal Conic A projected coordinate system used for east-west data. (p. 19)

Landsat A long-running U.S. remote sensing program that had its first satellite launched in 1972 and continues today. (p. 304)

Landsat scene A single image obtained by a Landsat satellite sensor. (p. 304)

large-scale map A map with a larger value for its representative fraction. Such maps will usually show a smaller geographic area. (p. 76)

LAS dataset An ArcGIS file structure used for storing, viewing, and analyzing one or more LAS files. (p. 396)

LAS file A standard file format for holding lidar data. (p. 393)

last return The final reflection of the lidar laser pulse, which usually indicates the elevation height of the ground. (p. 401)

layer A single dataset used in ArcGIS. (p. 6)

layer package One or more ArcGIS layers combined together into a single file for portability or distribution via ArcGIS Online. (p. 124) Also, a file containing a series of layers that can be shared via ArcGIS Online. (p. 431)

layout The assemblage and placement of various map elements used in constructing a map. (p. 73)

least-cost path The set of grid cells that represents the lowest value for accumulated cost to move from a source to a destination. (p. 448)

legend A graphical device used on a map as an explanation of what the various map symbols and colors represent. (p. 78)

lidar A method for measuring elevation values using laser pulses sent to the ground from an aircraft. (p. 393)

line One-dimensional vector object. (p. 13)

lineage The data sources and processes used for creating a dataset. (p. 164)

linear interpolation A method used in geocoding to place an address location among a range of addresses. (p. 240)

line-of-sight (LOS) The visibility in a straight line between two points. (p. 352)

line-on-polygon overlay A type of GIS overlay that results in a new line layer where the lines have the attributes of the polygons within which they are. (p. 224)

local interpolation A method that uses a subset of known values when approximating an unknown value. (p. 320)

locator package A single file created in ArcGIS that allows for the sharing of Address Locators. (p. 249)

logical consistency A measurement for examining if the same rules were used throughout a dataset. (p. 177)

many-to-one join When many records in an attribute table can have one other record from another table matched to them during a join. (p. 42)

map A visual representation of geospatial data that conveys some sort of message to its reader. (p. 59)

Map Algebra Operation performed on rasters, such as overlaying two or more grids. (p. 463)

map document (.mxd) File that contains information about where all of the data layers used in a session are located, as well as their appearance and settings. (p. 5)

Map Package (.mpk) A single file that contains the map document, all data layers used in the map document, as well as their appearance. (p. 28)

map scale A metric used to determine the relationship between measurements made on a map and their real-world equivalents. (p. 76)

map service Another term for "tiled map service." (p. 95)

map template A pre-made arrangement of items in a map layout. (p. 75)

Maplex Label Engine A toolset in ArcMap that allows for a greater degree of flexibility and options for labeling features. (p. 86)

mass points The selection of points used in creating a TIN. (p. 376)

merge A geoprocessing operation that combines two or more layers of data together into a single layer. (p. 211)

metadata Descriptive information about data. (p. 133)

middle infrared (MIR) Electromagnetic energy with wavelengths between 1.3 and 3.0 micrometers. (p. 302)

model An ArcGIS tool that combines data layers, geoprocessing tools, and outputs in a workflow. (p. 487)

model element Each of the items (Variables, Tools, and Connectors) that can be placed into the ModelBuilder interface. (p. 491)

model parameters The userdefined inputs or variables of a model. (p. 492)

ModelBuilder The interface of ArcGIS used in laying out workflows and creating models. (p. 467)

mosaic A raster created by joining several smaller rasters. (p. 293)

mosaic dataset An ArcGIS file structure designed for managing one or more rasters or collections of raster data. (p. 292)

MSS The Multispectral Scanner onboard Landsat 1 through 5. (p. 304)

multipatch The ArcGIS data format that allows for fully 3D objects and structures to be used. (p. 412)

multipoint A geodatabase feature class that can hold very large amounts of point data. (p. 408)

multispectral Sensing several bands of energy at once. (p. 302)

NAD27 North American Datum of 1927. (p. 18)

NAD83 North American Datum of 1983. (p. 18)

nadir The location under the sensor or camera in remote sensing. (p. 295)

NAIP The National Agriculture Imagery Program maintained by the USDA. (p. 313)

National Elevation Dataset (NED) A digital elevation model of the entire United States, provided by the USGS. (p. 347)

National Geospatial Program (NGP) A federal U.S. initiative for managing and distributing geospatial resources. (p. 113)

National Land Cover Dataset (NLCD) A 30 meter raster dataset of land cover for the entire United States. (p. 282)

The National Map An online resource for distributing U.S. geospatial data. (p. 114)

Natural Breaks A data-classification method that selects class break levels by searching for spaces in the data values. (p. 66)

NAVD 88 The North American Vertical Datum, a commonly used vertical datum for United States digital terrain models. (p. 341)

near infrared (NIR) Electromagnetic energy with wavelengths between 0.7 and 1.3 micrometers. (p. 302)

neatline A border that can be placed around some or all map elements in a layout. (p. 85)

network A series of junctions and edges connected together for modeling concepts such as streets and routes. (p. 253)

Network Analyst The ArcGIS extension used with transportation networks. (p. 257)

Network Dataset An ArcGIS structure of a series of connected junctions and edges, commonly used for transportation-related problems. (p. 257)

NoData A raster data cell that contains a null value. (p. 282)

nominal data A type of data that is a unique identifier of some kind. If numerical, the differences between numbers are not significant. (p. 38)

non-spatial data Descriptive information that does not have location-based qualities. (p. 33)

normalize Altering count data values so that they are at the same level of representing the data (such as using them as a percentage). (p. 63, 484)

north arrow A graphical device used to show the orientation of the map. (p. 81)

NOT (Map Algebra) The Boolean operator used with a negation of a grid or expression. (p. 477)

NOT (Query) A Boolean operator used to negate the function or query. (p. 48)

oblique image A remotely sensed image acquired at an angle. (p. 295)

off-nadir viewing The capability of a sensor to observe areas other than the ground directly underneath it. (p. 295)

offset Used to raise an object above the ground surface. (p. 421)

OLI The Operational Land Imager instrument onboard Landsat 8. (p. 304)

one-to-one join When a single record is linked to another single record during a join. (p. 42)

OpenStreetMap An opensource collaborative mapping project whose data are also available as a basemap in ArcGIS. (p. 128)

OR (Map Algebra) The Boolean operator used with a union of two grids. (p. 475)

OR (Query) The Boolean operator wherein the chosen features are those that meet one or the other (or both) of the criteria in the query. (p. 48)

ordinal data A type of data that refers solely to a ranking of some kind. (p. 39)

Ordinary Kriging A Kriging method that assumes an unknown mean and a lack of trend in the data.(p. 338)

orthoimage (orthophoto) A spatially referenced aerial photo with uniform scale. (p. 311)

overlay The combining of two or more layers together in GIS. (p. 207)

overshoot When a vertex of a line goes beyond its target location. (p. 172)

panchromatic Black-and-white imagery acquired by sensing the entire visible portion of the spectrum at once. (p. 302)

pan-sharpening The technique of fusing a higher resolution panchromatic band with lower-resolution multispectral bands to improve the clarity and detail seen in an image. (p. 310)

parsing Breaking an address into its component parts. (p. 239)

path distance A cost-distance function that allows for a variety of other frictions (such as horizontal or vertical factors) to be included in the analysis. (p. 461)

personal geodatabase A geodatabase that can hold a maximum of 2 GB of data and has the file extension .mdb. (p. 142)

perspective view Viewing GIS data at an oblique angle that causes it to take on a "three-dimensional" appearance. (p. 359)

point cloud The processed elevation points of a lidar dataset. (p. 393)

point digitizing The digitizing mode in which the user places a vertex by clicking the mouse button. (p. 153)

point Zero-dimensional vector object. (p. 13)

point-in-polygon overlay A type of GIS overlay that results in a new point layer where the points have the attributes of the polygons within which they are. (p. 223)

polygon Two-dimensional vector object. (p. 13)

positional accuracy How closely the spatial features of a dataset match their real world locations. (p. 170)

pour point A feature class that represents the outlet location for the drainage of water. (p. 536)

precision How consistent or exact a measurement is. (p. 163)

project To change a layer from one coordinate system into another. (p. 20)

projected coordinate system (PCS) A set of measurements made on a flat grid system, initially derived from a GCS. (p. 18)

pseudo-3D The term used to describe the perspective view of a digital terrain model because it is often a 2.5D model, not a 3D model. (p. 359)

publish Placing data or content onto a cloud server. (p. 94)

pyramids Lower-resolution versions of an image that are used when displaying the image at different scales in order for images to be quickly rendered on the screen. (p. 298)

Python An object-oriented, open-source programming language that is used for developing scripts for ArcGIS. (p. 226)

Quantiles A data classification method that attempts to place an equal number of values in each class. (p. 67)

query The conditions used to retrieve data from a database or table. (p. 45)

radiometric resolution A sensor's ability to determine fine differences in a band of energy measurements. (p. 302)

Raster Calculator ArcGIS tool used for Map Algebra operations. (p. 472)

raster data model A conceptualization of representing geospatial data with a series of equally spaced and sized grid cells. (p. 277)

ratio data A type of numerical data in which the difference between numbers is significant, but there exists a fixed, non-arbitrary zero point associated with the data. (p. 40)

Reclassify The ArcGIS tool that allows you to create a new grid by assigning new grid-cell values to an existing raster. (p. 456)

records The rows of an attribute table. (p. 33)

red Electromagnetic energy with wavelengths between 0.6 and 0.7 micrometers. (p. 302)

reference data The base street layer used as a source for geocoding. (p. 229)

reference map A map that serves to show the location of features, rather than thematic information. (p. 59)

region A set of contiguous grid cells that have the same value. (p. 285)

regularized A Spline method that results in an overall smoother surface but may also result in values that are beyond the range of the known point values. (p. 333)

relate A type of join that establishes a connection between tables but does not append fields from one table to another table. (p. 42)

relational operator One of the six operators (equal, not equal, greater than, greater than or equal to, less than, less than or equal to) used when building a Raster Calculator expression. (p. 471)

relational operators One of six connectors (=, <>, >, <, >=, <=) used to build a query. (p. 45)

remote sensing The process of collecting information related to the reflected or emitted electromagnetic energy from a target by a device a considerable distance away from the target onboard an aircraft or spacecraft platform. (p. 295)

Representative Fraction (RF) A value indicating how many units of measurement on the map are the equivalent of a number of units of measurement in the real world. (p. 76)

resolution The ground area represented by a single grid cell in a raster. (p. 291)

return The reflection of energy from a lidar laser pulse. (p. 401)

RMSE The Root Mean Square Error, an error measure used in determining the accuracy of the overall transformation of the unreferenced data. (p. 316)

SaaS (Software as a Service) A cloud structure wherein the software being used is stored on a server at another location and accessed on-demand via the Internet. (p. 90)

satellite imagery Digital images of Earth acquired by sensors onboard orbiting spaceborne platforms. (p. 295)

scale bar A graphical device used to represent scale on a map.(p. 76)

scale text Descriptive information used to represent scale on a map. (p. 77)

scene document (.sxd) The ArcScene equivalent of an ArcMap map document—a file that contains information about where all of the data layers used in an ArcScene session are located, as well as their appearance and settings. (p. 366)

script A short section of computer code. (p. 226)

segment A single digitized line. (p. 153) Also a single part of a line that corresponds to one portion of a street. (p. 229)

Select By Location An ArcGIS tool used to select features from one or more layers based on their spatial relationship to another layer. (p. 195)

selection When certain records or features are chosen and set aside from the remainder of the records or features. (p. 44)

semivariogram A model of the variance between points and their distance apart. (p. 337)

service A format for GIS data and maps to be distributed (or "served") to others via the Internet. (p. 94)

service area A polygon boundary created around sections of a network to determine which areas of the network are within a certain distance of a location. (p. 271)

shapefile A series of files (with extensions including .shp, .shx, and .dbf) that make up one vector data layer. (p. 13, 143)

share Distributing data, content, maps, or applications across the Internet. (p. 104)

shortwave infrared (SWIR) Another term used for electromagnetic energy with wavelengths between 1.3 and 3.0 micrometers. (p. 302)

Shreve method A stream-ordering method that allows for the stream-order ranking to be computed by adding the ranks of the two streams flowing into each another. (p. 530)

simple query A query that only contains one operator. (p. 45)

sinks Low areas in a digital-elevation model surrounded on all sides by high areas. (p. 522)

SketchUp A 3D design software owned and distributed by Trimble. (p. 431)

slice Grouping several grid-cell values together according to a classification scheme. (p. 452)

slope A measurement of the rate of elevation change at a location found by dividing the vertical height (the rise) by the horizontal length (the run). (p. 349)

small-scale map A map with a lower value for its representative fraction. Such maps will usually show a large geographic area. (p. 76)

snapping When two vertices link together to become a single vertex. (p. 171)

snapping tolerance How closely a vertex must be placed within a feature to activate snapping. (p. 172)

soft breaklines Enforced lines in a TIN that do not represent discontinuities. (p. 377)

source (Select By Location) The layer containing the features for which selections will be made in relation to when using the Select By Location tool. (p. 195)

spatial analysis Examining the characteristics or features of spatial data, or how features spatially relate to one another. (p. 183)

Spatial Analyst The extension that enables several raster functions and tools in ArcGIS. (p. 280)

spatial autocorrelation A measure of the degree of clustering of objects and their data values. (p. 204)

Spatial Data Transfer Standard (SDTS) A neutral file format used for the distribution of GIS data. (p. 132)

spatial interpolation The process of determining an unknown value at a location based on known values at other locations. (p. 319)

spatial join Appending attributes from one layer to another based on the spatial relationship of the layers. (p. 188)

spatial query Selecting records or objects from a layer based on their spatial relationships with other layers rather than their attributes. (p. 183)

spatial resolution The size of the area on the ground represented by one pixel's worth of energy measurement. (p. 299)

Spline An interpolation method that uses a mathematical process to fit the surface exactly through the known points. (p. 333)

SQL The Structured Query Language—a formal setup for building queries. (p. 45)

Standard Deviation A data-classification method that computes break values by using the mean of the data values and the average distance a value is away from the mean. (p. 68)

Standard toolbar ArcMap toolbar that contains functions such as opening, saving, and printing map documents, as well as options to open other windows such as ArcToolbox or Catalog. (p. 6)

State Plane The SPCS (State Plane Coordinate System) projected coordinate system that divides states into zones with measurements made in feet or meters. (p. 19)

stops Destinations to visit on a network. (p. 261)

story map A Web application designed to convey a specific theme or concept to the user using a special set of Web templates. (p. 108)

Strahler method A stream ordering method that allows for stream order to increase in rank only if two streams of the same order of magnitude flow into each another. (p. 530)

stream digitizing The digitizing mode in which vertices are automatically generated at a pre-determined distance of moving the mouse. (p. 153)

stream order A computed ranking of streams based on how they flow into one another. (p. 529)

street centerline A file containing line segments representing roads. (p. 233)

subset A raster created by removing a set of grid cells from a larger raster. (p. 293)

suitability index A ranking of locations according to a set of criteria. (p. 505)

surface A raster that contains a value for some phenomena at all locations. (p. 319)

symbology The appearance of a data layer. (p. 25)

Symmetrical Difference A type of GIS overlay that retains all of the features from both layers except for the features that they have in common. (p. 221)

Table of Contents Component of ArcGIS that shows all layers being used in a map document. (p. 6)

target (Select By Location) The layers from which features will be selected when using the Select By Location tool. (p. 195)

temporal accuracy The time period and currentness of a dataset. (p. 166)

tension A Spline method that results in a less smooth surface than a regularized Spline but that will contain values closer to the range of the known point values. (p. 333)

terrain dataset An ArcGIS file structure that can hold very large amounts of elevation data and render them quickly at a variety of scales and resolutions. (p. 408)

thematic map A map that displays a particular theme or feature. (p. 59)

thermal infrared (TIR) Electromagnetic energy with wavelengths between 3.0 and 14.0 micrometers. (p. 302)

three-dimensional (3D) A model of the terrain that allows for multiple z-values to be assigned to each x/y coordinate. (p. 359)

TIGER/Line Topologically Integrated Geographic Encoded Referencing boundary and road network GIS data created by the U.S. Census Bureau. Also refers to a file that contains (among other things) the line segments that correspond with roads in the United States. (p. 132, 233)

tiled map service A hosted service that sets up GIS data as a series of image tiles that can be displayed, but not queried or edited. (p. 95)

TIN Triangulated irregular network, a terrain model formed from non-overlapping triangles that allows for unevenly spaced elevation points. (p. 369)

TIRS The Thermal Infrared Sensor onboard Landsat 8.

TM The Thematic Mapper instrument onboard Landsats 4 and 5. (p. 304)

Tool (ModelBuilder) Specific term used in ModelBuilder for one of the geoprocessing tools, models, or scripts (by default they appear as yellow rectangles). (p. 491)

Tools toolbar ArcMap toolbar that contains functions such as zooming in or out, and selecting or identifying features. (p. 6)

topographic map A map created by the USGS to show landscape and terrain as well as the location of features on the land. (p. 59, 390)

topology How objects relate or connect to one another independent of their coordinates. (p. 180)

trace The problem solving techniques used with a geometric network. (p. 275)

transformation A process in which data are altered from unreferenced to having spatial reference. (p. 316)

Transverse Mercator A projected coordinate system used for north-south data. (p. 19)

trend Very general patterns or overriding processes that affect measurements. (p. 320)

true color composite An image arranged by placing the red band in the red channel, the green band in the green channel, and the blue band in the blue channel. (p. 303)

TSP The Traveling Salesman Problem, a mathematical process involving determining the optimal configuration of rearranging a series of stops on a network. (p. 266)

turns Information used by ArcGIS to determine information about valid flows along a network. (p. 257)

two-and-a-half dimensional (2.5D) A model of the terrain that allows for a single z-value to be assigned to each x/y coordinate. (p. 359)

undershoot When a vertex of a line falls short of its target location. (p. 172)

Union A type of GIS overlay that retains all of the features from both layers. (p. 221)

Universal Kriging A Kriging method that assumes a trend within the data (and removes the trend in order to perform interpolation). (p. 338)

Update A type of GIS overlay that retains all features from the second layer as well as those features from the first layer that are not in common with the second. (p. 221)

US Topo A digital topographic map series created by the USGS to allow multiple layers of data to be used on a map in GeoPDF file format. (p. 391)

UTM Universal Transverse Mercator projected coordinate system that divides the world into 60 reference zones with measurements in meters. (p. 19)

Value Attribute Table (VAT) The attribute table of a raster dataset. (p. 285)

Variable (ModelBuilder) Specific term used in ModelBuilder for a data layer or value input (by default they appear as blue ovals in ModelBuilder). (p. 491)

variables Different factors that are combined into an overall analysis. (p. 487)

vector data model Model that represents geospatial data with a series of vector objects. (p. 13)

vector objects Points, lines, and polygons that are used to model real-world phenomena using the vector data model. (p. 13)

vertex The beginning and ending points of a segment. (p. 153)

vertical datum A baseline used in measuring elevation values (above or below this level). (p. 341)

vertical exaggeration A 3D visualization technique that alters the vertical scale but keeps the horizontal scale the same. (p. 383)

vertical image A remotely sensed image in which the camera is looking down at the landscape. (p. 295)

viewshed The visibility within 360 degrees around a location. (p. 352)

visibility analysis Techniques used in GIS to determine what areas can be seen or not seen from a vantage point. (p. 352)

visible light Electromagnetic energy with wavelengths between 0.4 and 0.7 micrometers. (p. 302)

visual hierarchy How features are displayed on a map to emphasize their level of prominence. (p. 84)

Volunteered Geographic Information (VGI) User-generated geospatial data. (p. 128)

watershed The area for which all locations within it flow to a common drainage area. (p. 515)

Web application A structure that uses shared content or data within a specific ArcGIS Online template. (p. 105)

Web map An interactive online representation of GIS data, which can be accessed via a Web browser. (p. 89)

Web Mercator auxiliary sphere The projected coordinate system used by ArcGIS Online. (p. 95)

Web Scene An interactive version of an ArcScene scene that can be viewed and used through a Web browser. (p. 385)

WebGL The Web Graphics Library, a Web standard for viewing 3D graphics. (p. 385)

weighted overlay The process of combining two or more grids by assigning a relative importance to each grid. (p. 484)

WGS84 World Geodetic System datum of 1984. (p. 18)

workflow The sequence of actions performed using GIS layers and tools. (p. 464)

XOR (Map Algebra) The Boolean operator used with an "exclusive or" of two grids. (p. 476)

Zonal Statistics A method of calculating statistical values based on the raster grid-cell values that fall within a boundary. (p. 331)

zone One of several different values assigned to a raster grid cell. (p. 282)

z-value The elevation assigned to an x/y coordinate. (p. 341)

INDEX